Handbook of
Endocrine Research Techniques

Handbook of Endocrine Research Techniques

Edited by

Flora de Pablo
Centro de Investigaciones Biologicas
Consejo Superior de Investigaciones Científicas
Madrid, Spain

Colin G. Scanes
Department of Animal Sciences
Rutgers – The State University of New Jersey
New Brunswick, New Jersey

Bruce D. Weintraub
Consulting Editor
National Institutes of Health
Bethesda, Maryland

Academic Press, Inc.
A Division of Harcourt Brace & Company

San Diego   New York   Boston   London   Sydney   Tokyo   Toronto

This book is printed on acid-free paper.

Copyright © 1993 by ACADEMIC PRESS, INC.
All Rights Reserved.
No part of this publication may be reproduced or transmitted in any form or by any means, electronic or mechanical, including photocopy, recording, or any information storage and retrieval system, without permission in writing from the publisher.

Academic Press, Inc.
1250 Sixth Avenue, San Diego, California 92101-4311

United Kingdom Edition published by
Academic Press Limited
24–28 Oval Road, London NW1 7DX

Library of Congress Cataloging-in-Publication Data

Handbook of endocrine research techniques / edited by Flora de Pablo,
  Colin Scanes, Bruce Weintraub.
      p.   cm.
    Includes bibliographical references and index.
    ISBN 0-12-209920-6
    1. Hormones--Analysis,   2. Hormone receptors--Analysis,
  3. Molecular endocrinology--Technique,   I. De Pablo, Flora,
  II. Scanes, C. G.   III. Weintraub, Bruce.
    [DNLM: 1. Endocrinology--methods. 2. Research--methods,   WK 20
  H236 1993]
  QP571.H345   1993
  591. 19' 27' 0724--dc20
  DNLM/DLC
  for Library of Congress   93-17185
                                                                    CIP

PRINTED IN THE UNITED STATES OF AMERICA
93   94   95   96   97   98   BC   9 8 7 6 5 4 3 2 1

Contents

Contributors xv
Foreword xix
Preface xxiii

PART I
Hormone Assays

1 Radioimmunoassay and Radioreceptor Assay: Past, Present, and Future
   Sonia M. Najjar and Bruce D. Weintraub

   I. History and Development of Radioimmunoassay 4
   II. Principles of Radioimmunoassay 5
   III. Principles of Immunoradiometric Assays 10
   IV. Application of Antibody Fractionation in RIA 15
   V. Radioreceptor Assays 16
   VI. Conclusions and Future Perspectives 17
   VII. Appendix: Mathematical Aspects of Radioligand Assays 18
       References 21

2 Immunoradiometric Assays
   M.R. Pandian and D.A. Fisher

   I. Introduction 26
   II. Principles of Immunoradiometric Assays 27
   III. Solid-Phase Systems 30
   IV. Selection of Antibodies 31
   V. Variations in Immunoradiometric Assays 36
   VI. Development of an Immunoradiometric Assay 39
   VII. Factors Influencing Immunoradiometric Assays 49
   VIII. Advantages and Disadvantages of Immunoradiometric Assays 50
   IX. Development of Nonradioactive Assays 51
   X. Conclusions 52
       References 53

v

3 Enzyme-Linked Immunosorbent Assays
Endre V. Nagy and Kenneth D. Burman

    I. Background   55
    II. Overview of the General Principles Involved with ELISA   56
    III. Details of the ELISA Technique   58
    IV. Related Techniques   63
        References   64

4 Development and Characterization of Polyclonal and Monoclonal Antibodies
Joan Rener

    I. Introduction   67
    II. Immunogen Preparation   69
    III. Antisera Development   70
    IV. Monoclonal Antibody Development   74
    V. Monoclonal Antibody Production   86
    VI. Purification of Antisera and Monoclonal Antibodies   87
    VII. Characterization of Antisera and Monoclonal Antibodies   88
        References   90

5 Isolation of Polypeptide Hormones: General Separation Methods and Some Applications for Thyroid Hormones
Jorge Alemany, Jorge Garcia de Ancos, and Enrique Mendez

    I. Introduction   93
    II. Purification Techniques and Strategy   94
    III. Improvement in HPLC Fractionation of Thyroxine-Containing Peptides by Prior Acell Ion-Exchange Column Chromatography   99
    IV. Identification of T3- and T4-Containing Peptides by HPLC with Photodiode Array (PDA) Detectors   101
        References   105

6 Reverse Hemolytic Plaque Assays: Applications to Endocrine Problems
L. Stephen Frawley

    I. Introduction   107
    II. Procedures Common to All Plaque Assays   110
    III. Standard Plaque Assay   113

       IV. Sequential Plaque Assay   118
        V. Simultaneous Plaque Assay   121
       VI. Other Variations and Applications of Plaque Assays   121
      VII. Summary   124
           References   125

7 Capillary Electrophoresis Coupled to Fluorescence Detection for the Determination of Multiple Neuropeptides
Juan P. Advis, Khurshid Iqbal, A. Waseem Malick, and Norberto A. Guzman

        I. Introduction   127
       II. A Capillary Electrophoresis-Based Assay   132
      III. Comments and Future Perspectives   141
           References   143

8 Immunoreactive Heterogeneity in Peptide Measurement
I. Valverde and M.L. Villanueva-Peñacarrillo

        I. General Introduction   145
       II. Proglucagon-Derived Peptides   146
      III. Conclusion   154
           References   155

9 Why Use a Flow Cytometer for Endocrine Cell Analysis?
Frank M. Perez, Daniel R. Deaver, and Wesley C. Hymer

        I. Introduction   157
       II. Capabilities of a Flow Cytometer   158
      III. Identification of Pituitary Cells by Flow Cytometric Immunofluorescence   159
       IV. Sorting of Pituitary Cells by Light-Scatter Properties   169
        V. Sorting of Pituitary Cells by Fluorescence   174
       VI. Concluding Remarks   178
           References   179

10 Identification and Characterization of Insulin-like Growth Factor-Binding Proteins
Yvonne W-H. Yang and Matthew M. Rechler

        I. Introduction   181
       II. Detection   184

III. Identification of IGFBPs   193
IV. Conclusions   203
References   203

PART II
Histological and *in Situ* Approaches

11 Localization of Peptides: Double Labeling Immunohistochemistry

José E. García-Arrarás

    I. Preparation of Tissues   208
   II. Single Labeling   212
  III. Double Labeling   214
  IV. Triple Labeling   223
      References   224

12 Electron Microscopic Immunocytochemical Approaches to the Localization of Ligands, Receptors, Transducers, and Transporters

Robert M. Smith and Leonard Jarett

    I. Introduction   227
   II. Use of Nonimmunologic Electron-Dense Hormone Complexes to Localize Receptors and Ligands   228
  III. General Consideration for Successful Immunocytochemical Electron Microscopy   230
  IV. Localization of Hormones, Receptors, Transducers, and Transporters   237
   V. Future Prospects for Electron Microscopy   259
  VI. Summary and Conclusion   260
     References   261

13 *In Situ* Hybridization Histochemistry

Carolyn A. Bondy, Jian Zhou, and Wei-Hua Lee

    I. Applications   266
   II. Methodological Issues   271
  III. *In Situ* Hybridization Protocols   277
     References   285

PART III
Techniques for Receptors and Signal Transductors

14 Plasma Membrane Isolation Strategies for Cell Surface Receptors: Application for the Insulin Receptors
Maxine A. Lesniak, Joshua Shemer, and Phillip Gorden

    I. Introduction   289
    II. Preparation of Plasma Membranes   290
    III. General Considerations for Radioreceptor Assays   296
    IV. Binding of ^{125}I-Insulin to Plasma Membranes   297
    References   299

15 Analysis of Autophosphorylation and Substrate Phosphorylation by Receptor Tyrosine Kinases
Robert S. Garofalo

    I. Introduction   301
    II. Methods   306
    References   317

16 Involvement of Protein Tyrosine Phosphatase Activity in the Regulation of Insulin's Signal Transduction
Dalit Hecht and Yehiel Zick

    I. Introduction   321
    II. Regulation of Protein Tyrosine Phosphatase Activity   322
    III. Cellular Functions Mediated by Protein Tyrosine Phosphatases   323
    IV. Protein Tyrosine Phoshatases as Mediators of Insulin Action   323
    V. Alterations in Insulin Receptor–Protein Tyrosine Phosphatase Activity in Insulin-Resistant States   324
    VI. Protein Substrates for Protein Tyrosine Phosphatases   325
    VII. Assay of Protein Tyrosine Phosphatase Activity   325
    References   335

17 Techniques for Evaluating Protein Kinase C in Intact Cells
Perry J. Blackshear

    I. Measuring Activation of Protein Kinase C in Intact Cells   340
    II. Down-Regulation of Protein Kinase C   352
    References   354

18 Identification and Quantification of G-Proteins
Ravi Iyenga

    I. Introduction   357
    II. Receptor-Stimulated GTPase   358
    III. Guanine Nucleotide-Binding Assays   358
    IV. Labeling of G-Proteins by Bacterial Toxins   360
    V. Immunoblotting with Sequence-Specific Antisera   362
        References   363

19 Determination of Adenylyl Cyclase Catalytic Activity
Roger A. Johnson and Yoram Salomon

    I. Considerations for Establishing Reaction Conditions   366
    II. Radioactive Substrates: [^{3}H]ATP vs [α-^{32}P]ATP   372
    III. Stopping the Reaction   374
    IV. Chromatographic Alternatives   378
    V. Data Analysis   386
        References   388

20 Intracellular Mediators of Peptide Hormone Action: Glycosyl Phosphatidylinositol/Inositol Phosphoglycan System
Isabel Varela-Nieto, Luis Alvarez, and José M. Mato

    I. Introduction   391
    II. Methods for the Purification and Characterization of the Glycosyl Phosphatidylinositol/Inositol Phosphoglycan System   393
    III. Biological Activity of Inositol Phosphoglycan   400
        References   404

21 Free Calcium Measurements: Down to the Single-Cell Level
Antonio Sanchez-Bueno and Peter H. Cobbold

    I. Introduction   407
    II. Single Cell vs Population to Measure Ca^{2+} 408
    III. Properties of Aequorin   408
    IV. The Aequorin Technique   410
    V. Examples of Cytosolic Ca^{2+} Measurements in Single Cells Using Aequorin   415
        References   417

PART IV
Molecular Techniques and Specific Model Systems

22 Approaches for the Purification, Quantitation, and Analysis of Hormone and Receptor mRNAs

Martin L. Adamo, Bethel Stannard, Derek LeRoith, and Charles T. Roberts, Jr.

    I. General Introduction   421
   II. RNA Extraction   422
  III. Northern Blot Hybridization Analysis   432
  IV. Solution Hybridization/RNase Protection Assay   441
   V. Primer Extension   449
     References   454

23 Polymerase Chain Reaction: Applications to Endocrine Research

Alan R. Shuldiner and Riccardo Perfetti

     I. Introduction   457
    II. Polymerase Chain Reaction: Theory of the Method   459
   III. General Protocol: The Polymerase Chain Reaction   462
   IV. Comments: General Polymerase Chain Reaction Protocol   463
    V. Reverse Transcription–Polymerase Chain Reaction   466
   VI. Protocol for Reverse Transcription   467
  VII. Comments: Reverse Transcription   468
 VIII. Quantitative Reverse Transcription–Polymerase Chain Reaction   469
   IX. Contamination and Polymerase Chain Reaction False Positives   469
    X. RNA Template-Specific Polymerase Chain Reaction   472
   XI. Subcloning Polymerase Chain Reaction Products   474
   XII. Direct Sequencing of Polymerase Chain Reaction Products   478
  XIII. Some Commonly Encountered Polymerase Chain Reaction Artifacts   480
  XIV. Conclusions   481
     References   482

24 Detection of Mutations in Hormone Receptor Genes

Domenico Accili, Fabrizio Barbetti, Takashi Kadowaki, and Simeon I. Taylor

   I. Introduction   487
  II. Southern Blotting   488

III. Sequencing of Polymerase Chain Reaction-Amplified Genomic DNA and cDNA   492
IV. Allele-Specific Oligonucleotide Hybridization of Polymerase Chain Reaction Products   495
V. Denaturing Gradient Gel Electrophoresis   497
VI. Single-Stranded Conformation Polymorphism (SSCP)   500
References   503

25 Techniques to Study Regulation of Gene Expression
Fredric E. Wondisford

I. RNA Analysis   505
II. Direct Measurement of Transcriptional Rates   506
III. Indirect Measurement of Transcriptional Rates   508
References   514

26 Transgenic Animals as a Tool in Endocrinology: Structure/Function Studies of Peptide Hormones Employing Transgenic Mice
John J. Kopchick and Wen Y. Chen

I. Introduction   515
II. Plasmid Construction and Mutagenesis   517
III. Transgenic Mouse Production   518
IV. IGF-I Radioimmunoassay   522
V. Immunoblotting Analysis   522
VI. Radioreceptor-Binding Assay   524
VII. Comments   525
References   526

27 Use of Nonmammalian Animal Models in Endocrine Investigation
Ian P. Callard, Lynn Riddiford, and Colin G. Scanes

I. Introduction   530
II. Vertebrate Models for Hypothalamic Pituitary Interaction and Behavior   534
III. Vertebrate Reproduction   537
IV. Vertebrate Hormones and Metabolism   541
V. Invertebrate Models   546
References   549

28 Proliferation Induced by Growth Factors in Mammalian Fetal Cells

Manuel Benito and Margarita Lorenzo

    I. Introduction   555
   II. Isolation and Plating of Cells   556
  III. Establishment of Cell Quiescence   557
  IV. Induction of Cell Proliferation   560
      References   564

29 Expression of Hormones/Growth Factors and Their Receptors in Early Embryogenesis

Flora de Pablo, Gilbert A. Schultz, and Susan Heyner

    I. Introduction   567
   II. The Mammalian Preimplantation Embryo   569
  III. The Chicken Embryo   580
      References   586

Index   589

Contributors

Numbers in parentheses indicate the pages on which the authors' contributions begin.

Domenico Accili (487), Diabetes Branch, National Institute of Diabetes, and Digestive and Kidney Diseases, National Institutes of Health, Bethesda, Maryland 20892

Martin L. Adamo (421), Diabetes Branch, National Institute of Diabetes, and Digestive and Kidney Diseases, National Institutes of Health, Bethesda, Maryland 20892

Juan P. Advis (127), Department of Animal Sciences, Rutgers – The State University of New Jersey, New Brunswick, New Jersey 08903

Jorge Alemany (93), PharmaGen S.A., 28760 Madrid, Spain

Luis Alvarez (391), Instituto de Investigaciones Biomedicas, Consejo Superior de Investigaciones Científicas, 28029 Madrid, Spain

Fabrizio Barbetti (487), Istituto S. Raffaele, 20132 Milan, Italy

Manuel Benito (555), Departamento de Bioquimica y Biologia Molecular, Centro Mixto CSIC/UCM, Facultad de Farmacia, Ciudad Universitaria, 28040, Madrid, Spain

Perry J. Blackshear (339), The Howard Hughes Medical Institute Laboratories, and Section of Diabetes and Metabolism, Departments of Medicine and Biochemistry, Division of Endocrinology, Metabolism, and Nutrition, Duke University Medical Center, Durham, North Carolina 27710

Carolyn A. Bondy (265), Developmental Endocrinology Branch, NICHD, National Institutes of Health, Bethesda, Maryland 20892

Kenneth D. Burman (55), Endocrine – Metabolic Service, and Kyle Metabolic Unit, Walter Reed Army Medical Center, Washington, D.C. 20307

Ian P. Callard (529), Department of Biology, Boston University, Boston, Massachusetts 02215

Wen Y. Chen (515), Department of Biological Sciences, Molecular and Cellular Biology Program, and Edison Animal Biological Center, Ohio University, Athens, Ohio 45701

Peter H. Cobbold (407), Department of Human Anatomy and Cell Biology, Liverpool L69 3BX, United Kingdom

Daniel R. Deaver (157), Department of Dairy and Animal Sciences, The Pennsylvania State University, University Park, Pennsylvania 16802

Flora de Pablo (567), Centro de Investigaciones Biológicas, Consejo Superior de Investigaciones Científicas, 28006 Madrid, Spain

D.A. Fisher (25), Nichols Institute, San Juan Capistrano, California 92690

L. Stephen Frawley (107), Department of Cell Biology and Anatomy, Medical University of South Carolina, Charleston, South Carolina 29425

José E. García-Arrarás (207), Department of Biology, University of Puerto Rico, Rio Piedras, Puerto Rico 00931

Jorge Garcia de Ancos (93), Serono, 28760 Madrid, Spain

Robert S. Garofalo (301), Department of Anatomy and Cell Biology, State University of New York, Health Science Center at Brooklyn, Brooklyn, New York 11203

Phillip Gorden (289), National Institute of Diabetes, and Digestive and Kidney Diseases, National Institutes of Health, Bethesda, Maryland 20892

Norberto A. Guzman (127), Pharmaceutical Research and Development, Hoffmann–La Roche, Nutley, New Jersey 07110

Dalit Hecht (321), The Department of Chemical Immunology, The Weizmann Institute of Science, Rehovot 76100, Israel

Susan Heyner (567), Department of Ob-Gyn, University of Pennsylvania Medical Center, Philadelphia, Pennsylvania 19104

Wesley C. Hymer (157), Department of Molecular and Cell Biology, The Pennsylvania State University, University Park, Pennsylvania 16802

Khurshid Iqbal (127), Pharmaceutical Research and Development, Hoffmann–La Roche, Nutley, New Jersey 07110

Ravi Iyengar (357), Department of Phamacology, Mount Sinai School of Medicine, City University of New York, New York, New York 10029

Leonard Jarett (227), Department of Pathology and Laboratory Medicine, University of Pennsylvania School of Medicine, Philadelphia, Pennsylvania 19104

Roger A. Johnson (365), Department of Physiology and Biophysics, Health Sciences Center, State University of New York at Stony Brook, Stony Brook, New York 11794

Takashi Kadowaki (487), The Third Department of Internal Medicine, University of Tokyo Medical School, Tokyo, Japan

John J. Kopchick (515), Department of Biological Sciences, Molecular and Cellular Biology Program, and Edison Animal Biological Center, Ohio University, Athens, Ohio 45701

Derek LeRoith (421), Diabetes Branch, National Institutes of Health, Bethesda, Maryland 20892

Wei-Hua Lee (265), Developmental Endocrinology Branch, NICHD, National Institutes of Health, Bethesda, Maryland 20892

Maxine A. Lesniak (289), Diabetes Branch, National Institute of Diabetes, and Digestive and Kidney Diseases, National Institutes of Health, Bethesda, Maryland 20892

Margarita Lorenzo (555), Departamento de Bioquimica y Biologia Molecular, Centro Mixto CSIC/UCM, Facultad de Farmacia, Ciudad Universitaria, 28040 Madrid, Spain

A. Waseem Malick (127), Pharmaceutical Research and Development, Hoffmann–La Roche, Nutley, New Jersey 07110

José M. Mato (391), Instituto de Investigaciones Biomedicas, Consejo Superior de Investigaciones Científicas, 28029, Madrid, Spain

Enrique Mendez (93), Servicio de Endocrinologia, 28034 Madrid, Spain

Endre Nagy[1] (55), Endocrine–Metabolic Service, and Kyle Metabolic Unit, Walter Reed Army Medical Center, Washington, DC 20307

Sonia M. Najjar (3), National Institute of Diabetes, and Digestive and Kidney Diseases, National Institutes of Health, Bethesda, Maryland 20892

M.R. Pandian (25), Nichols Institute, San Juan Capistrano, California 92690

Frank M. Perez (157), Department of Molecular and Cell Biology, The Pennsylvania State University, University Park, Pennsylvania 16802

Riccardo Perfetti (457), National Institute on Aging, National Institutes of Health, Bethesda, Maryland 20892

Matthew M. Rechler (181), Growth and Development Section, Molecular and Cellular Endocrinology Branch, National Institute of Diabetes, and Digestive and Kidney Diseases, National Institutes of Health, Bethesda, Maryland 20892

Joan Rener (67), Hazleton Washington, Inc., Vienna, Virginia 22182

Lynn Riddiford (529), Department of Zoology, University of Washington, Seattle, Washington 98195

Charles T. Roberts, Jr. (421), Diabetes Branch, National Institute of Diabetes, and Digestive and Kidney Diseases, National Institutes of Health, Bethesda, Maryland 20892

Yoram Salomon (365), Department of Physiology and Biophysics, Health Sciences Center, State University of New York at Stony Brook, Stony Brook, New York 11794

Antonio Sanchez-Bueno (407), Department of Human Anatomy and Cell Biology, Liverpool L69 3BX, United Kingdom

Colin G. Scanes (529), Department of Animal Sciences, Rutgers – The State University of New Jersey, New Brunswick, New Jersey 08903

Gilbert A. Schultz (567), University of Calgary, Calgary, Alberta, Canada T2N 4N1

Joshua Shemer (289), Heller Institute of Medical Research, Sheba Medical Center, Tel-Hashomer 52621, Israel

Alan R. Shuldiner (457), Johns Hopkins University, School of Medicine, Baltimore, Maryland 21224

---

[1] Present address: Department of Medicine, University School of Debrecen, Debrecen pf. 19, 4-4012, Hungary.

Robert M. Smith (227), Department of Pathology and Laboratory Medicine, University of Pennsylvania School of Medicine, Philadelphia, Pennsylvania 19104

Bethel Stannard (421), Diabetes Branch, National Institute of Diabetes, and Digestive and Kidney Diseases, National Institutes of Health, Bethesda, Maryland 20892

Simeon I. Taylor (487), Diabetes Branch, National Institute of Diabetes, and Digestive and Kidney Diseases, National Institutes of Health, Bethesda, Maryland 20892

I. Valverde (145), Departamento Metabolismo, Nutrición y Hormonas, Fundación Jiménez Diaz, 28040 Madrid, Spain

Isabel Varela-Nieto (391), Instituto de Investigaciones Biomédicas, Consejo Superior de Investigaciones Científicas, 28029 Madrid, Spain

M.L. Villanueva-Peñacarrillo (145), Departamento Metabolismo, Nutrición y Hormonas, Fundación Jiménez Diaz, 28040 Madrid, Spain

Bruce D. Weintraub (3), National Institute of Diabetes, and Digestive and Kidney Diseases, National Institutes of Health, Bethesda, Maryland 20892

Fredric E. Wondisford (505), Case Western Reserve University Medical School, Department of Medicine, Division of Endocrinology and Hypertension, University Hospitals of Cleveland, Cleveland, Ohio 44106

Yvonne W-H. Yang (181), Growth and Development Section, Molecular and Cellular Endocrinology Branch, National Institute of Diabetes, and Digestive and Kidney Diseases, National Institutes of Health, Bethesda, Maryland 20892

Jian Zhou (265), Developmental Endocrinology Branch, NICHD, National Institutes of Health, Bethesda, Maryland 20892

Yehiel Zick (321), The Department of Chemical Immunology, The Weizmann Institute of Science, Rehovot 76100, Israel

---

[2] Present address: Osaka Bioscience Institute, 6-2-4 Furuedai, Suita, Osaka 565, Japan.

Foreword

I am pleased that the editors invited me to compose a foreword to their new volume, *Handbook of Endocrine Research Techniques*. This book is devoted to current up-to-date techniques and methods. The emphasis throughout the volume is on the contemporary, except in this essay where I muse over the past, the history of books, techniques, and endocrine research.

Books: Whether on papyrus, clay, animal skin, or paper, books have always been the dominant commodity of libraries. In English, as in many languages, the word for library stems from the word for book. In the latter part of this century, biomedical libraries have undergone an inversion (or perhaps a diversion) whereby books have become subordinated to journals. In most other fields books are still primary.

When one enters a grand national library like the Library of Congress in Washington, or a major urban library like Boston's Public Library, or a great university library such as Butler Library at Columbia, books dominate; magazines, other periodicals, and newspapers are clearly less important. However, in biomedical libraries, journals and other periodicals are paramount. Traditionally, science is rooted in books and, until recently, biomedical libraries also had a heavy emphasis on books. Now journals have assumed dominance. An expansive and expensive list of journals is standard in most libraries. A core set of journals is kept complete for the whole of this century. When journals are lost from a library, they are usually replaced. When books are missing they are often not replaced. Books often go out of print without notice; their circulation and distribution are quite idiosyncratic.

Is this phenomenon temporary or permanent? Has the extraordinary tradition of the book that arose with the invention of writing 5000 years ago, strengthened through 500 years of printing, been permanently lost in biomedicine?

Techniques: Advances in techniques and methods have classically been the basis for most major advances in biology and medicine. The development of compound lenses in the 17th century, the mastery over spherical aberration in the 19th century, and the invention of the electron microscope in the 20th century together with a continuing succession of new methods to obtain, prepare, preserve, stain, and visualize tissues, cells, organelles, and molecules have provided an ongoing series of advances in knowledge. Endocrinology in its century as a discipline, like molecular biology in its few decades of existence, is very heavily rooted in methodology. Yet, methods are not given the respect they deserve. It is very difficult to have a methods paper accepted for publication in a standard first-line journal. Often, papers that emphasize new methods are rejected from journals with the pejorative comment, "It is only a methods paper." Furthermore, with most articles in standard journals, editors exert great effort to constrict the methods section, but otherwise ignore it. Reflecting the editors' and reviewers' neglect of methods, authors often invest little in that part of the paper; carelessness and obscurity synergize to ruin it. This seems paradoxical. After all, when authors publish a paper, they are warranting to the reader that when the experiments are carried out with the reagents and experimental protocols that are described, the results will be reproduced. Yet the methods sections of published journal articles are so typically truncated and terse as to be obscure. It may be difficult to understand even in outline what the authors did, much less to use their printed work to reproduce their experiments. In some ways, the techniques are much more important than the results; the methods are widely applicable while the results are highly particular. Yet the results totally dominate and the methods are highly restricted.

Endocrine Research: Endocrinology today may appear to be a narrow specialty that studies hormones and their target cells, with a canon of endocrine glands established in the 19th and 20th centuries based largely on the links of specific organs to specific diseases. In fact, endocrinology over the past few decades has been a focus of seminal advances in both methods and concepts for all of biology. Intercellular communication, formerly the private purview of endocrinology, has spread to all specialties including oncology, immunology, hematology, cardiology, infectious disease, and neuroscience. Many of the most widely applied methods had their origins in endocrine research. The radioimmunoassay was first devised for measurements of circulating insulin. Adenylate cyclase and cyclic AMP, indeed the whole concept of signal transduction as we know it today, were inaugurated by the search to understand how epinephrine and glucagon stimulate glycogen breakdown in the liver. Cell surface receptors were first measured for ACTH and other hormones and the first diseases discovered involving receptors were those for hormones. The G-proteins were discovered in an attempt to reconcile glucagon-stimulated cAMP formation with glucagon binding to its receptor. The role of phosphorylation and dephosphorylation in cellular regulation was discovered for a glycogenolytic enzyme and brought to the fore by studies of hormone-sensitive pathways. Pioneer observations on the importance of calcium as an intracellular messenger were made during the study of the secretion of adrenaline from the adrenal medulla. In addition, the first peptides whose structures were uncovered and synthesized were the hormones of the posterior pituitary. The first protein whose sequence was elucidated was that of insulin; indeed it was the demonstration of the precise amino acid sequence of insulin that clarified the notion that a protein had a single unique structure rather than being a series of closely related vaguely defined components. Whole areas of modern biology have their roots in the endocrine research of the last half century.

Prospect: In the Foreword, I have looked backward and covered the past. The rest of this book dwells on the present. What does the future hold? I believe that computers will correct many of today's distortions and restore books to their rightful place — books will be universally accessible, unlosable, and fully indexed. Computers will also permit regular emendations. Books will be of greater value than ever. Indeed books and journals will merge into a single series of continuous and overlapping formats.

I can envision how methods and techniques will be fully indexed, presented in exquisite detail, and updated regularly via computer. The focus on methods will be sharpened.

I welcome this volume as we rapidly approach the threshold of the new era for books, techniques, and endocrine research.

Jesse Roth

Preface

Any biological problem in the field of endocrinology can be approached using biochemical, cellular, and molecular biological techniques that provide complementary answers. Often, the researcher must consult multiple books and articles to decide which approach might be the most successful to clarify the specific question proposed. This handbook was planned as a source of up-to-date methods and strategies particularly useful in the field of endocrinological research. Because its purpose is to help both the novice and the experienced investigator, each chapter includes an introduction to the area, general concepts, detailed protocols, and extensive references.

We have not attempted to be comprehensive; the focus of most chapters is on polypeptide hormones, their receptors, modulators, and mediators of their action. Areas not covered are steroids and thyroid hormones and their cytoplasmic/nuclear receptors, although some of the methods described in the book can be applied equally well to these types of hormones and receptors.

The emphasis is placed on methods that have been developed or have evolved recently. Less information is included about traditional techniques that have multiple well-documented sources (specific books on radioimmunoassay, bioassay, immunocytochemistry, etc.). Each chapter has been written by experts in the field with hands-on experience, using specific systems as examples. Some overlap between themes has been allowed; with the help of cross-references, this should help the reader view the same topic from slightly different angles.

The demonstration of the presence of hormones/growth factors and their receptors in very diverse cellular and physiological states and the understanding of their normal or perturbed mechanisms of action have progressed enormously in the past few decades with the availability of extremely sensitive techniques for measuring minute amounts of proteins and their mRNAs and for observing subcellular compartments. Methodological advances in the past 20 years, largely with the advent of molecular biology, have reduced the lag of many decades that used to occur between the recognition of a signaling factor (or its receptor) and the synthesis/expression of the pure molecule in the laboratory to a few years. Breakthrough techniques such as the polymerase chain reaction (PCR) have provided insights into the expression of hormones and growth factors that help enormously in our understanding of their function.

The book begins with a chapter on general principles of the most traditional of all endocrinological techniques, the radioimmunoassay (also radioreceptor?). The book continues with other hormone assays developed more recently. In Part II we have grouped the histochemical and *in situ* approaches to the analysis of peptide factors, hormonal or locally produced, and some mediators. In Part III receptors and transducers are analyzed more biochemically, while in Part IV the most common molecular techniques are discussed. The last few chapters shift to a few useful specific model systems. The applicability of protocols is generally broad and care has been taken to provide references that will allow the reader to explore in depth a particular topic.

We dedicate this handbook to the students in our laboratories. If their initial "immersion" in a research project is facilitated by this set of strategies and protocols, our main goal will be achieved.

Flora de Pablo
Colin G. Scanes

PART I

Hormone Assays

Radioimmunoassay and Radioreceptor Assay: Past, Present, and Future

Sonia M. Najjar and Bruce D. Weintraub

> "From 1950 until his untimely death in 1972, Dr. Solomon Berson was joined with me in this scientific adventure and together we gave birth to and nurtured through its infancy radioimmunoassay, a powerful tool for determination of virtually any substance of biologic interest. . . ."
>
> Dr. Rosalyn S. Yalow
> December 8, 1977
> Nobel Lecture

I. History and Development of Radioimmunoassay
II. Principles of Radioimmunoassay
   A. Conventional Radioimmunoassay
   B. Two-Step (Sequential, Saturation, Nonequilibrium) Radioimmunoassay
   C. Cooperative Immunoassay (CIA)
III. Principles of Immunoradiometric Assays (IRMA)
   A. Direct Immunoradiometric Assay
   B. Two-Site Immunoradiometric Assay
   C. One-Step, Two-Site IRMA
   D. Indirect Assay
IV. Application of Antibody Fractionation in RIA
V. Radioreceptor Assays
VI. Conclusions and Future Perspectives
VII. Appendix: Mathematical Aspects of Radioligand Assays
   A. Theory
   B. Laws of Mass Action and Scatchard Equation
References

I. History and Development of Radioimmunoassay

The concept of radioimmunoassay (RIA) was introduced by Yalow and Berson in the late 1950s following their discovery and quantitation of anti-insulin antibodies in patients with non-insulin-dependent diabetes mellitus (Berson et al., 1956). The two subsequently developed the theory and practical application of RIA, based on the competitive inhibition of the binding of a labeled antigen to its specific antibody by an unlabeled antigen (Berson and Yalow, 1959). Since these early pioneering studies, which were initially applied to the measurement of insulin and growth hormone (Glick et al., 1963), RIA has been applied to the measurement of a wide variety of peptide, steroidal, and thyroidal hormones as well as nonhormonal substances (Yalow and Yalow, 1980).

In 1968, Miles and Hales modified RIA by introducing the use of a labeled antibody instead of an antigen. The immunoradiometric assay (IRMA) was thus described. This method involves the purification of the labeled antibody using an immunoadsorbent column and the subsequent incubation of an excess of the purified antibody with an unknown antigen. Antigen-bound antibody is then counted after removal of free labeled antibody by precipitation with an immunoadsorbent-coupled antigen (see below). In 1971, Addison and Hales extended the modifications of IRMA by introducing the use of antibody immunoadsorbents to concentrate the antigen and then allowing the labeled antibody to bind (Addison and Hales, 1971).

Unlike RIA where high-affinity antibodies were usually employed at high dilution (1 : 100,000 to 1 : 10,000,000), IRMA assays consumed large amounts of antibodies and, thus, were not widely applied until the development of mouse monoclonal antibodies by Köhler and Milstein (1975). In an attempt to spare such consumption of antibody in immunoradiometric assays (direct and two-site), Beck and Hales developed in 1975 the indirect labeling technique of the antibody (Beck and Hales, 1975). This concept is based on the fact that anti-IgG from the same species as the animal in which the specific antibody was raised can be immunologically recognized by the antibody. Therefore, one can label a large amount of anti-IgG and allow it to bind to certain sites on the first antibody. This produces a labeled double-antibody complex that may be generally used for antigen detection in multiple types of either immunoradiometric assay.

Moreover, as researchers began to appreciate the limitations of using radioactivity as label, enzymatic, immunofluorescent, or chemiluminescent labeling of either antigen or antibody, directly or indirectly, became widespread. In general RIA using radioactive label, particularly ^{125}I, remains the most popular method for many research applications. However, the two-site or sandwich immunoassay has theoretical and practical superiority to RIA since their sensitivity is not dependent on the affinity constant of the antibody (which is limiting in RIA) and specificity can be enhanced by the use of two immunologic determinants (see later). Since two-site assays have employed a variety of labeling methods, nonradioactive labels are increasingly popular especially in automated assays used in large hospital or commercial laboratories.

II. Principles of Radioimmunoassay

A. Conventional Radioimmunoassay

The fundamental concept of RIA is based on observations that high-affinity, high-titer anti-hormone antibodies are able to bind ^{131}I- or ^{125}I-labeled hormone in a reversible manner and that such binding is competitively inhibited by unlabeled hormone (Fig. 1). To assess the validity of RIA, standards that are immunochemically identical to unknown biologic samples must be simultaneously assayed, such that the dilution curve of the unknowns is superimposed on the dilution curve of the standards over a wide concentration range (Yalow, 1973). Complete separation of the labeled antibody-bound and unbound antigen is required in order to obtain an accurate measurement of the free and bound labeled antigens, allowing for the determination of the hormone concentration in an unknown sample (cf. Appendix: Mathematical Aspects of Radioligand Assays).

In the remaining part of this section, a simplified version of the preparation of the reagents used in RIA is given. For a more detailed description, the reader is strongly referred to the comprehensive reviews cited in the bibliography (Dwenger, 1984; Edwards, 1985).

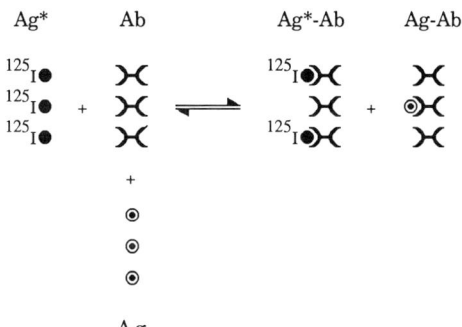

Figure 1 Schematic diagram of conventional radioimmunoassay. ◉, Unlabeled antigen (Ag); ●, labeled antigen (*);)—(, antibody (Ab).

Traditionally, polyclonal antibodies are used in RIA. These heterogenous antibodies offer an enhanced antigenicity permitting a sufficient dilution so that only the binding sites with the highest affinity or equilibrium constant are occupied by the antigen. This increases the sensitivity of RIA. In some cases, however, optimal sensitivity is not required, and the use of the monospecific monoclonal antibodies to enhance the specificity of the assay may be more advantageous (cf. Section IV).

As expected, purity of the labeled antigen is required to eliminate the potential recognition of labeled contaminants by nonspecific antibodies. The high specific activity and high counting efficiency of ^{125}I renders it the radionuclide of choice for the labeling of most antigens. Radioiodine readily labels the tyrosine or histidine residues via a variety of methods such as the Chloramine-T oxidation technique (Hunter, 1962; Heber *et al.*, 1978) as well as more gentle enzymatic methods such as lactoperoxidase (Schiller *et al.*, 1978). When labeling of these residues is not accessible, iodine labeling of the free amino groups such as the terminal and ϵ-NH$_2$ groups by the Bolton–Hunter reagent is recommended (Bolton and Hunter, 1973).

In principle, the antibody–antigen complexes might undergo spontaneous precipitation. However, the concentration of the reagents used in RIA is so low that the antibody–antigen complex remains soluble. Hence, the mechanical separation of the antibody-bound antigen from the free unbound antigen is required. A variety of methods is now available: (1) absorption of the free antigen or the antibody to a solid-phase system such as charcoal, cellulose, or ion-exchange resins (Schuurman and De Ligny, 1978a); (2) precipitation of the antibody–antigen complex by a second antibody (double-antibody) (Kumar *et al.*, 1976), salting-out, charcoal (Odell, 1980), organic solvents (Magyar *et al.*, 1979), or a solid-phase anti-γ-globulin reagent such as protein A of *Staphylococcus aureus* (Figenschau and Ulstrup, 1974).

The most subtle pitfall of classical RIA probably is the degradation of the labeled antigen by proteolytic enzymes in blood and other biologic fluids. Thus, appropriate controls must be simultaneously assayed during the determination of peptide hormone concentrations in serum or plasma. Moreover, hormone standards should be optically diluted in hormone-free serum at concentrations identical to those used for unknown samples to control for a variety of "nonspecific" effects which may inhibit binding of antibody to labeled hormone.

B. Two-Step (Sequential, Saturation, Nonequilibrium) Radioimmunoassay

Many substances and, especially, thyroid and steroid hormones circulate in the blood either as free particles or bound to carrier proteins. While these specific binding proteins may play a role in facilitating the cell entry of some substances (i.e., cholesterol), they principally function as a storage zone for thyroid and steroid hormones in the body fluids. Moreover, bound hormones are "biologically inert" and, therefore, the diagnosis of a thyroid or steroid hormone-related disease should not be made solely on the basis of the levels of bound hormones. Since conventional RIA provided the measurement of total hormone levels, bound and free, the development of an assay that measures free serum hormones irrespective of the bound fraction became essential for the diagnosis of thyroid and steroid hormone diseases. Based on the initial findings by Hales and Randle (1963), Ellis and Ekins developed in the early 1970s a direct RIA (Ekins *et al.*, 1970; Ellis and Ekins, 1973; Harvey *et al.*, 1970) to measure the concentrations of free thyroid fT_3 and fT_4 hormones in the biologic fluids. This assay can be classified into a one- or two-step RIA (Ekins, 1983). In the two-step radioimmunoassay, an equilibrium is initially reached between the unlabeled ligand and the antibody prior to the addition of either the labeled ligand or the labeled antibody (Fig. 2). As can be deduced, the labeled ligand-dependent two-step assay measures the fractional occupancy of antibody following preexposure of the latter to serum. Labeled antibody-dependent assay directly quantifies the occupancy by the hormone of the solid-phase antibody. It is noteworthy that a nonequilibrium two-step system can be generated by the preincubation to equilibrium of the labeled ligand to the antibody, followed by the addition of the unlabeled ligand. However, the interaction of the unlabeled ligand dose level and the rate of approach to equilibrium results in a decreased sensitivity in a nonequilibrium assay that includes preincubation with labeled ligand (Rodbard *et al.*, 1971), so that preincubation with unlabeled ligand is more commonly employed. In the one-step direct assay, however, an equilibrium state is reached upon the simultaneous addition of labeled and unlabeled antigens to the antibody. Moreover, the labeled antigen used is actually a labeled hormone analog which is assumed to be unreactive with serum-binding proteins. Although quicker and more practical, the one-step direct RIA is limited by the use of analogs which may abnormally bind to carrier proteins, such as albumin, in the serum of patients with thyroid or steroid disorders (Csako *et al.*, 1989). Since the two-step assay does not depend on hormone analogs that may unpredictably bind to some sera-binding proteins, it may be more reliable than its single-step counterpart in the diagnosis of endocrine disorders in which thyroid and steroid hormones are implicated (Csako *et al.*, 1986; Beckett *et al.*, 1991).

Berson and Yalow (1959) based their mathematical analysis of the equilibrium state of conventional RIA on many assumptions (see later), some of which are (1) the concentration of the antibody largely exceeds that of the antigen, and (2) the labeling of the antigen does not alter its affinity constant. In reality, however, these assumptions are not always upheld. In fact, the concentrations of the antibody and the antigen are comparable in most RIAs,

Step One:

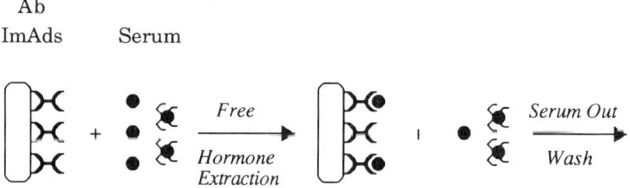

Step Two: Labeled Ligand

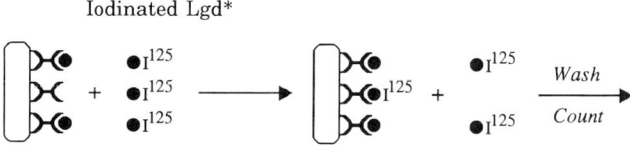

Step Two: Labeled Antibody

Figure 2 Schematic diagram of the nonequilibrium, two-step RIA. ImAds, immunoadsorbent;)—(, specific antibody (Ab); ●, free hormones; ⚭, bound hormones; Lgd*, iodinated ligand (hormone).

and the dose–response curves for RIA change when the labeled and unlabeled antigen are added at different times (Ekins *et al.*, 1970). These observations constitute the basis for the complicated kinetics of the two-step (nonequilibrium, sequential) relative to the one-step (equilibrium, simultaneous) RIA (Keilacker *et al.*, 1991). Comparisons by kinetic analysis and computer simulation predicted an improved sensitivity by a delayed addition of the labeled antigen as in the case of the two-step reaction (Rodbard *et al.*, 1971). Theoretically, the enhanced sensitivity by the two-step assay may stem from the fact that the antiserum used is more concentrated in the two-step than in the one-step assay in order to have the same percentage binding of the tracer. The increase of sensitivity by the two-step nonequilibrium RIA assay (a lower detection limit than that measured in a one-step assay carried out using the same original conditions for the antibody) was experimentally validated by Schuurman and De Ligny (1978b) for human serum immunoglobulin A and by Dale *et al.* (1983) for human granulocyte leukocyte antigen (L1). Despite the observation of a slightly compromised precision in the assay, the nonequilibrium RIA remains the assay of choice for large substances that are abundant in body fluids. Its main advantage is that the labeled tracer is exposed to the incubation mixtures for a shorter period, thereby reducing tracer damage during incubation.

C. Cooperative Immunoassay (CIA)

Faced with a scarcity of antisera with high association constants against peptide hormones with small molecular weights such as adrenocorticotropic hormone (ACTH), Matsukura *et al.* (1971) accidentally identified a paradoxical increase, instead of the conventional RIA decrease, in the bound-to-free ACTH ratio with an increment increase in the amounts of antigen up to 1 ng, beyond which the conventional decrease in labeled antigen binding to its antibody occurred. Although the authors encouraged the use of such an assay for antigens that exist at very low concentrations in unextracted plasma, they did not explain the precise mechanism for this phenomenon. Subsequently, Weintraub *et al.* (1973) reported a similar binding reaction in the RIA of human chorionic gonadotropin (hCG). The authors noted that mild papain digestion of the anti-hCG produced a divalent $F(ab)^2$ that, in contrast to univalent $F(ab)$, resulted in enhanced antigen binding. This enhanced binding at low amounts of antigens was found to be the result of a positive cooperative binding between the two antigenic binding sites of the multivalent antibodies. The exact mechanism of this positive cooperative binding was further attributed by Niederer (1974) to antigen-induced conformational change of the binding sites leading to increased affinity of the antibodies, but not to an allosteric effect (Ehrlich *et al.*, 1982). As monoclonal antibodies became more commonly employed in RIA, Moyle *et al.* (1983) used a mixture of monoclonals, each directed to different epitopes on the hCG, to develop mathematical models of the system. It was thus determined that the increase in the affinity of binding by mixing several monoclonals relative to each individual antibody for the antigen is caused by the formation of a circular complex between two antigen molecules and one of each type of antibody. Furthermore, Ehrlich and Moyle (1983) determined that the cooperative immunoassay (CIA) that involves two monoclonal antibodies to hCG is at least 100-fold more sensitive than a competitive conventional RIA employing each antibody alone. The only limitation on the sensitivity of the assay is the tracer specific activity.

It is noteworthy that the specificity of the assay could be either decreased or increased depending on the monoclonal antibodies used in the assay. Thus, the validity of mixing these antibodies in this assay largely depends on well-optimized standards (Ehrlich and Moyle, 1986).

III. Principles of Immunoradiometric Assays

A. Direct Immunoradiometric Assay

Conventional RIA was initially modified by Miles and Hales (1968a) by substituting labeled antibody for antigen. The purpose of the IRMA thereby introduced is to allow the antigen to react with an excess of labeled antibody to ensure that all the antigen in the sample or standard is bound. A solid-phase system consisting of antigen coupled to an immunoadsorbent is then added to bind to and, hence, to precipitate out the unbound antibody instead of the antigen as is the case in conventional RIA (see Fig. 3). The labeled bound fraction is then counted. The counts in the bound fraction are related directly to the total amount of antigen present in the assay. This differs from RIA in that there is an inverse relationship between the percentage of counts present

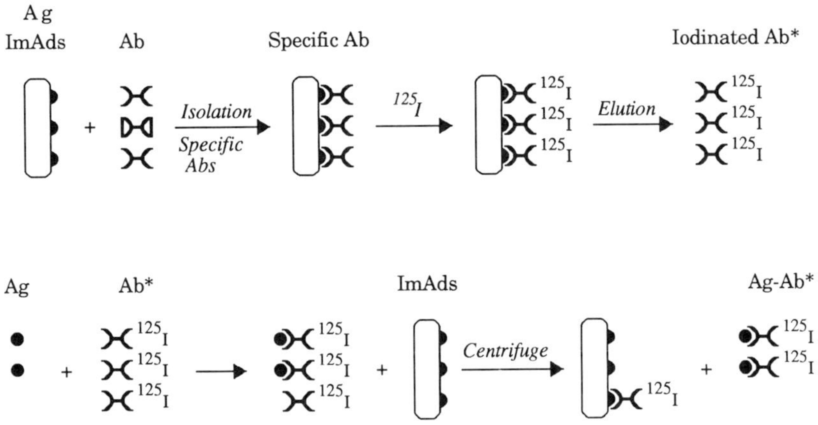

Figure 3 Schematic diagram of the immunoradiometric assay (IRMA). ImAds, immunoadsorbent;)—(, specific antibody (Ab); D—◁, nonspecific antibody; ●, antigen (Ag); *, iodinated species.

in the bound fraction and the initial amount of unlabeled antigen. In contrast to RIA, IRMA does not follow the laws of mass action (cf. Appendix: Mathematical Aspects of Radioligand Assays).

As described earlier, the increasing use of IRMA has led credence to its many advantages over the conventional RIA. (1) Many peptide hormones have proved difficult to iodinate due either to the lack of tyrosine residues or to induced conformational change in the iodinated molecule (Miles and Hales, 1968a). Moreover, iodinated hormones may be subject to altered immunogenicity. In contrast, the many tyrosine residues on the immunoglobulin render the molecule more readily iodinated. (2) Iodinated peptides are susceptible to proteolytic cleavage by plasma enzymes that are active during long incubation periods. In contrast, iodinated immunoglobulins are relatively more stable and uniform, rendering the results of IRMA more reliable and more correctly interpreted than those of RIA. (3) Since the labeling process of the antibody involves the purification of the antibody over an antigen-immunoadsorbent system, the antibody could be actually stored bound to this solid-phase system for at least 6 months (Woodhead *et al.*, 1974).

The question of enhanced sensitivity by IRMA over RIA has not yet been entirely resolved. In their mathematical analysis, Rodbard and Weiss (1973) indicated that the absolute sensitivity of each method depends on how closely in practice it approaches the ideal situation. Moreover, the relative sensitivity of the two systems depends largely on the specific activity of the labeled species. Theoretically, however, measurements of high-molecular-weight antigens by IRMA may be more sensitive than those of conventional RIA since several molecules of labeled antibody per molecule of antigen may bind (Hales, 1972). However, in most cases the sensitivity of IRMA has surpassed that of RIA, and such sensitivity gains are further enhanced by the use of nonradioactive labels (see later) which can be added at a much higher molar ratio than iodide to antibodies, producing very high specific activity label. RIA may display problems of specificity and IRMA displays certain advantages in this regard.

Because there may be stabilization of cross-reacting antigens on the immunoadsorbent in IRMA, the difference in the binding affinity of the specific and the less-specific cross-reacting antibodies toward the antigen may increase, leading to a better separation and, thus, to fewer contaminants (Miles, 1971).

The larger amounts of antibody needed for the assay and the amount of time it takes to prepare the reagents constitute the main disadvantages of IRMA compared to conventional RIA. Regardless, IRMA has gained a wide application in endocrinology and, especially, in studies of hormones that exist at very low concentrations in biologic fluids such as follicle-stimulating hormones in premenstrual females (Miles, 1971). Other applications involve insulin (Miles and Hales, 1968b), growth hormone (Miles and Hales, 1968c), parathyroid hormone (Addison et al., 1971), and calcitonin (Colt et al., 1971).

B. Two-Site Immunoradiometric Assay

In 1971, Addison and Hales modified the antibody-labeling-based IRMA into a two-site assay by concentrating and purifying the antigen on an antibody immunoadsorbent that consists of the immunoglobulin fraction of the immune sera coupled to the solid phase (Addison and Hales, 1971). Following the extraction of the antigen from the antibody immunoadsorbent, the iodinated antibody is then added, allowing it to bind to the other immunoreactive site of the antigen which is fixed onto the immunoadsorbent complex. The precipitate containing the antigen sandwiched between two antibody molecules, one of which is labeled and the other unlabeled and coupled to a solid phase, is then washed and counted (Fig. 4).

The concept of the two-site IRMA has made the use of antibody-coated tubes as the solid phase practical, simplifying the technical procedures and allowing large amounts of sera to be screened for antigen (Ling and Overby, 1972; Cheung et al., 1981). This constitutes one of the bases for the theoretical advantages of speed that this assay possesses. Moreover, relative to conventional RIA where sensitivity of the assay largely depends on the binding of the antibody to the highest affinity constant, the extraction of a highly specific antigen by antibody immunoadsorbents lowers the nonspecific radioactivity and, thus, enhances the sensitivity of the two-site IRMA with good precision over a wide concentration range since the reaction kinetics are independent of the antibody affinity constants (Ekins, 1978). Furthermore, the requirement that the antigen possesses at least two antigenic determinants improves the specificity of the assay and renders it the assay of choice for diagnosis and screening for many endocrine disorders such as neonatal hypothyroidism (Sutherland et al., 1982). Moreover, the assay has proved particularly useful in

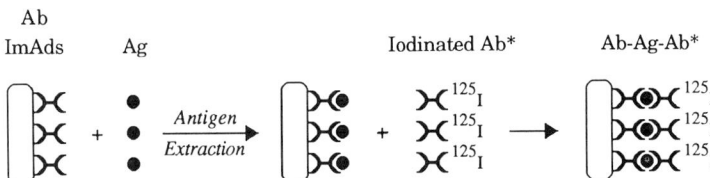

Figure 4 Schematic diagram of the two-site immunoradiometric assay. ImAds, immunoadsorbent; Ab, specific antibody; *, iodinated species; ●, antigen (Ag).

distinguishing intact hormone from metabolites or fragments, such as the carboxyterminal fragment of the parathyroid hormone (O'Riordan *et al.*, 1972).

The major limitation of the two-site IRMA, however, is the "high-dose hook effect," so described because the dose–response curve has a bell shape, characterized by a positive slope up to a limiting concentration, followed by a negative slope that is probably due to the dissociation of the antigen from the solid-phase antibody (Miles, 1977). This fall in the curve remains paradoxical despite attempts to analyze it both mathematically (Hoffman *et al.*, 1984) or at the bench. Heterogenous antibody with low affinity and incomplete washing of the immunoadsorbent–antigen complex have been considered likely candidates in explaining the fall in the curve. Regardless, the many advantages of the two-site IRMA far outweigh its disadvantages and render it the assay of choice for the detection of substances that exist at very low concentrations in biologic fluids.

C. One-Step, Two-Site IRMA

The advent of the production of large amounts of monoclonal antibodies in the mid 1970s (Köhler and Milstein, 1975, 1976) and the enhancement of their antigen-binding affinity by binding to two different epitopes in the same antigen molecule (Holmes and Parham, 1983) have extended the use of monoclonal antibodies in the two-site IRMA. In this fashion, one can use two different monoclonal antibodies directed to different determinants on the same antigen. The immunoadsorbent-coupled unlabeled monoclonal binds to the bivalent antigen at one immunoreactive site, leaving the other site accessible for the binding of the other labeled monoclonal. Thus, the use of monoclonal antibodies has greatly improved the specificity of the two-site IRMA in the measurement of many hormonal or nonhormonal peptide substances in the biologic fluids. Moreover, due to the different specificities recognized by two monoclonal antibodies, the antigen to be measured could be simultaneously incubated with an antibody fixed to an immunoadsorbent and another antibody that is labeled. This allows for the omittance of the washing step before the addition of labeled antibody as required in the standard two-site IRMA. This one-step procedure has rendered the two-site IRMA quicker and widely applicable in nonhormonal and hormonal systems. Among the nonhormonal systems are human atrial natriuretic factor (Lewis *et al.*, 1989), subnormal levels of erythropoietin in anemic and polycythemic patients (Andre *et al.*, 1992), and factor VIII-related antigen (VIIIRAg) in von Willebrand's disease (Sultan *et al.*, 1984). The application of this monoclonal antibody-based two-site IRMA has been even more widespread in hormonal systems. For instance, the many structural similarities between luteinizing and follicle-stimulating hormone result in elevated cross-reactivity between their respective polyclonal antibodies. This may contribute to the complexity and the relative invalidity of RIA in the determination of the fluctuating levels of serum gonadotropins from infancy to adulthood. Using specific monoclonal antibodies, De Hertogh *et al.* (1989) reported that the two-site IRMA is more reliable in the determination of lower serum gonadotropins in childhood. Similarly, the presence of substances in plasma that share with insulin some structural similarities such as proinsulin (Roth *et al.*, 1968) and split proinsulin intermediates (Kemmler *et al.*, 1971) may contribute to cross-reactivity in RIA that depends on polyclonal antibodies, lowering the accuracy of plasma insulin measurements in noninsulin-dependent diabetes. Thus, it seems reasonable that the use of the more specific monoclonal antibodies in a two-site IRMA would enhance the specificity and lead to a more accurate determination of plasma insulin levels (Temple *et al.*, 1992). Indeed, the use of monoclonal antibodies in a two-site IRMA has been greatly appreciated in the detection of low serum levels of other peptide hormones such as human chorionic gonadotropin (Cheong *et al.*, 1990), calcitonin in the medullary carcinoma of the thyroid (Guilloteau *et al.*, 1990), high-molecular-weight growth hormone (Luthman *et al.*, 1990), ACTH (Gibson *et al.*, 1989), and intact human parathyroid hormone (Frölich *et al.*, 1990).

D. Indirect Assay

The difficulty of iodinating a small amount of antibodies directly by conventional methods has made the immunoradiometric assay a large consumer of antisera. In order to decrease the amount of labeled antibody used to detect the antigen, Beck and Hales (1975) advocated indirect labeling of the first (antigen-detective) antibody through the use of a labeled antibody to the immunoglobulin (anti-IgG) of the same animal species as that in which the first antibody is raised (Fig. 5). Basically, the antibody used to detect the antigen and the anti-IgG antibody are separately purified onto antigen or IgG immunoadsorbents, simultaneously. The immunoadsorbent–IgG complex is then iodinated, and the iodinated anti-IgG is then eluted off. Since the first antibody was raised in the same species as IgG, it possesses immunological sites that recognize and bind to labeled IgG. These sites are different from those that are bound to the immunoadsorbent-coupled antigen. The double-antibody (antibody-IgG)-labeled complex is eluted off the antigen immunoadsorbent to be employed in either the direct conventional or the two-site immunoradiometric assay (Hales and Woodhead, 1980).

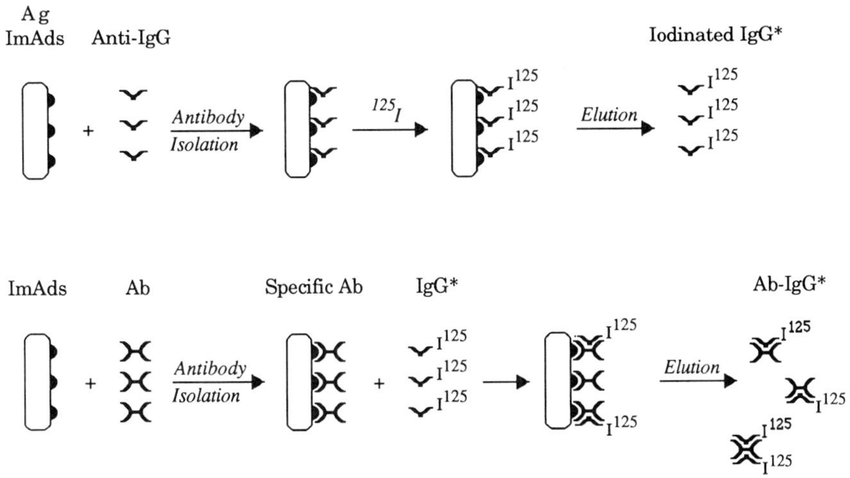

Figure 5 Schematic diagram of indirect labeling of antibody: double-antibody. Imads, immunoadsorbent; ●, antigen (Ag);)–(, specific antibody (Ab); ⌄⌄, anti-IgG.

The apparent advantages of the double-antibody labeling are numerous: (1) it is very practical to use a universal reagent such as anti-IgG, (2) it is more economical to label large amounts of readily available anti-IgG; and (3) it may enhance the specificity of the assay since one molecule of antibody may bind more than one molecule of labeled IgG.

IV. Application of Antibody Fractionation in RIA

In the conventional RIA, polyclonal or monoclonal antibodies are usually used as specific antisera. Polyclonal antisera normally contain a population of antibodies with different affinities for the antigen. Antibodies with low affinity reduce the sensitivity and the specificity of the assay. Monoclonal antibodies reveal high specificity for a single antigenic site, but have low affinity. Thus, it became essential to separate antibodies with high affinity from the heterogeneous polyclonal antisera in order to enhance the sensitivity and/or specificity of conventional RIAs. Immunoadsorption of antisera into populations with varying specificities was initially reported by Silman and Katchalski (1966) and Miles (1971). However, isolation of immunoglobulin populations with different affinities for antigen was first described by Weintraub and Kadesky (1971) in an attempt to enhance the sensitivity of RIA for human chorionic somatomammotropin. With the advent of antibody fractionation, it became apparent that the method by which the antibody fractions with high affinity constants were eluted off the immunoadsorbants was of great importance. These methods are pH gradient (Miles, 1977), denaturing solutes (Weintraub and Kadesky, 1971), aqueous acetonitrile (Hodgkinson and Lowry, 1981), and isoelectric focusing (Endo et al., 1985). All of these methods share limitations in terms of low yield, contamination by relatively low-affinity fractions, the quantity of the initial antisera used, and time required to obtain the antibody fractions. Regardless, their use in either RIAs or IRMA with labeled antibody has been widely appreciated (Kemball-Cook and Barrowcliffe, 1986). Despite their more specific nature, monoclonal antibodies have also undergone similar problems. In fact, Seghers et al. (1989) have shown that monoclonal antibodies encounter interferences by circulating anti-mouse immunoglobulins in human serum, falsifying the immunoassays. Furthermore, fractionation of serum by protein A chromatography has reduced their effect.

V. Radioreceptor Assays

Upon binding of the hormone to its receptor, the hormone–receptor complex must transduce a further signal in order to elicit a physiological response of the cell to the hormone. This property is usually referred to as the intrinsic activity of the hormone. Radioreceptor assay (RRA) measures the affinity of the hormone to its receptor in the initial step of the cascade. It is thus a functional assay that is based on the intrinsic ability of the hormone to bind to its specific receptor on the cell surface or other subcellular components. It is important to note that the assay estimates biological activity only if the hormone is known from independent assessment to have full intrinsic activity.

RRA was initially introduced for the measurement of ACTH in plasma (Lefkowitz et al., 1970). Eventually, its application in most polypeptide (Gorden et al., 1976) and steroid hormones (Lan and Baxter, 1982) became widespread. For a more comprehensive review on the historical development of the assay for peptide hormones and for the cholinergic and the β-adrenergic neurotransmitters, the reader is strongly encouraged to read the comprehensive review by Kahn (1976).

The apparent advantages of such an assay are numerous: (1) understanding the biological activity and measuring the physiologically active species of the hormone in plasma and other body fluids (Gavin et al., 1975), (2) studying the structure–function relationships of synthetic analogues of hormones, especially peptide hormones (Loumaye et al., 1982), and (3) estimating hormone receptor autoantibodies that are implicated in many endocrine disorders, such as myasthema gravis (Blecher and Bar, 1981), diabetes mellitus (Taylor et al., 1985; Boden et al., 1988), and Graves disease (Adams and Purves, 1956; Zacharija and McKenzie, 1978).

RRA is essentially an *in vitro* test. It measures an activity, not a specific structure. Molecules such as binding proteins in the biologic fluids may interact with the hormone receptor and be detected and measured by this technique. This binding may be nonspecific and, therefore, may interfere with the specific binding of the hormone to its receptor, thus altering the specificity of the assay. For these reasons, RRA is mainly a research tool rather than a routinely performed test in the diagnosis of endocrine-related diseases. Regardless, it often provides information additional to that obtained by standard radioimmunoassays and, thus, RIA and RRA combined may give insights into the mechanisms of the development of most endocrine diseases. Further details about the practical application of radioreceptor assays are beyond the scope of this chapter but can be found in several reviews (De Meyts, 1976; Gorden and Weintraub, 1992).

VI. Conclusions and Future Perspectives

Since their introduction in the late 1950s, radioimmunoassays have constituted the gold standard of basic and clinical endocrine research for the measurement of hormones in biologic fluids. The significant impact they have in the field of endocrinology justifies continuous efforts to improve underlying techniques and methodology. New assay innovations would be based on either the detection method used or the theoretical aspects of the assay itself. New technologic advances in nonradioactive detection systems have granted these methods increasing importance in improving the specificity and sensitivity of the assays. Over the past few years, cDNA technologies have made it possible to clone and sequence virtually all hormone receptors. Subsequent transfection of eukaryotic cells by plasmids carrying a receptor cDNA would allow for the expression of high levels of pure receptors in mammalian cells. Site-directed mutagenesis could lead to the synthesis of different analogs of the receptor that vary in terms of affinity and specificity by comparison to the natural receptor. This may prove an important tool in distinguishing agonists from antagonists which remains virtually impossible in assays currently used. Expression of the hormone receptor in various high-level expression systems (O'Reilly et al., 1992), for example, would result in the production of large amounts of pure receptors or receptor analogs. Such receptors or receptor analogs could be used to develop novel assays such as a combined receptor and antibody sandwich assay for a hormone. In such an assay the antibody might be coupled to a solid matrix and the receptor or receptor analog would be labeled. To complete the sandwich the hormone would have to have both specific immunologic determinants as well as receptor-binding determinants. More generally, it is likely that advances in molecular biology will lead to the development of a variety of synthetic binding proteins which will be exploited in novel assays.

VII. Appendix: Mathematical Aspects of Radioligand Assays

The principles described in this section are based on competitive radioreceptor assays, but are also applicable to competitive radioimmunoassays and any radioligand assay. Such principles are particularly applicable to a radioreceptor assay which is practically used not only to measure hormone, but also to measure the number and affinity of hormone receptors (or receptor sites) in various tissues and cells. For the latter application, familiarity with these principles, including the Scatchard (Scatchard, 1949) equation is mandatory (see later). Several assumptions are made in order to render the competitive radioreceptor and other radioligand systems suitable for mathematical analysis. These are (1) the hormone and the receptor are homogenous so that the reaction possesses a single equilibrium constant, K; (2) the hormone and the receptor are monovalent so that they bind to form a single species of the bound complex; (3) the bound hormone is perfectly separated from the free hormone; (4) the labeling of the hormone does not interfere with the formation of the hormone–receptor complexes; and (5) true equilibrium is reached in the reaction.

The effect of cooperativity in radioreceptor and radioligand binding on graphic analysis of the Scatchard plot is not discussed in this chapter. For more information, the reader is referred to studies by Freychet *et al.* (1971), De Meyts and Roth (1975), and Rodbard (1979).

A. Theory

$$H^* + R \longleftrightarrow H^*R$$

$$(F) + (B)$$

$$H \updownarrow HR$$

B. Laws of Mass Action and Scatchard Equation

1. Definitions

$H = F$ = free hormone concentration at equilibrium
R = free receptor site concentration at equilibrium
$HR = B$ = bound hormone or bound receptor sites at equilibrium
R^0 = total receptor sites
H^0 = total hormone (labeled + unlabeled)
k_1 = forward or association rate constant
k_2 = backward or dissociation rate constant
K_a = equilibrium association constant
$K_d = 1/K_a$ = equilibrium dissociation constant
B/F = ratio of bound to free hormone at equilibrium
$B/T = B/(B + F)$

2. Hormone-Receptor Kinetics

$$H + R \xrightarrow{k_1} HR \qquad \text{association rate} = k_1[H][R]$$

$$HR \xrightarrow{k_2} H + R \qquad \text{dissociation rate} = k_2[HR]$$

At equilibrium

$$k_1[H][R] = k_2[HR]$$

$$\frac{[H][R]}{[HR]} = \frac{k_2}{k_1} = K_d$$

$$K_d = \frac{1}{K_a}.$$

3. Derivation of Scatchard Equation

$$\frac{[HR]}{[H][R]} = \frac{k_1}{k_2} = K_a$$

$$\frac{[HR]}{H} = B/F = K[R]$$

$$B/F = K[R^0 - B]$$

$$B/F = -KB + KR^0 = \text{Scatchard equation.}$$

4. Scatchard Plot of Equilibrium Radioreceptor Assay Data B/F (y Variable) Is Plotted against B (x Variable for Different Amounts of H^0.

B is calculated $= [H^0] \times [B/T]$

$H^0 = H^0$ labeled $+ H^0$ unlabeled (where H^0 labeled is constant)

a. Example

y variable		$H°$	$H°$		x variable
B/F	B/T	labeled	unlabeled	$H°$	B
1.0	0.5	1×10^{-10} M	0 M	1×10^{-10} M	0.5×10^{-10} M
0.5	0.33	1×10^{-10} M	1×10^{10}	2×10^{-10} M	0.66×10^{-10} M
0.25	0.2	1×10^{-10} M	3×10^{10}	4×10^{-10} M	0.8×10^{-10} M

5. Types of Scatchard Plots

$$y \text{ intercept} = KR^0$$
$$x \text{ intercept} = R^0$$

In test tube, multiply by appropriate dilution factor to obtain R^0 in original extract or cell pellet and divide by receptor valence to convert receptor sites to receptor molecular concentration.

$$\text{slope} = -K \text{ in units } M^{-1} = L/M$$

a. Downward Curving Decreasing apparent K caused by either heterogeneity of receptor sites or negative cooperativity. Only kinetic studies can distinguish these possibilities.

$$x \text{ intercept} = R^0$$

Tangents to curve may estimate limiting high and low K, but there may be an infinite number of intermediate K's.

Do *not* extrapolate tangents to the x intercept in an attempt to obtain a concentration of high- or low-affinity sites.

These data are best analyzed by Sips or Hill plots.

b. Upward Curving Increasing apparent K caused by positive cooperativity (after artifacts have been ruled out).

$$x \text{ intercept} = R^0$$

These data are best analyzed by Sips or Hill plots.

References

Adams, D. D., and Purves, H. D. (1956). *Proc. Univ. Otago Med. Sch.* **34,** 11–12.
Addison, G. M., and Hales, C. N. (1971). *Horm. Metab. Res.* **3,** 59–60.
Addison, G. M., Hales, C. N., Woodhead, J. S., and O'Riordan, J. L. H. (1971). *J. Endocrinol.* **49,** 521–530.
Andre, M., Ferster, A., Toppet, M., Fondu, P., Dratwa, M., and Bergmann, P. (1992). *Clin. Chem.* **38,** 758–763.
Beck, P., and Hales, C. N. (1975). *Biochem. J.* **145,** 607–616.
Beckett, G. J., Wilkinson, E., Rae, P. W. H., Gow, S., Wu, P. S. C., and Toft, A. D. (1991). *Ann. Clin. Biochem.* **28,** 335–344.
Berson, S. A., and Yalow, R. S. (1959). *J. Clin. Invest.* **38,** 1996–2016.
Berson, S. A., Yalow, R. S., Bauman, A., Rothschild, M. A., and Nuverly, K. (1956). *J. Clin. Invest.* **35,** 170–190.
Blecher, M., and Bar, R. S. (1981). *In* "Receptors and Human Disease" (M. Blecher and R. S. Bar, eds.), pp. 237–257. Williams & Wilkins, Baltimore.
Boden, G., Fujita-Yamaguchi, Y., Shimoyama, R., Shelmet, J. J., Tappy, L., Rezvani, I., and Owen, O. E. (1988). *J. Clin. Invest.* **81,** 1971–1978.
Bolton, A. E., and Hunter, W. M. (1973). *Biochem. J.* **133,** 529–538.
Cheong, H. S., Chang, J. S., Park, J. M., and Byun, S. M. (1990). *Biochem. Biophys. Res. Commun.* **173,** 795–800.
Cheung, N.-K. V., Reid, M. J., Page, A. R., and Lewiston, N. J. (1981). *Ann. Allergy* **46,** 132–136.
Colt, E. W. D., Miles, L. E. M., Becker, K. L., and Shah, N. J. (1971). *J. Clin. Endocrinol. Metab.* **32,** 285–287.
Csako, G., Zweig, M. H., Benson, C., and Ruddel, M. (1986). *Clin. Chem.* **32,** 108–115.
Csako, G., Zweig, M. H., Glickman, J., Ruddel, M., and Kestner, J. (1989). *Clin. Chem.* **35,** 1655–1662.
Dale, I., Fagerhol, M. K., and Frigård, M. (1983). *J. Immunol. Methods* **65,** 245–255.
De Hertogh, R., Wolter, R., Van Vliet, G., and Vankrieken, L. (1989). *Acta Endocrinol. (Copenhagen)* **121,** 141–146.
De Meyts, P. (1976). *In* "Methods in Receptor Research" (M. Blecher, ed.), pp. 301–383. Dekker, New York.
De Meyts, P., and Roth, J. (1975). *Biochem. Biophys. Res. Commun.* **66,** 1118–1126.
Dwenger, A. J. (1984). *Clin. Chem. Clin. Biochem.* **22,** 883–894.
Edwards, R. (1985). "Immunoassay." Heinemann, London.
Ehrlich, P. H., and Moyle, W. R. (1983). *Science* **221,** 279–281.
Ehrlich, P. H., and Moyle, W. R. (1986). *In* "Immunochemical Techniques," Part I (J. Langone and H. Van Vunakis, eds.), Methods in Enzymology, Vol. 121, pp. 695–702. Academic Press, Orlando, Florida.

Ehrlich, P. H., Moyle, W. R., Moustafa, Z. A., and Canfield, R. E. (1982). *J. Immunol.* **128**, 2709-2713.
Ekins, R. P. (1978). *In* "Radioimmunoassay and Related Procedures in Medicine," Vol. 1, pp. 241-275. IAEA, Vienna.
Ekins, R. P. (1983). *In* "Immunoassays in Clinical Chemistry" (W. M. Hunter and J. E. T. Corrie, eds.), 2nd Ed., pp. 319-339. Churchill-Livingstone, Edinburgh.
Ekins, R. P., Newman, G. B., and O'Riordan, J. L. H. (1970). *In* "Statistics in Endocrinology" (J. W. McArthur and T. Colton, eds.), p. 345. MIT Press, Cambridge, Massachusetts.
Ellis, S., and Ekins, R. P. (1973). *Acta Endocrinol. (Copenhagen), Suppl.* No. 177, 106.
Endo, Y., Miyai, K., Hata, N., and Ichihara, K. (1985). *Anal. Biochem.* **144**, 41-46.
Figenschau, K. J., and Ulstrup, J. C. (1974). *Acta Pathol. Microbiol. Scand., Sect. B* **82**, 422-428.
Freychet, P., Roth, J., and Neville, D. M., Jr. (1971). *Proc. Natl. Acad. Sci. U.S.A.* **68**, 1833-1837.
Frölich, M., Walma, S. T., Paulson, C., and Papapoulos, S. E. (1990). *Ann. Clin. Biochem.* **27**, 69-72.
Gavin, J. R., III, Kahn, C. R., Gorden, P., Roth, J., and Neville, D. M., Jr. (1975). *J. Clin. Endocrinol. Metab.* **41**, 438-445.
Gibson, S., Pollock, A., Littley, M., Shalet, S., and White, A. (1989). *Ann. Clin. Biochem.* **26**, 500-507.
Glick, S. M., Roth, J., Yalow, R. S., and Berson, S. A. (1963). *Nature (London)* **199**, 784-787.
Gorden, P., and Weintraub, B. D. (1992). *In* "Williams Textbook of Endocrinology" (D. Foster and J. Wilson, eds.), 8th Ed., pp. 1647-1661. Saunders, Philadelphia.
Gorden, P., Lesniak, M. A., Eastman, R., Hendricks, C. M., and Roth, J. (1976). *J. Clin. Endocrinol. Metab.* **43**, 364-373.
Guilloteau, D., Perdrisot, R., Calmettes, C., Baulieu, J. L., Lecomte, P., Kaphan, G., Milhaud, G., Besnard, J. C., Jallet, P., and Bigorgne, J. C. (1990). *J. Clin. Endocrinol. Metab.* **71**, 1064-1067.
Hales, C. N. (1972). *Diabetologia* **8**, 229-235.
Hales, C. N., and Randle, P. J. (1963). *Biochem. J.* **88**, 137-146.
Hales, C. N., and Woodhead, J. S. (1980). *In* "Immunochemical Techniques," Part A (H. Van Vunakis and J. Langone, eds.), Methods in Enzymology, Vol. 70, pp. 334-355. Academic Press, New York.
Harvey, R. F., Williams, E. S., Ellis, S., and Ekins, R. P. (1970). *Acta Endocrinol. (Copenhagen)* **63**, 527-532.
Heber, D., Odell, W. D., Schedewie, H., and Wolfsen, A. R. (1978). *Clin. Chem.* **24**, 796-799.
Hodgkinson, S. C., and Lowry, P. J. (1981). *Biochem. J.* **199**, 619-627.
Hoffman, K. L., Parsons, G. H., Allerdt, L. J., Brooks, J. M., and Miles, L. E. M. (1984). *Clin. Chem.* **30**, 1499-1501.
Holmes, N. J., and Parham, P. (1983). *J. Biol. Chem.* **258**, 1580-1586.
Hunter, W. M. (1962). *Nature (London)* **194**, 495-496.
Kahn, C. R. (1976). *J. Cell Biol.* **70**, 261-286.
Keilacker, H., Besch, W., Woltanski, K.-P., Diaz-Alonso, J. M., Kohnert, K.-D., and Ziegler, M. (1991). *Eur. J. Clin. Chem. Clin. Biochem.* **29**, 555-563.
Kemball-Cook, G., and Barrowcliffe, T. W. (1986). *Br. J. Haematol.* **63**, 425-434.
Kemmler, W., Peterson, J. D., and Steiner, D. F. (1971). *J. Biol. Chem.* **246**, 6786-6791.
Köhler, G., and Milstein, C. (1975). *Nature (London)* **256**, 495-497.
Köhler, G., and Milstein, C. (1976). *Eur. J. Immunol.* **6**, 511-519.
Kumar, M. S., Safa, A. M., and Deodhar, S. D. (1976). *Clin. Chem.* **22**, 1845-1849.
Lan, N. C., and Baxter, J. D. (1982). *J. Clin. Endocrinol. Metab.* **55**, 516-523.
Lefkowitz, R. J., Roth, J., and Pastan, I. (1970). *Science* **170**, 633-635.

Lewis, H. M., Ratcliffe, W. A., Stott, R. A. W., Wilkins, M. R., and Baylis, P. H. (1989). *Clin. Chem.* **35**, 953–957.
Ling, C. M., and Overby, L. R. (1972). *J. Immunol.* **109**, 834–841.
Loumaye, E., Naor, Z., and Catt, K. J. (1982). *Endocrinology (Baltimore)* **111**, 730–736.
Luthman, M., Jónsdóttir, I., Skoog, B., Wivall, I.-L., Roos, P., and Werner, S. (1990). *Acta Endocrinol. (Copenhagen)* **123**, 317–325.
Matsukura, S. M., West, C. D., Ichikawa, Y., Jubiz, W., Harada, G., and Tyler, F. H. J. (1971). *J. Lab. Clin. Med.* **77**, 490–500.
Maygar, D. M., Elsner, C. W., Nathanielsz, P. W., Lowe, K. C., and Buster, J. E. (1979). *Steroids* **34**, 111–119.
Miles, L. E. M. (1971). *J. Clin. Endocrinol. Metab.* **33**, 399–408.
Miles, L. E. M. (1977). In "Clinical and Biochemical Analysis, Vol. 5: Handbook of Radioimmunoassay" (G. E. Abraham, ed.), pp. 131–177. Dekker, New York.
Miles, L. E. M., and Hales, C. N. (1968a). *Nature (London)* **219**, 186–189.
Miles, L. E. M., and Hales, C. N. (1968b). *Biochem. J.* **108**, 611–618.
Miles, L. E. M., and Hales, C. N. (1968c). *Lancet* **ii**, 492–493.
Moyle, W. R., Lin, C., and Corson, R. L. (1983). *Mol. Immunol.* **20**, 439–452.
Niederer, W. (1974). *J. Immunol. Methods* **5**, 77–82.
Odell, W. D. (1980). In "Immunochemical Techniques," Part A (H. Van Vunakis and J. Langone, eds.), Methods in Enzymology, Vol. 70, pp. 274–279. Academic Press, New York.
O'Reilly, D. R., Miller, L. K., and Luckow, V. A. (1992). "Baculovirus Expression Vectors, A Laboratory Manual." Freeman, New York.
O'Riordan, J. L. H., Addison, G. M., Woodhead, J. S., Keutmann, H. T., and Potts, J. T. (1972). *Proc. Int. Symp. Endocrinology, 3rd, London, 1971* pp. 386–392.
Rodbard, D. (1979). *Am. J. Physiol.* **237**, E203–E205.
Rodbard, D., and Weiss, G. H. (1973). *Anal. Biochem.* **52**, 10–44.
Rodbard, D., Ruder, H. J., Vaitukaitis, J., and Jacobs, H. S. (1971). *J. Clin. Endocrinol.* **33**, 343–355.
Roth, J., Gordon, P., and Pastan, I. (1968). *Proc. Natl. Acad. Sci. U.S.A.* **61**, 138–145.
Scatchard, G. (1949). *Ann N.Y. Acad. Sci.* **51**, 660–672.
Schiller, H. S., Kulchinski, L., and Luthy, D. A. (1978). *Clin. Chem.* **24**, 275–279.
Schuurman, H. J., and De Ligny, C. L. (1978a). *Clin. Chim. Acta* **89**, 191–207.
Schuurman, H. J., and De Ligny, C. L. (1978b). *Clin. Chim. Acta* **89**, 209–219.
Seghers, J., Schrurs, F., De Nayer, P., and Beckers, C. (1989). *Eur. J. Nucl. Med.* **15**, 194–196.
Silman, I. H., and Katchalski, E. (1966). *Annu. Rev. Biochem.* **35**, 873.
Sultan, Y., Avner, P. H., Maisonneuve, P., Arnaud, D., and Jeanneau, C. H. (1984). *Thromb. Haemostasis* **52**, 250–252.
Sutherland, R. M., Ratcliffe, J. G., and Chapman, R. S. (1982). *Clin. Chim. Acta* **124**, 1–11.
Taylor, S. I., Underhill, L. H., and Marcus-Samuels, B. (1985). In "Hormone Action," Part I (L. Birnbaumer and B. O'Malley, eds.), Methods in Enzymology, Vol. 109, pp. 656–667. Academic Press, Orlando, Florida.
Temple, R., Clark, P. M. S., and Hales, C. N. (1992). *Diabetic Med.* **9**, 503–512.
Weintraub, B. D., and Kadesky, Y. M. (1971). *J. Clin. Endocrinol. Metab.* **33**, 432–435.
Weintraub, B. D., Rosen, S. W., Andrew McCammon, J., and Perlman, R. L. (1973). *Endocrinology (Baltimore)* **92**, 1250–1255.
Woodhead, J. S., Addison, G. M., and Hales, C. N. (1974). *Br. Med. Bull.* **30**, 44–49.
Yalow, R. S. (1973). *Pharmacol. Rev.* **25**, 161–178.
Yalow, R. S., and Yalow, A. A. (1980). *Trans. N.Y. Acad. Sci.* **40**, 253–266.
Zacharija, M., and McKenzie, J. M. (1978). *J. Clin. Endocrinol. Metab.* **47**, 249–254.

2

Immunoradiometric Assays

M. R. Pandian and D. A. Fisher

I. Introduction
II. Principles of Immunoradiometric Assays
III. Solid-Phase Systems
IV. Selection of Antibodies
    A. Monoclonal vs Polyclonal Antibodies
    B. Purification of Antibodies
V. Variation in Immunoradiometric Assays
    A. Direct Sandwich
    B. Indirect Sandwich
    C. Simultaneous Assay vs Delayed Assay
VI. Development of an Immunoradiometric Assay
    A. Examples of Polyclonal and Monoclonal Antibody IRMA
    B. Sensitivity of IRMA
    C. Precision of IRMA
    D. Specificity of IRMA
    E. Data Reduction for IRMA
    F. Quality Control of IRMA
VII. Factors Influencing Immunoradiometric Assays
    A. Matrix Interferences
    B. Hook Effect
VIII. Advantages and Disadvantages of Immunoradiometric Assays
    A. Methodologic Advantages
    B. Methodologic Disadvantages
    C. Clinical Advantages
IX. Development of Nonradioactive Assays
X. Conclusions
    References

I. Introduction

Quantitative analytical methods involving antibodies and antigens as primary reagents are now integral to many immunoassays used in clinical, pharmaceutical, and basic scientific investigations. Clinically important analytes, proteins, peptides, steroids, or amino acid derivatives (including hormones, oncofetal antigens, other tumor markers, pathogen antigens, and specific antibodies) are quantified by the immunoassays. Even where an analyte can feasibly be determined by other standard procedures (chromatographic, colorimetric, etc.), quantitative immunoassays are often used because of their speed, simplicity, reproducibility, and relatively low cost. The widespread dissemination of extremely sensitive and specific radioimmunoassay (RIA) methods revolutionized many aspects of biology and clinical chemistry (Miles, 1977; Morris et al., 1987).

In radioimmunoassay, a small amount of the analyte to be quantified must be available in pure form. This pure substance is radiolabeled and mixed with the sample, which contains the same material in unlabeled, impure form. Specific antibody is added, the antigen–antibody complex is precipitated, and the radioactivity in the precipitate is determined. The RIA depends on competition of labeled and unlabeled antigen for the binding sites in the antibody. Thus, the higher the concentration of unlabeled ligand in the sample, the less radioactivity (labeled antigen–antibody complex) will be precipitated. The major strength of the RIA method is the specificity of the antibodies (Pandian et al., 1982). The principal disadvantages are the need for highly purified analyte to be radiolabeled and the fact that the antibody has to have high affinity to achieve high sensitivity. Hence, polyclonal antibodies are popular for RIA; low-affinity monoclonal antibodies are not as well suited.

The need to precipitate the antigen–antibody complex from solution is an additional inherent difficulty with RIA. To avoid the need for purified ligand as label, immunoassay systems were developed using labeled antibody instead of labeled antigen. This immunoradiometric assay (IRMA) is subject to interference from antibodies against constituents of the mixture other than the substance of interest. This problem is partially ameliorated by using monoclonal antibodies. The increased specificity of monoclonal antibodies, as compared to polyclonal antisera, confers on the IRMA the desired specificity (Hales and Woodhead, 1980). The IRMA is generally carried out as a solid-phase procedure, with a capture antibody immobilized on a solid carrier. Another antibody against the antigen is employed to bind the captured antigen and serve as the radiolabeled detection reagent ("readout antibody" or "labeled antibody," Fig. 1).

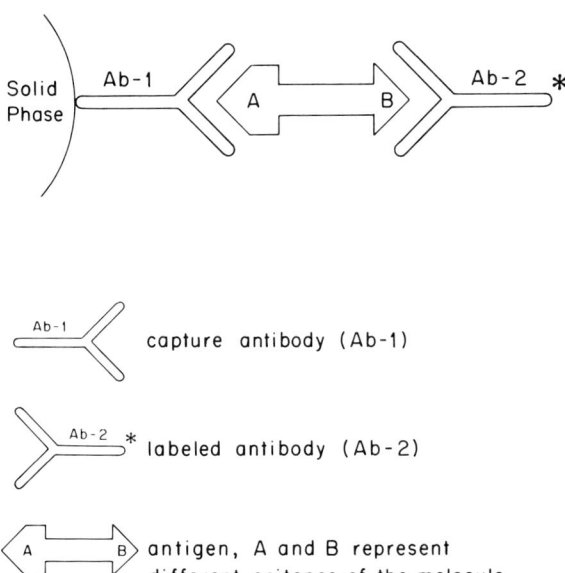

Figure 1 Direct IRMA. Schematic diagram of the antigen sandwiched between two antibodies, Ab-1 and Ab-2. The capture antibody is attached directly to the solid-phase system. The Ab-2 is the signal antibody with the radioactive label.

II. Principles of Immunoradiometric Assays

The use of labeled antibodies to measure soluble antigen was first described in 1968 by Miles and Hales (1968). This procedure was given the name immunoradiometric assay. Later publications described a variation of the IRMA called "two-site" or "sandwich" IRMA (Al-Shawi et al., 1981). Certain fundamental and theoretical advantages of the sandwich IRMA over competitive radioimmunoassay were immediately recognized, but various technical and practical obstacles tended to limit the realization of these advantages and consequently limit the use of the sandwich procedure. However, in recent years, numerous publications have appeared describing sandwich IRMAs with high sensitivity, specificity, and speed (Al-Shawi et al., 1981; Seth et al., 1984; Nussbaum et al., 1988). Careful optimization of these assays has allowed for tremendous enhancements in specificity for hormones, cancer markers, and infectious disease antigens. Diagnostic applications for this methodology have expanded tremendously. Recent advances include a 10-min quantitative assay for human parathyroid hormone (PTH) (Nussbaum et al.,

1988) and a highly sensitive second-generation thyrotropin (TSH) assay (Seth *et al.*, 1984) for use in the diagnosis of hyperthyroidism.

IRMA methodology has been referred to by many names:

1. Immunoradiometric assay
2. Radioimmunometric assay
3. "Sandwich" assay
4. Two-site assay/two-antibody assay/double-antibody assay
5. Labeled antibody assay/radiolabeled antibody assay
6. Second-generation immunoassay
7. Capture/signal antibody assay

The fundamental requirement for IRMA is a multivalent antigen, capable of being bound by at least two antibodies simultaneously. One antibody is usually attached to a solid-phase support (test tube or bead) or is in some other way made precipitable, such as by addition of a species-specific precipitating antibody or coupling with microcellulose or magnetic particles. The second antibody is labeled with a tracer material such as radioiodine (^{125}I). The antigen

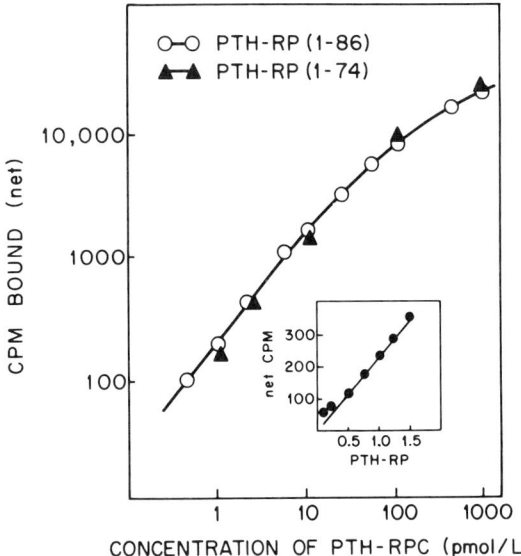

Figure 2 Dose–response curve for PTH-RP using two standard preparations, PTH-RP (1–86) and PTH-RP (1–74). The log concentration of PTH-RP is plotted against the response (log CPM bound to the bead). The CPM bound at 0 standard (NSB) is subtracted from all the signals for various doses. The inset is the dose–response curve at a low concentration of PTH-RP. **Based on Fig. 2 in Pandian *et al.*, 1992.**)

is incubated with both antibodies either simultaneously or sequentially, forming a sandwich (Fig. 1). Separation of labeled antibody bound in the sandwich from free labeled antibody is accomplished by washing (in solid-phase assays) or by precipitation (centrifugation or magnetic separation in the other methods). The IRMA uses excess amounts of antibodies both on a solid phase (to facilitate separation) and for labeling. There is a direct, linear antigen–antibody reaction, because the limiting reactant is the antigen. The bound signal is thus directly proportional to the concentration of antigen, since label can only bind to solid phase via attachment to antigen. Figure 2 shows a typical standard curve, plotting the dose of the standard (PTH-related protein) versus the signal (CPM bound). The linear response is manifest until the "excess"

Table I
Comparison of RIA and IRMA

	IRMA	RIA
1. Radiolabelled material	Antibody	Antigen
2. Principle of the method	Sandwich (of antigen with two antibodies)	Competitive inhibition
3. Antibodies	a. Purified antibodies are recommended	Crude antibody could be used
	b. Two antibodies required. Capture antibody (Ab-1) and radiolabeled antibody (Ab-2)	Single polyclonal antiserum commonly used
	c. Monoclonal (low affinity) as well as polyclonal antibodies have been successfully used	High-affinity antiserum necessary for best results
	d. Large amount of antibody required when compared to RIA	Small quantity of antibody required
4. Purity of the antigen	Crude antigen can be used	Purified antigen required for radiolabeling
5. Size of the antigen	Successfully used for large molecules (peptides, proteins); providing at least two epitopes	Useful for small as well as large molecules
6. Separation system	Solid phase is the separation system of choice	Liquid-phase system commonly used; solid phase can also be used
7. Standard curve	Wide range; linear relationship between the concentration of standard and bound counts (highest counts with highest standard)	Limited range; high sample has to be diluted; inverse relationship between concentration of standard and bound counts

concentration of the labeled antibody begins to affect the reaction; that is, it is no longer in true excess. If more labeled antibody were present, the effective linearity of the system could be extended. Practically, safety precautions limit the use of excess radioactive material, and it is recommended that radioactivity be limited to 0.2 μCi/tube.

IRMA has many similarities to RIA with regard to reagent requirements and the kinetics of antibody and antigen interaction. IRMA and RIA are compared in Table I.

III. Solid-Phase Systems

A number of solid supports are available commercially (Valkirs and Barton, 1985). These are useful for specific adsorption and concentration of antigen or antibodies, and they aid in the separation of specific analytes from biological fluids (serum, urine, etc.). An additional advantage is that the separation of the antigen–antibody complex from free analytes can be achieved by simple washing or filtration eliminating the need for centrifugation. This separation procedure has enhanced the availability of home-testing products (Norman *et al.*, 1985). A list of solid support systems commonly used follows:

1. Tubes (polystyrene, polypropylene, and glass)
2. Beads (polystyrene) in different sizes
3. Cellulose particles
4. Magnetic particles
5. Polystyrene particles (sizes from 0.018 μm (18 nm) to 1000 μm (1 mm)
6. Membranes (nylon, cellulose)

The surface chemistry of solid-phase systems differs depending on the material and its preparation. For instance, plain polystyrene is hydrophobic; but the surface chemistry can be altered to produce very hydrophilic surfaces by adding a wide variety of functional groups (Catarero *et al.*, 1980). These include aldehydes (CHO), aliphatic amines (CH_2 NH_2), amides (CO NH_2), aromatic amines (O NH_2), carboxylic acid (COOH), chloromethyl groups (CH_2 Cl), hydrazide (NH NH_2), hydroxyl groups (OH), sulfates (SO_4), and sulfonates (SO_3).

The linking of analytes (antibody or antigen) to a solid phase can be achieved by simple adsorption or by covalent linkage. A wide variety of solid-phase materials are available. A common problem of solid-phase materials is nonspecific attachment or adsorption of labeled analytes (high nonspecific binding). This often can be minimized by altering the solid-phase functional group. Methodologies for adsorption of analytes to solid surfaces have been described in a variety of publications (Wood and Missler, 1990). Methods for specific assay systems are included in Section VI.

Adsorption of protein and other biochemicals onto the solid supports is due to hydrophobic binding forces largely independent of pH (Catt and Tregear, 1967). Nonetheless, for protein adsorption, charges do play a role. The conformation of protein is affected by charge, and the formation of the most compact monolayer of proteins is generally maximum at the isoelectric point of the protein (pH 7.8 for human IgG). Nonionic detergents (Triton X-100, etc.) depress the binding of proteins, mainly by hydrating the proteins and hindering hydrophobic binding. Other factors which influence adsorption are ionic strength of the buffer, temperature, time, and analyte concentration.

IV. Selection of Antibodies

A. Monoclonal vs Polyclonal Antibodies

A polyclonal antiserum contains a spectrum of antigen-directed antibodies; therefore, in the presence of the several antibody species, the observed concentration of antigen–antibody will be the sum of all the separate antigen–antibody reactions. These reactions are affected by the binding strength (affinity constant, K_a) and the concentration of each antibody. In the monoclonal antibody reaction, the combining site—where antibody reacts with antigen—is the same for every molecular event resulting in homogeneous reactivity. Thus, linear immunometric assays (IRMAs) depend on two criteria: excess reagent and homogeneity of reactivity. Polyclonal antisera display neither while monoclonal antibodies display both characteristics. The availability of antibodies, mono- or polyclonal, dictates the use of either monoclonals and/or polyclonals in the respective systems. When polyclonal antibodies are selected for an IRMA it is understood that a large volume of antiserum is available (source could be sheep, goat, donkey, horse, etc.) and that the laboratory is equipped for purification methodologies.

Two monoclonal antibodies should provide better results than a mixture of monoclonal and polyclonal, or purely polyclonal, antibodies. In practice, this is not always the case. Many monoclonal antibodies have low-affinity constants when compared with their polyclonal counterparts. Many monoclonal antibodies are "sensitive" to handling (i.e., immobilization or labeling causes loss of activity). In the case of polyclonal antibodies, the clones which lose binding activity during coating or labeling are compensated by those which are "resistant" to these processes. In screening of polyclonal antisera, all antibodies which show poor binding after coating or labeling, or show high nonspecific effects, are eliminated from the list of "potentials." For monoclonal antibodies, this sorting must take place for each antibody.

B. Purification of Antibodies

Antibodies must be purified prior to use either from ascites fluid (monoclonal) or from serum (polyclonal). Simple purification steps include ammonium sulfate or polyethylene glycol (PEG) precipitation or column separation methods (molecular sieving or ion-exchange column methods, DEAE-cellulose). For the purification of monoclonal antibodies, protein A or protein G is commonly used (Martin and Helening, 1991; Nussbaum et al., 1987).

1. Protein A/Protein G Separation Methods

Protein A is a cell wall constituent of *Staphylococcus aureus*. The purified protein, 42 kDa, exhibits a markedly extended overall shape featuring four homologous, globular units in the N-terminal region (Martin and Helening, 1991). Each of these units is capable of binding the Fc portion of certain mammalian IgG molecules. Commonly available protein A is a formalin-fixed and heat-killed *S. aureus*, Cowan I strain. This strain has been found to express more than 50% of its protein as protein A. The resulting reagent is a high-capacity, stable solid-phase immunoadsorbent. The binding capacity is greater than 1.6 mg human IgG/ml 10% cell suspension.

Protein G is a cell wall constituent of *Streptococcus* species. Like protein A, this protein will bind the Fc portions of certain mammalian IgG molecules. However, numerous studies have shown that protein G reacts with a broader range of mammalian IgGs and has stronger binding in many cases. The immunoglobulins to be purified bind to the columns, protein A Sepharose 4B or protein G Sepharose 4B. The column-bound immunoglobulins are eluted with glycine HCl buffer and neutralized immediately.

2. Affinity Purification of Antibodies

Affinity purification is commonly used for polyclonal antibody fractionation, particularly for isolating specific antibodies from the serum pools. Affinity purification of analyte specific antibodies from a polyclonal antibody pool or a mixture of antibodies can be achieved using columns to which the specific protein is bound. Nussbaum et al. (1987) separated amino terminal specific antibodies (anti-PTH 1–34) and carboxy terminal specific antibodies (anti-PTH 39–84) from a polyclonal pool of antisera developed in goats using

PTH 1–84. This was possible since the primary structure of PTH was known and synthetic fragments (1–34 and 39–84) were available. Similar fractionation of polyclonal antibodies has been accomplished for ACTH (Zahradnik et al., 1989) and PTH-related protein (PTH-RP) (Pandian et al., 1992). Such fractionation is not possible if the molecular structure of the antigen is not known. Hence, this technique may not be applicable to crude antigens or antigens not fully characterized.

Various factors have to be considered in the use of affinity columns including choice of matrix, choice of coupling method, column parameters, wash buffers, elution buffers, and reuse of the column.

a. Choice of Affinity Matrix A variety of affinity supports are available commercially (Morrissey, 1981) (Sepharose from Pharmacia; Affi-gel from Bio-Rad Laboratories). Variations include the composition of the solid support, the addition of "spacer" arms which keep the coupled antibody away from the solid support matrix (and so improve access), and the chemical groups provided for coupling with antibody. The ideal support matrix will have minimal binding of proteins nonspecifically. Nonspecific binding of proteins is more likely if the support matrix contains charged groups or hydrophobic spacearms. The final choice among the commercially available materials is largely a matter of personal preference.

b. Choice of Coupling Method Some of the support materials are available for use with several alternative conjugation methods (Cuatrecasas et al., 1968). There is limited evidence for the superiority of one method over another. However, immunoabsorbents based on ether linkages (derived from epoxy-derivatized agarose) are believed more stable than those prepared using cyanogen bromide-activated agarose or thiol-based materials. However, cyanogen bromide-activated agarose is commonly used because of simplicity and reproducibility (Cuatrecasas et al., 1968). Coupling reactions usually involve a reaction in which electron-rich atoms (N or S) on the protein react with the derivatized matrix. Ammonium, Tris salts, or azide inhibit the reaction and should not be included in the coupling buffer. In some cases the solid support matrix is available commercially and the derivatization is carried out in the laboratory (cyanogen bromide-activated Sepharose 4B). Unless large-scale work is planned, it usually is more economical to buy already-derivatized support material.

c. Column Parameters The size of the column will depend on the amount of antigen bound and the amount of antibody to be prepared. Typically, 1 to 3 mg of purified antigen can be bound per milliliter of support bed (Sepharose CL-4B, Pharmacia). The resulting column will have a capacity approximating 0.5 mg of immunoglobulin per milliliter of bed. Using a larger column than is necessary increases nonspecific absorption and decreases yields.

With immunoaffinity purification the antibody is bound reversibly to the column and retained during washing. The bound antibody is dissociated from the antigen on the column by the elution buffer. The size of the column is important if the antibody has low affinity. Conversely, if the antibody has high affinity, the column dimensions are less important.

A typical setup consists of a column containing the gel coupled with antigen. The column should be precycled immediately before use, by washing with the elution buffer (to remove any impurity which may be eluted later and contaminate the immunoglobulin) and then equilibrated in the sample buffer. The column is run preferably at 4°C., unless adsorption and elution are temperature dependent. A slow flow rate (20 to 30 ml/hr for a 5-ml column) is used; the antiserum to be purified is applied followed by 5 to 10 column volumes of sample buffer. One-milliliter fractions are collected, and all fractions are kept for subsequent analysis.

d. Wash Buffers Having applied the antibody solution to the column, it is necessary to wash off unbound material. This may be done using the sample buffer, but it may be advantageous to use acidic buffer (pH 5.0) to remove nonspecifically bound material (i.e., material other than the antibody). This assumes that such binding (due to electrostatic interaction, hydrophobic interaction, and possibly, cross-reactivity with the antigen) will be of a lower affinity than antigen–antibody binding. Further washing is monitored to determine whether the antibody is eluted; a sequence of washes suitable for the antibody is determined.

e. Elution Buffer Elution of bound antibody requires a change in conditions such that the antigen–antibody complex dissociates (Zola, 1987). These conditions involve high or low pH, addition of agents which reduce the polarity of the solvent (for example, dioxane) or detergents, or the use of high salt concentrations. Certain ions are more conducive to dissociating complexes than are others; these are referred to as "chaotropic" ions. Thiocyanate is one of the most chaotropic anions. Thus potassium thiocyanate or lithium thiocyanate at high concentration (3 M) often is used as eluant. An acidic pH (2.5) is used commonly for γ-globulin elution. However, the elution buffer needs to be optimized for each particular antibody.

Damage to the eluted immunoglobulin can be reduced by rapid neutralization or dilution of the eluent. The eluate may be collected into a high-molarity neutral buffer, (0.5 M sodium phosphate; Pandian *et al.*, 1992), or a pH-stat may be used to neutralize the eluate. The elution should be monitored (for example, by using radioactive antigen), but the immunoglobulin usually elutes in 1 to 2 column volumes if the eluting buffer does not allow antibody–antigen reassociation. Elution effectiveness will also depend on physical parameters. Higher temperature favors dissociation but accelerates any reactions which damage the antibody. Elution may be more effective if the immunoabsorbent is removed from the column and eluted in a larger volume, with mixing (Zola, 1978).

f. Regeneration and Reuse of the Column Effective antibody purification may be achieved with a single use of the column, but in most studies cost effectiveness is improved if the column is reused. Elution conditions sometimes reduce the capacity of the column; this is more likely if elution is achieved using extreme pH (Oldham, 1984). The column should be restored to a neutral pH and chaotropic ions washed off as soon as possible by passing through 8 to 10 column volumes of neutral buffer. The covalent bonds linking the antigen to the matrix are subject to hydrolysis and there will be some spontaneous loss of antigen. This is minimized by storing the column in the cold, in buffer containing a bacteriostatic agent. Probably the most important factor limiting the life of an immunoabsorbent column is contamination with particulate and fatty material from the antibody solution. It is important to filter the mixture or dilute the antiserum before applying it to the column. Columns which are running slowly because of contamination may be improved by removing the immunoabsorbent from the column and washing it extensively before repacking. If handled carefully, immunoabsorbents are stable for 1 year or longer and can be reused many times.

3. Epitope Mapping of Antibodies

Selection of specific antibodies from a pool of antibodies can be achieved successfully if epitope maps are available. To map the epitope of an antibody, one has to know the amino acid sequence of the antigen. Based on this information, five–eight amino acid peptides can be synthesized to cover the full length of the molecule. As shown in Fig. 3, the peptide sequences are overlapped (e.g., the sequential peptides could be AVSEHQ, VSEHQL, SEHQLL, etc.). These peptides are adsorbed to microtiter plates or attached to a polyethylene pin where the top binds the specific population of antibodies. The antibodies are quantified by enzyme-linked immunosorbant assay (ELISA) testing using peroxidase-labeled second antibody (Mason *et al.*, 1982). By plotting the optical density (OD) vs the sequence, it is possible to identify the specific amino acid sequences (epitopes) with which the antibody has maximal reactivity (Fig. 3). Epitope mapping is helpful in selecting antibodies which

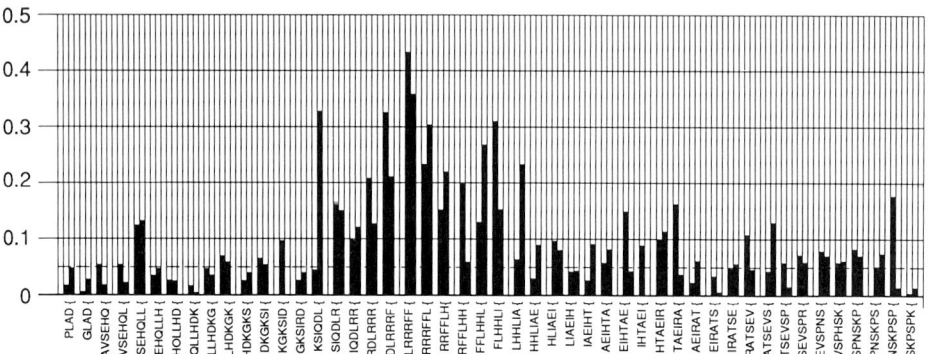

Figure 3 Epitope mapping of a polyclonal antibody (goat) for PTH-RP (1–86). The diluted antibody is adsorbed to the peptide (six amino acids) linked to the pins. The synthesized peptides represent the sequential amino acids of PTH-RP with overlapping sequences. The adsorbed antibody to the pins (bound to peptide) is washed and quantified using rabbit anti-goat γ-globulin conjugated with horseradish peroxidase. The horseradish peroxidase is quantified using O-phenylenediamine in a spectrophotometer at 405 μ. The optical density (OD) is directly proportional to the concentration of γ-globulin. The OD at 405 μ is plotted against the peptide sequence. The first two peptides with four amino acids (PLAG and GLAG) are the controls. The area around the sequence of LRRRFF has the maximal OD indicating the primary epitope for the antibody is in the overlapping sequence close to LRRRFF (*Source*: Dr. S. Hutchison, Nichols Institute Diagnostics).

produce minimal steric hindrance when the sandwich is formed. Screening for suitable monoclonal antibodies is described in detail for TSH in Section VI.

V. Variations in Immunoradiometric Assays

The principle of IRMA is illustrated in Fig. 1. Two forms of solid-phase IRMA are possible, direct sandwich and indirect sandwich assays (Boscato *et al.*, 1989). And each of these can be performed as simultaneous assays (one step) or delayed assays (two step) (Hoffman, 1985).

A. Direct Sandwich

Figure 1 illustrates a direct IRMA assay in which the antigen (Ag) has at least two different epitopes, A and B. Two antibodies recognize different epitopes; one utilized as the capture antibody (Ab-1) recognizes epitope A while the second, labeled antibody (Ab-2), recognizes epitope B. The capture antibody (Ab-1) is attached to the solid phase either covalently or noncovalently. The signal antibody (Ab-2) is radioactively labeled, commonly with ^{125}I. When all the analytes (solid-phase Ab-1, Ab-2, and the antigen) are mixed, a "sandwich" of antigen–antibodies complex (Ab_1–Antigen–Ab_2) is formed. In this IRMA, the capture antibody is attached directly to a solid phase and there is a possibility of steric hindrance when the reaction has to occur close to the solid surface. Indirect IRMA alleviates the steric hindrance, enhances reaction kinetics, and improves assay sensitivity.

B. Indirect Sandwich

The indirect sandwich format is a modification of the direct IRMA (Boscato *et al.*, 1989). To facilitate the kinetic reaction, the capture antibody is attached indirectly to the solid phase through a bridge (Figs. 4A and B). In Fig. 4A, the solid base attachment is mediated through the avidin–biotin reaction taking advantage of high affinity of biotin with avidin (K_d 10^{-15}) (Odell *et al.*, 1986). The solid phase has avidin on the surface; biotin is covalently attached to the capture antibody (Ab-1). Detailed description of the avidin–biotin-mediated indirect sandwich for TSH is described later in this chapter (Section VI,A). By analogy with the biotin–avidin system a first antibody–second antibody system (γ-globulin–anti-γ-globulin) also can be used in an indirect sandwich system (Fig. 4B). A detailed description of this system using rabbit γ-globulin and anti-rabbit γ-globulin is described for the PTH-RP assay (Section VI,A).

The indirect sandwich approach also can facilitate signal amplification. In this format, the signal antibody (labeled antibody) can be amplified by using second antibodies which are radiolabeled. One signal antibody (Ab-2) can accommodate many radiolabeled second antibody molecules (Fig. 4C). Similarly, an avidin–biotin system could be used. For this purpose, the biotin is radiolabeled and the resultant complex will be solid phase–Ab-1–Antigen–Ab-2–biotin–^{125}I–avidin. Since many biotin molecules could be linked to Ab-2, they can form a sandwich with radiolabeled aviden and, thus, increase the signal. The indirect sandwich also can be formatted with amplification at both ends, capture and signal (Berman and Bosch, 1980).

C. Simultaneous Assay vs Delayed Assay

Direct or indirect assays can be accomplished by simultaneous addition of all the analytes and the solid phase. This "one-step" assay procedure is

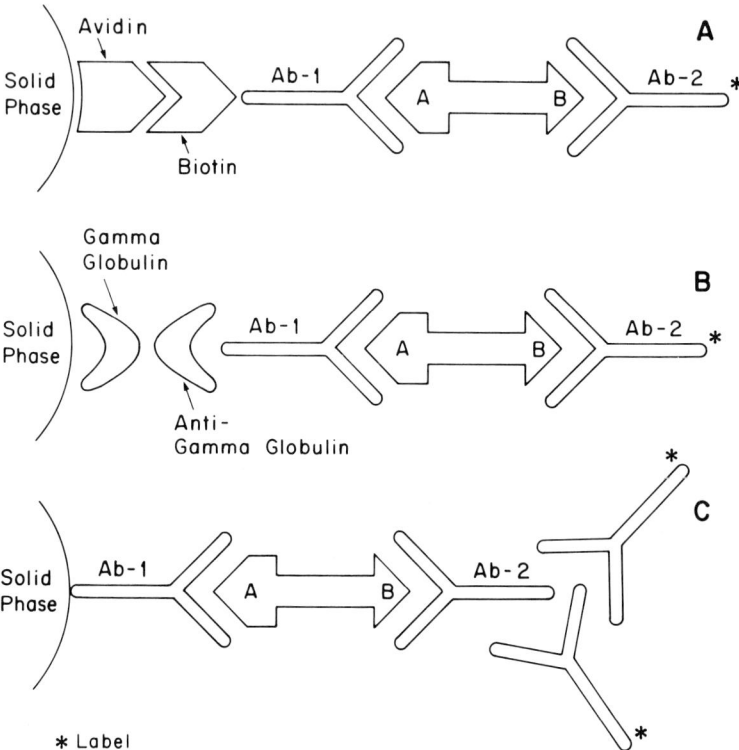

Figure 4 Schematic diagrams of indirect (multistep) IRMA sandwich assay reactions. The antigen vs antibodies interactions are similar to the direct sandwich as in Fig. 1, except (A) adsorption of the capture antibody (Ab-1) to the solid phase is accomplished through the intermediate avidin–biotin bridge; (B) adsorption of Ab-1 to the solid phase is accomplished through a γ-globulin and anti-γ-globulin bridge; and (C) the signal antibody is amplified by radiolabeled second antibody (anti-immunoglobulin antibody).

popular because of simplicity and minimal procedural steps. Interference by serum protein or steric hindrance of the reactants may occur resulting in poor precision and limited assay sensitivity (Boscato and Stuart, 1988). A sequential assay approach (two step or multistep in indirect sandwich) can be used to minimize these problems. For a sequential assay, the sample is incubated with the solid phase first (capture Ab) and washed prior to the addition of labeled antibodies. In some cases the antibodies (capture and labeled) are incubated with antigen before the solid-phase separation system is added. The sequential assay has been used successfully to minimize a hook effect (Rodbard *et al.*, 1978), to extend the standard curve range, to increase the sensitivity by immunoconcentration, and to improve the reproducibility by minimizing interference by nonspecific sample components (Newman *et al.*, 1989).

VI. Development of an Immunoradiometric Assay

A. Examples of Polyclonal and Monoclonal Antibody IRMA

Linearity of immunoradiometric assays depends on two criteria: excess of antibody and homogeneity of reactivity. Hence, monoclonal antibodies are commonly used in IRMA assays. Polyclonal antisera have been used in some cases. It is possible to build a two-site assay using two monoclonal antibodies (mono–mono system) (Odell *et al.*, 1986), two specific polyclonal antibodies (poly–poly system) (Pandian *et al.*, 1982), or one monoclonal and one polyclonal antibody (mono–poly system) (Schwarz *et al.*, 1985). We describe two systems in detail, a poly–poly system for PTH-RP and a mono–mono system for human thyrotropin (hTSH).

1. IRMA for PTH-RP

a. General Description A humoral factor responsible for hypercalcemia-associated cancer has been isolated from several solid tumors. The gene for this factor has been cloned and characterized as a PTH-RP. An IRMA for PTH-RP was originally described by Burtis *et al.* (1990); Pandian *et al.* (1992) modified the IRMA to improve sensitivity and specificity. The assay uses affinity-purified polyclonal antibodies from rabbits; one is N-terminal specific, the other C-terminal specific.

b. Affinity Purification of Antibodies The circulating form(s) of PTH-RP is not well defined. By analysis of the messenger RNA from the tumor tissues, at least three polypeptides of different lengths (1 to 139, 1 to 141, and 1 to 173) have been predicted. Availability of intact PTH-RP (1–141) is limiting, but the 1–74 and 1–84 peptides are available commercially. Rabbits were immunized with PTH-RP peptide (1–74) conjugated with keyhole limpet hemocyanin (Burtis *et al.*, 1990). The antibodies were purified through an affinity column into amino terminal antibodies (anti-PTH-RP 1–36) and carboxy terminal antibodies (anti-PTH-RP 37–74). Human PTH-RP (1–36) and human PTH-RP (37–74) immunoaffinity resins were prepared by coupling 1 mg of human PTH-RP (1–36) or human PTH-RP (37–74) to 1 g of cyanagen bromide-activated Sepharose 4B. Protein A affinity columns were prepared by coupling protein A to Sepharose 4B (Bacquet and Twumasi, 1984).

Polyclonal rabbit antiserum developed by immunizations with human PTH-RP (1–74) was affinity purified into two classes: one specific for PTH-RP (1–36), the other specific for PTH-RP (37–74).

In a typical purification procedure, 4 ml of polyclonal anti-PTH-RP (rabbit) was applied to a PTH-RP (1–36) affinity column. The antiserum was cycled through the column five times and collected for application to a PTH-RP (1–74) affinity column. The column was washed with 20 ml of 0.01 M phosphate-buffered saline (PBS), pH 7.4, containing 0.1% bovine serum albumin (BSA); 10 × 2 ml fractions were collected in 12 × 75-mm glass tubes. A series of washings/elutions were conducted with 0.2 M glycine, 0.1% BSA at progressively decreasing pH (pH 4.0, 3.5, 3.0, 2.5). After each washing 6 × 2-ml fractions were collected in 12 × 75 tubes with 500 μl of 0.2 M sodium phosphate buffer, pH 7.4, and 30 μl phenol red (Pandian et al., 1992). Phenol red was the pH indicator and phosphate buffer was used to neutralize the pH. After collection, pH was adjusted with 0.4 M NaOH. For subsequent use the column was washed with 20 ml PBS and stored at 4°C.

The fraction eluted at pH 4.0 was discarded. Dilutions (1:10 and 1:100) of the remaining collected fractions were prepared in assay buffer and reacted with ^{125}I-labeled PTH-RP (1–36) to check binding and with ^{125}I-labeled PTH-RP (37–74) to check cross-reactivity. Fractions with optimal binding were pooled and the pool was filtered through 0.20-μm filters. These immunoglobulin preparations were further purified through a protein A Sepharose 4B (PAS) column to remove BSA and denatured immunoglobulins. The preparation eluted at pH 2.5 had higher affinity than that eluted at pH 3.5. Eluted antiserum that failed to bind PTH-RP (1–36) was adsorbed onto an affinity column with immobilized human PTH-RP (37–74). The adsorbed antibodies were eluted with glycine–HCl and further purified with PAS as described earlier. The high-affinity antibodies were then assessed for sensitivity and yield using ^{125}I-labeled PTH-RP (37–74). Specificity was confirmed by showing failure to bind to ^{125}I-labeled Tyr^0–human PTH-RP (0–36). The purified anti-PTH-RP (37–74) was quantified spectrophotometrically (at 280 nm) and diluted in PBS to the required concentration.

c. Coating of the Polystyrene Beads The antibodies can be adsorbed directly to the solid-phase bead or indirectly through protein layers. Indirect adsorption is preferred to minimize the steric hindrance in the reaction kinetics between antibodies and antigen. This approach also reduces the amount of antibody used. The plastic beads were coated indirectly with affinity-purified anti-PTH-RP (37–74). Plastic beads (8 mm) from Clifton Plastics, Inc., were used for coating. The beads were washed with running distilled water for 10 min and incubated with rabbit IgG solution (15 mg/l) in PBS (250 ml for 1000 beads) for 20 hr at room temperature (23°C). The adsorption of antibodies to the beads is a noncovalent interaction. After washing with deionized water, the beads were incubated for 20 hr with goat anti-rabbit γ-globulin solution diluted 1:100 with 0.1% BSA in PBS. The beads were then washed with distilled water, incubated with anti-PTH-RP immunoglobulin (500 μg/l), and stored at 4°C in 0.1% BSA/PBS until use.

d. Radiolabeling of Anti-PTH-RP The affinity-purified immunoglobulins were radioiodinated with ^{125}I using the Chloramine-T method (Hunter and Greenwood, 1962). The specific activity of the iodinated immunoglobulin ranged from 100 to 400 μCi/μg. Lower specific activity results in poor assay sensitivity; high specific activity leads to poor stability of the radiolabeled material.

e. Development of PTH-RP IRMA The anti-PTH-RP (1–36) and anti-PTH-RP (37–74) antibodies were used in combination, one as capture and the other as radiolabeled antibody. The IRMA format is shown schematically in Fig. 5. The assay was performed sequentially (two-step IRMA) using anti-PTH-RP (37–74) antibodies immobilized onto polystyrene beads as the capture antibodies and ^{125}I-labeled anti-PTH-RP (1–36) as the tracer. In a typical assay, we incubated 200 μl of sample or standard with rotation at 4°C for 16 hr with an antibody-coated bead in a 12 × 75-mm polystyrene tube. We then washed the bead with Triton X-100, 1 ml/l, in PBS and added 200 μl of radiolabeled anti-PTH-RP (1–36) in assay buffer (70,000 CPM). One liter of assay buffer contained 10 mmol of sodium phosphate buffer (pH 7.4), 154 mmol of sodium chloride, 1 g of BSA, 1 ml of Triton X-100, 50 ml of normal rabbit serum, 39.4 mg of aprotinin, and 2.5 mg of leupeptin (Pandian *et al.*, 1992). Incubation continued for another 16 hr at 4°C with rotation. After washing the beads, the radioactivity is determined in a gamma counter.

Human PTH-RP (1–86) was used routinely as the standard in the assay. We routinely diluted the standard in the assay buffer to 0.5, 1.0, 2.0, 10, 50, 100, 500, and 1000 pmol/l. The standard curve is shown in Fig. 2. There was a linear relationship between the concentration of PTH-RP and CPM bound. PTH-RP peptides containing 74 amino acids or more (1–74, 1–86, or 1–141) had similar binding characteristics. Smaller fragments (PTH-RP 1–36, 37–74) did not result in sandwich indicating that the two epitopes must be in the same molecule for a successful IRMA. This IRMA has been optimized for sensitivity, precision (reproducibility), ease in handling, and parallelism of the samples (Pandian *et al.*, 1992).

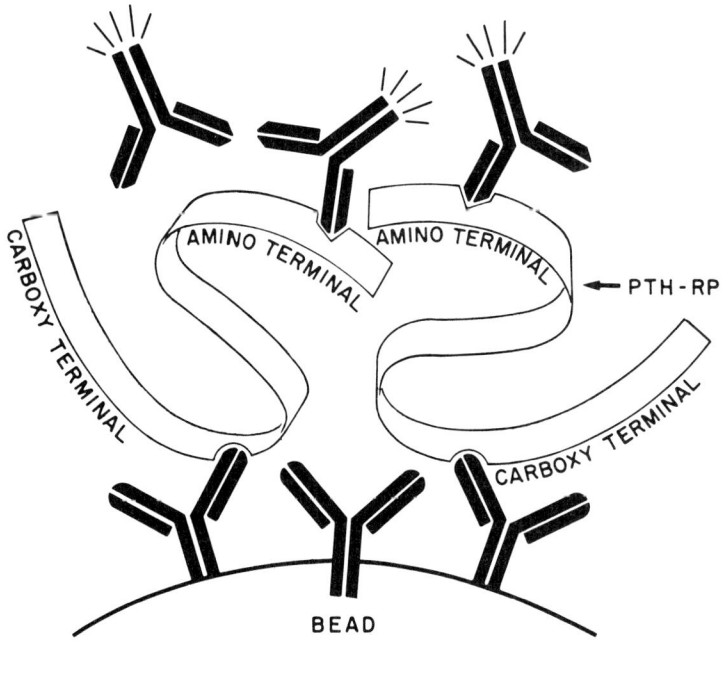

^{125}I Anti PTH-RP (1-36)   Anti PTH-RP (37-74)

Figure 5 Schematic diagram of IRMA for PTH-RP. See text for details.

2. IRMA for TSH

a. General Comments As indicated earlier, linear immunometric assays depend on two criteria: excess antibody and homogeneity of reactivity by the antibodies. Monoclonal antibodies display both characteristics, since they can be produced in large quantities and each reacts with one epitope of a molecule. Odell *et al.* (1986) developed a sensitive, specific, noncompetitive sandwich

IRMA for hTSH using two monoclonal antibodies. The assay was constructed as an indirect IRMA using the avidin–biotin method of separation.

b. Biotinylation of Monoclonal Antibody The antibody for biotinylation was dissolved in PBS. N-hydroxysuccinimidobiotin (NHS–biotin; Sigma Chemical Co., St. Louis, MO) was added to the antibody solution (44% of the weight of the protein), and the NHS–biotin, dissolved in N,N-dimethylformamide, was added to the dissolved protein. After mixing for 2 min and incubating at room temperature for 16–24 hr, the reactant was dialyzed in 0.9% sodium chloride solution. The dialyzed, biotinylated antibody was diluted with the antibody buffer (0.4 mol of potassium phosphate, 0.4 mol of sodium phosphate, 10 ml of horse serum, 5 g of bovine serum albumin, 10 ml of normal mouse serum, 1 g of sodium azide, and 8.77 g of sodium chloride per liter, pH 7.4) to a working dilution.

c. Preparation of Avidin-Coated Beads The coating was performed at room temperature similar to the procedure described for PTH-RP. The beads were rinsed with distilled water, allowed to stand with a 1% glutaraldehyde solution in distilled water for 16–24 hr, and rinsed 10 times with distilled water to remove any unbound material. The mixture then was incubated for 6–8 hr with a 30 mg/ml solution of biotinylated bovine serum albumin in PBS. After washing, the beads were incubated for 16 to 24 hr in 30 mg/ml of avidin and 1 mg/ml of BSA in PBS, rinsed in distilled water, and coated by incubating for 20 min with phosphate-buffered saline containing 1 g of bovine serum albumin and 25 g of sucrose per liter. They were either lyophilized to dryness or spread on adsorbent paper and allowed to dry thoroughly. When dry, the beads were placed in plastic containers, sealed tightly, and stored in the $-70°C$ freezer.

d. Screening of Monoclonal Antibodies Odell et al. (1986) purified a total of 10 monoclonal antibodies; all were iodinated and biotinylated. Using the solid-phase assay, each iodinated antibody was tested in the presence of hTSH and biotinylated antibody. All possible combinations of iodinated antibodies and biotinylated antibodies were tested (in a χ^2 design) in the presence or absence of TSH. TSH-free "0" standard was used to determine nonspecific binding. A low concentration of TSH was used to test sensitivity and a high standard for capacity. The combinations of antibodies having the higher affinities for hTSH were chosen for assay development, one as the biotinylated capture antibody and the other as radiolabeled antibody.

e. Assay Procedure Biotinylated monoclonal antibody was diluted with the radioiodinated antibody buffer to a concentration of 0.5 μg of biotinylated antibody and 300,000 CPM of radiolabeled antibody in a total volume of

100 μl. The assay was performed by combining 100 μl of standard or serum. The tubes were rotated for 2 hr at room temperature. Following incubation, the liquid was aspirated from the bead, after which the bead was washed twice with 3 ml of PBS containing 0.1% of Triton X-100, and radioactivity counted in a gamma counter. The assay was designed as a second-generation TSH assay to screen patients with hyperthyroidism from normals as well as to identify hypothyroid patients. The standard curve ranged from 0.1 to 100 mIU/liter (Fig. 6). The assay had a signal/noise ratio of 2 at 0.1 mIU/liter.

B. Sensitivity of IRMA

In a sandwich assay, sensitivity usually is defined as that concentration of analyte corresponding to a mean signal plus two standard deviations at zero analyte concentration. The sensitivity is calculated by performing assays on 20 tubes with 0 standard and determining the concentration related to mean +2 SD CPM (Rodbard and Feldman, 1978). Both theory and observation demonstrate that sandwich assays are more sensitive by about an order of magnitude than competitive RIA methods. The direct dose–response curve of

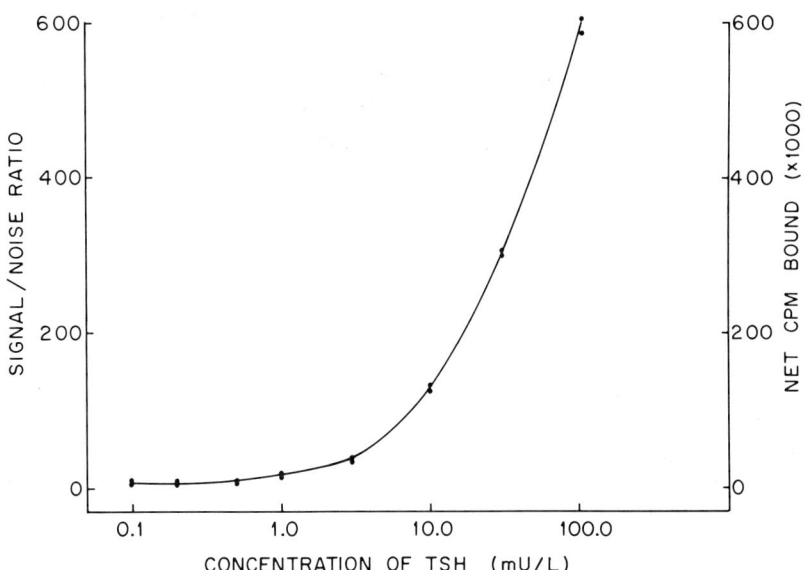

Figure 6 Dose–response curve for TSH. The log concentration is plotted against signal/noise ratio (CPM bound at a TSH standard/CPM bound at "0" standard). Net CPM is the actual CPM bound to the bead-CPM bound at 0 standard (NSB).

the sandwich assay is much less affected by antibody association constants. This is because sandwich assays employ a relative excess of antibody. The two variables that most influence sensitivity in sandwich assays are nonspecific binding (NSB) and specific activity of the labeled antibody. A practical limit to increasing specific activity for purposes of improved sensitivity is that NSB could increase, or antibody activity and stability may be lost. Ideal sensitivity should be the lowest analyte concentration in the assay which is reproduced within 10–15% coefficients of variation (CVs).

C. Precision of IRMA

The fact that sandwich assays use a relative excess of antibody accounts for some of the improvement in precision, since minor variations in the amount of antibody per test will not result in appreciable changes in the binding. In an optimized competitive assay the lowest CVs are obtained around the midpoint of the dose–response curve, with substantially lower precision at high and especially low concentrations. IRMAs have a relatively low and fairly constant variation profile throughout the dose–response curve but still show some worsening of precision at the dose extremes (Fig. 7). Poor

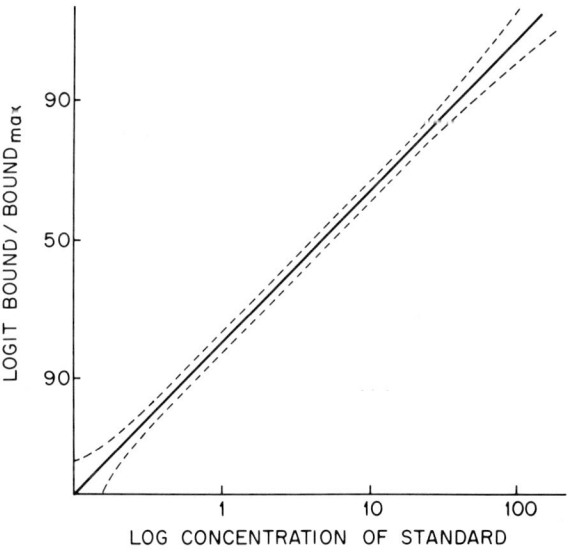

Figure 7 Log–logit transformation of a standard curve in an IRMA. The dotted lines represent the interassay variations expected in the standard curve. Note the heteroscedasticity at the low and high doses of standard.

reproducibility is expected at low dose due to low signal. The noise (CPM at 0 standard), due to radioactive decay, influences the reproducibility at that level. At high antigen concentrations, when antibody is less in excess, imprecise addition of antibody causes more of a problem (Hoffman, 1985). Furthermore, linearity of the dose–response tends to be diminished at high analyte concentrations. It is always possible to optimize IRMA assays in the most precise region of the standard curve since the analytical range is high.

D. Specificity of IRMA

For a cross-reacting molecule to be falsely detected as analyte in an IRMA would require binding by both antibodies. Therefore, the specificity of sandwich assay is much better than if either antibody were used separately. In practice the cross-reactivity of a sandwich assay is approximately equal to the product of cross-reactivities of both antibodies in competitive assays. For example, if one antibody has 10% and the other 1% cross-reactivity in competitive assays, the sandwich assay cross-reactivity would be 0.1% (Hoffman, 1985). This enhancement of specificity has been used to great advantage in assays for gonadotropic hormones, tumor markers, and infectious disease-related antigens.

The cross-reactivity in sandwich assays can also manifest as a decrease in the observed analyte concentrations. This can be seen in assays where one antibody is very specific and the other has high cross-reactivity with a substance present in high concentration relative to the intended analyte. In this case the highly cross-reactive antibody is not in excess relative to the cross-reacting substance, resulting in competition between the analyte and the cross-reacting material. The other antibody is specific and, hence, high concentration of the cross-reactant will inhibit sandwich formation, resulting in an underestimation of the true analyte levels.

Experimental data to illustrate this manifestation of cross-reactivity are shown in Table II. In this simultaneous follicle-stimulating hormone (FSH) sandwich assay, one antibody recognizes the α common subunit of the gonadotropins while the other is highly specific for the β subunit of FSH. When the potentially interfering substances TSH, leutinizing hormone (LH), and human chorionic gonadotropin (hCG) were added to a pool containing FSH, no increase in apparent FSH was seen (data not shown). However, when added to standard containing 50 mIU/ml of FSH, these cross-reactants had the effect of decreasing the measured FSH concentrations (Table II). Consequently, it is recommended that cross-reactivity be evaluated in the presence of known amounts of analyte (ideally at diagnostic cutoff levels) using the highest concentration of cross-reacting substance likely to be encountered. This type

Table II
Specificity of an IRMA for FSH

Cross-reactant	Cross-reactant dose studied	Change in FSH concentration (mIU/ml)
hLH	50 mIU/ml	−0.7
	102 mIU/ml	−0.5
hCG	10,000 mIU/ml	−9.6
	30,000 mIU/ml	−12.1
	100,000 mIU/ml	−13.6
hTSH	70 μIU/ml	−2.4
	150 μIU/ml	−4.2
	252 μIU/ml	−5.1

Source. Direction insert for FSH kit from Nichols Institute Diagnostics.
Note. The cross-reactivity of human luteinizing hormone (hLH), human chorionic gonoadotropin (hCG), and human thyroid stimulating hormone (hTSH) were determined in the Allegro FSH immunoassay by addition of each analyte to a FSH standard (50 mIU/ml). The resulting measured FSH concentrations attributed to each cross-reactant are shown in the table.

of interference can be overcome by developing a two-step or sequential sandwich assay (Hoffman, 1985).

E. Data Reduction for IRMA

Dose–response curves for an immunoradiometric assay can be curve fitted by a variety of data reduction methods. Haven *et al.* (1987) demonstrated the influence of data reduction on a TSH assay. Using the data from the same assay, but calculated using a different data reduction program, they observed clinically relevant differences in TSH concentrations. The data reduction programs also can influence sensitivity and precision of the assays. Hence, selection of the data reduction method is as critical as the technical validation of an IRMA assay. Many methods of data reduction have been described: point-to-point method, spline/modified spline curve fit, four-parameter logistic function, linear square fit (logit–log program, probit–log program, or log–linear plot), nonlinear square fit, etc. (Rodbard and Huff, 1974).

Automated data reduction programs using computers in the laboratory and automation of IRMA assays have produced improvements in sandwich assays. These advances make it possible to convert a nonequilibrium and even nonlinear assay into a clinically valid assay in terms of precision and accuracy. Most of the curve-fitting techniques applied to competitive assays have been successfully adapted to sandwich assays (Dudley *et al.*, 1985). Perhaps the most important parameter to consider in any sandwich assay data reduction scheme is the treatment of the nonspecific signal since this can greatly impact accuracy at low concentrations. While certain methods may be more mathematically correct, the practical considerations of maximum analytical range, low-end accuracy, and optimal precision tend to dictate which is used. The direct dose–response nature of sandwich assays means that with proper optimization, a very linear dose–response curve can be achieved over an adequate analytical range. Thus, for most analytes a simple logit–log transformation will suffice.

Various four-parameter logistical equations are more versatile and will accommodate some assay nonlinearity. An unfortunate problem with these methods is that they may not accurately report doses between the zero and the first standard. The sensitivity, defined earlier as precision around the zero, is typically at doses well below the first standard. Sandwich assays are inherently precise and the cubic spline fitting method will provide very good low-dose accuracy, meaning a realization of assay sensitivity, while accommodating high-dose nonlinearity. A combination of four-parameter and spline has been used. In this method the four-parameter approach is used to fit the curve between the standards, while spline is used to determine the dose–response line between the zero and the first standard. Spline/modified spline curve fit also is commonly used and easily adaptable to automated instruments. The point-to-point method is most popular in research settings where manual computations are common. Four-parameter logistic functional calculation also can be used in automated immunoassay systems.

F. Quality Control of IRMA

The reproducibility of an IRMA can be monitored by running controls with known values (preferably throughout the standard curve) routinely in each assay. An interassay variation of less than 10% is graded as very good. The intraassay variation should be lower than the interassay variation. Intraassay variation data are helpful in making the decision whether to run a sample in duplicate or not and to determine acceptability criteria for duplicate values. It is also important to monitor nonspecific binding (noise or signal at 0 standard) to indicate performance of the radiolabeled antibody. Changes in signal at the highest concentrations of standard indicate a problem with the radiolabeled antibody and/or capture antibody. Monitoring of the standard curve (CPM bound or signal/noise ratio at various doses), slope, and back calculation of the standard are helpful in troubleshooting a problem assay.

VII. Factors Influencing Immunoradiometric Assays

A. Matrix Interferences

In general, sandwich assays show only limited effects from sample matrix differences (such as variations in protein concentration, lipemia, and hemolysis and differences between serum and plasma). In competitive assays such interference effects are manifest as an apparent change in the association constant of the antibody. In sandwich assays the antibody excess means that changes in the association constant will have limited impact on the assay. Interference caused by constituents from the patient's serum (such as "heterophilic antibodies" to the species of immunoglobulin used in the assay, or rheumatoid factor) are usually overcome by the use of nonimmune animal serum in the assay (Kajubi *et al.*, 1981). Other approaches include using different species for each antibody, the sequential assay format, solid-phase separation procedures, and/or modification of the antibodies such as $F(ab)$ and $F(ab)^2$ fragments. Boscato and Stuart (1986) have cautioned in their study with hCG that nonspecific immunoglobulin added to an assay system may not always be enough to block interference in all samples.

B. Hook Effect

The phenomenon of a falling dose–response curve at very high analyte concentrations is referred to as "high-dose hook effect" (Gosling, 1990). Various mechanisms have been proposed to explain the hook effect, but the usual problem is that the antibodies are not in sufficient excess for high analyte concentrations. In the case of analytes (like hCG) present in a very wide concentration range, the hook effect can be a very serious issue and the solution to the problem often involves compromise. The effect is usually overcome only at the expense of sensitivity, assay range, or procedural modifications such as sequential analysis, which increase the steps and overall incubation time (Ryall *et al.*, 1982).

VIII. Advantages and Disadvantages of Immunoradiometric Assays

A. Methodologic Advantages

Advantages of the IRMA methodology include the following:

1. Labeling of the antigen is avoided. Radiolabeled antibodies are stable, compared to most of the radiolabeled antigen.
2. The assay can be applied to "hard to label" antigens (e.g., C-peptide of insulin, a case where no tyrosine is in the molecule).
3. It is ideal for antigens in short supply.
4. It is particularly adapted to monoclonal antibodies. No affinity purification is required and monoclonals are low-affinity antibodies. Antibody affinity has less influence on the sensitivity of the assay.
5. IRMA is more specific — only the intact analyte is measured, not the fragments of a protein.
6. Antibodies to different antigenic sites can enhance affinity and increase sensitivity.
7. The first antibody permits immunoconcentration and can thus increase sensitivity.

B. Methodologic Disadvantages

Disadvantages of IRMA can be summarized as follows:

1. Preparation of materials is time-consuming and requires a high level of expertise (e.g., protein purification).
2. Consumption of reagents is high due to purification and coupling losses, in addition to the use of excess antibodies.
3. Antibody iodination sometimes can result in a substantial reduction of affinity.
4. IRMAs are particularly liable to high-dose hook effects at high analyte concentrations. (This can sometimes be eliminated by sequential addition of antibodies or using large quantities of the "capture" antibody.)
5. The presence of heterophilic antibodies results in artifactually elevated results.
6. IRMA requires antigens capable of binding two antibodies simultaneously. Hence, IRMA is not adaptable to small molecules like steroid hormones, thyroid hormones, catecholamines, etc.

C. Clinical Advantages

As indicated, IRMA has many advantages over competitive-binding radioimmunoassays: increased sensitivity, extended standard range, shortened incubation time, enhanced precision, improved specificity, and technical simplicity. An example of improved sensitivity is an IRMA for TSH. The TSH by IRMA is a second-generation assay with a sensitivity of 0.1 mU/liter (Odell *et al.*, 1986). This improved sensitivity results in improved clinical correlations so that the assay is useful to differentiate hyperthyroid from euthyroid patients as well as hypothyroid from euthyroid individuals. The differentiation of hyperthyroid from euthyroid patients was not possible using the first-generation RIA. Similarly, improvement in sensitivity of the IRMA for ACTH (Zahradnik *et al.*, 1989), PTH (Nussbaum *et al.*, 1987), PTH-RP (Pandian *et al.*, 1992), LH (Gariboldi *et al.*, 1991), or thyroglobulin (Piechaczyk *et al.*, 1989) resulted in improved clinical correlations. The improved LH sensitivity allows differentiation of LH levels from normal children vs children with an abnormal tempo of sexual maturation, precocious puberty, and delayed puberty (Gariboldi *et al.*, 1991).

IRMA also has enhanced specificity for the analyte. The PTH assay by IRMA does not recognize fragments of PTH and measures only intact hormone. Carboxyterminal fragments are not biologically active but react in first-generation radioimmunoassays to falsely elevate measured hormone levels. IRMA PTH assays are therefore useful to differentiate chronic renal failure patients with high concentrations of nonbioactive PTH fragments (Nussbaum *et al.*, 1987). Using RIA, samples with high concentrations of reacting analytes have to be repeated after dilution since the standard curve range is limited. However, using the IRMA assay, the standard curve range is increased. Another example is the hCG assay: the standard curve range for the hCG RIA is 3.1–200 U/liter; for the hCG IRMA it is 1.0 to 100,000 U/liter (Norman *et al.*, 1990). The use of solid-phase separation in IRMA results in improved reproducibility and technical simplicity, with the possibility for automation.

IX. Development of Nonradioactive Assays

The use of radioactivity creates a variety of problems in both research and clinical laboratories. Some of the gamma-emitting isotopes are short lived (e.g., half life of ^{125}I is 60 days) which limits the shelf life of the assays. Not only are the isotopes biologically harmful, but they are costly to purchase and maintain. Further, users and radiation safety officers must monitor experiments constantly and keep detailed records to ensure safety. Finally, radioisotope disposal is a growing environmental problem. A variety of nonradioactive techniques have been used as alternative methods of detection. The immunometric methods are easily adaptable to various nonradioactive procedures by exchanging the radioactivity (^{125}I) with nonisotopic labeling materials. Some of the labels for nonradioactive application are listed below (Schall and Tenoso, 1981; Jackson and Ekins, 1986; Sturgess *et al.*, 1986):

Technique	Labels
Enzyme immunoassay (EIA)	Horseradish peroxidase alkaline phosphatase β-galactosidase
Chemiluminescent assay or immunochemiluminometric assay (ICMA)	Acridinium esterase Alkaline phosphatase Isoluminol
Fluoroimmunoassay (FIA)	Fluorescein Ethidium Rhodamine
Time-resolved fluorescense assay	Europium

X. Conclusions

Recent advances in immunoassay methods, especially involving monoclonal antibodies and affinity-purified polyclonal antibodies, have allowed the development of high-performance immunoradiometric assays. The advantages of IRMA, including sensitivity, specificity, precision, wide analytical range, and rapid performance, have led to increased application of this technique in endocrinology, cancer diagnosis, and measurement of infectious disease-related antigens. IRMA also is a gateway technology for various nonisotopic immunoassay techniques, leading to complete automation.

Acknowledgments

The authors express their appreciation to Ms. Kathy Tomczik and Ms. Mary Fenton for their help in the preparation of the manuscript. We thank Dr. Scott Hutchison for useful discussion.

References

Al-Shawi, A., Ali, M., Houts, T., and Alahuddin, A. R. (1981). *Ligand Q.* **4**, 43.
Bacquet, C., and Twumasi, D. Y. (1984). *Anal. Biochem.* **136**, 487.
Berman, J. W., and Bosch, R. S. (1980). *J. Immunol. Methods* **36**, 335.
Boscato, L. M., and Stuart, M. C. (1986). *Clin. Chem.* **32**, 1491.
Boscato, L. M., and Stuart, M. C. (1988). *Clin. Chem.* **34**, 27.
Boscato, L. M., Egan, G. M., and Stuart, M. C. (1989). *J. Immunol. Methods* **117**, 221.
Burtis, W. J., Brady, T. G., Orloff, J. J., Ersbak, J. B., Warrell, R. P., Olson, B. R. Woth, Mitnick, M. E., Broadvs, A. E., and Stewart, A. F. (1990). *N. Engl. J. Med.* **322**, 1106.
Catarero, L. A., Butler, J. E., Osborne, J. W. (1980). *Anal. Biochem.* **105**, 375.
Catt, K., and Tregear, G. W. (1967). *Science* **158**, 1570.
Cuatrecasas, P., Wilcheck, M., and Anzinsen, C. B. (1968). *Proc. Natl. Acad. Sci. U.S.A.* **61**, 636.
Dudley, R. A., Edwards, P., Ekins, R. P., Finney, O. J., McKenzie, I. A. M., Raab, A. M., Rodbard, D., and Rodgern, R.P.C. (1985). *Clin. Chem.* **31**, 1264.
Gariboldi, R., Picco, S., Magier, S., Chevli, R., and Aceto, T. (1991). *J. Clin. Endocrinol. Metab.* **72**, 888.
Gosling, J. P. (1990). *Clin. Chem.* **36**, 1408.
Hales, C. N., and Woodhead, J. S. (1980). *In* "Immunochemical Techniques," Part A (H. Van Vunakis and J. Langone, eds.), Methods in Enzymology, Vol. 70, p. 334. Academic Press, New York.
Haven, M. C., Orsulak, P. J., Arnold, L. L., and Crowley, G. (1987). *Clin. Chem.* **33**, 1207.
Hoffman, K. L. (1985). *J. Clin. Immunoassay* **8**, 237.
Hunter, W. M., and Greenwood, F. C. (1962). *Nature (London)* **194**, 495.
Jackson, T. M., and Ekins, R. P. (1986). *J. Immunol. Methods* **87**, 13.
Kajubi, S. K., Yang, R. K., Li, H. R., and Yalow, R. S. (1981). *Ligand Q.* **4**, 63.
Martin, K., and Helening, A. (1991). *Cell* **68**, 117.
Mason, D. Y., Jacqueline, L., Cordell, J. L., Abdulaziz, Z., Naiem, M., and Bordenave, G. (1982). *J. Histochem. Cytochem.* 30, 1114.
Miles, L. E. M. (1977). *In* "Handbook of Radioimmunoassay" (G. E. Abraham, ed.), pp. 131–177. Dekker, New York.
Miles, L. E. M., and Hales, C. N. (1968). *Nature (London)* **219**, 186.
Morris, B. L. A., Clifford, M. N., and Jackman, R., eds. (1987). "Immunoassay for Veterinary and Food Analysis—1." Elsevier, London.
Morrissey, J. H. (1981). *Anal. Biochem.* **117**, 307.
Newman, E. S., Moskie, L. A., Duggal, R. N., Goldenberg, D. M., and Hansen, H. J. (1989). *Clin. Chem.* **35**, 1743.
Norman, R. J., Lowings, C., and Chard, T. (1985). *Lancet* **i**, 19.
Norman, R. J., Buck, R. H., and DeMedeiros, S. (1990). *Ann. Clin. Biochem.* **27**, 183.
Nussbaum, S. R., Zahradnik, R., Lavigne, J., Brennan, G. L., Nozawa-Ung, K., Kim, L. Y., Keutmann, H. T., Wang, C. A., Potts, Jr., J. T., and Segre, G. V. (1987). *Clin. Chem.* **33**, 1364.
Nussbaum, S. R., Thompson, A. R., Hutcheson, K. A., Gaz, R. D., and Wang, C. A. (1988). *Surgery* **104**, 1121.
Odell, W. D., Griffin, J., and Zahradnik, R. (1986). *Clin. Chem.* **32**, 1873.
Oldham, R. J. (1984). *J. Biol. Response Modif.* **3**, 229.
Pandian, M. R., Horvat, A., and Said, S. I. (1982). *In* "Vasoactive Intestinal Polypeptide" (S. I. Said, ed.), pp. 35–50. Raven, New York.

Pandian, M. R., Morgan, C. H., Carlton, E., and Segre, G. V. (1992). *Clin. Chem.* **38**, 302.
Piechaczyk, M., Baldet, L., Pau, B., and Bastide, J. M. (1989). *Clin. Chem.* **35**, 422.
Rodbard, D., and Feldman, Y. (1978). *Immunochemistry* **15**, 71.
Rodbard, D., and Huff, D. M. (1974). "Radioimmunoassays and Immunoradiometric (Labelled Antibody) Assays," Vol. 1. IAEA, Vienna.
Rodbard, D., Feldman, Y., Jaffe, M. L., and Miles, L. E. M. (1978). *Immunochemistry* **15**, 77.
Ryall, R. G., Story, C. J., and Turner, D. R. (1982). *Anal. Biochem.* **127**, 308.
Schall, R. F., Jr., and Tenoso, H. J. (1981). *Clin. Chem.* **27**, 1157.
Schwarz, S., Berger, P., and Wick, G. (1985). *Clin. Chem.* **38**, 1322.
Seth, J., Kellett, H. A., Caldwell, G., Sweeting, V. M., Beckett, G. J., Gow, S. M., and Toft, A. D. (1984). *Br. Med. J.* No. 289, 1334.
Sturgess, M. L., Weeks, I., Mpoko, C. N., Laing, I., and Woodhead, J. S. (1986). *Clin. Chem.* **32**, 532.
Valkirs, G. E., and Barton, R. (1985). *Clin. Chem.* **31**, 1427.
Wood, W. G., and Missler, V. (1990). *In* "Luminescence Immunoassay" (K. V. Dyke and R. V. Dyke, eds.), Chap. 9. CRC Press, Boston.
Zahradnik, R., Brennan, G., Hutchison, J. S., and Odell, W. D. (1989). *Clin. Chem.* **35**, 804.
Zola, H. (1978). *J. Immunol. Methods* **21**, 51.
Zola, H., ed. (1987). "Monoclonal Antibodies: A Manual of Techniques," Chap. 6. CRC Press, Boca Raton, Florida.

3

Enzyme-Linked Immunosorbent Assays[1,2]

Endre V. Nagy and Kenneth D. Burman

I. Background
II. Overview of the General Principles Involved with ELISA
III. Details of the ELISA Technique
   A. Helpful Hints
   B. Interpretation of the Data
IV. Related Techniques
References

I. Background

Because of its versatility, the concept of antibody binding to immobilized protein molecules has been exploited for decades in the detection and quantitation of antibodies against known antigens. In practice, protein antigens have been immobilized on the surface of red blood cells (agglutination) or in gels (immunoelectrophoresis). The discovery that many biomolecules can covalently bind to plastic materials under certain conditions simplified the immunoassays and made it possible to perform a large number of measurements in a short time period at a substantially lower cost. When enzyme-linked immunosorbent assay (ELISA) is mentioned, the term usually implies the performance of a test in a 96-well ELISA plate. However, synthetic

---

[1] The number of publications and books on this topic is enormous. In the present discussion we concentrate on the technique itself, predominantly from the viewpoint of an endocrinologist.

[2] The opinions or assertions contained herein are the private views of the authors and are not to be construed as official or reflecting the views of the Army or the Department of Defense.

membranes, plastic beads, or tubes may replace the 96-well plate in some systems. The prerequisite of any accurate measurements in such systems is the availability of the protein in a relatively pure form (Baker et al., 1983). With the increasing availability of recombinant proteins and synthetic peptides this issue is becoming less relevant. Crude cellular extracts may not be appropriate and seemingly reliable readings with such extracts may result in misleading data, influenced by the characteristics of the "sticky" mixture of the wide range of cellular components attached to the plastic.

II. Overview of the General Principles Involved with ELISA

The classical ELISA detects and quantitates an antibody (Engvall and Perlmann, 1972) which is present in a solution (serum, ascites, culture supernatant, etc.). The principle of the method is that the antibody which is to be measured (the so-called first antibody) binds to the antigen immobilized on the plastic surface and, after the unbound antibodies are washed away, a second antibody is added which recognizes the constant region of the bound first antibody. This second antibody carries an enzyme, and in the last step of the assay the enzyme turns its colorless substrate into a visible product. The color intensity is proportionate to the amount of the bound first antibody (Fig. 1).

For the practicing endocrinologist, an understanding of the ELISA principles and their application to the measurement of serum antibodies is very important. In a wide array of events, autoimmune attack against endocrine gland components is involved in the disease process, and the detection of these antibodies is useful in the diagnosis of these diseases as well as in following the effectiveness of the treatment. However, from a broader endocrinological view, detection and quantitation of serum hormone levels are usually the most important laboratory information desired. The classical ELISA is the method for antibody detection (Voller et al., 1980), but can be conveniently modified for measurement of other molecules, including hormones (Weeks et al., 1984; Tseng et al., 1985). In this case, plates are coated with a monoclonal antibody specific for the hormone in question. Serum dilutions are then added, and the hormone, if present, binds to the immobilized monoclonal antibody. An enzyme-coupled, usually polyclonal, hormone-specific antibody is applied in the next step. The antigen (in this case, hormone) is inserted between two immunoglobulin molecules ("antibody sandwich ELISA"). This "sandwich", before addition of the substrate, is depicted in Fig. 2. Alternatively, the enzyme

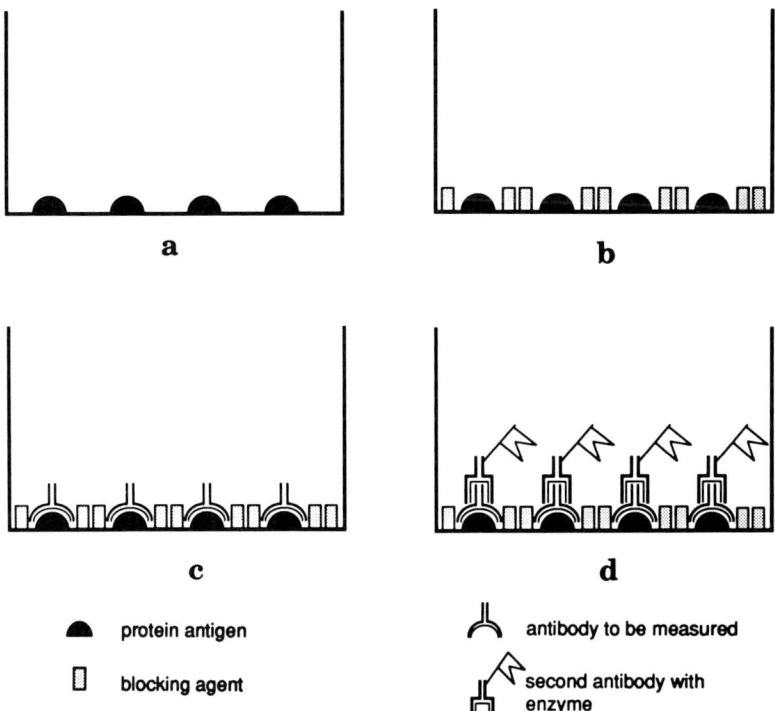

Figure 1 Schematic representation of the ELISA principle. One well of a 96-well plate is shown, depicting the situation at the end of each main step of the procedure. (a) During the coating step, the protein binds covalently to the plastic surface. (b) The blocking agent fills up unused sites on the plate and prevents nonspecific binding during subsequent steps. (c) If there is an antibody in the serum against the plastic-bound protein, it will bind to the coating protein, while other antibodies are removed by washing. (d) The second antibody, linked to an enzyme, identifies the presence of the first antibody. Excess second antibody-enzyme complexes will be washed away prior to substrate being added. When the entire complex is present, the enzyme will turn color after the substrate is added. The intensity of the color change correlates with the concentration of bound second and, consequently, first antibody.

may be carried by a third antibody, specific for the constant region of the second (polyclonal) antibody; this technique generally enhances the sensitivity of the test. The final step is the color development using the substrate, with subsequent quantitation.

Serum hormone levels may be measured by the traditional radioimmunoassay (RIA) or by ELISA. The rationale to use one assay over another relates

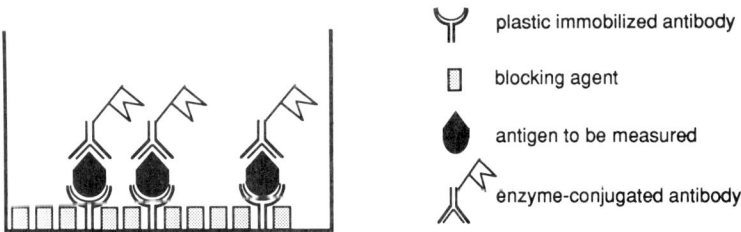

Figure 2 Schematic representation of the "sandwich" when ELISA is used to quantitate protein or hormone. The immobilized antibody is usually monoclonal.

to laboratory features, environment, availability, and expense of reagents (Burman, 1991). A comparison showing the advantages–disadvantages for these two methods is made in Table I. The basic chemical interactions are comparable, even in the competition-type RIAs, and in some cases only the detection systems are different.

Below, we provide a flow sheet for the "classical" ELISA; this procedure can be accommodated to specific individual requirements.

III. Details of the ELISA Technique

Equipment

1. *Washing and pipetting equipment.* The fastest way to wash plates is the use of automatic plate-washing devices. Less convenient, but only slightly

Table I
ELISA and RIA: Comparison of Advantages

	ELISA	RIA
Isotope usage	No	Yes
Reagent shelf life	Long	Limited
Equipment needed	ELISA reader	Nuclear lab
Sensitivity	High	High
Cost per test	Low	Acceptable
Convenience	Excellent	Good

slower, are the multichannel pipettors; a six- or eight-channel one with a 50- to 300-μl range will perform this function well.

2. *Solution basins for multichannel pipettors.*

3. *Constant temperature 37°C incubator, 4°C refrigerator.* Special ELISA incubators are available which provide even heat distribution, but any thermostat with 37°C setting is sufficient.

4. *ELISA reader.* Inexpensive models offer several wavelengths and will print out the actual readings. If several measurements are performed daily, a more sophisticated reader which is faster, makes the background calculations, and calculates the ELISA indices using appropriate software is preferable.

Materials and Reagents

1. *ELISA plates (96 well).* High-binding flat-bottom polystyrene plates, such as Dynatech's Immulon 2, give good results even with relatively short peptides. Occasionally, the binding conditions for some proteins, antibodies, glycolipids, glycoproteins, and lipoproteins may not be predicted and only experimental trials with several types of plastic, plate pretreatment, and coating conditions (see below) will allow a decision regarding optimal systems. There had been no methods developed by which the coating protein can be removed from microwells once bound, and plates must be discarded after use. Recently, a procedure has been described which removes everything from the microwells, with only the antigen coat remaining in place and retaining its antigenic activity (Baunoch *et al.*, 1992). This newer method should be considered if a coating protein is involved which is difficult or expensive to produce.

2. *Coating buffer.* Carbonate–bicarbonate buffer 0.05 M, pH 9.6.

3. *Coating protein.* Prepare solutions in coating buffer in the 1 to 20 μg/ml range. Short peptides below 20 amino acids may require higher concentrations or may not bind sufficiently for assay. In future tests, the smallest concentration which still results in coating should be employed. For most proteins, 2–5 μg/ml results in sufficient coating.

4. *Serum and antibody diluent.* Phosphate-buffered saline, pH 7.4 (PBS).

5. *Wash solution.* PBS containing 0.05 % Tween-20 (PBS-T).

6. *Blocking solution.* One percent bovine serum albumin in PBS (see Table II for alternatives).

7. *Samples.* Dilute samples in PBS (without Tween-20). A typical dilution for sera is 1:1000; dilutions under 1:100 may result in unreliable results due to high nonspecific binding. For titer determinations, serial dilutions are

Table II

Blocking agents
Bovine serum albumin (BSA) 1% w/v
Gelatin 1% w/v
Nonfat dried milk 3% w/v

tested between 1:100 and 1:32000. Culture supernatants with low antibody content may require lower dilutions to obtain satisfactory results.

8. *Second antibody.* If the sample tested is human serum and you want to measure IgG against the antigen coated on the plate, the second antibody should be anti-human IgG, conjugated to alkaline phosphatase, raised in another species, e.g., goat (for other possible enzymes, see Table III). If the working dilution of the antibody for ELISA is not specified by the manufacturer, the 1:500 to 1:10,000 range should be tested initially. A typical working dilution is 1:3000.

9. *Substrate.* p-Nitrophenyl phosphate (pNP).

10. *Diethanolamine buffer for substrate solution.* To 800 ml of distilled water, add NaN_3, 200 mg; $MgCl_2(6H_2O)$, 100 mg; diethanolamine, 97 ml. Adjust pH 9.6 with 1 N HCl. Bring up to 1 liter with distilled water. Store at

Table III
Common Enzymes and Substrates for ELISA

Enzyme	Substrate/indicator[a]
Alkaline phosphatase	p-Nitrophenyl phosphate (pNPP) 4-Methylumbelliferone phosphate (MUP)[b]
Peroxidase	2,2'-Azino-bis(3-ethylbenzthiazoline-6-sulfonic acid) (ABTS) o-Phenylenediamine (OPD) 3,3',5,5'-Tetramethylbenzidine (TMB) 5-Aminosalicylic acid (5-AS)
Urease	Urea/bromcresol purple
Beta-galactosidase	o-Nitrophenyl-β-D-galactopyranoside (ONPG) 4-Methylumbelliferone-β-D-galactopyranoside (MUG)[2]

[a] For ELISA purposes, only water soluble end products are considered.
[b] Fluorescent substrate.

4°C in the dark. Bring to room temperature before use and prepare 1 mg/ml solution of pNP.

11. *Stop solution.* 3 M NaOH.

DETAILS

1. Draw a schematic diagram of your plate(s) in advance, showing the arrangement of your samples. Do not use the outer rows; these lanes tend to give higher readings. Different plastic characteristics and uneven heat distribution might account for this "side lane phenomenon," which is empirically observed to occur but not adequately explained. One has to be sure in advance which wells or which column will be used as blank by the ELISA reader (usually the first column); be sure to include appropriate control wells (see below). Duplicate or triplicate samples are preferable.
2. Pipette 50 μl of the coating solution into the wells of a 96-well ELISA plate. Incubate at 4°C overnight. Always use lids or plastic foil to cover plates while incubating.
3. Pour out the fluid by inverting the plate and empty completely by hitting firmly against a flat surface covered by a paper towel or cloth. The same method of emptying the plate(s) will be used throughout the procedure. Wash the wells twice with 200 μl or more PBS-T, for 3 min each. The firmer you spray the wash solution into the wells the more efficient the washing.
4. Add 200 μl of the blocking solution to the wells and incubate for 1 hr at 37°C, covered. As an alternative, the blocking agent (1% BSA) may be added to the samples in step (6) below, and steps (4) and (5) omitted in turn. This will shorten the procedure by an hour, but it has to be individually determined if this shortcut will work in a given system.
5. Wash three times with PBS-T as described under (3) above.
6. Add 100 μl of the prepared samples to the wells according to your plate plan. Incubate for 2 hr at 37°C.
7. Wash three times with PBS-T as described under (3) above.
8. Add 100 μl of the enzyme-conjugated second antibody to the wells. Incubate for 1 hr at 37°C.
9. Wash three times with PBS-T as under (3) above.
10. Add 100 μl of substrate solution to the wells. Follow color development at room temperature. When marked differences can be seen (5 to 60 min), stop reaction by adding the stop solution, 100 μl per

well. Some substrates are sensitive to light and the color reaction may deepen when the stop solution has been added; pNP is not among these.

11. Read plates at 405 nm within 2 hr.

A. Helpful Hints

1. How to Set Up Control Wells

The inclusion of appropriate control wells is extremely important; these wells represent the basic test of reliability of the results obtained in an ELISA. For this purpose, one component of the ELISA sandwich for each control well is systematically omitted and replaced by buffer only; each control is preferred in duplicate. The "no coat" well evaluates if there is antibody binding to the blocking agent. High reading in this well means that the blocking agent has to be replaced. The "no sample" well indicates the background binding (i.e., the nonspecific binding of the second antibody to the coat). Including a "no second antibody" well is useful as its positivity might suggest that the coating material intrinsically contains enzyme activity similar to the enzyme coupled to the second antibody. For example, the thyroid microsomal antigen is the enzyme thyroid peroxidase and consequently it contains intrinsic peroxidase activity; should this example turn out to be the case, another enzyme system is suggested. Also, buffer blanks (wells containing buffer only) should be included and arranged according to the requirement of the ELISA photometer. At least one positive control sample and one negative control sample are needed for each plate, and these samples are processed identically to the unknown samples. Instead of a single positive control, a series of known samples may be included. As color development is not linear with time, setting up a standard curve with known samples will allow interassay comparisons.

If any of the wells gives a reading above the range of the machine, dilute samples and transfer 100 μl into a new plate and read again.

2. The Second Antibody

Affinity-purified second antibodies directed against single or multiple immunoglobulin classes and subclasses are commercially available. When testing for an IgM response, the selection of an Fc-specific second antibody instead of one directed against the whole IgM molecule will substantially diminish background binding. Second antibodies are diluted in PBS before use; however, the stock solution containing 50% glycerol can be kept at 4°C for at least 1 year. Thimerosal 0.01% is also a good preservative; NaN_3 interferes with peroxidase and EDTA interferes with alkaline phosphatase activity.

3. Choosing the Enzyme and the Substrate

Inexpensive enzymes most commonly used are probably horseradish peroxidase and alkaline phosphatase (Table III). A wide selection of antibodies is available as their conjugate. The end product of the substrate reaction has to be water soluble; a list of possible substrates is provided in Table III.

B. Interpretation of the Data

After determining the means of duplicate or triplicate wells, the usual manner to express results is the calculation of ELISA indices. An ELISA index (EI) indicates how many times greater the reading in a test well is than in the control well:

$$EI = \frac{\text{actual reading}}{\text{negative control reading}}.$$

The negative control reading, at least in the case of sera, is usually higher than the background (buffer control) reading.

IV. Related Techniques

Utilizing the same general principles, several alternative methods can be employed. Using a non-enzyme-coupled second antibody followed by an enzyme-coupled third antibody generally makes this method very sensitive. Another possibility to increase sensitivity is the use of the avidin–biotin system: biotin is coupled to the second antibody (biotinylation) and, in the next step, avidin-coupled enzyme is added (Green et al., 1971; Bayer et al., 1979). Alternatively, avidin can be inserted between the biotinylated second antibody and a biotinylated enzyme. Streptavidin, as opposed to egg white avidin, has no glycoprotein component and seems to give less nonspecific binding. This system increases sensitivity because of the very high affinity of avidin to biotin.

Sensitivity may also be enhanced if a fluorogenic substrate is used (Shalev et al., 1980); this "FELISA" or "FIA" method requires a fluorescent microplate reader. Instead of enzyme or radionuclide, other assays utilize a chemiluminescent tracer or compound (Davis et al., 1984) linked to either antigen or antibody; sensitivity can be enhanced as much as 100-fold.

When proteins or peptides with differing abilities to bind to plastic are used for coating, a special, 96-well nitrocellulose filter-bottomed plate might help to achieve better binding and diminish differences caused by varying binding (Davis *et al.*, 1984). Also, the length of each step is related to the time of pipetting the fluid and eluting through the bottom. However, a special plate suction apparatus is needed, and samples have to be transferred into a regular ELISA plate to read the color. Alternatively, this system works as a dot assay if water-insoluble substrate is used and the color of the membranes quantitated. Some membranes have to be prewetted in methanol before use and no detergent (Tween-20) is allowed when washing.

With "cell-ELISA," the relative quantitation of cell surface antigens *in situ* is possible by using an adherent cell monolayer as antigen in the wells (Posner *et al.*, 1982; Nagy *et al.*, 1991); after cells are fixed with aldehyde, consecutive steps are similar to the classical ELISA (steps 3–11 in Details, earlier). Recently, the immobilization of nonadherent cells became possible for this purpose (Bishop and Hwang, 1992).

The ELISPOT technique (Czerkinsky *et al.*, 1983) enables enumeration of antibody-secreting cells with a given specificity. Antigen-coated membranes are incubated with living B cells on their surface. After the cells are removed, the second antibody–enzyme complex will detect the spots where the specific lymphocytes were sitting on the membrane if a water-insoluble substrate is added. The method can be modified to identify cells secreting a given substance (hormone, lymphokine). This method utilizes techniques considered intermediate between an ELISA and the dot blot–Western blot analysis.

Acknowledgments

Dr. Nagy is an NIH Fogarty Research Fellow (1 F05 TW04412). Mrs. Alice Jacobson provided editorial assistance; the figures are the work of Mrs. Annabelle Wright.

References

Baker, J. R., Jr., Lukes, Y. G., Smallridge, R. C., Berger, M., and Burman, K. D. (1983). *J. Clin. Invest.* **72**, 1487–1497.
Baunoch, D. A., Das, P., Browning, M. E., and Hari, V. (1992). *BioTechniques* **12**, 412–417.
Bayer, E. A., Skutelsky, E., and Wilchek, M. (1979). *In* "Vitamins and Coenzymes," Part D (D. McCormick and L. Wright, eds.), Methods in Enzymology, Vol. 62, pp. 308–315. Academic Press, New York.
Bishop, G. A., and Hwang, J. (1992). *BioTechniques* **12**, 326–330.
Burman, K. D. (1991). *In* "Encyclopedia of Human Biology" (R. Dulbecco, ed.), pp. 351–356. Academic Press, San Diego.

Czerkinsky, C., Nilsson, L. A., Nygren, H., Ouchterlony, O., and Tarkowski, A. (1983). *J. Immunol. Methods* **65**, 109–121.
Davis, J. W., Angel, J. M., and Bowen, J. M. (1984). *J. Immunol. Methods* **67**, 271–278.
Engvall, E., and Perlmann, P. (1972). *J. Immunol.* **109**, 129–135.
Green, N. M., Konieczny, L., Toms, E. J., and Valentine, R. C. (1971). *Biochem. J.* **125**, 781–791.
Nagy, E. V., Kalman, K., Szabo, J., and Bako, G. (1991). *Immunobiology* **182**, 405–413.
Posner, M. R., Antoniou, D., Griffin, J., Schlossman, S. F., and Lazarus, H. (1982). *J. Immunol. Methods* **48**, 23–31.
Shalev, A., Greenberg, A. H., and McAlpine, P. J. (1980). *J. Immunol. Methods* **38**, 125–139.
Tseng, Y.-C., Burman, K. D., Baker, J. R., Jr., and Wartofsky, L. (1985). *Clin. Chem.* **31**, 1131.
Voller, A., Nidwell, D., and Bartlett, A. (1980). *In* "Manual of Clinical Immunology" (N. Rose and H. Friedman, eds.), 2nd Ed., p. 359. Am. Soc. Microbiol., Washington, D.C.
Weeks, I., Sturgess, M., Siddle, K., Jones, M. K., and Woodhead, J. S. (1984). *Clin. Endocrinol.* **20**, 489.

4

Development and Characterization of Polyclonal and Monoclonal Antibodies

Joan Rener

I. Introduction
II. Immunogen Preparation
   A. Conjugation
   B. Adjuvants
III. Antisera Development
   A. Animals
   B. Immunization
IV. Monoclonal Antibody Development
   A. Animals
   B. Immunization
   C. Myeloma Cells
   D. Fusion Protocols
   E. Hybridoma Evaluation
   F. Cloning
   G. Cryopreservation
V. Monoclonal Antibody Production
   A. In Vitro
   B. In Vivo
VI. Purification of Antisera and Monoclonal Antibodies
   A. Antisera
   B. Monoclonal Antibodies
VII. Characterization of Antisera and Monoclonal Antibodies
   A. Specificity
   B. Affinity
   References

I. Introduction

Immunoglobin-based assays have played a central role in the detection, measurement, and characterization of steroidal and nonsteroidal hormones for over 35 years (Berson *et al.*, 1956; Yalow and Berson, 1959). During this time, a wide range of antibody-based assay techniques has been developed which utilize electrophoresis, fluorescence, chemiluminescence, laser nephelometry, solid-phase immunoadsorbents, and enzyme conjugates (Steward and Lew, 1985). Today, affinity-purified antisera and monoclonal antibodies (MAb) are the basis of an ever-growing industry devoted to immunological reagents, automated analyzers, and diagnostic kits.

The specificity and sensitivity of these assay formats depend directly on the properties of the immunoreagents that are used. Hybridoma technology (Köhler and Milstein, 1975) took immunoassays into a new era in which preselected antibodies could be developed and produced in large quantities as homogenous, characterized reagents. Technology for improving immunoglobulins is still evolving as a result of continuing advances in genetic engineering. Through genetic engineering it is possible to avoid cell fusions altogether, and at the same time increase the chances of developing antibodies against unique epitopes. Immunoglobulin light and heavy chains, encoded in DNA libraries in λ phage, can be recombined in every theoretical permutation and expressed in *Escherichia coli* for rapid screening and production (Huse et al., 1989). However, at this time the technology is not freely available to researchers and its added value to hormonal assay development remains to be proven.

Reagent development for immunoassays should be approached with certain questions in mind: What is the availability of the antigen? What is its molecular weight? To what level has it been purified? What level of sensitivity is needed for the assay? What problems might be expected with cross-reactivity? What will the assay be used for? The answers to these questions will determine what immunoreagents should be developed and suggest ways to design an efficient screening program to choose the best antisera and/or MAb.

Most MAb developed for hormonal assays are of mouse origin. Although reliable techniques have been published for rat (Kilmartin et al., 1982) and human MAb (Ernst and Sonneborn, 1990; James and Bell, 1987; Engleman et al., 1985), they are less successful, as a rule, than protocols based upon the original Köhler and Milstein (1975) methods. This chapter focuses on techniques which can be used for the development of rabbit antisera and murine hybridomas. Unless noted, the immunological and biological reagents referred to in this chapter are available from commercial sources that are listed in Linscott's Directory (1992–1993).

NOTE The *in vivo* procedures described in this chapter are designed to minimize pain and discomfort to animals. Investigators should contact their institutional animal care and use committee to verify that all animal procedures meet approved standards.

II. Immunogen Preparation

A. Conjugation

Antisera and MAb development begin with the same challenge: eliciting a strong, specific immune response in animals. Low molecular weight analytes (less than 10,000 Da), such as steroids, small polypeptides, and synthetic peptides should be treated as haptens. Haptens are nonimmunogenic compounds that acquire immunogenicity when they are covalently bound to a carrier (Erlanger et al., 1959). Commonly used carriers include proteins such as bovine serum albumin (Kohen et al., 1982), keyhole limpet hemocyanin (Tateishi et al., 1980), bovine thyroglobulin (Robinson, 1980), synthetic polymers, such as poly-L-lysine (Brown et al., 1970), and X-linking agents, such as glutaraldehyde (Kagan and Glicks, 1979).

The conjugation method for linking a hapten and a protein is critical to success. While the carrier provides ancillary epitopes to trigger T-cell recognition, its immunogenicity can impede a specific response unless the hapten is oriented toward the outside of the complex. For this reason, a spacer is frequently incorporated to move the hapten away from the carrier's surface.

The choice of a conjugation procedure depends on the stability and solubility properties of the hapten, the active sites on the hapten that are available for conjugation, and the particular mode of attachment to the protein that is desired (Parker, 1976). Most coupling methods rely on the presence of free amino, sulfhydryl, phenolic, or carboxylic acid residues on the protein carrier (Harlow and Lane, 1988; Carraway, 1979). Hapten modifications usually involve blocking unwanted sites to achieve site-directed conjugation (Kohen et al., 1982). In most cases, a hapten:carrier molar ratio of 5–15:1 is desirable. Higher ratios may result in stearic hindrance and decreased specific immune response.

Conjugation, with or without a spacer, frequently results in the formation of new epitopes which may be more immunogenic than the hapten. For MAb development, prepare two conjugates, utilizing different protein carriers, and, if possible, two different conjugation methods. Immunize with one preparation and screen for specificity with the other.

B. Adjuvants

The response to an immunogen can be enhanced by presenting it in a mixture with an adjuvant. Adjuvants protect antigen from rapid degradation and nonspecifically stimulate an immune response. Always use adjuvant with the first inoculation, even with antigens that are known to be immunogenic. Freund's complete adjuvant (FCA) is usually recommended, although alum and *Bordetella pertussis* are also used (Kenney *et al.*, 1989). The active ingredient in FCA is a mycobacterium preparation which causes a severe inflammatory response. To avoid undue discomfort to animals, the National Institutes of Health Guideline (Institute of Laboratory Animal Resources, 1988) recommends that FCA be used only for the initial immunization and Freund's incomplete adjuvant (FIA) for subsequent boosts. The Guideline recommends that no more than 0.05 ml of adjuvant be injected per subcutaneous site, at a concentration of no greater than 0.1 mg mycobacterium per 1 ml of inoculum. Apparatuses, such as the "Mulsi Churn," are available commercially (Thomas, 1989), but complete mixing of Freund's adjuvants and antigen is easily accomplished using a double-hubbed needle or a stopcock connector between two glass syringes. Plastic syringes may be used provided the plunger is prelubricated with mineral oil.

Polysaccharides and copolymer adjuvants are increasingly popular since they are less irritating and seem to have good immunomodulator effects (Hunter *et al.*, 1989). At this time, however, FCA and FIA remain the reagents of choice for most investigators. Freund's adjuvants given intraperitoneally in mice may adversely affect the spleen, potentially complicating hybridoma development.

NOTE Freund's adjuvants should never be given intravenously.

ANTIGEN/FREUND'S ADJUVANT PREPARATION

1. Place antigen preparation in one glass syringe and an equal amount of adjuvant in another (1:1, v:v).
2. Connect the syringes and inject the aqueous antigen solution into the adjuvant.
3. Emulsify the mixture by repeated transfer between the barrels. Continue until the emulsion is thick and creamy. The emulsion should remain as one phase when a small amount is added to water.

III. Antisera Development

A. Animals

Antiserum for research purposes is generally raised in rabbits because they are inexpensive and easy to handle. Pasteurella-free New Zealand White rabbits, 3–4 kg, are recommended because of their availability and the ease with which they can be housed, handled, and bled. Maintain rabbits according to the National Institutes of Health "Guide for the Care and Use of Laboratory Animals" (U.S. Department of Health, 1985). Upon receipt, quarantine the animals for 1 week, allowing them to recover from shipping stress. During this time, observe them for general health.

B. Immunization

Immunogen which is absorbed slowly and presented in small amounts over time seems to be most effective in eliciting a strong immune response (Parker, 1971; Vaitukaitis *et al.*, 1971). This is usually accomplished through multiple intradermal inoculations (mid) with amounts of antigen ranging from 0.25 mg to 0.50 mg/dose, administered at 10 to 40 sites on the animal's back. Higher doses in the range of 10 mg/dose have been used with weak immunogens, but as a rule, excess antigen should be avoided since it may result in tolerance and suppressed response. Intramuscular (im), subcutaneous (sc), intravenous (iv), and lymph node routes of administration have all been used successfully (Harlow and Lane, 1988). The methods and schedules described in this section have been used to produce high-affinity antisera ($K_a > 10^{10}$ liter mol^{-1}) against a variety of steroidal and polypeptide hormones and peptides. Due to the variation in immune responses between animals, many rabbits may need to be immunized before an acceptable antiserum is developed. Immunize at least three animals, using *B pertussis* tuberculosis to stimulate the animals' immune systems prior to immunization.

INTRAMUSCULAR INJECTION

1. Prepare immunogen in a glass syringe fitted with a 25-gauge needle. Use a new sterile syringe and needle for each animal.
2. Place the animal in a restraining device, or work with a partner who can wrap it tightly in a towel and hold its leg in an accessible position.
3. Insert the needle into the gluteal muscle on either hind leg. Withdraw the plunger slightly; there should be resistance and an absence of blood.

4. Administer the immunogen in a slow steady motion.
5. Dispose of needle and materials appropriately.

MULTISITE INTRADERMAL INJECTIONS

The method of Vaitukaitis (1981) is followed:

1. Prepare immunogen in a glass syringe fitted with a 25-gauge needle. Use a new sterile glass syringe and needle for each animal.
2. Restrain the animal.
3. Shave the rabbit's back and swab it with alcohol.
4. Insert the needle, bevel side up, under the skin. Inject no more than 100 μl/injection site. The inoculum will form a small bump or blister.
5. Compress the skin as the needle is withdrawn to prevent leakage.
6. Repeat at another site.

INOCULA PREPARATION (FOR THREE RABBITS)

1. *B. pertussis* suspension
    a. Suspend *B. pertussis* (20×10^{10} killed cells per ampule) in 2 ml saline.
    b. Four days prior to primary inoculation, administer, im, 6×10^{10} cells (0.6 ml) per rabbit.
2. Primary inocula (Day 1)
    a. Emulsify the following until thick and creamy:

    1.5 ml   Antigen solution
    1.5 ml   FCA

    b. Administer, mid, 1.0 ml (0.50 mg antigen) per rabbit, distributed between 30–40 injection sites.
3. Booster inocula
    a. Emulsify the following until thick and creamy:

    0.75 ml   Antigen solution
    0.75 ml   FIA

    b. Administer, mid, 0.5 ml (0.25 mg antigen) per rabbit distributed at 30–40 injection sites.

Immunization Regimen for Antisera Development

Day	Procedure
−4	Prebleed
	Inject *B. pertussis*, im
1	Initial immunization, mid, antigen + *M. tuberculosis* + FCA
21	Boost, mid, antigen + FIA
31	Test bleed
42	Boost, mid, antigen + FIA
52	Test bleed
63	Boost, mid, antigen + FIA
73	Production bleed

Boost every 3 weeks; test bleed 10 days after boosts.

COLLECTION OF BLOOD VIA MEDIAL AURICULAR ARTERY

1. Label blood collection tubes.
2. Restrain the animal.
3. Locate the medial artery and rub it vigorously with 70% alcohol on gauze.
4. When the artery dilates, puncture it at approximately midlength. The needle should penetrate only 1 to 2 mm into the artery.
5. Allow blood to drip into a tube. When using a blood collection set, push an evacuated tube onto the needle when blood appears in the hub.
6. Stroke the artery as necessary to ensure blood flow.
7. Remove the needle and immediately apply pressure to the collection site.
8. Allow blood to clot at 37°C for 1 hr. Use a Pasteur pipette to release the clot from the sides of the tube and incubate the tube at 4°C overnight.
9. Centrifuge the tube at 1500 rpm for 10 min.
10. Remove the serum, aliquot it into sterile tubes, and hold at 4°C, prior to testing or at −20°C for long-term storage.

Monitor the specific antibody response during the immunization regimen by evaluating sera for rising titer. Initiate test bleeds (approximately 10 ml of blood/bleed) several days prior to immunization to establish a baseline for each animal. Production bleeds of 45 ml of blood/animal are initiated as needed once sufficient titer is reached.

IV. Monoclonal Antibody Development

A. Animals

Balb/c mice are the animals of choice for hybridoma development, since most of the cell lines used as fusion partners were derived from a Balb/c myeloma tumor (Potter, 1972); syngeneic fusion partners seem to yield the highest number of stable hybridomas. In addition, the hybridomas can be grown as Balb/c ascites tumors for antibody production.

Mice should be purchased from a breeder who will supply adequate documentation concerning their disease-free status and a serologic profile for common murine pathogens (Weisbroth, 1984). Latent viruses, such as mouse hepatitis virus and Sendii virus, can compromise immune functions, inhibit hybridoma cell growth after fusion, and decrease antibody production in ascites fluid. Animals should be housed according to the National Institutes of Health's, "Guide for the Care and Use of Laboratory Animals" (U.S. Department of Health, 1985).

B. Immunization

Techniques for immunogen preparation are the same as those for antisera development. Approximately 200 to 500 μg of purified antigen is sufficient to complete an immunization regimen and to evaluate immune sera. Additional antigen will be required to evaluate culture supernatants as monoclonal antibody development proceeds. Immunize at least six animals.

SUBCUTANEOUS INOCULATION

1. Calculate the amount of antigen so that the total volume to be injected is 0.3 ml/mouse (use 0.15 ml antigen/PBS and 0.15 ml adjuvant).
2. Completely mix the antigen/PBS and adjuvant through the stopcock and the disposable syringes (when emulsified, the consistency will be like mayonnaise).
3. Holding mouse by the skin of the neck and base of tail, position mouse so that abdomen is accessible.

4. Swab abdomen with alcohol.
5. Using a 21-gauge needle and a 3-ml disposable syringe, insert the needle just below the skin along, but above, muscle wall at the groin.
6. The needle should be able to move slightly so that it is not intradermal or into the abdomen wall.
7. Inject 0.3 ml of antigen/PBS/adjuvant subcutaneously.
8. Remove needle and swab site of injection with alcohol to be certain that no antigen leaks out.

When only small amounts of antigen are available, alternative methods of *in vivo* immunization, such as intrasplenic inoculation (Spitz *et al.*, 1984; Donley, 1989) or subcutaneous implantation of antigen adsorbed to nitrocellulose discs, have been used successfully (Harlow and Lane, 1988). In most cases, a final iv boost via the tail vein is performed 3 days prior to fusion.

Picomole amounts of hypothalamic growth hormone-releasing factor have been used successfully with *in vitro* immunization techniques (Luben *et al.*, 1982). However, *in vitro* immunization is usually less successful for the development of high-affinity antibodies and is not addressed in this chapter.

INTRAVENOUS INOCULATION VIA TAIL VEIN

Intravenous injection requires skill and patience. Practice administering saline to naive mice, iv, prior to boosting an immunized animal. Precious antigen can be lost by missing the vein or by inadvertently injecting an air bubble into the vein which will kill the animal.

1. Prepare antigen without adjuvant in a sterile 1-ml syringe fitted with a 27 g, 0.25-in. needle. Hold the syringe upright and remove any air bubbles.
2. Dilate the tail veins by warming the mouse under a heat lamp for a few minutes, being careful not to cause heat stress, or vigorously rub the dorsal surface of the tail with 70% alcohol.
3. Restrain the mouse.
4. Hold the tail at the tip and insert the needle, bevel side up, near the distal end of one of the two lateral tail veins. If this is unsuccessful, try again, working up the tail.
5. Push the plunger slowly, administering a volume of no more than 0.25 ml/mouse.
6. Withdraw the needle slowly and apply pressure to the injection site to prevent bleeding.

Immunization Regimen for Monoclonal Antibody Development

Day	Procedure
−1	Earmark six mice, prebleed, store sera at −20°C
1	Immunize six mice, 20 μg antigen + adjuvant, inoculated sc
21	Boost, 20 μg antigen + adjuvant, inoculated sc
31	Test bleed; prebleed and test sera evaluated by ELISA
42	Boost, 20 μg antigen + adjuvant, inoculated sc
52	Test bleed; test sera evaluated by ELISA
63	Boost, 20 μg antigen + adjuvant, inoculated sc
73	Test bleed; test sera evaluated by ELISA
76	Boost animal with highest titer, iv, with 10 μg antigen

C. Myeloma Cells

Several cell lines, all descended from the Balb/c MOPC tumor-derived cell line, P3K (Horibata and Harris, 1970), are reliable fusion partners. Three such lines are P3-X63-Ag8.653 (Kearney et al., 1979); NSO/1 (Galfré and Milstein, 1981); and a hybridoma derived from a P3-X63-Ag8 fusion, Sp2/0-Ag14 (Shulman et al., 1978). These cell lines lack the enzyme hypoxanthine guanine phosphoribosyl transferase (HGPRT) and are resistant to killing by 8-azaguanine. They will die in selection medium containing hypoxanthine, aminopterin, and thymidine (HAT), which blocks the biosynthetic pathways for nucleotide synthesis but allows HGPRT+ cells to survive via the salvage pathway (Littlefield, 1964). Only hybrids between these myeloma cells (HGPRT−) and lymphocytes (HGPRT+) will survive in HAT medium. Obtain the cell line from a cell repository such as the American Type Culture Collection (Rockville, MD) to assure its identity and freedom from mycoplasma contamination. Mycoplasma interferes with fusions and has been implicated in decreased antibody secretion rates.

Myeloma cells should be maintained in an established tissue culture laboratory by technicians experienced in good cell culture techniques. Take care to maintain the cells in a healthy condition with viability over 90%. Cryopreserve and store an adequate number of ampules of cells at −170°C, in vapor-phase liquid N_2 (see Section IV, G).

Most myeloma cell lines grow well in cell culture and can be adapted to several media such as RPMI 1640 or Iscove's modified Dulbecco's medium (IMDM). The following sections on hybridoma development and maintenance utilize supplemented or complete IMDM (CIMDM) as the basic medium (Brown, 1985). Use a biological safety cabinet, level 2, and aseptic technique to perform all cell culture manipulations. Cell culture reagents and materials must be sterile to prevent contamination.

PREPARATION OF CIMDM

Materials

1. Premade liquid IMDM stored at 4°C
2. Fetal bovine serum (heat inactivated)
3. L-Glutamine (200 mM)
4. Sodium pyruvate (100 mM)
5. Pen/Strep (5000 U penicillin/ml; 5000 µg streptomycin/ml)

Directions

To a volume of IMDM, add the following:
1. 10% Fetal bovine serum
2. 1% Sodium pyruvate
3. 1% L-Glutamine
4. 1% Pen/Strep

CIMDM is used to maintain myeloma and hybridoma cell lines. Established cell lines can be weaned off serum and grown in serum-free medium (Brown, 1987; Kawamoto et al., 1983; McHugh et al., 1983). It should be noted that some hybridoma cell lines cultured in serum-free or low-serum media may have decreased antibody production (Ozturk and Palsson, 1990) or changes in glycosylation patterns. The use of antibiotics in cell culture should be minimized, but may be used at critical steps such as fusion and clonings until stable cell lines have been developed.

D. Fusion Protocols

MYELOMA CELL PREPARATION

1. To assure a homogenous population of HGPRT− cells, passage the myeloma cells several times in CIMDM supplemented with 10^{-4} M 8-azaguanine which will kill any HGPRT+ revertants.
2. Two passages prior to fusion, switch the cells to CIMDM (without 8-azaguanine).
3. On the day of the fusion, wash approximately 5×10^7 to 1×10^8 cells, in log phase with over 90% viability, in serum-free medium and suspend them in 1 ml of serum-free medium.

COLLECTING IMMUNE SERUM FROM THE BRACHIAL ARTERY

1. Based on immune sera titers, choose the mouse that will be used as the donor animal and boost it, iv, 3 to 4 days before the fusion.
2. On the day of the fusion, work at a hood used for animals and anesthetize the mouse with CO_2 inhalation; place it on its back with one of its front paws extended.
3. Make a small incision in the flap of skin under the arm, forming a pocket to collect the blood.
4. Cut the brachial artery, deep within the armpit, and use a Pasteur pipette to collect the blood as it pools. Approximately 1 to 2 ml can be collected for use as positive controls in screening assays.

SPLENOCYTE PREPARATION

1. Place the exsanguinated animal on its right side and spray it with 70% alcohol.
2. Using sterile instruments and gloves, make a small nick in the skin on the lower left side just below the ribs.
3. Using both hands, gently pull back the skin from each side of the nick, exposing the spleen which is visible through the peritoneal wall.
4. Spray again with 70% alcohol.
5. Using fresh instruments, cut the wall and grasp the spleen with forceps. Lift it out and cut it free from underlying connective tissue.
6. Place the spleen in a 50-ml centrifuge tube containing 20 ml of cold, serum-free IMDM supplemented with 2% Pen/Strep (SF-IMDM) on ice.
7. Transfer the tube to a tissue culture hood.
8. Pour the spleen into a petri dish and discard most of the medium.
9. Using a sterile No. 10 scalpel, cut off one tip of the spleen and split the spleen lengthwise.
10. Using the flat side of the scalpel, massage out the freely separating cells, mincing as necessary to release cells from the more fibrous margins.
11. Transfer the cells and medium to a 50-ml centrifuge tube with 10 ml of SF-IMDM.
12. Wash the petri dish with additional media to collect all the cells and add to the tube.

13. Allow any clumps to settle out of suspension for approximately 5 min.
14. Transfer the cell suspension to a fresh 50-ml centrifuge tube, leaving the clumps behind, and wash three times with SF-IMDM.
15. Resuspend the cells in 10 ml SF-IMDM and transfer to a T-25 flask.
16. Incubate the flask at room temperature for 30 min allowing fibroblasts to adhere to the bottom.
17. Transfer the cell suspension to a new 50-ml centrifuge tube and perform a viable cell count utilizing the trypan blue dye exclusion method (Phillips, 1973). Approximately 10^8 cells are obtained from a mouse spleen.

FUSION BETWEEN SPLENOCYTES AND MYELOMA CELLS

Solutions

1. *Stock solution of 50× HAT* (Gustafsson, 1990). Dissolve 68 mg of hypoxanthine and 19.5 mg of thymidine in 100 ml of distilled water in a water bath for 30 min. Add a few drops of 1.0 M NaOH if the hypoxanthine does not dissolve easily. Adjust pH to 7.2 with HCl after reagents dissolve. Add 0.9 mg aminopterin and filter through a 0.2-μm filter. Store in aliquots at $-20°C$ protected from light.

NOTE Aminopterin is poisonous. Wear gloves and a mask while preparing solutions.

2. *HAT medium.* Add 2 ml of 50× HAT solution to 100 ml CIMDM.
3. *50% polyethylene glycol (PEG) 4000 solution.* While PEG is usually purchased as a waxy preparation, there is less toxicity associated with ultrapure PEG that is manufactured as chromatography-grade crystals. Add 25 g PEG 4000 to 20 ml IMDM. Place the bottle in a 80°C water bath until the PEG melts. Bring the volume to 50 ml with IMDM and filter through a 0.2-μm filter. Aliquot and store at 37°C until use.

DIRECTIONS

1. Combine the spleen and myeloma cells in a ratio of 4:1 in a single 50-ml centrifuge tube and centrifuge at 200 g for 5 min.
2. Pour off the supernatant and remove residual media with a Pasteur pipette.

3. Place the tube in a 37°C water bath in the hood.
4. For every 1.6×10^8 total cells, add 1 ml warm PEG solution. Using a 1-ml pipette, let the PEG run down the side of the tube from approximately 1 in. above the pellet while slowly turning the tube.
5. Leave the tube undisturbed for 1 min.
6. Dilute the fusion mixture by adding 1 ml IMDM, drop by drop, over 1 min.
7. Add a second 1-ml IMDM over the second minute.
8. Slowly add 10 ml IMDM with gentle mixing and centrifuge at 150 g for 5 min.
9. Discard the supernatant and resuspend the cells in 44 ml HAT selection media (CIMDM with HAT). This is sufficient for 600 microtiter wells, adding one drop per well.
10. Using a 10-ml pipette, distribute the cell suspension, adding one drop per well to the 60 inner wells of each of ten 96-well cell culture plates. Each drop, approximately 80 µl, will contain approximately 2.5×10^5 cells.
11. Add an additional two drops of HAT selection media to the same wells.
12. In an additional plate, plate parental myeloma cells in HAT selection media.
13. Fill the outer wells of all the plates with media or sterile water to prevent evaporation.
14. Place them in a dry CO_2 incubator, at 37°C, 5% CO_2:95% air, without additional moisture. Using a dry incubator reduces the chance for contamination.
15. Feed the cells twice a week by gently removing half of the spent supernatant and replacing it with fresh HAT selection media.

The control cells should die within 6 days. By Day 10, hybridomas should become visible and the media should be turning yellow. Colonies are selected for assays when they cover approximately one-third of the bottom of a well.

E. Hybridoma Evaluation

The strategy for choosing the right hybridomas for expansion should be established well before the first immunizations are performed. Although a useful monoclonal antibody is defined by the assay in which it will be utilized, the amount of work necessary to screen all of the developing colonies usually dictates a multiphase screening approach.

1. Initial Assays

The first assays should be rapid and suitable for screening large numbers of supernatants easily. Many different assays are used for this purpose (Harlow and Lane, 1988), but a solid-phase enzyme-linked immunosorbent assay (ELISA) is usually the method of choice (see Section VII, A). Use 96-well plates for the assay to facilitate the transfer of supernatants from 96-well microculture plates to wells with the same designation. Precharged assay plates, such as Nunc-Immulon 2 and Corning ELISA plates, are available from several sources for easy adsorption of conjugates, proteins, and peptides to the surfaces of the wells. Carbohydrate antigens (Jackson et al., 1984), whole cells (Kennett, 1980; King, 1989), and cell lysates (Kelleher et al., 1983) can also be used as coating agents.

2. Hapten Evaluation

If the target antigen is a hapten, screen with a hapten conjugate other than the one used for immunization. This will avoid false positives due to cross-reactivity with the protein carrier or with new epitopes formed during conjugation. Screen supernatants that are positive in this first assay in a second ELISA, to demonstrate that antibody binding to the hapten conjugate is blocked in the presence of free hapten (competitive ELISA).

3. Confirmatory Assays

Once supernatants have been identified as positive by ELISA, further tests are needed to verify the antibody's usefulness in the assay format for which it is intended. Use a competitive ELISA to verify that the antibody will recognize the native form of a protein if it is to be detected in body fluids rather than adsorbed to a solid phase. If the antibody will be used in immunohistostaining procedures, verify that the target epitope for binding is retained after tissue/cell manipulations or exposure to various fixatives. Western blotting, or immunoblotting, is an excellent procedure for visualizing antibody specificity for antisera (Ramlau, 1988). MAbs will bind only if they recognize a primary amino acid sequence that does not undergo denaturation during electrophoresis.

4. Isotyping

Knowing the immunoglobulin isotype is important for assay development and affinity purification. Several isotyping kits are available that utilize ELISA technology for rapid determinations. Isotyping should be done with supernatants from cloned cells rather than from ascites fluid or uncloned cells which can contain irrelevant antibodies.

F. Cloning

Clone colonies that are positive for specific antibody production immediately to avoid overgrowth by nonproducing or irrelevant hybrids. Cloning by limiting dilution is rapid and requires no additional techniques. While a feeder layer of thymocytes can be used to increase cloning efficiency (Rener *et al.*, 1985), commercial products containing interleukin 6 are excellent substitutes (Sugasawara, 1989). One such product, Origen®, is available from Igen, Inc., (Rockville, MD).

Because each hybrid varies in cloning efficiency and health, plate the cells at several theoretical densities ranging from 4 to 0.25 cells/well to assure recovery. Using this method, hybrids can be cloned directly into CIMDM containing 30% Origen or thymocyte feeder layers. Cryopreserve the cells that are left over as back up stock (see Section IV, G). Clones should appear within 2 weeks. As colonies arise and supernatants are assayed for specific activity, choose colonies for expansion from wells that were seeded at the lowest density. Repeat clonings until all of the supernatants from a single plate are positive. Recloning can usually be done at less than 1 cell/well. Transfer positive clones to 2-ml wells for further expansion and cryopreservation. Cryopreserve cells from each round of cloning to assure back up stock in case clones or antibody production are lost. Hybridomas that are maintained in culture over long periods of time should be recloned periodically to avoid changes in antibody characteristics or secretion rates.

G. Cryopreservation

Minimize loss of important hybridomas to contamination, changes in antibody characteristics and secretion rates, or cell death by establishing reliable techniques for cryopreservation and recovery of frozen cells. Freeze ampules of cells from each line in development, starting with the remaining cells from the original fusion well and cells from every round of cloning; split a stock of 20 ampules or more of the final subclone between two freezers to assure security.

Two techniques are useful for freezing and recovering cell lines during development: conventional cryopreservation of cells harvested from suspension cultures and *in situ* cryopreservation of colonies of new hybridomas and clones growing in 96-well microtiter plates (Wells and Price, 1983). The latter technique is invaluable if a fusion results in too many colonies to be evaluated easily or if the screening assay is cumbersome or time consuming. Once the cells are frozen, supernatants can be tested over a longer period of time. In both cases, cells should be stored in vapor-phase L N_2.

NOTE Use gloves and face mask for protection from L N_2 and defective vials.

CRYOPRESERVATION FROM SUSPENSION CULTURES

1. Harvest cells from exponential growth-phase cultures almost at confluency. Confluence can be judged visually when the culture looks slightly cloudy and the pH has dropped, turning the medium slightly orange/yellow. Cell concentration should be approximately 1×10^6/ml.
2. Concentrate the cells by centrifugation and resuspend them in cold CIMDM containing 10% dimethyl sulfoxide (DMSO) to approximately $5 \times 10^6 - 1 \times 10^7$ cells/ml. DMSO is toxic to cells and freeze media should be prepared just before use and kept cold.
3. Add 1-ml aliquots of cell suspension to 2-ml cryotubes on ice.
4. Transfer the vials to a $-70°C$ freezer overnight and then place in vapor-phase L N_2 (approximately $-170°C$) for long-term storage. For maximum viability, freeze the vials in a programmable, controlled-rate freezer before transferring to L N_2.
5. To recover frozen cells, remove a vial from the cryorepository and immediately place it in a 37°C water bath. If necessary, put the vial on dry ice while transporting it from the repository to the cell culture laboratory.
6. Swirl the vial gently until the cells are just thawed.
7. Wipe the outside of the vial with 70% alcohol and slowly add the contents to a centrifuge tube containing 10 ml CIMDM.
8. Spin the cells gently at 200 g for 5 min, decant, and resuspend the cells in prewarmed CIMDM.
9. Transfer the cells to a T-25 flask and incubate.

IN SITU CRYOPRESERVATION

New hybridomas (DeLeij et al., 1983), and clones can be frozen directly in 96- or 24-well plates avoiding the need to assay or expand many cultures at one time (Wells and Price, 1983; Price, 1985). Utilizing this method, recovery of over 90% of the wells can be expected (Walker et al., 1989).

1. Remove most of the medium from each well, being careful not to disrupt the cells. If the supernatants are retained for screening, transfer them to a fresh plate in the same orientation as the original wells for easy identification.
2. Add 80 μl of freeze medium to each well.
3. Wrap each plate with parafilm and hold on wet ice until all of the plates are completed.
4. Transfer to a $-20\,°C$ freezer for 2 hr and then to $-70\,°C$ where they can remain for storage up to 3 months. For longer storage, transfer to L N_2 vapor phase. For best results, freeze the plates to $-60\,°C$ in a programmable controlled-rate freezer and immediately transfer to L N_2 vapor phase.
5. Prior to recovering plates, put a flat sponge in a $45\,°C$ water bath and fill the bath with enough water to come to the top of the sponge (approximately $\frac{1}{2}$ in.).
6. Take the plates from the freezer and put them in an insulated freezer box on dry ice for transportation.
7. Remove the parafilm from a plate and place the plate directly on the sponge.
8. Gently push the plate down several times into the sponge until the wells are just thawed.
9. Using a 10-ml pipette, immediately add 1 drop of warm ($37\,°C$) FBS to each well and incubate at $37\,°C$ for 5 min.
10. Using an 18-gauge needle connected to a vacuum apparatus, gently remove the media, being careful not to disrupt the cells.
11. Add two drops of CIMDM to each well and return the plate to the incubator.

Table I
Summary of Media Described in the Preceding Sections on Hybridoma Cell Line Development

	Complete IMDM	HGPRT-selection medium	Wash medium SF-IMDM	PEG fusion medium	HAT selection medium	Cloning medium	Cryopreservation medium
Fetal bovine serum	10%	10%	—	—	10%	10%	10%
Sodium pyruvate	1%	1%	—	—	1%	1%	1%
L-Glutamine	1%	1%	—	—	1%	1%	1%
PEN/STREP	1%	1%	2%	—	1%	1%	1%
8-Azaguanine	—	$10^{-4} M$	—	—	—	—	—
50X HAT	—	—	—	—	2 ml/100 ml	—	—
Polyethylene glycol	—	—	—	25 g/50 ml	—	—	—
Origen®	—	—	—	—	—	30%	—
Dimethyl sulfoxide	—	—	—	—	—	—	10%

Note. Media formulations: Each formula is based on additions to presterilized, liquid 1X Iscove's Modified Dulbecco's Medium (IMDM).

V. Monoclonal Antibody Production

Large quantities of MAb can be produced in bulk from a variety of cell culture vessels or from mouse ascitic fluid.

A. In Vitro

Supernatants from confluent cultures of hybridomas grown in T-flasks or cell culture factories usually yield an optimal antibody concentration of approximately 10–20 µg/ml. Allowing the cultures to overgrow and exceed maximal cell densities can increase the concentration twofold since antibodies are relatively resistant to proteases released from lysed cells (Harlow and Lane, 1988). Simple spinner flasks, shaker cultures, and roller bottles can be used to increase the volume of supernatant that can be handled easily, while improving culture conditions. Concentrate and purify antibody simultaneously from supernatant by ultrafiltration.

A variety of bioreactors for large-scale production are available that monitor and control pH, O_2, nutrients, CO_2, and waste removal. These systems have the advantage of maintaining cultures for long periods of time at optimal culture conditions. Many systems incorporate antibody concentration and harvesting as on-line features (Clark *et al.*, 1990).

B. In Vivo

The easiest way to make high-titer antibody in a short period of time is to grow hybridoma cell lines as ascites tumors in pristane-primed syngeneic mice. A gram of antibody can be produced using approximately 100 animals. Although studies have been done to determine optimal conditions for ascites production (Brodeur *et al.*, 1984; Hoogenraad *et al.*, 1983), success seems dependent on intrinsic characteristics of each hybridoma cell line. Specific antibody content in ascitic fluid can vary from 1 to over 20 mg/ml; the volume of fluid collected from one harvest can vary from 0 to 5 ml. Total volume of fluid collected from a single animal can be as high as 15 ml (Rener, 1985).

ASCITES PRODUCTION

At least 10 to 14 days prior to injecting the cells, pretreat adult mice of the same genetic background as the hybridoma with pristane (2,6,10,14-tetramethylpentadecane) injected into the interperitoneal cavity (ip). FIA has also been used with equal success and in some cases has been superior (Gillette, 1987). These reagents cause increased number of monocytes to congregate in the peritoneum and secrete high levels of cytokines which support hybridoma cell growth. Use primed Balb/c mice within 10 weeks, since pristane can produce spontaneous plasmacytomas within 3 months (Potter *et al.*, 1972).

1. *Pristane priming.* Using a 22-gauge, 1½ in. needle, inject each mouse with 0.5 ml pristane, ip. Insert the needle no more than 1 cm and at an upward angle through the lower abdominal wall to avoid damaging internal organs.
2. *Passaging cells.* Using a 22-gauge, 1½ in. needle, inject 1×10^6 log-phase hybridoma cells in 0.5 ml buffer, ip, into a pristane-primed mouse as above. Swelling should be apparent within 10 to 20 days.
3. *Collecting ascitic fluid.* Tap the mouse as soon as swelling becomes apparent. Swab the abdomen with 70% alcohol. Insert an 18-gauge, 1½ inch needle upward through the lower abdominal wall and drain ascites directly into a 50-ml centrifuge tube.
4. *Concentrating cells.* Concentrate the cells by centrifugation at 3000 rpm for 5 min in a refrigerated centrifuge.
5. *Transferring ascitic fluid.* Transfer the ascitic fluid to storage tubes and hold at 4°C for further processing.

The mouse can be tapped again on successive days before it is sacrificed by CO_2 inhalation or exsanguination. Cells harvested from ascitic fluid can be passaged again in mice for further production. Reclone a cell line after three successive passages in mice to avoid changes in cell line and antibody characteristics.

Hybridomas that are not of Balb/c origin can be grown in F1 generation, or X-irradiated animals using the same procedures. More easily, the same procedures can be used with equal success in athymic (nude) mice if minimal precautions are taken to prevent spurious infections. Since mice are held for short periods of time during ascites production, good animal husbandry is usually sufficient to maintain the health of athymic mice. Wipe their abdomens with 70% alcohol before any invasive technique is begun; house them in polycarbonate cages with filter bonnets.

VI. Purification of Antisera and Monoclonal Antibodies

Depending upon their final application, antisera and MAb are processed to varying degrees of purification. All antibody preparations can be concentrated by salt fractionation with 50% saturated ammonium sulfate and dialyzed with appropriate buffer. This may be an end in itself or only the first step since extraneous proteins, including irrelevant antibodies, may remain which can interfere with assay specificity and sensitivity. Purification by simple salt fractionation can be improved by prior precipitation with caprylic acid (Reik et al., 1987). The choice of further purification methods depends upon many factors.

A. Antisera

For many assay applications, antisera are used without purification. Nonspecific cross-reactivity can usually be removed by affinity purification if sufficient antigen is available.

B. Monoclonal Antibodies

In addition to conventional ion-exchange chromatography, mouse MAb are frequently purified by DEAE Affi-gel blue (Bruck et al., 1982), affinity chromatography using immobilized antigen, or protein A Sepharose (Ey et al., 1978; Harshman, 1985). Protein A Sepharose will bind all Ig classes except IgM (some IgG1 antibodies may not bind well). Stepwise elution with buffers at pH 6.5, 5.5, 4.5, and 3.0 will elute IgG1, IgG2a, IgG2b, and IgG3, respectively. IgM antibodies can be purified by ammonium sulfate precipitation followed by hydroxylapatite chromatography (Bukovsky and Kennett, 1987; Harshman, 1989). The level of immunoglobulin purity can best be evaluated by SDS-PAGE, sodium dodecyl sulfate–polyacrylamide gel electrophoresis (Laemmli, 1970).

VII. Characterization of Antisera and Monoclonal Antibodies

A. Specificity

Traditionally, antiserum is evaluated for its ability to bind to pure, radiolabeled antigen in a precipitation assay. This test is of limited use for evaluating MAb; depending on the distribution and density of the target epitope, many high-affinity MAbs may not work in a precipitation assay. Both antisera and MAb can be evaluated in a 96-well ELISA format simply by changing the secondary immunoreagent from anti-mouse IgG to anti-rabbit IgG (Prigent et al., 1990). Hybridoma supernatant is tested undiluted; mouse immune sera or rabbit antisera are diluted 1:100–1:1000. Always include the appropriate controls. Negative controls should include reagents such as buffer, myeloma cell line supernatant, prebleed, or normal mouse or rabbit sera; positive controls might include other antisera or mouse immune sera.

ELISA EVALUATION OF ANTISERA AND MAB

1. Dilute pure antigen in 0.1 M carbonate buffer, pH 9.6.
2. Add 50 μl of diluted, pure antigen to each well. If the antigen preparation is partially purified, or conjugated to a carrier protein, add 2 μg/well. Purified protein can be added to each well at nanogram levels (50–100 ng/well).
3. Cover the plates and incubate overnight at 4°C.
4. Wash the plates 4× with phosphate-buffered saline containing 0.05% Tween 20 (PBS-T), blotting each time on a paper towel.
5. Add 200 μl of phosphate-buffered saline with 0.05% gelatin (PBS-G) to each well to block nonspecific binding.
6. Incubate 60 min at 37°C.
7. Decant and wash 4× with PBS-T.
8. Add 50 μl/well of test fluid diluted in PBS-G.
9. Cover the plates and incubate them for 1 hr at room temperature on a rotary shaker.
10. Decant and wash plates 4× with PBS-T, blotting each time.
11. Add 50 μl/well of peroxidase-conjugated anti-murine IgG serum or peroxidase-conjugated anti-rabbit IgG serum diluted according to manufacturers specifications, in PBS-G.
12. Cover and incubate the plates for 1 hr at room temperature on a rotary shaker.
13. Decant and wash the plates 6× with PBS-T, blotting each time.
14. Add 100 μl of freshly prepared substrate solution/well. Add 10 mg O-phenylenediamine (OPD) and 4 μl 30% hydrogen peroxide to 10 ml 0.1 M sodium citrate/citric acid buffer, pH 4.5.

NOTE OPD is carcinogenic. Wear gloves during preparation.

15. Cover the plates with foil-lined lids to protect from light and incubate 30 min at room temperature on rotary shaker.
16. Monitor absorbance at 450 nm or visually compare color intensities to controls.
17. The reactions can be stopped by the addition of 25 μl of 8 N sulfuric acid to each well prior to storing the plates.

Assay sensitivity can be enhanced by substituting biotinylated antimouse or anti-rabbit IgG reagents combined with a streptavidin horseradish peroxidase conjugate. Immunoglobulins which must discriminate between homologous proteins should be evaluated by performing assays simultaneously against the target antigen and potential cross-reacting compounds.

B. Affinity

The affinity constants of immunoglobulin reagents directly affect immunoassay sensitivity (Steward and Lew, 1985). Antisera and MAbs being developed for hormonal assays should have affinity constants in the range of $10^9 - 10^{10}$ liter mol^{-1} to detect low concentrations in body fluids. Measuring the binding of constant trace amounts of labeled antigen by increasing dilutions of culture supernatants, ascites fluid, or antisera is a rapid method of ranking immunoreagents early in development. Choose antisera or MAb for detailed characterization based upon the best titer. Running the assays at different pHs will also give some indication of best binding interactions (Van Heyningen et al., 1983). Scatchard analysis (Scatchard, 1949; Thakur and Rodbard, 1979) may be used to get a first approximation of the binding affinities of candidate antisera and MAb which can then be refined by nonlinear least-squares estimation techniques. Computer programs are available for such analyses (Munson and Rodbard, 1980; Thakur, 1990).

Acknowledgments

The author thanks Steve Harshman, Norman Beaudry, James Noel, Bruce Brown, and Ajit Thakur for their critical appraisals of the manuscript. Special thanks are due Florence Nicholson, Regina Donley, Tracey Walker, Tom O'Brien, Terry King, and Sharon Saturn for excellent technical assistance over the years as these procedures were developed and implemented in our laboratories.

References

Berson, S. A., Yalow, R. S., Bauman, A., Rothschild, M. A., and Nuverly, K. (1956). *J. Clin. Invst.* **35**, 170–190.
Brodeur, B. R., Tsang, P., and Larose, Y. (1984). *J. Immunol. Methods* **71**, 265–272.
Brown, B. L. (1985). *J. Tissue Cult. Methods* **9**(3), 137–140.
Brown, B. L. (1987). In "Commercial Production of Monoclonal Antibodies. A Guide for Scale-Up" (S. Seaver, ed.), pp. 35–48. Dekker, New York.
Brown, B. L., Ekins, R. P., Ellis, S. M., and Reith, W. S. (1970). *Nature (London)* **220**, 359.
Bruck, C., Portetelle, D., Glineur, C., and Bollen, A. (1982). *J. Immunol. Methods* **53**, 313–319.

Bukovsky, J., and Kennett, R. H. (1987). *Hybridoma* **6**(2), 219–228.
Carraway, R. E. (1979). *In* "Methods of Hormone Radioimmunoassay" (B. M. Jaffee and H. R. Behrman, eds.), 2nd Ed., pp. 139–169. Academic Press, New York.
Clark, S. A., Griffiths, J. B., and Morris, C. B. (1990). *In* "Animal Cell Culture" (J. W. Pollard and J. M. Walker, eds.), Methods in Molecular Biology, Vol. 5, Chap. 52. Humana Press, Clifton, New
DeLeij, l., Poppema, S., and The, T. H. (1983). *J. Immunol. Methods* **62**, 69–72.
Donley, R. (1989). *J. Tissue Cult. Methods* **12**(3), 91–92.
Engleman, E. G., Foung, S. K. H., Larrick, J., and Raubitschek, A. (1985). "Human Hybridomas and Monoclonal Antibodies." Plenum, New York.
Erlanger, B. F., Borek, F., Beister, S. M., and Lieberman, S. (1959). *J. Biol. Chem.* **228**, 713–727.
Ernst, M., and Sonneborn, H. (1990). *Hum. Antibodies Hybridomas* **1**(3), 122–125.
Ey, P. L., Prowse, S. J., and Jenkin, C. R. (1978). *Immunochemistry* **15**, 429–436.
Galfré, G., and Milstein, C. (1981). *In* "Immunochemical Techniques," Part B (J. Langone and H. Van Vunakis, eds.), Methods in Enzymology, Vol. 73, pp. 1–45. Academic Press, New York.
Gillette, R. W. (1987). *J. Immunol. Methods* **88**, 21–23.
Gustafsson, B. (1990). *In* "Animal Cell Culture" (J. W. Pollard and J. M. Walker, eds.), Methods in Molecular Biology, Vol. 5, Chap. 48. Humana Press, Clifton, New Jersey.
Harlow, E., and Lane, D. (1988). "Antibodies: A Laboratory Manual." Cold Spring Harbor Lab. Press, Cold Spring Harbor, New York.
Harshman, J. S. (1985). *J. Tissue Cult. Methods* **9**(3), 183–186.
Harshman, J. S. (1989). *J. Tissue Cult. Methods* **12**(3), 115–118.
Hoogenraad, N., Helman, T., and Hoogenraad, J. (1983). *J. Immunol. Methods* **61**, 317–320.
Horibata, K., and Harris, A. W. (1970). *Exp. Cell Res.* **60**, 61–77.
Hunter, R. L., Bennett, B., Howerton, D., Buynitzky, S., and Check, I. J. (1989). *In* "Immunological Adjuvants and Vaccines" (G. Gregoriadis, A. C. Allison, and G. Post, eds.), pp. 133–144. Plenum, New York.
Huse, W. D., Lakshmi, S., Iverson, S. A., Angray, S. K., Alting-Mees, M., Burton, D. R., Benkovic, S J, and Lerner, R. A. (1989). *Science* **246**, 1275–1281.
Institute of Laboratory Animal Resources. (1988). *ILAR NEWS* **30**(2), 9.
Jackson, S., Folks, T. M., Wetterskog, D. L., and Kindt, T. J. (1984). *J. Immunol.* **133**, 1553–1557.
James, K., and Bell, G. T. (1987). *J. Immunol. Methods* **100**, 5–40.
Kagan, A., and Glicks, S. (1979). *In* "Methods of Hormone Radioimmunoassay" (B. Jaffe and H. Behrman, eds.), 2nd Ed., Chap. 15. Academic Press, New York.
Kawamoto, T., Sato, J. D., Le, A., McClure, D. B., and Sato, G. H. (1983). *Anal. Biochem.* **130**, 445–453.
Kearney, J. F., Radburch, A., Liesegang, B., and Rajewsky, K. (1979). *J. Immunol.* **123**, 1548–1550.
Kelleher, P. J., Mathews, H. L., Woods, L. K., Farr, R. S., and Minden, P. (1983). *Cancer Immunol. Immunother.* **14**, 185–190.
Kennett, R. H. (1980). *In* "Monoclonal Antibodies: Hybridomas, A New Dimension" (R. H. Kennett, T. J. McKearn, and K. B. Bechtol, eds.), pp. 376–377. Plenum, New York.
Kenney, J. S., Hughes, B. W., Masada, M. P., and Allison, A. C. (1989). *J. Immunol. Methods* **121**, 157–166.
Kilmartin, J. V., Wright, B., and Milstein, C. (1982). *J. Cell Biol.* **93**, 576–236.
King, T. (1989). *J. Tissue Cult. Methods* **12**(3), 107–109.
Köhler, G., and Milstein, C. (1975). *Nature (London)* **256**, 495–497.
Kohen, F., Lichter, S., Eshhar, Z., and Lindner, H. R. (1982). *Steroids* **39**, 453–459.

Laemmli, U. K. (1970). *Nature (London)* **227**, 680-685.
"Linscott's Directory of Immunological and Biological Reagents." Seventh Ed. (1992-1993). 4877 Grange Road, Santa Rosa, California 95404.
Littlefield, J. W. (1964). *Science* **145**, 709.
Luben, R. A., Brazeau, P., Bohlen, P., and Guillemin, R. (1982). *Science* **218**, 887.
McHugh, Y. E., Walthall, B. J., and Steimer, K. S. (1983). *BioTechniques* Jun./July, 72-77.
Munson, P. J., and Rodbard, D. (1980). *Anal. Biochem.* **107**, 220-239.
Ozturk, S. S., and Palsson, B. O. (1990). *Hybridoma* **9**(2), 167-175.
Parker, C. W. (1971). *In* "Principles of Competitive Protein-Binding Assays" (W. D. Odell and W. H. Daughaday, eds.), pp. 25-56. Lippincott, Philadelphia.
Parker, C. W. (1976). *In* "Radioimmunoassay of Biologically Active Compounds" (A. G. Osler and L. Weiss, eds.), pp. 4-23. Prentice-Hall, Englewood Cliffs, New Jersey.
Phillips, H. J. (1973). *In* "Tissue Culture Methods and Applications" (P. F. Kruse and M. K. Patterson, Jr., eds.), pp. 406-408. Academic Press, New York.
Potter, M. (1972). *Physiol. Rev.* **52**, 631-719.
Potter, M., Pumphrey, J. G., and Walters, J. L. (1972). *J. Natl. Cancer Inst.* **49**, 305-308.
Price, P. (1985). *J. Tissue Cult. Methods* **9**(3), 167-169.
Prigent, S. A., Stanley, K. K., and Siddle, K. (1990). *J. Biol. Chem.* **265**, 9970-9977.
Ramlau, J. (1988). *In* "Handbook of Immunoblotting of Proteins" (O. J. Bjerrum and N. H. H. Heegaard, eds.), Vol. 1, pp. 151-158. CRC Press, Boca Raton, Florida.
Reik, L. M., Maines, S. L., Ryan, D. E., Levin, W., Bandiera, S., and Thomas, P. E. (1987). *J. Immunol. Methods* **100**, 123-130.
Rener, J. (1985). *J. Tissue Cult. Methods* **9**(3), 187-190.
Rener, J., Brown, B. L., and Nardone, R. M. (1985). *J. Tissue Cult. Methods* **9**(3), 175-177.
Robinson, I. C. A. F. (1980). *J. Immunoassay* **1**, 323.
Scatchard, G. (1949). *Ann. N.Y. Acad. Sci.* **51**, 660-672.
Shulman, M., Wilde, C. D., and Köhler, G. (1978). *Nature (London)* **276**, 269-270.
Spitz, M., Spitz, L., Thorpe, R., and Eugui, E. (1984). *J. Immunol. Methods* **70**, 39-43.
Steward, M. W., and Lew, A. M. (1985). *J. Immunol. Methods* **78**, 173-190.
Sugasawara, R. (1989). *J. Tissue Cult. Methods* **12**(3), 93-95.
Tateishi, K., Hamaoka, T., Takatsu, K., and Hayashi, C. (1980). *J. Steroid Biochem.* **13**, 951-959.
Thakur, A. K. (1990). *In* "Handbook of the Laboratory Diagnosis and Treatment of Infertility" (B. A. Keel and B. W. Webster, eds.), pp. 271-289. CRC Press, Boca Raton, Florida.
Thakur, A. K., and Rodbard, D. (1979). *J. Theor. Biol.* **80**, 383.
Thomas, G. (1989). *In* "Hormonal Assay Techniques," 15th Training Course Syllabus, pp. 6-9. Endocrine Soc., Bethesda, Maryland.
U.S. Department of Health and Human Services. (1985). "Guide for the Care and Use of Laboratory Animals," NIH Publ. 85-23. Natl. Inst. Health, Bethesda, Maryland.
Vaitukaitis, J. L. (1981). *In* "Immunochemical Techniques," Part B (J. Langone and H. Van Vunakis, eds.), Methods in Enzymology, Vol. 73, pp. 46-52. Academic Press, New York.
Vaitukaitis, J. L., Robbins, J. B., Nieschlag, E., and Ross, G. T. (1971). *J. Clin. Endocrinol. Metab.* **33**, 988-991.
Van Heyningen, V., Brock, D. J., and Van Heyningen, S. (1983). *J. Immunol. Methods* **62**, 147-153.
Walker, T. L., Whitlow, C. L., and Brown, B. L. (1989). *J. Tissue Cult. Methods* **12**(3), 97-98.
Weisbroth, S. (1984). *Lab. Anim.* **13**, 13-25.
Wells, D. E., and Price, P. J. (1983). *J. Immunol. Methods* **59**, 49-52.
Yalow, R. S., and Berson, S. A. (1959). *Nature (London)* **184**, 1648-1649.

5

Isolation of Polypeptide Hormones: General Separation Methods and Some Applications for Thyroid Hormones

Jorge Alemany, Jorge Garcia de Ancos, and Enrique Mendez

    I. Introduction
   II. Purification Techniques and Strategy
      A. Purification Techniques
      B. Purification Strategy
  III. Improvement in HPLC Fractionation of Thyroxine-Containing Peptides by Prior Acell Ion-Exchange Column Chromatography
      A. Protocol for Filtration of Tryptic Digests
  IV. Identification of T3- and T4-Containing Peptides by HPLC with Photodiode Array (PDA) Detectors
      A. Protocol for Peptide Detection Using PDA
      B. Advantages of This Approach
      References

I. Introduction

The techniques and methods used for the isolation of polypeptide hormones are the same as those used for the separation and purification of proteins and peptides in general. The improvement in reagents, technology, and instrumentation in the past two decades has made peptide purification more predictable and controllable, but it is still more an "art than a science." Therefore it is essential to have some degree of familiarity with the full range of available procedures, including knowledge of their strengths and pitfalls.

In general, for the isolation of peptides, one should start with extraction methods (i.e., homogenization, centrifugation, microfiltration), followed by low-selectivity fractionation techniques (i.e., "salting-out"/precipitation techniques, batch adsorption), and finally purification methods with a high degree of selectivity and efficiency (i.e., ultrafiltration, chromatographies). This approach is appropriate for the isolation of peptides whose chemical properties are not completely defined or when the peptide has to be purified from biological fluids. In this case a single component has to be separated out from thousands of substances (many of them chemically closely related) in the biological fluid. Of particular importance are the carrier proteins bound to some peptide hormones which can interfere with the isolation procedure. Additional complexity is encountered when one considers that the substance of interest is often present in small amounts. The isolation of a biological substance, such as a peptide, requires the use of procedures that will prevent the modification or loss of structural and/or functional properties.

Despite the sophisticated equipment available, difficulties for peptide isolation still remain, especially concerning two aspects: (1) finding the optimum conditions for extraction and sample pretreatment (avoidance of denaturation, inactivation, hydrolysis) and (2) choosing a suitable method for monitoring the concentration of the peptide and its biological activity during and after the purification process. These principles extend to the purification of peptides and proteins that are produced by chemical syntheses and by genetically engineered microorganisms or cultured eukaryotic cells.

The proliferation of conventional and high-performance techniques for purification of peptides has made choosing the most appropriate, in each particular case, a matter of considerable difficulty. For this purpose there are several useful books (Scopes, 1982; Janson and Ryden, 1989) and a review by Gavin (1991) that should be useful in designing a strategy. In the present chapter we mention briefly the most utilized techniques for peptide purification and describe a rapid and sensitive peptide identification method for monitoring peptides by high performance liquid chromatography (HPLC).

II. Purification Techniques and Strategy

A. Purification Techniques

The extraction and purification of a peptide ("downstream processing") can be divided into three steps:

1. *Initial fractionation.* The purpose of this step is to obtain a solution suitable for chromatography.
2. *Purification.* This step separates the peptide from contaminants and related molecules.

3. *Final "polishing."* This step removes aggregates and degradation products and leaves the purified peptide prepared in a suitable solution for final formulation.

1. Initial Fractionation

There are several techniques to clarify the initial cell extract:

a. Centrifugation and Microfiltration The clarification of any cell homogenate is usually no problem on a laboratory scale, where refrigerated high-speed centrifuges covering operating speeds from 20,000 to 75,000 rpm (40,000 to 500,000 g) can be used. During the past few years tangential or cross-flow microfiltration has received increased attention, especially for large-scale applications.

b. Ultrafiltration Ultrafiltration is a gentle, fast, and inexpensive method that is widely used in preparative biochemistry. Ultrafiltration membranes are available with different cutoff limits for separation of molecules from 1000 to 300,000 Da. This method is excellent for the separation of salts and other small molecules from a higher molecular weight protein fraction and at the same time ultrafiltration concentrates the protein.

c. Adsorption Chromatography The early work by Zechmeister and Cholnoky (1942) has been applied for decades to the separation of peptides (Tiselius *et al.*, 1956) using hydroxyapatite (HA) columns. This calcium/phosphate matrix is useful for the initial separation of crude mixtures. The binding occurs by nonspecific attraction between protein positive charges and HA and by specific interaction between protein carboxyl groups and calcium ions on the gel (Gorbunof, 1984a,b). As the HA has a high adsorption capacity, it is useful for batch fractionation before further purification by more refined methods. The elution of the adsorbed proteins is achieved by increasing ionic strength in the eluent buffer, and this can be manipulated to achieve more selective elutions. This technique has been applied to the purification of follicle-stimulating hormone (FSH) (Steelman, 1958) and thyrotrophin (Wynston *et al.*, 1960).

d. Precipitation Fractional precipitation is a classical means for a second clarifying step purification. First, bulk proteins in the solution are precipitated together with residual particulate matter and then the peptide of interest is precipitated from the resulting supernatant solution. Sometimes the peptide of interest is allowed to remain in the "mother" liquor solution for direct application to chromatographic columns, i.e., hydrophobic interaction (ad-

Table I
Chromatographic Separation Methods

Method	Separation criteria
Gel filtration	Particle size
Ion exchange	Particle charge
Hydrophobic interaction and reverse phase	Hydrophobicity
Affinity chromatography	Biospecific interaction

sorption of proteins in ammonium sulfate solution) or ion exchange (absorption in polyethylene glycol). The most common precipitating agents are ammonium sulfate, ethanol, and acetone.

2. Purification Step

At this stage the main technique used is chromatography and if the amount of protein needed is low one can also use electrophoresis. Both techniques have been extensively discussed previously (Scopes, 1982; Janson and Ryden, 1989; Deutscher, 1990). Various chromatographic methods available for protein purification are summarized in Table I.

a. Ion-Exchange Chromatography Peptides and proteins have a net charge whose size and sign (+ or −) depend on pH. This property is utilized in ion-exchange chromatography for selective adsorption onto a carrier derivatized with charged groups. There are anion exchangers, with positively charged groups like diethylaminoethyl (DEAE) and quaternary ammonium (QA), and cation exchangers like carboxymethyl (CM) and sulphopropyl (SP). Those exchangers can be either weak (DEAE, CM) or strong (QA, SP). The sample components bind to the exchanger, at a suitable pH, at low ionic strength, and can be selectively eluted by increasing the ionic strength or varying the pH.

When the ion-exchange chromatography is completed, the technique of election for further purification is either hydrophobic interaction affinity chromatography or gel filtration chromatography.

b. Affinity Chromatography In affinity chromatography, the product to be purified is specifically adsorbed on an affinity ligand coupled to a matrix. This is analogous to the specific binding processes that are necessary for the biological function of peptides, but with the difference that the immobilized ligands do not undergo subsequent conversion. Well over 100 ready-to-use media are available with different starting and elution conditions. They can be divided in three groups according to use: (1) specific ligands (e.g., monoclonal antibodies, protein A, enzyme inhibitors) for single proteins; (2) the so-called general or group-specific ligands such as immobilized metal ions; and (3) artificial ligands such as textile ligands.

Because of the high selectivity of the affinity media, very large volumes of the product can be processed. The sample mixture is generally applied to the column at low ionic strength and in the neutral pH range. After the nonbound impurities are washed out, the product can be eluted by using a gradient of increasing ionic strength or by reducing the pH.

c. Hydrophobic Interaction Chromatography (HIC) Most peptides (not only "hydrophobic" peptides) possess hydrophobic regions on their surface which allow interaction with a hydrophobically derivatized matrix in aqueous solutions. The forces responsible for adsorption to a carrier material of this type depend on the increase in entropy arising from the displacement of hydrophobic molecular regions from the polar surrounding. These forces are increased at high salt concentrations (salting-out).

The most common ligands used in HIC are phenyl, butyl, and octyl residues. The hydrophobic properties of phenyl matrixes are suitable for purifying most proteins under mild conditions. Chromatographic techniques using low ionic strength as the starting condition can be utilized after HIC.

d. Reverse-Phase Chromatography Hydrophobic interactions are also used in the reverse-phase chromatography (RPC). However, whereas HIC is a nondenaturing method, peptides are usually denatured by the nonpolar solvents used for elution from the highly hydrophobic RPC matrices. The preparative use of RPC is limited to the fractionation of peptides up to about the size of insulin (≈ 6000 Da). The use of RPC together with gel filtration and ion exchange has already proved to be of value in the production of peptides such as oxytocin and vasopressin (Larsson, 1985).

e. High-Pressure Liquid Chromatography This procedure represents one of the true recent advances in chromatographic separation of peptides. In HPLC all the chromatographic techniques described above can be applied, at higher speed and high resolution and with an automated operational format. The basic feature is that the mobile phase is liquid and the stationary phase (either solid or liquid) is bonded to or supported by a solid matrix. The velocity of the mobile phase is controlled by a pump and the column contains small-diameter packing material.

HPLC has been successfully used in isolating hormonal peptides, like insulin and insulin-like growth factor-I (IGF-I), from tissues or cultured cells

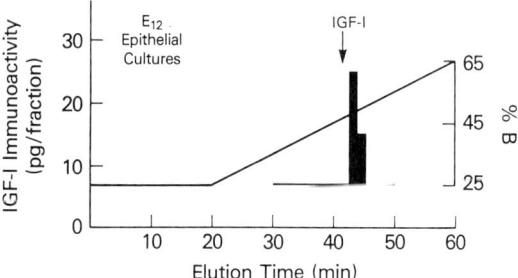

Figure 1 HPLC profile of insulin-like growth factor-I (IGF-I) from embryonic lens epithelial cells in primary culture. Cells from Day 12 embryo lenses were placed in culture for 24 hr. The cell extract was passed through C18-SepPak to remove IGF-I-binding proteins, as well as to desalt and concentrate. The eluate was resuspended in HPLC buffer and chromatographed using HPLC with a linear acetonitrile gradient. The percentage of solution B (80% acetonitrile/20% H_2O/0.1% trifluoroacetic acid) is plotted on the right vertical axis. Fractions 30 through 50 were analyzed in duplicate in an IGF-I RIA. The sensitivity of the RIA is indicated by a horizontal line. The position of an IGF-I standard is indicated by an arrow. (For experimental details see Caldés et al., 1991.)

(Caldés et al., 1991). Figure 1 shows the HPLC profile of IGF-I [identified by radioimmunoassay (RIA)] from epithelial lens cells in primary culture. Prior to chromatography and after homogenization, the cells were passed through a C18 SepPak (hydrophobic interaction-type) cartridge to separate the IGF-I-binding proteins. The reverse-phase HPLC was performed using a Vidac C-18 hydrophobic interaction TP104 column as described (Scavo et al., 1989).

B. Purification Strategy

The purification protocol of a peptide should be designed on the basis of as much information as possible about the chemical properties of the peptide to be isolated and, if possible, the characteristics of the impurities and contaminants. Useful data for the separation techniques to be used are the molecular weight, pI, presence of carbohydrates, presence of -SH groups, and the degree of hydrophobicity. This information can be derived either from the DNA level if the nucleotide sequence is available or by preliminary trials using crude extracts. It is also very important to know the stability range of the protein. Thus, one should collect information about the parameters which affect the protein structure and function such as pH, temperature, and presence of organic solvents, oxygen, heavy metals, etc.

Other important aspect to consider when designing a purification protocol are as follows:

What is the starting material? If the starting material is very precious one should favor high yield over speed and convenience.

What is the peptide to be used for? The purity requirements for a peptide hormone for research purposes are different than those for therapy.

What is the required amount? It is important to be aware of the scale-up ultimately expected, since some protocols useful for the purification of a few micrograms cannot be applied to high-scale preparations.

How will the protein be monitored? A sensitive and precise assay is necessary to monitor the location and losses of the peptide during the whole process. The most common techniques for this purpose are electrophoresis, immunological techniques (RIA, Western blot, enzyme-linked immunosorbent assay), and HPLC; new detectors such as the photodiode array have made this last technique even more powerful. We should also mention the capillary zone electrophoresis (CZE, see Chapter 7) that appears to have enormous applicability (Deyl and Struzinsky, 1991).

III. Improvement in HPLC Fractionation of Thyroxine-Containing Peptides by Prior Acell Ion-Exchange Column Chromatography

The isolation and purification of peptides containing tyrosyl residues in thyroglobulin (Tg) is of great interest for the understanding of structure–function relationships in this iodoprotein. Tg is a very large glycoprotein (660 kDa, 19 S) composed of two identical subunits (330 kDa, 12 S). Each subunit has 72 tyrosyl (Tyr) residues which, considering their size, is not a high number but rather average. Of the 144 Tyr residues contained in the Tg molecule, only about 25% are accessible to iodination, and of those 36 are iodinated as mono- and diiodotyrosine (MIT and DIT, respectively). In fact a maximum of only 14 Tyr residues directly participate in hormone formation. By reduction alkylation of Tg from several species, hormone-rich peptides ("hormonogenic peptides") have been isolated (Dunn et al., 1981; Ratwitch et al., 1983). However, due to the high molecular size of Tg, the purification of these peptides is very laborious since, in order to separate them from the more abundant contaminating peptides, many HPLC runs with various mobile phases are required and the yield is very low. In trying to improve this procedure, one can subject the tryptic digests, prior to the HPLC fractionation, to Acell self-packed cation (CM)- and anion (QMA)-exchange chromatography, eluting with a volatile solvent (Bennet, 1986).

A. Protocol for Filtration of Tryptic Digests

Passage through ion-exchange columns results in the enrichment of T_4-containing peptides, greatly facilitating their subsequent fractionation and purification by HPLC. The enrichment of T_4-containing peptides is only relative. However, the use of ion-exchange chromatography still represents an improvement and it can be used as preliminary step to separate and purify T_4- and iodotyrosine-containing peptides or "acceptors". The latter act as donors in the coupling reaction to form the hormones T_4 and T_3. The use of ion-exchange columns facilitates the subsequent separation and purification by HPLC (Miguel *et al.*, 1988).

Reduced, carboxymethylated Tg tryptic digests are divided into three aliquots (approximately 1.1 mg each): (i) the control sample subjected directly to HPLC fractionation; (ii) the sample passed through a self-packed Acell CM cation-exchange (carboxymethyl) column (0.8–4 cm); and (iii) the sample passed through a self-packed Acell QMA anion-exchange (quarternary methylammonium) column (0.8–4 cm), both from Waters. The columns are equilibrated with 4.5 ml of 0.05 M ammonium acetate buffer, pH 5.0. The samples are subsequently eluted from the columns with 4.5 ml of 20, 50, and 80% acetonitrile in 0.05 M ammonium acetate buffer, pH 5.0. Finally, the columns are washed with 4.5 ml of 0.4 M NaCl. The eluates are desiccated on a rotavapor and subsequently subjected to HPLC fractionation.

Aliquots from the different pools are dissolved in 5% acetonitrile containing 0.7% ammonium bicarbonate (phase A) and are applied on a C18 μBondapack reverse-phase column using a chromatograph consisting of two Waters M6000 A pumps, a Waters 680 automated gradient controller, and a Waters 480 variable wavelength absorbance detector. The column is eluted over 110 min with a linear gradient of 5% phase B to 100% phase B (100% acetonitrile). The column is run at room temperature at a flow rate of 0.5 ml/min collecting 0.5-ml fractions.

Figure 2 summarizes the procedure followed in order to test whether ion-exchange column chromatography, prior to HPLC fractionation of the Tg tryptic digests, enhances the separation and subsequent purification of the T_4-containing peptides. In the CM column, 67.5% of the ^{125}I-labeled material was not retained and 23.4% eluted with 20% acetonitrile, accounting between them for more than 90% of the ^{125}I-labeled tryptic digest applied to the column. In the QMA column, however, 34.9% was not retained and 19.7% eluted with 20% acetonitrile, accounting between them for only 54.6% of the tryptic digest applied. While in the CM column almost nothing eluted with 0.4 M NaCl, in the QMA one, 34% eluted with this salt. In other words, it was T_4 enriched with respect to the control sample.

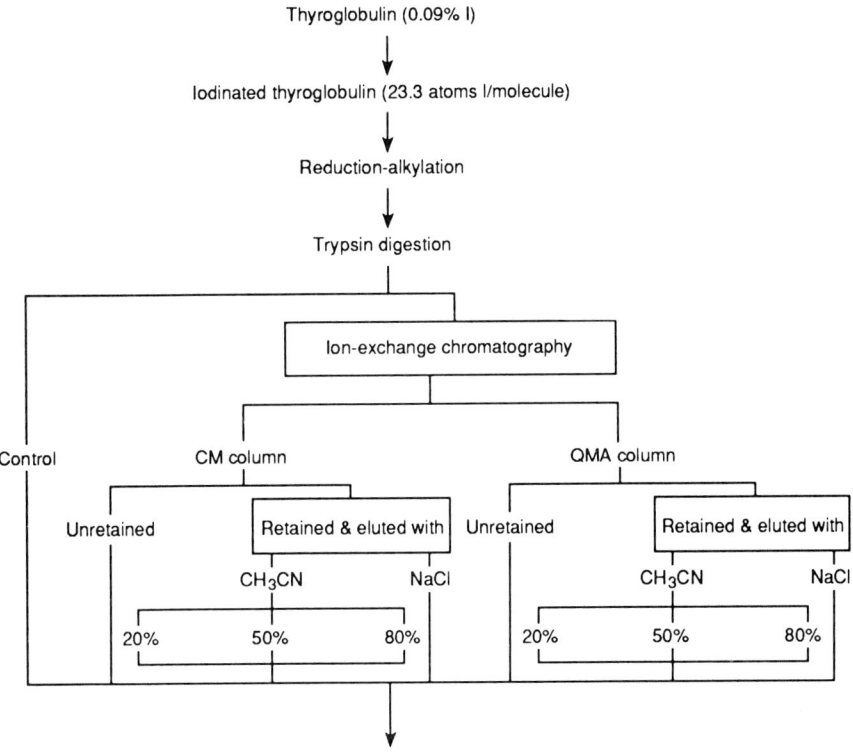

Figure 2 Outline of the procedure used for testing the improvement in separation of T_4-containing peptides from thyroglobulin tryptic digests by self-packed ion-exchange column chromatography. (For experimental details see Miguel et al., 1988.)

IV. Identification of T3- and T4-Containing Peptides by HPLC with Photodiode Array (PDA) Detectors

Since the introduction of HPLC, many different procedures for the identification of peptides have been described, such as RIA, ELISA, and PDA. In the past few years, PDA, ultraviolet [UV or ultraviolet-visible (UV-VIS)] detectors have been employed as components of HPLC systems. The advantage of these detectors is that they can continuously monitor all wavelengths of the spectrum and collect the data for future processing. After chromatography the data can be analyzed via software routines, such as spectral analysis for peak identification and purity confirmation or comparison of absorption spectra of separated peaks, to obtain specific information about the sample. Recently, metabolites of aromatic amino acids (Fell et al., 1984), mycotoxins (Frisvad and Thrane, 1987) and other fungal metabolites (Escribano et al., 1988), protein heterogeneous in charge, and several chromophores (Alfredson and Sheelan, 1986; Tejler and Grubb, 1976) have been analyzed with this detection system.

A. Protocol for Peptide Detection Using PDA

This section focuses on the characterization of peptides containing aromatic amino acids, cysteine, MIT, DIT, T_3, and T_4. We report a general procedure for identifying of all the above-mentioned peptides, using several simple data processing modes: "spectrum analysis," "second-derivative spectra," "spectrum index plot," and "multichromatogram analysis" (Escribano et al., 1990).

The chromatograph consists of two Waters 680 automated gradient controllers and a Waters 990 PDA detector with a dynamic range from the ultraviolet to the visible region (190–800 nm), based on an NEC APC III personal computer. All sample injections are performed with a Waters U6K universal injector.

Size exclusion HPLC is performed on a TSK 3000 SWG column (300 × 21.5 mm i.d.) (Toyo Soda, Tokyo, Japan), fitted with a TSK 3000 SWG guard column, by isocratic elution with 0.1 M ammonium acetate buffer (pH 5.0). The column is operated at room temperature at a flow rate of 1 ml/min.

Reverse-phase (RP)-HPLC is performed with a NovaPak column (150 × 3.9 mm i.d.) (Waters Assoc.), protected by a guard column packed with μBondapack C_{18}/Corasil (Waters Assoc.). The column is eluted with acetonitrile gradients containing 0.1% (v/v) TFA (pH 2.0). The TFA-insoluble material is solubilized with 6 M guanidinium chloride and eluted with an acetonitrile gradient containing 0.7% (w/v) ammonium hydrogen-carbonate (pH 8.0). The column is operated at room temperature at a flow rate of 0.5 ml/min.

The Tg are iodinated enzymatically *in vitro* in 0.067 M phosphate buffer (pH 7.0) at 37°C with 10^{-4} M iodine, labeled with radioiodine (I*, sp act 1.0 μCi/μat), 1.5 μg/ml lactoperoxidase, 1.0 mg/ml glucose, and 1.5 μg/ml glucose oxidase for 15 min. The number of atoms of iodine bound per mole of protein is calculated as described previously (Lamas et al., 1986).

The iodinated Tg is reduced with mercaptoethanol (ME) (100 mol ME/mol S-S in Tg) and subsequently S-cyanoethylated with acrylonitrile (2 mol/mol ME added), stopping the reaction with ME (2 mol/mol acrylonitrile added). Excess reagents are eliminated by passage through an Econopac 10DG disposable chromatographic column (Bio-Rad Labs., Richmond, CA). Tg

5. Isolation of Polypeptide Hormones

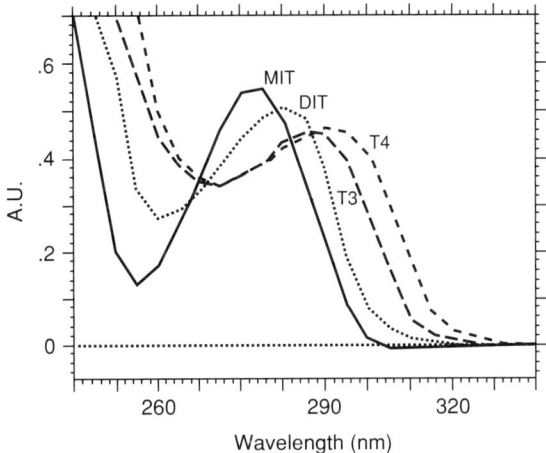

Figure 3 Comparative analysis from 240 to 350 nm of MIT, DIT, T_3, and T_4 standards. Spectra were normalized in order to eliminate concentration differences. (For experimental details see Escribano *et al.*, 1990.)

fractions are pooled and digested with TPCK-trypsin (5%, w/w) for 16 hr at 37°C. Aliquots from iodinated protein are digested with pronase and the iodoamino acid distribution is determined as described previously (Lamas *et al.*, 1986).

MIT, DIT, T_3 and T_4 standards are dissolved in 50% aqueous acetonitrile containing 0.1% (v/v) trifluoroacetic acid (TFA), in 0.7% aqueous ammonium hydrogencarbonate containing 0.1% (v/v) TFA, or in 52.5% acetonitrile containing 0.35% (w/v) ammonium hydrogen carbonate. The absorption spectrum is obtained using a Waters 990 PDA detector.

B. Advantages of This Approach

We use a routine procedure with a PDA detector adapted to our HPLC system for direct detection of iodoamino acids containing peptides, such as those derived from Tg, which is the Tyr-containing protein that forms thyroid hormones with the highest efficiency of iodination and couples to some of its Tyr residues, or protein human complex-forming glycoprotein (HC), which does not form thyroid hormones efficiently after *in vitro* iodination. Comparative spectral analysis of MIT, DIT, T_3, and T_4 standards injected into the HPLC system is shown in Fig. 3. Using the spectrum analysis program (Fig. 3), the absorbance maxima in TFA for MIT, DIT, T_3, and T_4 are 284, 290, 298, and 302 nm, respectively. These data show a gradual shift toward higher wavelengths with increasing number of iodine atoms in the molecule. Although the spectra of T_3 and T_4 have similar maxima around 300 nm, the spectrum of T_3 ends at about 334 nm whereas that of T_4 ends at about 340 nm. The clearly different absorption of each iodoamino acid allows their identification.

Based on these standards one can correlate those iodoamino acids containing peptides derived from proteolytic digestion of HC (Fig. 4) with MIT, DIT, T_3, and T_4. The HC digest is injected into the HPLC system and peaks (fractions) are analyzed using either the spectral analysis or the spectrum index plot program. It is possible to obtain automatically the maximum absorption spectrum at any wavelength range for each peak of the chromatogram (Fig. 4). In one example, peaks 47 and 87 contained 85.5 and 92.5% MIT,

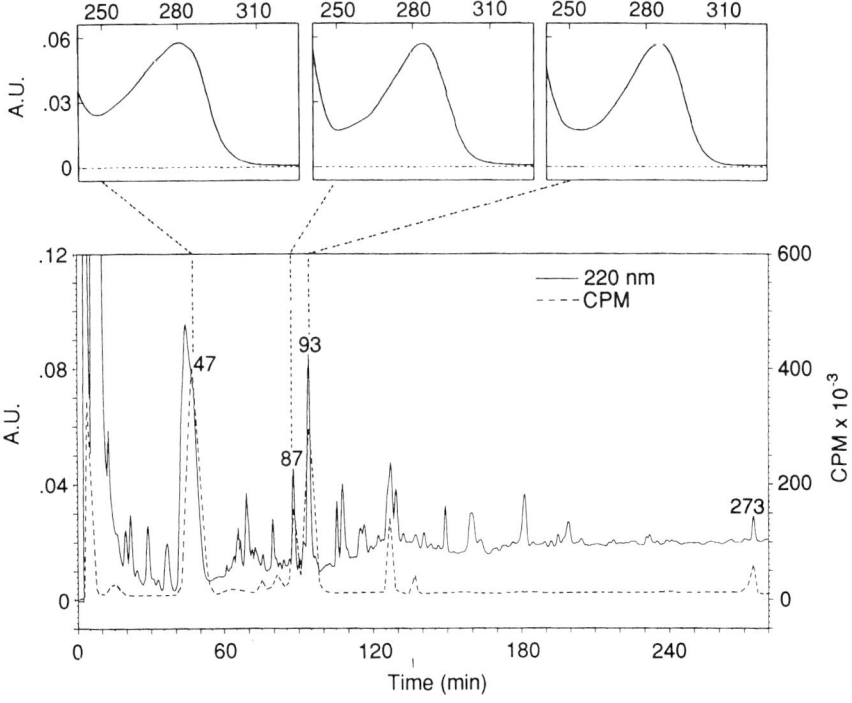

Figure 4 Fractionation of a pronase digest of urinary ^{125}I-labeled protein HC. Sample, 31.0 nmol; column, Novapack C_{18} (300 × 21.5 mm i.d.); flow rate, 0.5 ml/min. The column was equilibrated with 0.1% (v/v) aqueous TFA, and peptides were eluted at room temperature, using a linear gradient from 0 to 80% of acetonitrile containing 0.1% (v/v) TFA. Fractions of 2 ml were collected and the radioactivity was measured. The chromatogram was analyzed by monitoring the absorbance at 220 nm (bottom). Automatic spectra were acquired from the peak maxima from 240 to 330 nm (top). (For experimental details see Escribano et al., 1990.)

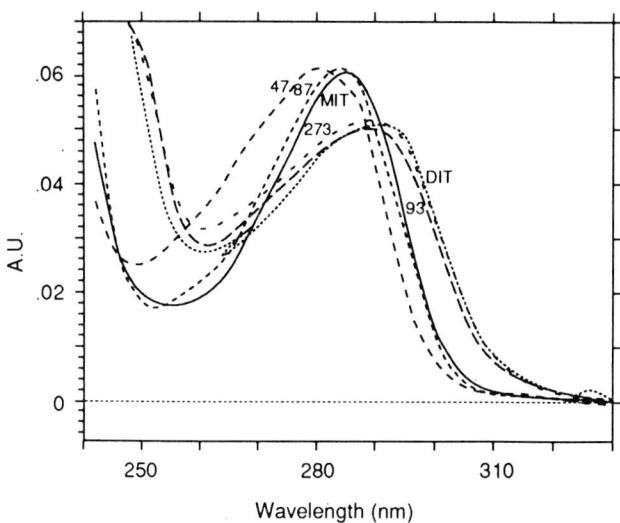

Figure 5 Comparative spectral analysis from 240 to 330 nm of protein HC pronase peptides 47, 87, 93, and 273 from Fig. 4. The spectra of MIT and DIT are included for comparison. Spectra were normalized in order to eliminate differences. (For experimental details see Escribano et al., 1990.)

respectively, whereas peaks 93 and 273 contained 90.0 and 82.2% DIT, respectively. This was confirmed by using the spectral analysis program as shown in Fig. 5, as the spectrum of fraction 87 coincided exactly with that of the MIT standard, and the spectra of fractions 93 and 273 coincided with that of the DIT standard. However, peak 47 does not coincide completely with either the MIT or the DIT standard by spectral analysis (Fig. 5), probably due to the contribution of aromatic amino acid contaminants, as the absorbance peak and the ^{125}I peak around that fraction do not coincide (Fig. 4).

References

Alfredson, T., and Sheelan, T. (1986). *J. Chromatogr. Sci.* **24,** 495.
Bennet, H. P. J. (1986). *J. Chromatogr.* **359,** 383.
Caldés, T., Alemany, J., Robcis, H., and de Pablo, F. (1991). *J. Biol. Chem.* **266,** 2786.
Deutscher, M. P., ed. (1990). "Guide to Protein Purification," Methods in Enzymology, Vol. 182. Academic Press, San Diego.
Deyl, Z., and Struzinsky, R. (1991). *J. Chromatogr.* **569,** 63.
Dunn, J. T., Dunn, A. D., Heppner, D. G., and Kim, P. S. (1981). *J. Biol. Chem.* **256,** 942.
Escribano, J., Matas, R., and Méndez, E. (1988). *J. Chromatogr.* **444,** 165.
Escribano, J., Asunción, M., Miguel, J., Lamas, L., and Méndez, E. (1990). *J. Chromatogr.* **512,** 255.

Fell, A. F., Clark, B. J., and Scott, H. P. (1984). *J. Chromatogr.* **297**, 203.
Frisvad, J. C., and Thrane, U. (1987). *J. Chromatogr.* **404**, 195.
Gavin, J. R. (1991). *In* "Introduction to Endocrine Investigation," pp. 93–101. Endocr. Soc., Bethesda, Maryland.
Gorbunof, M. J. (1984a). *Anal. Biochem.* **136**, 425.
Gorbunof, M. J. (1984b). *Anal. Biochem.* **136**, 433.
Janson, J.-C., and Ryden, L. (1989). "Protein Purification. Principles, High Resolution Methods and Applications." VCH, New York.
Lamas, L., Santisteban, C., Turmo, C., and Seguido, A. M. (1986). *Endocrinology (Baltimore)* **118**, 2131.
Larsson, K. B. A. (1985). *Symp. HPLC Proteins, Pept. Polynucleotides, 5th, Toronto.*
Miguel, J., Asunción, M., Marin, C., Seguido, A., Lamas, L., and Mendez, E. (1988). *FEBS Lett.* **232**, 399.
Ratwitch, A. B., Chernof, S. B., Litwer, M. R., Rouse, J. B., and Hamilton, J. W. (1983). *J. Biol. Chem.* **258**, 2079.
Scavo, L., Alemany, J., Roth, J., and de Pablo, F. (1989). *Biochem. Biophys. Res. Commun.* **162**, 1167.
Scopes, R. K. (1982). "Protein Purification. Principles and Practice." Springer-Verlag, New York.
Steelman, S. L. (1958). *Biochim. Biophys. Acta.* **27**, 405.
Tejler, L., and Grubb, A. O. (1976). *Biochim. Biophys. Acta* **439**, 82.
Tiselius, A., Hjerten, S., and Levin, J. G. (1956). *Arch. Biochem. Biophys.* **65**, 132.
Wynston, L. K., Free, C. H., and Pierce, J. G. (1960). *J. Biol. Chem.* **235**, 85.
Zechmeister, L., and Cholnoky, V. (1942). "Principles and Practice of Chromatography." Chapman & Hall, London.

6

Reverse Hemolytic Plaque Assays: Applications to Endocrine Problems

L. Stephen Frawley

I. Introduction
II. Procedures Common to All Plaque Assays
III. Standard Plaque Assay
    A. Validation
    B. Quantification and Applications
IV. Sequential Plaque Assay
    A. Validity
    B. Quantification and Applications
V. Simultaneous Plaque Assay
VI. Other Variations and Applications of Plaque Assays
VII. Summary
    References

I. Introduction

The reverse hemolytic plaque assay (RHPA) has become a powerful tool for endocrinologists because it enables the detection of hormone release at the single-cell level. The technique is based on antibody-directed, complement-mediated lysis of indicator cells (ovine red blood cells) in the vicinity of hormone-secreting cells (Neill and Frawley, 1983). To put it simply, one type of hormone secretor can be distinguished from other types because it induces the formation of a microscopically identifiable plaque (zone of hemolysis) around itself. This occurs when the secretory cell is incubated with indicator cells in the presence of hormone-specific antiserum and complement (Fig. 1). More specifically, protein A (Pr-A), which has an affinity for antibodies, is chemically coupled to ovine red blood cells. The protein A-coated ovine red blood cells (oRBCs) are then mixed with a suspension of monodispersed

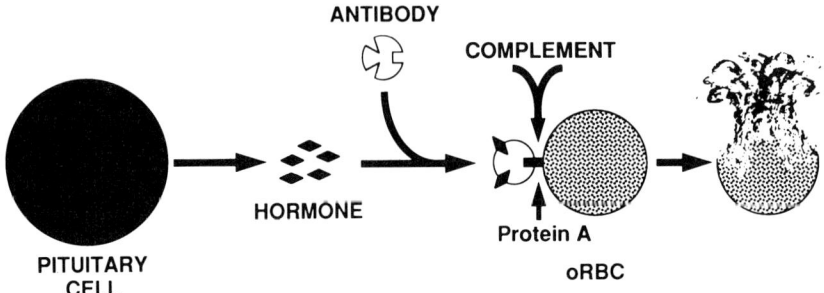

Figure 1 Principles underlying the reverse hemolytic plaque assay. Details relevant to this and the following figure are provided in the Introduction.

secretory cells (in our lab these are anterior pituitary cells). The suspension is infused into an assay chamber (Fig. 2) constructed using double-sided tape and coverslips pressed onto poly-L-lysine-treated slides. After a preincubation period, the monolayer is washed and flooded with a solution of medium containing antibody. Secretagogues can be used during the incubation step. Plaques are then developed during a subsequent incubation with a dilute solution of guinea pig serum (a source of complement). The binding of the hormone–antibody complex to the oRBCs facilitates complement fixation

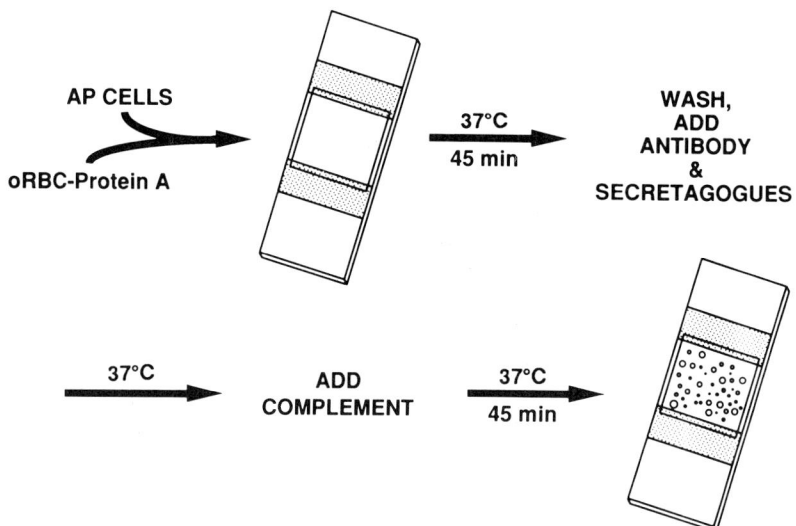

Figure 2 Protocol for performing the standard version of the reverse hemolytic plaque assay.

and eventual lysis of the oRBCs. A low-magnification photomicrograph of a monolayer containing a pituitary cell which formed a plaque after exposure to antibody and complement is shown in Fig. 3.

Although the first hormone-based plaque assay developed by Neill and Frawley might have seemed to appear rather abruptly in 1983, the methodology required for this application was developed by immunologists during the preceding 2 decades. The technology, as we currently know it, required several innovative modifications, four of which stand out in importance. The first of these was the adaptation by Jerne *et al.* (1974) of the agar plaque assay (originally used for enumerating bacterial viruses) to detect the release of immunoglobulin from individual B lymphocytes. In this system, the antibody was secreted by the cell under study whereas the antigen resided on the surface of the indicator erythrocyte. Next, Cunningham and Szenberg (1968) im-

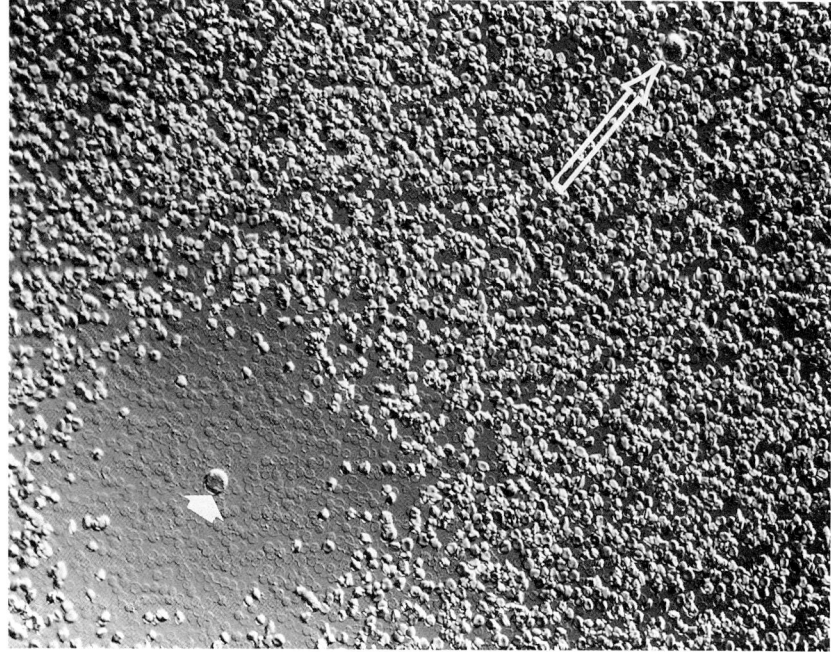

Figure 3 Photomicrograph (160×) of a plaque-forming cell. Shown here are two pituitary cells in a lawn of protein A-coated ovine erythrocytes. One of these (solid arrow) caused the lysis of neighboring erythrocytes when incubated with growth hormone (GH) antiserum and complement and is therefore identified as a somatotrope. The other pituitary cell (open arrow) is not a GH secretor as evidenced by the absence of the hemolytic reaction.

proved the sensitivity and logistics of performing the assay by adapting it to the monolayer condition and developing the slide chambers now commonly employed. An additional quantal step in development was accomplished by Molinaro *et al.* (1981), who reversed the relative positions of antigen and antibody on the erythrocytes. Thus, in their reverse hemolytic plaque assay a secretory cell released a protein which was detected by antibody chemically linked to the erythrocyte surface. Finally, Gronowicz *et al.* (1976), devised a "universal" indicator erythrocyte by attaching protein A to the surface of red blood cells. This innovation circumvented the requirement of antibody purification prior to coupling and dramatically reduced the amount of antibody consumed.

The aim of this brief review is to familiarize the reader with the principles and protocols for plaque assays and with the type of information they can provide. The organization plan will be to focus on the attributes and limitations of the three major versions of the plaque assay (standard, sequential, and simultaneous) that are used for quantifying hormone release from single cells. This will be followed by a section that illustrates how the power of the plaque assay is greatly augmented when the technique is used in combination with other analytical methods. First, however, detailed descriptions of reagents and procedures common to all three major types of plaque assays will be provided. This is intended to facilitate the presentation of specific plaque assays in more conceptual terms.

II. Procedures Common to All Plaque Assays

Reagents

1. 4% Bovine serum albumin (BSA)
   a. Gently place 4 g BSA (Sigma, frac. V) on 100 ml minimum essential medium (SMEM; see below).
   b. Allow this to sit for 1–2 hr and then invert gently.
   c. Add 1.1 ml penicillin–streptomycin solution (GIBCO).
   d. Sterile filter.
2. SMEM–0.1% BSA (for dispersion and retrypsinization)
   a. 100 ml SMEM (GIBCO)
   b. 1.1 ml penicillin–streptomycin solution
   c. 2.6 ml 4% BSA
3. Dulbecco's modified eagle medium (DMEM)–0.1% BSA
   a. 100 ml DMEM (GIBCO)

b. 1.1 ml penicillin-streptomycin solution
   c. 2.6 ml 4% BSA
4. DMEM-10% Horse Serum
   a. 90 ml DMEM-0.1% BSA
   b. 10 ml heat-inactivated horse serum (GIBCO)

PREPARATION OF MONODISPERSED CELLS FROM PITUITARY TISSUE

Pituitary tissue from rats is monodispersed with trypsin using a modification of the method described by Hymer et al. (1973). Each rat is killed by rapid decapitation (guillotine) and the adenohypophysis is obtained and freed of neurohypophyseal tissue by blunt dissection. The following procedures are then carried out under sterile conditions with the aid of a laminar flow hood.

1. Transfer anterior pituitary lobes to a silicone-treated petri dish containing 3-4 ml SMEM-0.1% BSA. Transfer using flamed forceps.
2. Decant and replace with 3-4 ml of fresh SMEM-0.1% BSA.
3. Repeat step 2.
4. Using two sterile No. 11 scalpel blades, cut anterior lobes into 1-mm^3 pieces.
5. Rinse pieces of tissue several times to remove blood cells.
6. Place 10 mg of trypsin in a tube, add 10 ml SMEM-0.1% BSA, and dissolve.
7. Transfer solution to silicone-treated Spinner's flask using a 10-cc sterile syringe and Millex-GV filter.
8. Pipette anterior lobe pieces to Spinner's flask using a 5-ml pipette and SMEM-0.1% BSA.
9. Replace top on flask and gas (95:5, $O_2:CO_2$) flask through side arm for approximately 20 sec.
10. Place on a magnetic stirrer at 37°C (water bath) so that the impeller bar gently agitates the fragments.
11. Triturate with a *flame-polished pipette* according to the following schedule:
    a. Adults: 50 min, 80 min, 100 min, 120 min (2 hr total)
    b. Five-day juveniles: 15 min, 30 min, 45 min (45 min total)
    c. Fetuses: 15 min, 30 min (30 min total)

NOTE Gas through side arm after each trituration. If cells are dispersed before elapsed time, go to step 12.

12. Transfer dispersed cells to sterile, 15-ml centrifuge tube, fill with SMEM–0.1% BSA to capacity, and centrifuge at 1500 (275 g) rpm for 10 min.
13. Decant. Add 1 ml SMEM–0.1% BSA per pituitary (or at least 5 ml), triturate with flame-polished pipette, and remove an aliquot for cell count (hemacytometer).
14. Fill tube to capacity with SMEM–0.1% BSA, centrifuge at 1500 rpm for 10 min.
15. Decant. Resuspend in DMEM–0.1% BSA, add 0.5–1.0 ml of cell suspension per poly-L-lysine-coated plastic petri dish (2-4 million cells/35mm plate).
16. Incubate 45 min at 37°C.
17. Decant DMEM–0.1% BSA, add 2 ml DMEM–10% horse serum per plate.
18. Culture at 37°C for 18–42 hr before use in plaque assay.

RETRYPSINIZATION OF PRIMARY PITUITARY CULTURES

1. Dissolve 2.5 mg trypsin/10 ml SMEM–0.1% BSA solution.
2. Remove media from petri dish.
3. Add approximately 2 ml of trypsin solution to petri dish and replace in incubator.
4. Incubate dish for 10 min total, tapping bottom of dish lightly at 5 min.
5. After 10 min incubation, triturate cells using flame-polished pipette.
6. Transfer contents of dish to 15-ml centrifuge tube, rinse dish with SMEM–0.1% BSA, and fill centrifuge tube to capacity with SMEM–0.1% BSA.
7. Spin at 1500 rpm for 10 min.
8. Decant.
9. Add 5 ml SMEM–0.1% BSA to pellet, triturate, and remove aliquot to count.
10. Fill centrifuge tube to capacity with SMEM–0.1% BSA. Spin at 1500 rpm for 10 min.
11. Decant and resuspend in DMEM–0.1% BSA to desired concentration of pituitary cells.

PREPARATION OF PROTEIN-A-COATED ERYTHROCYTES

The following method used in our laboratory is adapted from the method of Gronowicz et al. (1976).

1. Place 7.5 ml of blood from Colorado Serum (1 part whole blood: 1 part Alsevier's solution) in 15-ml centrifuge tube, spin 10 min at 1500 rpm.
2. Decant supernatant, add 10 ml normal saline, centrifuge again.
3. Decant supernatant, carefully remove buffy coat with Pasteur pipette, resuspend with 10 ml saline, centrifuge.
4. Repeat step 3, removing remaining buffy coat, and adjust packed-cell volume to 1 ml. Resuspend with 10 ml saline, centrifuge.
5. Decant supernatant; add IN ORDER, 5 ml saline, 1 ml protein A (0.5 mg/ml in saline, Sigma); resuspend with flame-polished pipette; then add 5 ml 0.02% chromium chloride · 6 H_2O (in saline). Invert tubes several times, incubate at 30°C for 1 hr.
6. Invert tubes twice every 15 min.
7. After 1 hr, centrifuge, decant supernatant, resuspend in 10 ml saline by inverting and tapping tubes, centrifuge.
8. Repeat step 7, then resuspend in DMEM–0.1% BSA, centrifuge.
9. Decant supernatant, resuspend in 5 ml DMEM–0.1% BSA with flame-polished pipette, then pool multiple tubes, divide equally, and bring each to 50 ml with DMEM–0.1% BSA. This yields a 2% solution of oRBCs. Store at 4°C.

NOTE Conjugate oRBCs within 2 weeks of collection date, and use conjugated blood within 1 week.

III. Standard Plaque Assay

The standard plaque assay is the simplest of the three versions considered in this review. However, all are conceptually similar in that they are based on the same principles outlined earlier. For the sake of clarity, this and the following section on sequential assays are subdivided as to protocol followed by validation and quantification/applications.

PROTOCOL

1. An incubation (Cunningham) chamber is constructed by using double-sided Scotch tape to attach a glass coverslip to a poly-L-lysine-

coated glass microscope slide. This coating imparts a positive charge to the floor of the incubation chamber which ensures rapid and virtually complete attachment of pituitary cells and oRBCs as a monolayer. The coating is achieved by immersing slides in a solution of poly-L-lysine ($> 300,000$ MW, Sigma, 0.50 mg/ml deionized H_2O) for 15 min followed by three washes of the same duration in deionized H_2O. A chamber constructed in this manner has a volume of approximately 25 µl and dimensions which favor diffusion of hormone laterally toward the oRBCs.

2. A suspension of monodispersed pituitary cells (6×10^4 cells/ml) in medium (DMEM – 0.1% BSA) is mixed with an equal volume of a 12% suspension (as assessed by packed cell volume) of oRBCs in the same medium. This mixture is infused by capillary action into the assay chamber.

3. The cells are allowed to attach to the floor of the incubation chamber for 45 min. The unattached cells are then washed from the chamber by placing fresh medium on one side and drawing it through to the other side by use of an absorbent piece of paper.

4. The chamber is then refilled in the same manner with antiserum diluted in medium. Secretagogues (when appropriate) are usually added at this step. The length of this reaction generally varies from 0.5 to 8.0 hr and depends on how the plaque assay will be quantified (*e.g.*, sizes of plaques formed *versus* percentages of plaque-forming cells *versus* rate of plaque development). This consideration is discussed in detail in Section III, B.

5. The chamber containing the monolayer is flushed with fresh medium and refilled with guinea pig complement (GIBCO) diluted in medium.

6. Fifty minutes later the plaque-forming reaction is completed, and the monolayers are fixed in place by filling the chamber with fixative (2.0% glutaraldehyde in isotonic saline).

7. The monolayers are stained by exposure to a saturated, filtered solution of toluidine blue in fixative for 30 min followed by several washes. This facilitates and ensures visualization of all pituitary cells in the oRBC mat. The chambers are then stored in fixative for up to several months.

8. The monolayers are viewed with the aid of a microscope and two parameters of plaque development can be quantified, as will be discussed in detail below. One can determine the percentage of all pituitary cells in the monolayer that form plaques. Alternatively, one can use an ocular micrometer to measure the sizes of plaques that formed.

A. Validation

The following minimal criteria must be satisfied during the initial development and at frequent intervals during the routine conduct of plaque assays. First and foremost, the antibody must be extremely specific for the hormone to which it was raised. This is critically important with plaque assays because antisera are generally used at very low dilutions. Accordingly, preabsorption of the antiserum with homologous hormone (but not heterologous hormone) should abolish plaque development. Second, plaques surrounding individual secretory cells should get larger as a function of incubation time with antibody; very small plaques that do not grow progressively with time are likely to be artifacts. Third, deletion of antibody or complement from the reaction mixture (or substitution with inactive reagents) should prevent plaque formation. Finally, hormone secretagogues should exert predictable effects on plaque development; agents that stimulate the release of a particular hormone by entire cultures of cells should also augment (on average) the quantifiable parameters of plaque development (see later), whereas inhibitory agents should evoke the opposite response.

Parenthetically, the aforementioned need to use antibodies at relatively low dilutions has created a practical problem that deserves consideration. Because the quantity of antibody required generally exceeds the amount one can reasonably expect to obtain as a gift, investigators interested in employing this methodology are compelled either to raise their own antisera or to purchase expensive commercial preparations. This single consideration has significantly restricted the more generalized use of plaque assays in endocrine research.

B. Quantification and Applications

Plaque development is generally quantified in two ways. First, one can determine the proportion of all cells in culture that form plaques after maximal plaque development has occurred (usually 2 to 6 hr for most systems). This approach provides information on the fractional abundance of hormone-*releasing* cells of a given type, much like immunocytochemistry provides complementary estimates of hormone-*containing* cells. A variation on this theme is to determine the percentage of plaque-forming cells (PFCs) after variable-length incubations with antibody, and then to use the rate of plaque development as an index of the rate of hormone secretion. The rationale here is that a threshold amount of antigen (hormone) is required around a cell to induce plaque formation. It follows, therefore, that the incubation time required to first detect a plaque around a single cell can be used as an indicator of hormone

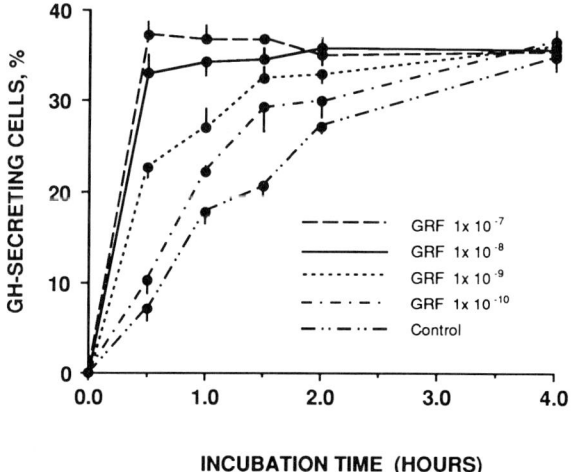

Figure 4 Effect of a stimulatory secretagogue on the rate of plaque formation. In this experiment, standard plaque assays for GH were performed by incubating pituitary cells from male rats with GH antibody in the absence or presence of GH-releasing factor (GRF) for 30, 60, 90, 120, or 240 min prior to the addition of complement. After terminating the reaction by infusion of fixative, the percentage of all pituitary cells that formed GH plaques was determined for each time point with the aid of a microscope. As shown, GRF accelerated the rate of plaque development in a dose-related manner, and this effect was most obvious at early time points. However, the percentage of cells that eventually formed plaques by 4 hr was the same in all treatment groups. On the basis of this response, we conclude that GRF accelerates the rate of rat GH secretion but does not acutely influence the absolute number of cells committed to GH release. (From Boockfor *et al.*, 1986b.)

release; those cells that release more hormone molecules per unit time form plaques faster then those that release fewer molecules. In short, a reduction (or increase) in the rate of plaque formation caused by a treatment provides evidence for inhibition (or stimulation) of hormone release. Utilization of this approach has provided important information about the functional heterogeneity of hormone-secreting cells (Fig. 4). For example, discontinuities in the rate of plaque formation among prolactin (PRL) cells from adult male rats revealed that only a small fraction were maximally responsive to the inhibitory actions of dopamine or the stimulatory effects of thyrotropin-releasing hormone (TRH) (Frawley and Clark, 1986). Likewise, rate studies demonstrated that PRL cells taken from different regions of the rat adenohypophysis were differentially responsive to these same two secretagogues (Boockfor and Frawley, 1987). Finally, insulin-like growth factor-I (IGF-I), which is known to feedback negatively on growth hormone (GH) release by a direct pituitary

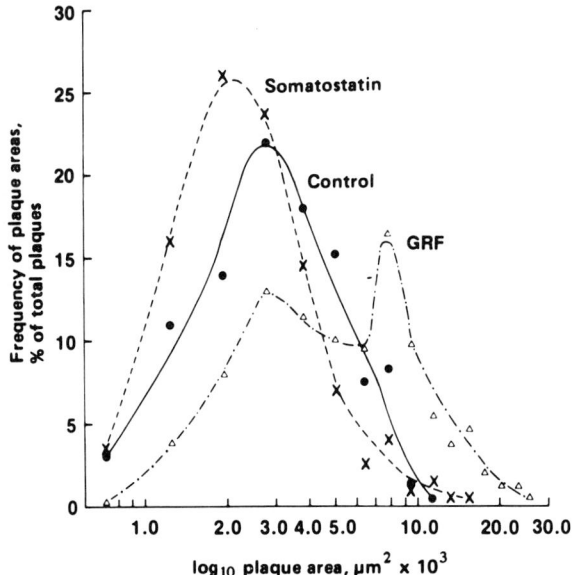

Figure 5 Frequency distribution of GH plaque areas measured after a 4-hr incubation of pituitary cells from male rats. By plotting the frequency of plaque areas against the sizes of plaques formed, it is possible to determine whether all cells respond similarly to inhibitory or stimulatory secretagogues. As shown here, somatostatin-treated monolayers exhibited unimodal frequency distributions that were shifted to the left (toward smaller plaques) relative to the controls. This type of result shows that all GH secretors were responsive to the inhibitory actions of somatostatin. In contrast, GRF-treated cells yielded a bimodal frequency distribution. Results of this latter type demonstrate the existence of a somatotrope population which is preferentially responsive to GRF. (From Frawley and Neill, 1984.)

action, was shown to inhibit the rate of plaque formation by only a subpopulation of GH cells from adult males (Hoeffler *et al.*, 1987).

An alternative method for quantifying plaque development is to measure the sizes of plaques that form, since this variable (plaque area) is proportional to the relative amount of hormone released. Such a relationship has been demonstrated by indirect as well as direct methods. For example, Smith *et al.* (1986) compared leutinizing hormone (LH) secretion in dishes (measured by radioimmunoassay) with that in Cunningham chambers (determined by plaque assay) and observed a linear relationship between plaque area and hormone released over a wide range of responses. A more direct analysis was performed by Allaerts *et al.* (1988) who developed a computation model that takes into account the spreading of secreted molecules and is based on the governing principles of diffusion and antigen absorption by immobilized antibodies.

Their attainment of close agreement between observed and predicted results for PRL plaque formation validated the assumptions underlying this approach. Thus, the results of both types of studies support the same conclusions: plaque area provides a reliable index of the relative amount of hormone released and all cells of a given type do not secrete the same amount of hormone. An example of the type of information that can be derived from plaque size measurements is provided in Fig. 5.

IV. Sequential Plaque Assay

In its simplest sense, the sequential assay might be envisioned as the serial performance of two standard assays for different hormones on the same pituitary cells. However, the technical modifications required to translate this concept to reality make this assay more onerous to perform than anticipated (Frawley *et al.*, 1985).

PROTOCOL

1. The primary difference between this system and the one described above is that only the pituitary cells are attached to the floor of the incubation chamber; the oRBCs are flushed into and out of the chamber during each sequence. Therefore, the day before conducting an assay, monodispersed pituitary cells are plated onto glass slides NOT treated with poly-L-lysine (thus, several hours are required for attachment).

2. Six hours later, the slides are flooded with DMEM – 10% horse serum with antibiotics and placed in the incubator overnight.

3. On the day of an experiment, an incubation chamber is constructed over the pituitary monolayer with the use of a glass coverslip photoengraved with a numbered/lettered grid pattern to facilitate reidentification of specific cells in various regions of the slide (Lin and Ruddle, 1981).

4. After a 45-min preincubation, the chamber is filled with a mixture of oRBCs, antibody, and complement. Within 10 min, the oRBCs settle to the floor of the chamber and form a mat, but do not attach due to the absence of poly-L-lysine.

5. After plaque formation is completed (generally within 2 hr), videocassette recordings are made of cells within particular grids.

6. The incubation chamber is then flushed with fresh medium and refilled with a second assay mixture containing oRBCs, complement, and a different antibody.
7. After an additional incubation, a second set of recordings is taken for comparison with the first. In this manner, cells secreting one or both hormones can be identified.
8. It should be noted that the final concentrations of antibody and complement used for the sequential assay are routinely higher than those used in the standard versions of the plaque assay. The reason for this is that the oRBCs must be present at a greater density to enable formation of a mat in the absence of poly-L-lysine.

A. Validation

There are three considerations central to the validation of the sequential plaque assay. The first of these relates again to the issue of antibody specificity because the potential for cross-reaction in the sequential system is magnified as a consequence of the higher concentration of antibody employed. Therefore, this variable should be rigorously examined not only in the "cold displacement" approach used for assessing suitability of antisera for radioimmunoassay purposes but also in the plaque assay *per se* by preabsorption studies with homologous and heterologous hormones. Having established antibody specificity, one should then seek to ensure that plaque formation in each sequence occurs independently. The lack of "carryover" effects between sequences can be established by the demonstration that plaque formation during the second sequence can be inhibited by omission of antibody or complement. Moreover, when different antibodies are used in each sequence the results should be the same regardless of which hormone is detected in the initial sequence. Finally, the assay should be repeatable from sequence to sequence (*i.e.*, the same cells should form plaques when reagents for the detection of the same hormone are applied in both sequences).

B. Quantification and Applications

The characteristics of the oRBC mat impose restrictions on the quantification of sequential assays that are not relevant to the standard versions of this system. For example, determination of the percentage of plaque-forming cells

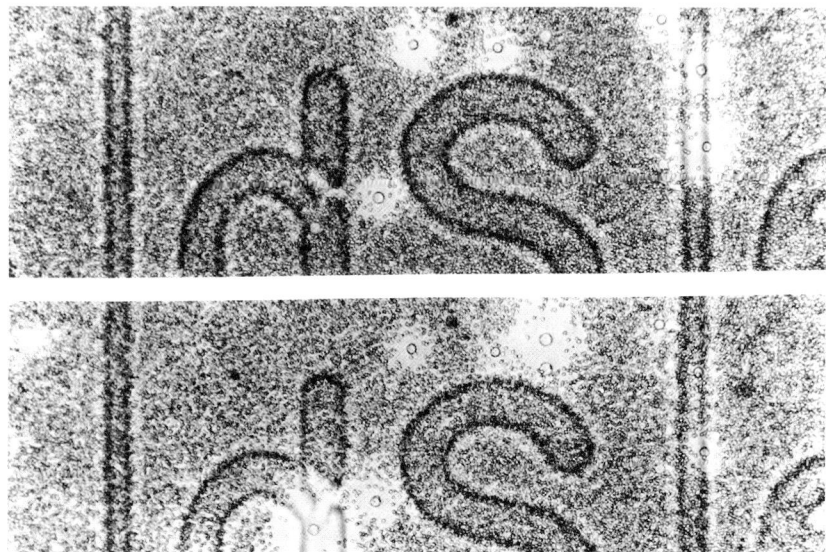

Figure 6 Results of a typical sequential plaque assay. Dispersed pituitary cells from male rats were assayed first for GH (top) and then for PRL (bottom). Note that some cells formed plaques in the presence of only GH antiserum (traditional somatotropes) or PRL antiserum (classical mammotropes) whereas others induced plaque development when incubated with either antiserum (mammosomatotropes). (From Frawley *et al.*, 1985.)

is complicated by the fact that the oRBC mat is usually more than a single cell layer deep and consequently one cannot always be confident about visualizing all non-plaque-formers. Likewise, analysis of plaque areas becomes problematic in the sequential system because plaque development occurs in three dimensions (due to the thicker oRBC mat), whereas the method of quantification takes only two of these into account. Nevertheless, this technique can provide extremely useful and unique information when the experimental question is judiciously posed with these constraints in mind. In this regard, the sequential assay is most powerful when used for analyzing the secretory behavior of individual cells in real time. Indeed, we have employed this system to obtain the first direct evidence that individual pituitary cells can concurrently release both GH and PRL (Fig. 6). Alternatively, the sequential assay can be employed for repeated measurement of the same hormone from the same cells. Such an approach was utilized by Neill *et al.* (1987) who observed that the sizes of plaques which formed around given mammotropes varied greatly over 4 days without any obvious pattern.

V. Simultaneous Plaque Assay

The simultaneous plaque assay (Porter *et al.*, 1990) is truly a hybrid of the standard and sequential versions both in terms of protocol and in the types of information that can be obtained. The simultaneous assay is carried out much like the standard assay in plain Cunningham chambers (no photoengraved coverslips) to which the secretory cells and oRBCs are affixed as a monolayer with poly-L-lysine. The major deviation from the standard system is that pituitary cells in companion chambers are incubated with either PRL or GH antiserum alone or with both antisera simultaneously. As a procedural control, preimmune or hyperimmune sera are included with each individual antiserum to maintain constant the total serum concentrations among the three types of assay conditions.

The rationale underlying the simultaneous assay is that a particular cell can be identified as a plaque former only once, whether it secretes one or both hormones. As a consequence, the chambers incubated with both antisera provide a valid estimate of the cumulative number of cells that release GH and/or PRL, whereas the arithmetic sum of the two single hormone assays yields an overestimate of this population. Thus, the degree of functional overlap (*i.e.*, the percentage of mammosomatotropes) can be calculated by subtracting the percentage of plaque formers measured when both antisera are used together from the arithmetic sum of the proportions obtained when single antisera are employed. Using the same logic, the relative abundance of cells that release only GH can be estimated by subtracting the percentage of PRL secretors (single antiserum assay) from the percentage obtained with both antisera used simultaneously. Likewise, the percentage of PRL-only cells can be derived in a comparable manner.

The advantage of the simultaneous assay over the standard version is rather obvious; only the former can provide information about the proportions of dual-hormone secretors as well as their monohormonal counterparts. Similar types of data can be (and have been) generated by employing the sequential plaque assay (Frawley *et al.*, 1985; Hoeffler *et al.*, 1985; Boockfor *et al.*, 1986a). However, the sequential system is considerably more "expensive" in terms of time and resources, and its use should be reserved for those situations in which information about specific, identified cells (not populations) is imperative.

VI. Other Variations and Applications of Plaque Assays

The combination of plaque assays with other contemporary methodologies enables the resolution of experimental questions that would be difficult or impossible to address with more conventional strategies. One such application is the *a priori* identification of a given type of hormone secretor for further study in the living state. This is possible because the hemolytic reaction will destroy the indicator erythrocytes but not the secretory cells, provided that the concentrations of hormone-specific antibody and complement are titrated appropriately. Using this paradigm, Lingle *et al.* (1986) employed the standard plaque assay to identify individual lactotropes in mixed cultures of pituitary cells. Plasma membrane currents in these identified cells were subsequently characterized by employing whole-cell- or patch-recording techniques. Screening with the plaque assay allowed these investigators and others (Croxton *et al.*, 1989) to record only from the specific cells of interest and thus circumvent the redundancy imposed by *post facto* identification, which is usually accomplished by immunocytochemistry. With a conceptually similar approach, Holl *et al.* (1988) simultaneously monitored GH release and intracellular calcium oscillations in individual, living pituitary cells. This was accomplished by combining the plaque assay with digital-imaging microscopy of cells loaded with the fluorescent calcium indicator Fura-2. Similar studies have been performed with LH-secreting cells (Leong and Thorner, 1991).

The *post facto* application of other analytic methods to hormone secretors identified by plaque assay can also amplify the information-generating potential of the technique. For example, the combined use of plaque assay and immunocytochemistry has led to the identification of secretory subpopulations that store a particular hormone but do not release it (Kineman *et al.*, 1990), or conversely, release the hormone without storing appreciable quantities (Boockfor *et al.*, 1985). Also, *in situ* hybridization for the mRNA encoding one hormone has been performed on cells that form plaques for a second hormone, thus demonstrating pluripotential secretors in mixed populations (Lyons *et al.*, 1992). Another application developed by Smith and Neill (1987) allows simultaneous quantification of hormone release (plaque assay) and secretogogue binding (receptor autoradiography) by individual gonadotropes treated with radiolabeled gonadotropin-releasing hormone (GnRH). A variation on this later theme is termed "surround immunoprecipitation" (Neill *et al.*, 1987). Here, pituitary cells are pulse labeled, and the release of PRL (a fraction of which is radiolabeled) is measured subsequently by plaque assay. Because the labeled hormone remains bound to lysed oRBC and can be microscopically visualized by autoradiography, a ratio of newly synthesized to total hormone release can be estimated. Thus, by establishing the relationship between plaque size and autoradiographic grains, it is possible to determine whether a given cell preferentially releases newly synthesized or older, stored hormone in response to a particular stimulus. Lastly, a method has been developed that enables electronmicroscopic examination of hormone secretors identified by plaque assay (Neill *et al.*, 1987). The potential of this

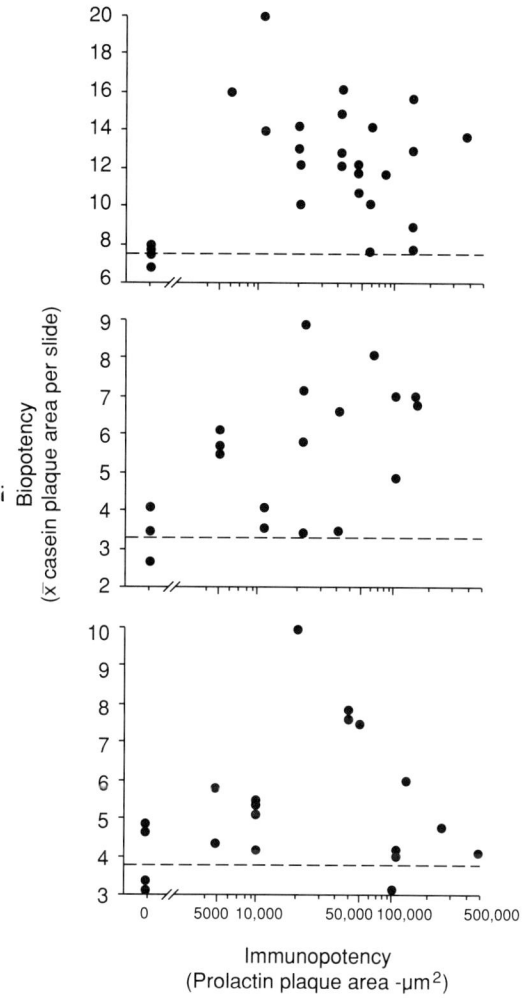

Figure 7 Example of how the plaque assay can be used to measure the bio- as well as the immunopotency of a particular hormone. Individual pituitary cells were incubated in microwells for 14 hr after which the medium was removed for assessment of PRL biopotency using a novel method in which PRL-induced casein release from mammary cells is measured by plaque assay. The same pituitary cells were then subjected to PRL–plaque assays to measure the release of immunoreactive hormone. The results (three separate experiments) show that individual mammotropes differ greatly in the bio- to immunopotency of secreted hormone. See the text for additional details. (From Frawley et al., 1986.)

combined approach is that it might be possible to finally correlate the morphological heterogeneity of various secretory types with the functional heterogeneity in the amount of hormone released.

The final application to be considered demonstrates how the incorporation of plaque assay methodology can improve the detectability of assay systems that are based on biological rather than immunological end points. In hormonal systems studied thus far, the degree of sensitivity exhibited by plaque assays has been consistently greater than that of corresponding radioimmunoassays, which in turn have historically outperformed biological assays. Given the exquisite sensitivity of plaque assays, it seemed reasonable to propose that their use might greatly augment the detectability of hormone-induced biological responses. To test this logic, our group developed a PRL bioassay based on the ability of the hormone to induce a dose-related release of casein from cultured mammary cells. Because the release of casein is measured by a standard plaque assay for the milk protein, we have termed this system the casein plaque bioassay (Frawley *et al.*, 1986). The technique is extremely sensitive (150 fg PRL/assay slide) and it enables us to evaluate the biopotency of PRL released from single mammotropes. In fact, by combining it with the standard plaque assay for PRL, it has been possible to estimate both the bio- and immunopotencies of hormone released from the same individual cells. These studies show clearly that some cells release PRL with extremely high biopotency and very low immunoreactivity and *vice versa* (Fig. 7). When viewed in light of reports that PRL is microheterogeneous and that its variant forms deviate in biological and immunological potencies, these findings indicate that mammotropes differ from one another in the molecular variant(s) of PRL released.

VII. Summary

On the basis of just the few examples provided, it should be clear that the reverse hemolytic plaque assay comprises an extremely versatile and powerful tool for analyzing hormone release at the single-cell level. From a qualitative standpoint, it enables the identification and enumeration of hormone secretors of a given type in mixed populations and thus circumvents the need to purify or enrich secretory cell types in order to study them. Quantitatively, it allows measurement of the relative amount of hormone released from single, living cells and by extension it enables the assessment of differential responsiveness of secretory subpopulations to stimulatory and inhibitory agents. The combined use of plaque assays with other contemporary methodologies should continue to allow investigators to address new levels of scientific questions. Its potential applications in this regard are limited only by the imagination.

References

Allaerts, W., Wouters, A., Van der Massen, D., Persoons, A., and Denef, C. (1988). *J. Theor. Biol.* **131**, 441-459.
Boockfor, F. R., and Frawley, L. S. (1987). *Endocrinology (Baltimore)* **120**, 874-879.
Boockfor, F. R., Hoeffler, J. P., and Frawley, L. S. (1985). *Endocrinology (Baltimore)* **117**, 418-420.
Boockfor, F. R., Hoeffler, J. P., and Frawley, L. S. (1986a). *Am. J. Physiol.* **250**, E103-E105.
Boockfor, F. R., Hoeffler, J. P., and Frawley, L. S. (1986b). *Neuroendocrinology* **42**, 64-70.
Croxton, T. L., Armstrong, W. McD., and Ben-Jonathan, N. (1989). *In* "Hormone Action, Part K: Neuroendocrine Peptides" (P. Conn, ed.), Methods in Enzymology, Vol. 168, pp. 144-166. Academic Press, San Diego.
Cunningham, A. J., and Szenberg, A. (1968). *Immunology* **14**, 599-600.
Frawley, L. S., and Clark, C. L. (1986). *Endocrinology (Baltimore)* **119**, 1462-1466.
Frawley, L. S., and Neill, J. D. (1984). *Neuroendocrinology* **39**, 484-487.
Frawley, L. S., Boockfor, F. R., and Hoeffler, J. P. (1985). *Endocrinology (Baltimore)* **116**, 734-737.
Frawley, L. S., Clark, C. L., Schoderbek, W. E., Hoeffler, J. P., and Boockfor, F. R. (1986). *Endocrinology (Baltimore)* **119**, 2867-2869.
Gronowicz, E., Coutinho, A., and Melchers, F. (1976). *Eur. J. Immunol.* **6**, 588-590.
Hoeffler, J. P., Boockfor, F. R., and Frawley, L. S. (1985). *Endocrinology (Baltimore)* **117**, 187-195.
Hoeffler, J. P., Hicks, S. A., and Frawley, L. S. (1987). *Endocrinology (Baltimore)* **120**, 1936-1941.
Holl, R. W., Thorner, M. O., Mandell, G. L., Sullivan, J. A., Sinha, Y. N., and Leong, D. A. (1988). *J. Biol. Chem.* **263**, 9682-9685.
Hymer, W. C., Evans, W. H., Kracier, J., Mastro, A., Davis, J., and Griswold, E. (1973). *Endocrinology (Baltimore)* **92**, 275-287.
Jerne, N. K., Henry, C., Nordin, A. A., Fuji, H., Koros, A. M. C., and Lefkovits, I. (1974). *Transplant. Rev.* **18**, 130-191.
Kineman, R. D., Faught, W. J., and Frawley, L. S. (1990). *Endocrinology (Baltimore)* **127**, 2229-2235.
Leong, D. A., and Thorner, M. O. (1991). *J. Biol. Chem.* **266**, 9016-9022.
Lin, P. F., and Ruddle, F. H. (1981). *Exp. Cell Res.* **134**, 485-488.
Lingle, C. J., Sombati, S., and Freeman, M. E. (1986). *J. Neurosci.* **6**, 2995-3005.
Lyons, C. E., Jr., Barrett, J. R., and Goth, M. I. (1992). *Program Annu. Meet. Endocr. Soc. 74th* Abstr. 1368.
Molinaro, G. A., Eby, W. C., and Molinaro, C. A. (1981). *In* "Immunochemical Techniques, Part B" (J. Langone and H. Van Vunakis, eds.), Methods in Enzymology, Vol. 73, pp. 326-388. Academic Press, New York.
Neill, J. D., and Frawley, L. S. (1983). *Endocrinology (Baltimore)* **112**, 1135-1137.
Neill, J. D., Smith, P. F., Luque, E. H., Munoz de Toro, M., Nagy, G., and Mulchahey, J. J. (1987). *Recent Prog. Horm. Res.* **43**, 175-229.
Porter, T. E., Hill, J. B., Wiles, C. D., and Frawley, L. S. (1990). *Endocrinology (Baltimore)* **127**, 2789-2794.
Smith, P. F., and Neill, J. D. (1987). *Proc. Natl. Acad. Sci. U.S.A.* **84**, 5501-5505.
Smith, P. F., Luque, E. H., and Neill, J. D. (1986). *In* "Hormone Action, Part J: Neuroendocrine Peptides" (P. Conn, ed.), Methods in Enzymology, Vol. 124, pp. 443-465. Academic Press, Orlando, Florida.

7

Capillary Electrophoresis Coupled to Fluorescence Detection for the Determination of Multiple Neuropeptides

Juan P. Advis, Khurshid Iqbal, A. Waseem Malick, and Norberto A. Guzman

I. Introduction
II. A Capillary Electrophoresis-Based Assay
   A. General Principles
   B. Instruments and Sample Preparation
   C. Separation and Optimization
III. Comments and Future Perspectives
References

I. Introduction

This chapter presents strategies, using a capillary electrophoresis (CE)-based assay as an alternative method (to traditional techniques), to increase the separation and sensitivity for the simultaneous determination of multiple peptides. As an example, *in vivo* release and tissue content samples containing neuropeptides from the median eminence of the ewe brain are used. Simultaneous determination of luteinizing hormone-releasing hormone (LHRH), neuropeptide Y (NPY), and β-endorphin (βEND) is obtained in nanoliter amounts of each of these samples, using CE and fluorescence detection.

Rapid progress in the determination of neuropeptides has enabled more stringent conditions for their analysis in tissues and biological fluids. At present, radioimmunoassays (RIAs) are the most common methods used to determine these biological active substances. However, antibodies used in RIAs are targeted to determine one specific neuropeptide at a time. Thus, sample volume might become a limiting issue when multiple neuropeptides must be assessed in the same sample by RIAs, especially when their sample concentration is low. Other drawbacks of RIAs are the use of radioactive material and the need to have access to high-titer-specific antibodies for each neuropeptide to be measured. In general, RIAs allow the determination of picogram levels of neuropeptides per assay tube. High-performance liquid chromatography (HPLC) is another commonly used technique that, in contrast to RIAs, enables the determination of several neuropeptides simultaneously. However, although HPLC is a technique with high separation power it is only able to detect neuropeptides with a sensitivity that is usually unacceptable even for measuring tissue level concentrations of these substrates, when coupled to UV detection at 214 nm. Furthermore, most reagents used in the fluorescence derivatization of amines have properties that limit their usefulness for HPLC detection of peptides when coupled to fluorescence detection, since their reaction systems tend to precipitate in the presence of HPLC solvents (Lingeman et al., 1985; Ogden and Foldi, 1987; Lunte and Wong, 1989; Newcomb, 1992; Wright, 1991). Finally, it is noteworthy to mention that during in vivo release of neuropeptides, as well as during preparation of samples extracted from biological fluids and/or tissues, some biochemical processing of neuropeptides (e.g., deamidation) might occur (Wright, 1991). These changes might yield peptide fragments that RIA cannot detect, but that might be related to changes in biological activity of the peptide, as for example in synaptic transmission or in the recently postulated volume transmission (Fuxe and Agnati, 1991). Most peptide fragments are usually resolved by HPLC.

Although capillary electrophoresis is currently been used as a separation technique in a large number of applications, for the purpose of this chapter we address only its possible use for the separation of brain neuropeptides, as an example of the capabilities of this novel technique. It is believed that analysis of brain perfusates might become the method of choice to determine in vivo reactions in the brain (Kendrick, 1989; Kasting and Martin, 1989; Lunte et al., 1991). For this, push-pull cannula (PPC) and microdialysis sampling are currently the two most prominent technologies. The main drawback, when attempting to measure multiple neuropeptides in perfusates obtained using these sampling approaches, is the low sample volume (20-100 μl) and the low neuropeptide concentration in these samples. The latter problem is further magnified when using microdialysis, because neuropeptide recovery through its probe range from 0.5 to 15% (Kendrick, 1989) depending on the neuropeptide and the length of the dialysis membrane used (2 vs 5 mm in length). In addition, microdialysis probes 5 mm long cannot be used when dealing with small target areas or areas adjacent to bone structures, as is true of the hypothalamic median eminence (ME). Neither recovery of neuropeptides nor position of the tip probe are problems when PPC sampling is used.

Capillary electrophoresis, a powerful analytical and separation technique, has the potential to assess multiple peptides in the same sample (Advis et al., 1989; Guzman et al., 1990a, 1991a), although large improvements in detection sensitivity still remain to be achieved. The basic principle of this technique using a single capillary is illustrated in Fig. 1. While the Fig. 1A is a diagram of a CE system attached to a UV detector, Fig. 1B is a diagram of a CE system attached to a fluorescence detector (for detailed description see Section II, A Capillary Electrophoresis-Based Assay). So far, CE has been successfully used to assess neurotransmitters in a single cell (Wallingford and Ewing, 1990), to reach zeptomole (10^{-21} mol/liter) levels of detectability (Hernandez et al., 1991), and to use femtoliters (10^{-15} liter) of sampling volume (Wallingford and Ewing, 1990). Thus, novel approaches using CE might be developed to assess multiple neuropeptides in brain perfusate samples (Advis et al., 1989; Guzman et al., 1990a, 1991a; Advis and Guzman, 1993). This will probably be achieved by using strategies to increase assay sensitivity. Recently, several approaches have been developed to increase sensitivity of capillary electrophoresis systems. Such improvements include the following:

1. The use of special lenses (Tsuda et al., 1990), for example, cylindrical lenses to concentrate the light beam to the capillary center, thus reducing light scattering and increasing sensitivity.

2. The development of a multirefractive system (Wang et al., 1991), for example, using a silver coating to amplify the light beam within the capillary, thus increasing capillary light output downstream and therefore increasing sensitivity.

3. The use of a Z-shaped capillary (Chervet et al., 1991), for example, by changing the spatial configuration of a portion of the capillary to increase the path length of the light beam within the capillary, thus increasing sensitivity.

4. The use of a bundle of capillaries (Guzman et al., 1991b) to increase the loading volume and in turn the detection of the sample, thus increasing sensitivity.

5. The use of an analyte concentrator (Guzman et al., 1991c) to increase the loading volume of an in-line affinity segment of the capillary located at the injection site, from which a specific analyte (e.g., if an antibody, lectin, or enzyme is used) or nonspecific analytes (e.g., if a C-18 support matrix or other chromatographic media is used) are captured until saturation of the binding sites and then eluted, thus increasing sensitivity.

6. The use of derivatization techniques (Guzman et al., 1992a,b), for example, the use of a chromophore to enhance the detection signal either in the UV and/or fluorescence region, thus increasing sensitivity. In this chapter we expand on this latter approach.

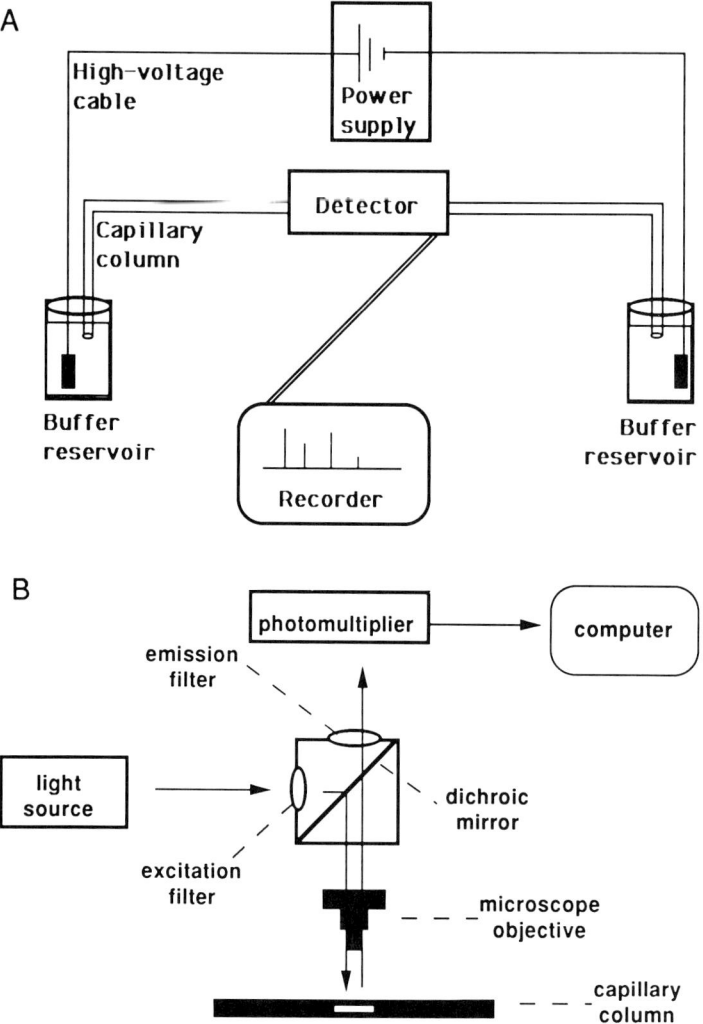

Figure 1 Schematic representation of the general principle of a capillary electrophoresis system and its main components. (A) With UV detection. (B) With fluorescence detection.

This chapter presents strategies using a capillary electrophoresis-based assay for the simultaneous determination of multiple neuropeptides. Specifically this approach uses CE, for the separation of various neuropeptides found in ME samples and their derivatization with the fluorogenic chromophore

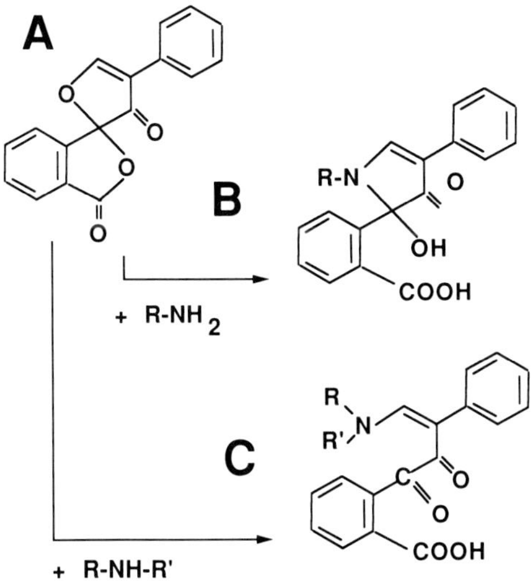

Figure 2 Schematic representation of the molecular structure of fluorescamine (A), the derivatized reaction product involving a reacting primary amine functional group-containing analyte (B), and the derivatized reaction product involving a reacting secondary amine functional group-containing analyte (C).

fluorescamine to amplify their signal. This chromophore has been applied by others to peptide fluorescence derivatization and shown to be of limited usefulness since it tends to precipitate in the presence of HPLC solvents used in liquid chromatography (Newcomb, 1992; Felix *et al.*, 1975; Rubenstein *et al.*, 1979; Tsuda *et al.*, 1988; Brown and Jenke, 1987). However, since our CE separation buffer does not have HPLC solvents, we have been able to use fluorescamine for the derivatization of neuropeptides found in *in vivo* ME release samples (obtained by push-pull perfusion in ewes) and in ME tissue samples. Fluorescamine reacts efficiently with primary amino acids to form intensely fluorescent substances and with secondary amino acids to form nonfluorescent aminoenone-type chromophores that are easily detected at the low UV region (see Fig. 2). Derivatization to fluorescamine also enhances the degree of resolution of the neuropeptides, enabling baseline separation of the derivatized analytes. The utility of CE for the analysis of neuropeptides has been previously demonstrated (Advis *et al.*, 1989; Guzman *et al.*, 1989, 1990a, 1991a; Advis amd Guzman, 1993). Because of its limitations in detection sensitivity, however, it is essential to develop CE conditions to maximize detection, in order to visualize small amounts of neuropeptides and other constituents of brain tissue. Thus, potential advantages of using CE as a quantitative analytical tool for the determination of neuropeptides are discussed in this chapter.

II. A Capillary Electrophoresis-Based Assay

A. General Principles

Capillary electrophoresis has received a great deal of attention due primarily to the high-power resolution and the significant sensitivity levels it can reach. It is surprising that in a tiny capillary, approximately the size of a human hair, high efficiencies can be achieved for sample separation allowing resolution of remarkably similar substances. In addition, samples can be analyzed in a nanoliter/picoliter range (about 10^9 picoliters are in a drop of water) detecting solute concentrations reaching zeptomole range (1 zeptomole = 10^{-21} mol or 600 molecules). Capillary electrophoresis permits the rapid and efficient separations of charged and neutral molecules in an instrumental format, with minimal sample volume required. In general terms, CE is the electrophoretic separation of a substance from (usually) a complex mixture within a narrow tube. Because small tubes dissipate heat efficiently and prevent disruption of separations by thermally driven convection currents, CE can use relatively large electric fields to separate components in very small samples, rapidly and effectively. High voltage (5–30 kV) and low current (20–200 μA) are normal conditions for separations using this technique. Table I summarizes some of the most important characteristics of CE. To date, many reviews on CE have been published (Wallingford and Ewing, 1990; Guzman et al., 1989, 1990b; Karger et al., 1989; Novotny et al., 1990; Deyl and Struzinsky, 1991; Kuhr and Monnig, 1992; Alvin et al., 1993; Karger and Foret, 1993). In open-tube CE, a buffer-filled capillary is generally suspended between two reservoirs containing the same buffer. A third reservoir contains the sample of interest (see Fig. 1A). A platinum–iridium electrode is placed in the third reservoir together with the anodic terminal (positive high voltage) of the capillary, and the sample is introduced into the capillary column for a very short time (usually 2–12 sec) by either electrokinetic or displacement injection methods. The other terminal of the capillary is connected to the cathodic terminal or electrically grounded end of the system. This terminal also contains a platinum–iridium electrode. Once the sample is introduced into the capillary (and the high-voltage circuit is disconnected), the anodic terminal of the capillary is transferred manually or automatically into a buffer-containing reservoir. At this stage, the high-voltage circuit is reopened and separation of the analyte(s) starts. A strong electroosmotic flow carries solutes, usually without regard to charge, from the positive end to the negative (grounded) electrode.

Table I
Characteristics of Capillary Electrophoresis

Nanotechnique	The amount of injected material usually is 1–20 nl, although it is possible to achieve picoliter and femtoliter quantities.
High speed	Normally substances are resolved in 10–40 min using 50 or 75 μm i.d. fused-silica capillaries; however, using less than 25 μm i.d. columns it is possible to achieve separations in 2–5 min.
High resolution	Higher plate counts are possible (several million theoretical plates have been achieved).
High reproducibility	When separation conditions have been optimized, it is possible to obtain migration times that can be repeated with less than 0.5% error and peak areas with less than 2.0% error (RSD).
High sensitivity	Subattomole levels of detection are possible. Using laser-induced fluorescence as a detection method, it has been possible to achieve limits of 1×10^{-15} M or 1×10^{-23} mol or less than 200 molecules.
No molecular weight limitation	In open-tube operation, smaller substances can be separated easily from complex macromolecules.
Mild conditions	Mobile phases usually range from slightly acidic to slightly basic aqueous buffers. However, nonaqueous buffers have also been used. Similarly, a variety of additives have been added successfully to the buffers to improve resolution and quantification.
Compatibility with biological systems	More than 95% of the original enzymes and their immunological activities have been recovered.
Instrumentation	From the introduction of the sample into the capillary to the collection of the separated components and data acquisition, every step can be controlled by semiautomatic to fully automatic instrumentation.
Micropreparative collection	Several methods are now available to collect separated components in the nanogram–microgram quantities. These methods include the use of multiple open-tube capillaries or the presence of a solid support matrix, such as agarose-acrylamide, or the presence of a solid support analyte concentrator.
Characterization by spectral analysis	Because the direction of migration can be reversed with the electrical polarity of the system, the complete spectrum of a substance can be measured every 1–2 nm after a single injection of a substance in which the substance is exposed in front of the detector as many times is necessary. Faster scanning systems, such as diode array detectors, can scan the sample on-line.
Minimal reagent consumption	Limited consumption of reagents makes capillary electrophoresis more economical.

B. Instruments and Sample Preparation

A laboratory-made capillary electrophoresis apparatus or a commercial unit can be used: (a) the laboratory-made apparatus consists of a capillary electrophoresis section and a fluorescent microscope with a photomultiplier linked to a computer as a detection system. One can use a microscope like the Nikon Optiphoto II model epiillumination fluorescence microscope (Nikon Instrument Group, Garden City, NY) attached to a computerized single-photon detector photomultiplier (Photon Technology International, South Brunswick, NJ) or similar. The excitation lamp is a 50-W high-pressure mercury lamp (model HV 1010 1AF); the bandpass filter a 380- to 425-nm filter; a 430-nm dichroic mirror is used; and the emission filter is a 450-nm filter (V-2A Nikon filter cube). A direct-current, high-voltage power supply is used (Spellman High Voltage Electronics, Plainview, NY). High voltage is applied through platinum–iridium electrodes for both electrokinetic injections and sample separations. Samples are injected electrokinetically at 25 kV for 2 min, and analytes are separated for 1 hr at 25 kV. (b) A commercially available CE instrument, as for example the P/ACE System 2000 (Beckman Instruments, Palo Alto, CA), contains a UV detection system. In this instrument, the capillary is housed in a cartridge constructed so as to allow a flow of recirculating liquid for Peltier-temperature control of the capillary column. In this system, samples are stored in a microapplication vessel assembly, consisting of a 150-μl conical microvial inserted into a standard 4-ml glass reservoir, and held in position for injection by an adjustable spring. To minimize evaporation of the sample volume (100 μl), about 1–2 ml of cool water can be added to the microapplication vessel housing the microvial. The external water serves as a cooling bath for the sample in the microvial and as a source of humidity to prevent sample evaporation and concentration. After insertion of the microvial, the microapplication vessel assembly is covered with a rubber injection septum and placed into the sample compartment of the CE instrument. Samples are injected into the capillary column by pressure. Peak visualization and data acquisition are performed using the UV detection system of the CE instrument and the System Gold Chromatography Software package (Beckman Instruments, San Ramon, CA). Data integration is also carried out with a Model D-2500 Chromato-Integrator (Hitachi Instruments, Inc., Danbury, CT).

1. Sample Preparation

Stock solutions of standard samples are individually prepared by dissolving LHRH (3.2 mg/ml), βEND (0.5 mg/ml), and NPY (0.5 mg/ml) in 0.1 M

sodium tetraborate (borax) buffer, pH 9.0. Median eminence tissue samples are homogenized in 0.1 N HCl and centrifuged at 3000 rpm in an Eppendorf microcentrifuge, and the supernatant is stored frozen until derivatized. Finally, *in vivo* release perfusate samples (PPC) are stored frozen until derivatized. All chemicals are obtained at the highest purity level available from the manufacturer and are used without additional purification. For example, sodium hydroxide, sodium phosphate (Na_2HPO_4), lithium chloride, borax ($Na_2B4O_7 \cdot 10H_2O$), and fluorescamine might be purchased from Sigma Chemical Co (St. Louis, MO). Acetone (HPLC grade), pyridine (Fisher Certified), and hydrochloric acid solution (12 M) might be obtained from Fisher Scientific (Fair Lawn, NJ). Neuropeptide standards (LHRH, NPY, βEND) might be purchased from Peninsula Laboratories (Belmont, CA). Reagent solutions and buffers are prepared using triply distilled and deionized water and are routinely degassed and sonicated under vacuum after filtration. Millex disposable filter units (0.22 μm) might be purchased from Millipore Corporation (Bedford, MA), and fused-silica capillary columns might be obtained from Scientific Glass Engineering (Austin, TX) and Polymicro Technologies (Phoenix, AZ).

2. Sample Derivatization

The molecular structure of fluorescamine and the chemical reactions it is involved with are shown in Fig. 2. Fluorescamine (A) is a nonfluorescent substance that reacts efficiently with primary amino acids to form intensely fluorescent substances (B). Additionally, it reacts with secondary amino acids to form nonfluorescent aminoenone-type chromophores (C), which are easily detected at the low UV region. For CE analysis without fluorescamine derivatization, assay samples are diluted to desired concentrations with sample dilution buffer (0.1 M sodium tetraborate buffer, pH 9.0) and directly transferred to the conical vial and then inserted into the microapplication vessel assembly on the CE instrument (Beckman) or to a microcentifuge tube and placed on the turntable of the CE instrument (laboratory-made). For CE analysis of fluorescamine derivatives, however, solutions of the respective analyte samples (concentration ranging from 8 to 128 μg, or from 6.7 to 108 nmol/100 μl reaction mixture) are transferred to a 500-μl microcentrifuge tube, and their total volume is adjusted to 70 μl by addition of sample dilution buffer. Derivatization is performed by the addition of 30 μl of fluorescamine solution (3 mg/ml fluorescamine in acetone, containing 20 μl pyridine) to the sample while continuously and vigorously vortexing. After approximately 2 min, the content of the microcentrifuge tube is transferred to the appropriate microvials for analysis.

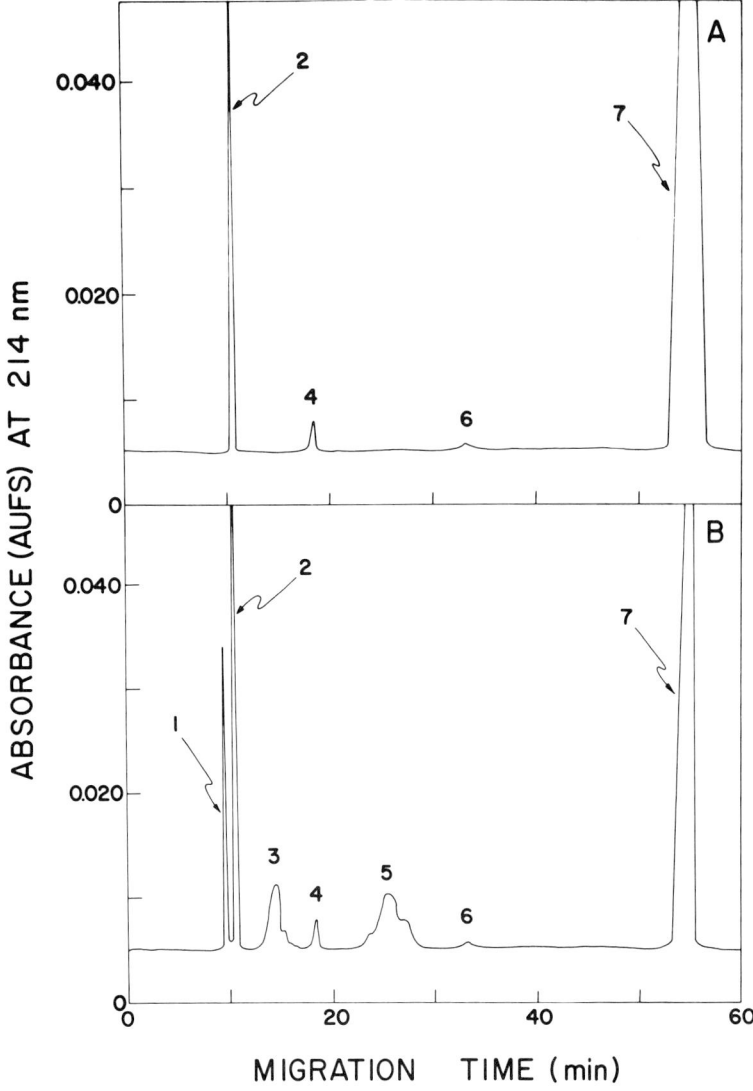

Figure 3 Capillary electrophoresis profile of fluorescamine-derivatized neuropeptide standards analyzed by UV detection at 214 nm. (A) Electropherogram of the fluorescamine solution control. Peak 2 represents acetone; peaks 4 and 6, reagent contaminants; and peak 7, fluorescamine reagent. (B) Electropherogram of LHRH, NPY, and βEND. Peak 1 represents LHRH; peak 2, acetone; peak 3, NPY; peak 4, reagent contaminants; peak 5, βEND; peak 6, reagent contaminant; and peak 7, fluorescamine reagent. The observed decreased peak area for the fluorescamine (B, peak 7), in comparison with the control value (A, peak 7), confirms that a significant amount of the reagent is immediately consumed in reaction with the analytes. Notably, LHRH under these conditions must be strongly positive since it migrated faster than the neutral acetone peak.

3. Running Conditions

Sample solutions for analysis in the appropriate microvials are placed into the sample holder of the respective instrument. The analysis program is initiated and the first sample automatically injected into the capillary by a positive nitrogen pressure of 0.5 psi (3500 Pa) for 20 sec (Beckman instrument), or electrokinetically at 25 kV for 45 sec (laboratory-made instrument). At the completion of each run, the capillary column is sequentially washed by injection of 2.0 N sodium hydroxide solution, 0.1 N sodium hydroxide solution, and distilled-deionized water, and then regenerated with running buffer. The CE separations reported are performed using 0.05 M sodium tetraborate buffer, pH 8.3, containing 0.025 M LiCl. The commercial CE instrument is equipped with a 57-cm (50 cm to the detector) × 75-μm i.d. capillary column and the laboratory-made instrument with a 125-cm (100 cm to the detector) × 75-μm i.d. capillary column. The CE separation is performed at 18 kV when using the Beckman instrument or at 25 kV when using the laboratory-made instrument. Capillary temperature for all experiments is maintained at 25°C during the run. Under these conditions, approximately 120 nl (6 nl/sec) is injected into the capillary column of the Beckman instrument (Harbaugh et al., 1990), or approximately 120 nl (2.7 nl/sec) into the column of the laboratory-made instrument. The analytes are monitored at a wavelength of 214 nm for the Beckman instrument and with fluorescence detection (475 emission) for the laboratory-made instrument.

C. Separation and Optimization

Good separation of LHRH, NPY, and βEND standards (Figs. 3 and 4), as well as of these neuropeptides from *in vivo* release samples and tissue content samples (Figs. 5 and 6, A and B, respectively), is obtained using the above-described conditions. For example, a good separation of a control fluorescamine solution and of a fluorescamine-derivatized mixture of LHRH, NPY, and βEND standards, as determined by UV detection at 214 nm, is shown in Fig. 3. In contrast, Fig. 4 shows electropherograms of fluorescamine-derivatized neuropeptide standards as determined by fluorescence detection at 475 nm emission. Figure 4A shows the fluorescamine-derivatized LHRH (peak 1) when LHRH is the only neuropeptide included in the reaction mixture. Peak 2 is presumably a reagent contaminant. Figure 4B shows an electropherogram obtained when all three neuropeptide standards were included in the reaction mixture. Peak 1 represents LHRH, peak 3 is NPY, and peak 4 is βEND. Peak 3 and peak 4 are not single peaks, and they probably represent either variants of each neuropeptide or impurity products. Detection sensitivity is considerably

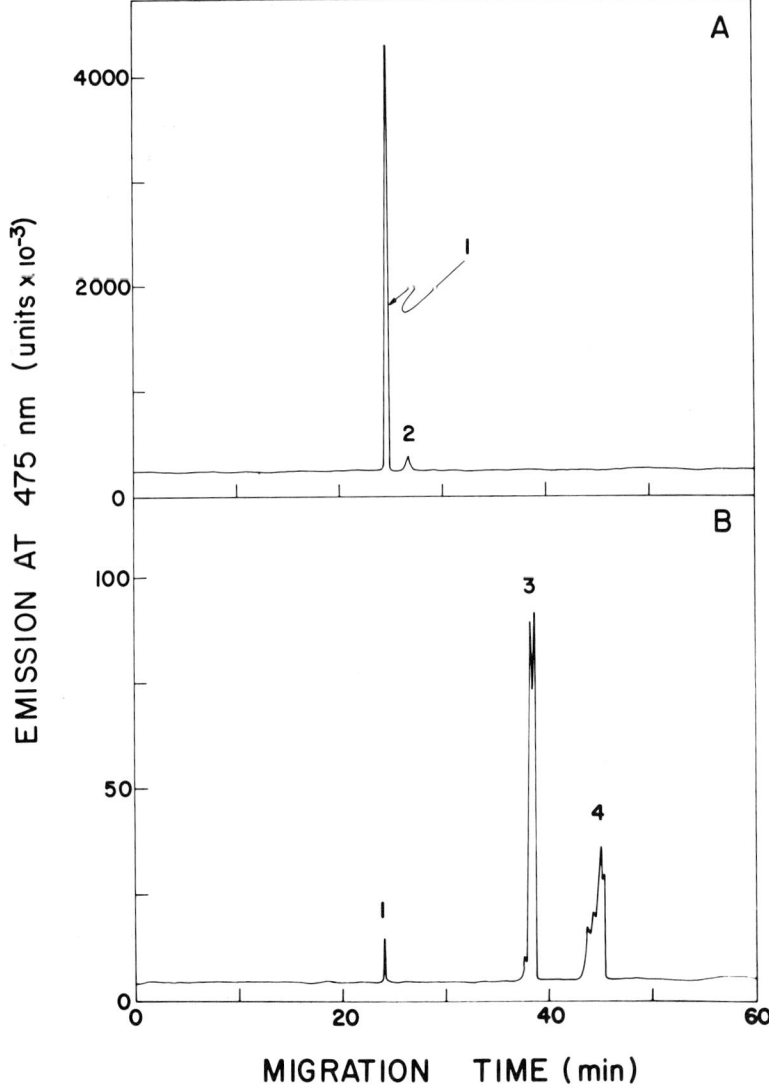

Figure 4 Capillary electrophoresis profile of fluorescamine-derivatized neuropeptide standards analyzed by fluorescence detection at 475 nm emission. (A) Electropherogram of LHRH standard. Peak 1 represents LHRH when LHRH is the only neuropeptide included in the reaction mixture and peak 2 probably represents a peptide impurity. (B) Electropherogram of a mixture of the three neuropeptides. Peak 1 represents LHRH; peak 3, NPY; and peak 4, βEND. Peak 3 and peak 4 are not single peaks, and they probably represent either variants of each neuropeptide or impurity products. It is worth noting that the quality control information for these neuropeptides standards provided by the manufacturer indicates the presence of a single peak for each one, as shown by HPLC and UV detection at 214 nm, a less-resolutive and less-sensitive system than the one we used.

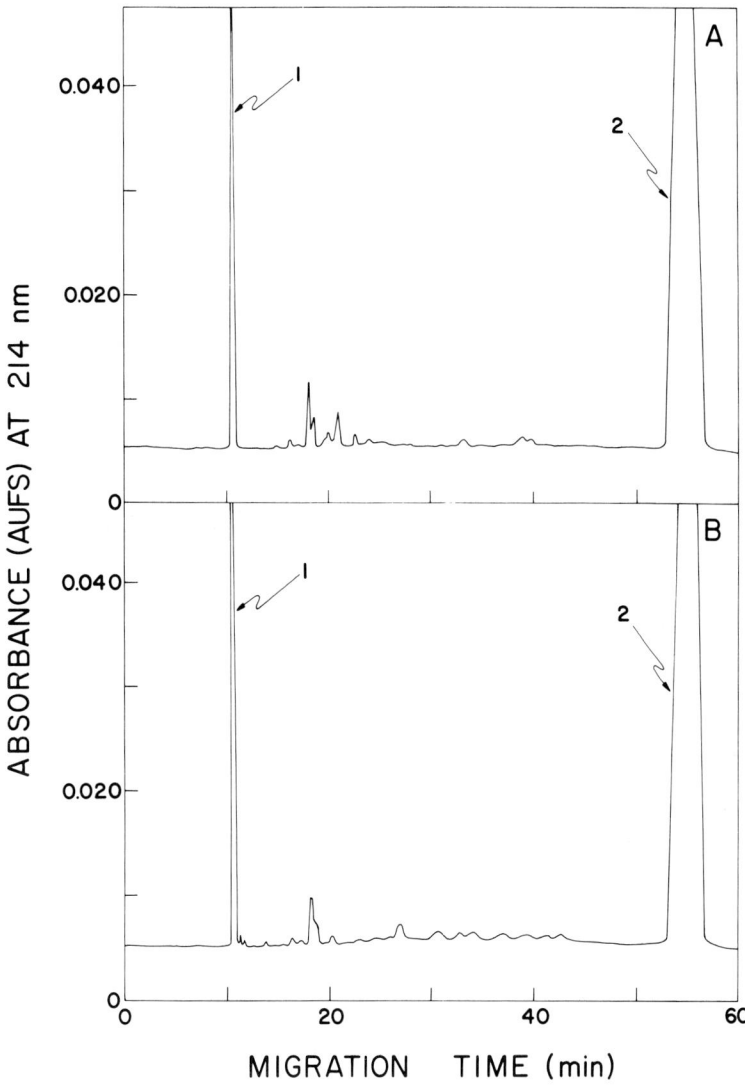

Figure 5 Capillary electrophoresis profile of fluorescamine-derivatized PPC and tissue samples analyzed by UV detection. (A) Electropherogram of a representative PPC sample. Peak 1 represents the acetone and pyridine organic solvents, and peak 2 represents the fluorescamine reagent. Other peaks, some of which comigrate with the previously mentioned reagent contaminants (Fig. 3A), are currently being identified. No LHRH, NPY, or βEND peaks were detected using UV detection in either of these samples, at the injected volumes (120 nl of a 100-μl reaction mixture, having either 40 μl of original PPC perfusate sample or 10 μl of original tissue supernatant sample). In contrast, fluorescence detection clearly allowed the visualization of these neuropeptides in these samples at the same injected volumes and conditions (see Fig. 6).

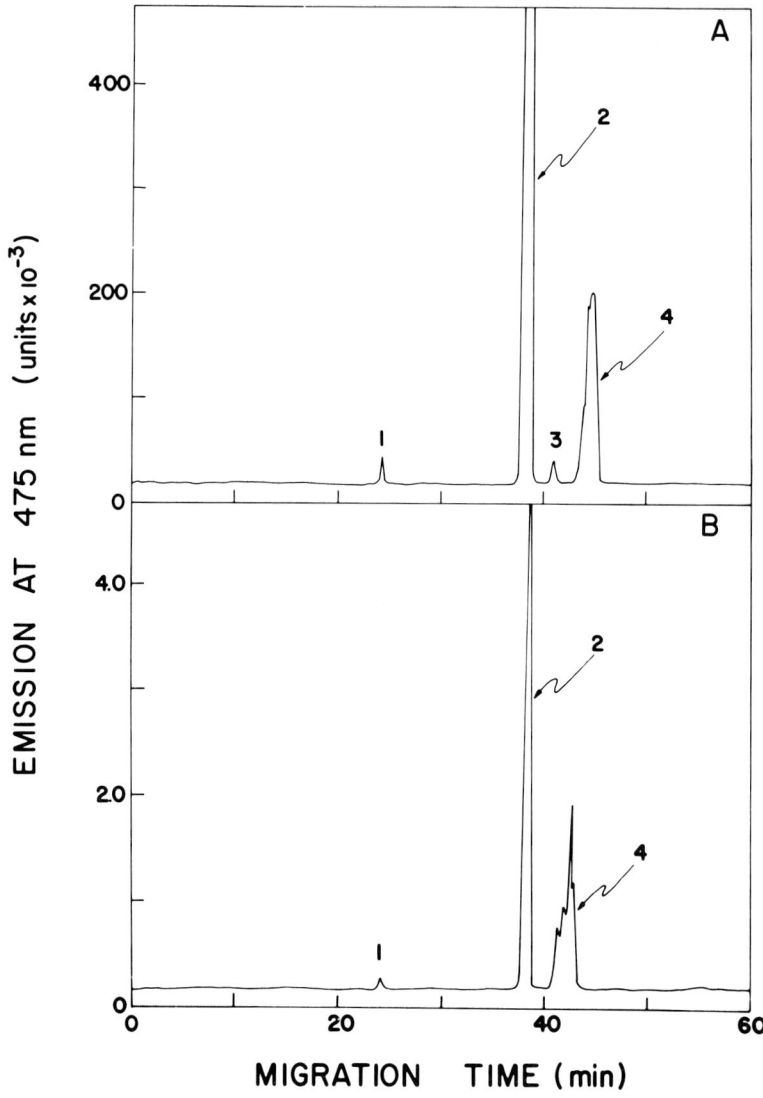

Figure 6 Capillary electrophoresis profile of fluorescamine-derivatized PPC and tissue sample analyzed by fluorescence detection. (A) Electropherogram of a representative PPC sample. Peak 1 represents LHRH; peak 2, NPY; peak 3, unknown; and peak 4, βEND. (B) Electropherogram of tissue sample. Peak 1 represents LHRH; peak 2, NPY; and peak 4, βEND. All three neuropeptides were well defined in the electropherogram of each sample (PPC and tissue). Some differences in migration time were observed, primarily for βEND in tissue samples (compare peak 4 in Figs. 6B vs 4B), which is probably related to the pH of the original tissue sample affecting the stochiometry of the derivatization reaction (tissues were homogenized in 100 μl of 0.1 N HCl).

increased when using fluorescence detection. For example, when using 30 μl standard/100 μl reaction mixture (96 μg of LHRH) or 115 ng (97.2 picomoles) in 120 nl of volume injected into the column, UV detection is 2×10^{-6} units and fluorescence detection is 71×10^{-6} units. That is a 34-fold increase in peak area. This increased sensitivity is important when dealing with endogenous samples since LHRH, NPY, and βEND peaks are not detected using UV detection (see Fig. 5), at the injected volumes (120 nl of a 100-μl reaction mixture, having either 40 μl of original PPC perfusate sample or 10 μl of original tissue supernatant sample). In contrast, fluorescence detection clearly allowed the visualization of *in vivo* release of these neuropeptides at the ME, as sampled by PPC, at the same injected volumes and conditions (see Fig. 6). Figure 6 shows an electropherogram of a PPC and tissue samples (6A and 6B, respectively) using fluorescence detection. Peak 1 represents LHRH, peak 2 is NPY, peak 3 is a reagent contaminant, and peak 4 is βEND. All three neuropeptides were well defined in the electropherogram of each sample (PPC and tissue). Some differences in migration time were observed, primarily for βEND in tissue samples (compare peak 4 in Figs. 6B vs 4B), which is probably related to the pH of the original tissue sample affecting the stochiometry of the derivatization reaction (tissues were homogenized in 100 μl of 0.1 N HCl).

III. Comments and Future Perspectives

The capillary electrophoresis system outlined in this chapter allows the simultaneous electrophoretic assessment of neuropeptides present in large amounts in PPC samples, such as NPY and βEND, which are found at levels between 10 and 100 pg/100 μl PPC perfusate/10 min as resolved by RIA (Advis *et al.*, 1990). However, simultaneous detection of *in vivo* release of a neuropeptide like LHRH present at levels between 0.1 and 1 pg/100 μl PPC perfusate/10 min from the hypothalamic median eminence (Advis *et al.*, 1989, 1990; Guzman *et al.*, 1990a, 1991a) will probably be achieved when laser-induced fluorescence detection of derivatized neuropeptides is coupled to additional sensitivity enhancers already in existence (e.g., special lenses, multirefractive systems, Z-shaped capillaries, capillary bundles, analyte concentrator). In the best circumstances, RIAs can reach detection limits of 10^{-9} to 10^{-10} M. Capillary electrophoresis coupled to laser-induced fluorescence detection is reaching levels of detection for concentrations ranging from 10^{-12} to 10^{-15} M. These values are the most sensitive concentrations so far reported in the literature (Hernandez *et al.*, 1991), and they have been obtained using a modification of the standard laser-induced fluorescence detection (Hernandez *et al.*, 1990). A model of the laser-induced fluoresence system is depicted in Fig. 7. This exceptional sensitivity level will be required to determine the electro-

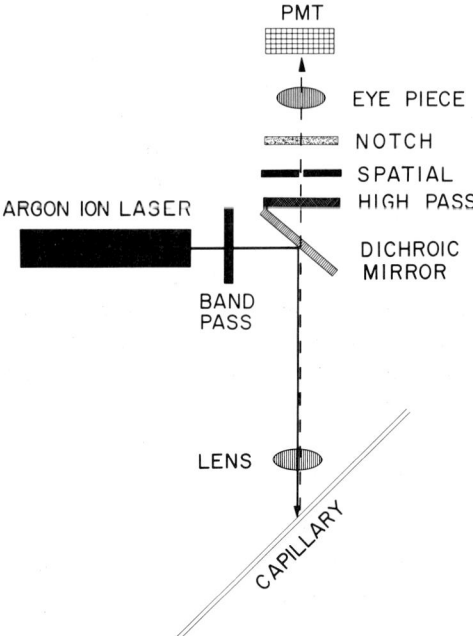

Figure 7 Diagram of a colinear, laser-induced fluorescence detection system. The main axis of the capillary is at a 90° angle with the laser beam and with the optical axis of the microscope. Adapted and reproduced by permission (Hernandez et al., 1991).

phoretic characteristics of neuropeptides being released in low concentrations from discrete brain sites, as for example LHRH from the median eminence. For comparison purposes it is interesting to note that the Comitée Internationale des Pois et Measures (Chenm et al., 1991) has recently accepted zeptomole and yoctomole as the smallest detection sensitivity values (1 zeptomole = 10^{-21} mol = 600 molecules; 1 yoctomole = 10^{-24} mol = 0.6 molecules). This is probably the lowest range possible for the determination of peptides found in minute amounts, for instance, at neuroendocrine controlling sites.

On the other hand, slight structural modifications of the neuropeptides being released (e.g., deamidation) might exist under different physiological conditions (Wright, 1991), thus giving origin to neuropeptide variants, which might affect their diffusion characteristics and therefore play a major role in a volume transmission context (Fuxe and Agnati, 1991). We should also keep in mind that slight modifications in neuropeptide structure might render it unrecognizable by specific antibodies and therefore by RIA. In this context, for example, it is interesting to note that multiple electrophoretic peaks are associated with NPY and βEND regardless of their origin (synthetic standards or ewe samples). The quality control information for the neuropeptides standards provided by the manufacturer indicates the presence of a single peak for each one, as shown by HPLC and UV detection at 214 nm, a less-resolutive and less-sensitive system than capillary electrophoresis and fluorescence detection. Since the profile associated with these multiple electrophoretic peaks of NPY and βEND is different in samples obtained during the follicular and luteal phases of the ewe estrous cycle (results not shown), the CE system might be resolving variants associated to these neuropeptides. The decapeptide LHRH was always resolved as a single peak.

The analysis of neuropeptides by capillary electrophoresis coupled to laser-induced fluorescence detection has the potential to become the method of choice for determining brain constituents. An appealing aspect of this technology are the simplicity of the methodology, its efficiency coupled to the need of low-volume samples, and, most important, its ability to separate (i.e., high resolution) and to detect (i.e., high sensitivity) simultaneously multiple neuropeptide components in a single sample. Great improvements have so far been achieved in the development of capillary electrophoresis, and at present this technique might almost be reaching its maximum detection sensitivity levels. Further enhancement in sensitivity will probably be achieved by advances in the optics of the instrumentation, as for example new laser-induced fluorescence detection systems, as well as advances in the derivatization chemistries.

Acknowledgments

Supported by USDA 89-37240-4587 to J. P. Advis.

References

Advis, J. P., Hernandez, L., and Guzman, N. A. (1989). *Pept. Res.* **2**, 389–394.
Advis, J. P., Conover, C. D., McDonald, J. K., and Kuljis, R. O. (1990). *Ann. N.Y. Acad. Sci.* **611**, 468–470.
Advis, J. P., and Guzman, N. A. (1993). *J. Liquid Chromatogr.* **16**, 2129–2148.
Alvin, M., Grossman, P. D., and Moring, S. E. (1993). *Anal. Chem.* **65**, 489A–497A.
Brown, D. S., and Jenke, D. R. (1987). *J. Chromatogr.* **410**, 157–168.
Chenm, D. Y., Swerdlow, H. P., Harke, H. R., Zhang, J. Z., and Dovichi, N. J. (1991). *J. Chromatogr.* **559**, 237–246.
Chervet, J. P., van Soest, R. E. J., and Ursem, M. (1991). *J. Chromatogr.* **543**, 439–449.
Deyl, Z., and Struzinsky, R. (1991). *J. Chromatogr.* **569**, 63–122.
Felix, A. M., Toome, V., Debernardo, S., and Weigle, M. (1975). *Arch. Biochem. Biophys.* **168**, 601–608.

Fuxe, K., and Agnati, L. F. (1991). *In* "Volume Transmission in Brain" (K. Fuxe and L. F. Agnati, eds.), pp. 1–9. Raven, New York.

Guzman, N. A., Hernandez, L., and Hoebel, B. G. (1989). *BioPharmacology* **2**, 22–37.

Guzman, N. A., Hernandez, L., and Advis, J. P. (1990a). *In* "Current Research in Protein Chemistry" (J. J. Villafranca, ed.), pp. 203–216. Academic Press, San Diego.

Guzman, N. A., Hernandez, L., and Terabe, S. (1990b). *In* "Analytical Biotechnology: Capillary Electrophoresis and Chromatography" (C. Horváth and J. G. Nikelly, eds.), ACS Symp. No. 434, pp. 1–35. Am. Chem. Soc., Washington, D.C.

Guzman, N. A., Trebilcock, M. A., and Advis, J. P. (1991a). *In* "Techniques in Protein Chemistry II" (J. J. Villafranca, ed.), pp. 37–54. Academic Press, San Diego.

Guzman, N. A., Trebilcock, M. A., and Advis, J. P. (1991b). *Anal. Chim. Acta* **249**, 247–255.

Guzman, N. A., Trebilcock, M. A., and Advis, J. P. (1991c). *J. Liq. Chromatogr.* **14**, 997–1015.

Guzman, N. A., Moshera, J., Iqbal, K., and Malick, A. W. (1992a). *J. Liq. Chromatogr.* **15**, 1163–1177.

Guzman, N. A., Moshera, J., Bailey, C. A., Iqbal, K., and Malick, A. W. (1992b). *J. Chromatogr.* **598**, 123–131.

Harbaugh, J., Collette, M., and Schwartz, H. E. (1990). *Beckman Instrum. Tech. Inf. Bull.* N°.TIBC-103 (4sp-890-10B).

Hernandez, L., Marquina, R., Escalona, J., and Guzman, N. A. (1990). *J. Chromatogr.* **502**, 247–255.

Hernandez, L., Joshi, N., Escalona, J., and Guzman, N. A. (1991). *J. Chromatogr.* **559**, 183–196.

Karger, B. L., Cohen, A. S., and Guttman, A. (1989). *J. Chromatogr.* **492**, 585–614.

Karger, B. L., and Foret, F. (1993). *In* "Capillary Electrophoresis Technology" (N. A. Guzman, ed.) pp. 3–63. Marcel Dekker Inc., New York.

Kasting, N. W., and Martin, J. B. (1989). *In* "Neuroendocrine Peptide Methodology" (P. M. Conn, ed.), pp. 253–264. Academic Press, San Diego.

Kendrick, K. M. (1989). *In* "Neuroendocrine Peptide Methodology" (P. M. Conn, ed.), pp. 229–252. Academic Press, San Diego.

Kuhr, W. G., and Monnig, C. A. (1992). *Anal. Chem.* **64**, 389R–407R.

Lingeman, H., Underberg, W. J. M., Takadate, A., and Hulshoff, A. (1985). *J. Liq. Chromatogr.* **8**, 789–874.

Lunte, S. M., and Wong, O. S. (1989). *LCGC* **7**, 908–916.

Lunte, C. E., Scott, D. O., and Kissinger, P. T. (1991). *Anal. Chem.* **63**, 773A–780A.

Newcomb, R. (1992). *LCGC* **10**, 34–39.

Novotny, M., Cobb, K. A., and Liu, J. (1990). *Electrophoresis* **11**, 735–749.

Ogden, G., and Foldi, P. (1987). *LCGC* **5**, 28–40.

Rivier, J., and McLintock, R. (1989). "The Use of HPLC in Receptor Biochemistry," pp. 77–105. Alan R. Liss, New York.

Rubenstein, M., Chen-Kiang, S., Stein, S., and Udenfriend, S. (1979). *Anal. Biochem.* **95**, 117–121.

Tsuda, T., Kobayashi, Y., Hori, A., Matsumoto, T., and Suzuki, O. (1988). *J. Chromatogr.* **456**, 375–381.

Tsuda, T., Sweedler, J. V., and Zare, R. N. (1990). *Anal. Chem.* **62**, 2149–2152.

Wallingford, R. A., and Ewing, A. G. (1990). *Adv. Chromatogr.* **29**, 1–76.

Wang, T., Aiken, J. H., Huie, C. W., and Hardwick, R. A. (1991). *Anal. Chem.* **63**, 1372–1376.

Wright, H. T. (1991). *Crit. Rev. Mol. Biol.* **26**, 1–52.

8

Immunoreactive Heterogeneity in Peptide Measurement

I. Valverde and M.L. Villanueva-Peñacarrillo

I. General Introduction
II. Proglucagon-Derived Peptides
   A. Glucagon Measurement
   B. Glicentin and Oxyntomodulin Measurements
   C. Truncated GLP-1 Measurement
III. Conclusion
   References

I. General Introduction

Radioimmunoassay (RIA; Yalow and Berson, 1960) was the tool that first allowed the detection and quantitation of hormones and peptides in circulation and tissues. The essential requirements for a RIA are a specific antibody with high affinity for the substance to be determined, a pure (cold and radiolabeled) native or synthetic substance, and a standardized method to separate the bound from the free portion of the substance. Today, there are commercial RIA kits available for a great number of hormones (polypeptides and steroids), which are used for clinical and even research purposes, and other commercial systems, like ELISA, which do not require radioisotopes.

Yalow (1974) first pointed out that an antiserum raised against a given peptide, which is specific in terms of not having cross-reactivity with other known peptides, can also detect other circulating forms of different molecular sizes containing the whole or part of the peptide. Similar findings in tissue extracts were the basis of the search for the precursor molecules of several peptides. An example of immunoreactive heterogeneity is proinsulin, which is present in plasma and pancreatic extracts; specific antibodies for insulin also react, variably, with proinsulin. The measurement of immunoreactive insulin (IRI) in a sample (plasma, pancreas perfusion medium, pancreatic islets, or pancreas extracts) reflects not only the amount of insulin, but also, in an undetermined proportion, the amount of proinsulin and potentially intermediate or altered forms of (pro)insulin. On the other hand, in a RIA for C-peptide (the connecting peptide, or the noninsulin part of the proinsulin molecule), insulin is not detected while proinsulin is.

The adequate quantitation of insulin or C-peptide requires separation of these molecules from proinsulin, prior to RIA. The traditional gel filtration method, although time consuming, was and still is the best choice to process large volumes of plasma or tissue extracts.

Alternative methods are available in which two site-specific antisera are used. For instance, by using both an anti-insulin and an anti-C-peptide, it is possible to develop a two-site RIA (Heding, 1977), a radioimmunometric assay (Rainbow et al., 1979), and an ELISA assay (Hartling et al., 1986). These can directly determine the concentration of proinsulin in a given sample, and it is then possible to estimate the insulin or C-peptide concentration by subtracting the proinsulin value from the IRI or C-peptide assay result. More recently, some monoclonal antibodies raised in guinea pigs, using proinsulin as immunogen, have been proven to have less than 1% cross-reactivity for insulin or C-peptide (Deacon and Conlon, 1985), and when purified by adsorption to insulin and C-peptide conjugated to Sepharose, the cross-reactivities to insulin and C-peptide decrease to less than 0.2% (Cohen et al., 1985). With these types of antisera, direct and specific RIAs for proinsulin have been developed (Cohen et al., 1986; Deacon and Conlon, 1985; Hampton et al., 1988; Yoshioka et al., 1988).

By sequencing the corresponding cDNA, it has been possible to infer the primary structure of the precursors of many peptides and hormones. Moreover, it is now clear that different parts of the precursor molecule can exist in the tissue and circulation. To detect these, efforts have been made to produce peptide region-specific antisera with the aim of excluding at least some of the molecular forms previously detected by other less-specific antisera. A good example is the measurement of the proglucagon-derived peptides.

II. Proglucagon-Derived Peptides

The primary structure of the proglucagon molecule has been deduced by the sequence analysis of the glucagon cDNA of several mammals and piscine species (reviewed in Bell, 1986). In all mammals studied, the preproglucagon is very similar; it consists of 180 amino acids, being the first 20 N-terminal residues, the signal peptide. The same proglucagon (PG) molecule is produced in the pancreatic A-cells and in the intestinal L-cells. However, different final products are produced in the A- and L-cell-types due to different post-translational processing (Novak *et al.*, 1987, and Fig. 1). In the pancreas, the precursor molecule is processed to glucagon (PG33–61). In addition, the following peptides are present in pancreas extracts, pancreas perfused media, and plasma: PG1–33 (glicentin-related polypeptide, GRPP), PG1–61, PG64–69, and PG72–160 (major proglucagon fragment, MPGF). In the small intestine, the processing of proglucagon produces glicentin (PG1–69), oxyntomodulin (PG33–69), PG78–107 amide (truncated glucagon-Like peptide-1, tGLP-1), and PG126–158 (glucagon-Like peptide-2, GLP-2).

A. Glucagon Measurement

The RIA for glucagon was first developed by Unger *et al.* (1961) soon after the description of that for insulin. Apart from glucagon, the Unger's group (Eisentraut *et al.*, 1968) also detected, with some glucagon antisera, the presence, in intestinal extracts and in plasma after oral glucose administration, of immunoreactive glucagon substances which were named GLI (glucagon-like immunoreactivity). The types of glucagon antisera not reacting with the intestinal peptides (consisting of antibodies directed to the C-terminal end of the glucagon molecule) were considered specific for glucagon. However, it was later demonstrated that these C-terminal glucagon antibodies could also detect a molecule larger than glucagon. This molecule was present in pancreatic extracts (Rigopoulou *et al.*, 1970) and plasma (Valverde *et al.*, 1974) and was increased in the plasma of glucagonoma patients and nephrectomized animals. This molecule has been identified as proglucagon 1-61 (Baldissera and Holst, 1986).

If antisera to the C-terminal of glucagon, including the 30K from R.H. Unger, are used to analyze glucagon-immunoreactive heterogeneity in plasma samples (human and animal), by gel filtration, the presence of four distinct glucagon-immunoreactive fractions is demonstrated. These four peaks consist of one co-eluting with the plasma proteins, named BPG (big plasma glucagon) or interference factor; a second 9000-Da peak later identified as PG1-61; a third, which corresponds to glucagon; and a final smaller-sized one (Valverde *et al.*, 1974; Valverde, 1983; Weir *et al.*, 1975). Only one antiserum, the RCS5 from Dr. S. Bloom, was found not to detect the protein-associated fraction (BPG) (Valverde and Rovira, 1983).

This high-molecular-sized glucagon-immunoreactive fraction (BPG) represents a widely variable percentage of the total immunoreactivity present in plasma, depending on the individual sample and on the antiserum used (from about 0 to about 100%). Some methods have been proposed to remove BPG prior to assay, such as an acid and/or ethanol precipitation. Alternatively, BPG can be measured as that remaining glucagon immunoreactivity in the sample treated with a charcoal–dextran mixture which absorbs the glucagon molecule (Weir et al., 1975). Nevertheless, these methods give only an approximation of the true glucagon concentrations in plasma because PG1–61 is still also present.

In conclusion, any RIA for glucagon, as it occurs with those for insulin, will detect not only the native hormone but also other hormone-containing forms (e.g., the large glucagon-containing peptide, PG1–61) present in pancreatic extracts and plasma. This makes the gel filtration of the sample obligatory to accurately quantitate glucagon.

1. Gel Filtration

This technique may be used as the only or as first step to isolate the peptide to be determined; in the case of glucagon, it should be sufficient if a C-terminal antiserum is used. Although Biogel P-10 and Sephadex G-50 are appropriate, Biogel P-30 is the most widely used matrix. The eluant could be a basic buffer, preferably the same as that used in the subsequent glucagon RIA, e.g., 0.2 M glycine, pH 8.8, containing 0.5% human serum albumin (HSA) and 500 IU/ml of Trasylol. Alternatively, the elution buffer can be 50 mM ammonium bicarbonate, pH 8.8, or 2M acetic acid; these facilitate concentration of the sample, by lyophilization, prior to the assay. In either case, albumin should be present in the elution buffer to prevent poor recovery of the peptide. Internal molecular size markers should be used in each run. In addition, a small quantity of [^{125}I]glucagon can be included, making certain that the radioactivity present will be negligible in the eluate and not interfere later with the assay. The whole procedure should be performed at 4°C. As an indication, up to 4 ml of plasma can be filtered in a 40 × 1-cm column, if a basic eluation buffer is being used. Figure 1A shows, as an example, the analysis of the C-terminal glucagon-immunoreactive components in the basal plasma of a diabetic subject. With acid eluants, no more than 1 ml of plasma should be loaded for columns this size to prevent possible precipitation of plasma proteins. For low glucagon concentration, the volume of plasma to fractionate must be increased; this requires a proportionally larger column and a subsequent lyophilization of the eluates prior to assay. Figure 1B shows the analysis of the C-terminal glucagon-immunoreactive components in the basal plasma of a pancreatectomized subject.

B. Glicentin and Oxyntomodulin Measurements

These two intestinal peptides are derived from the proglucagon molecule in the intestinal L-cell (Fig. 2). They contain in their structure the glucagon

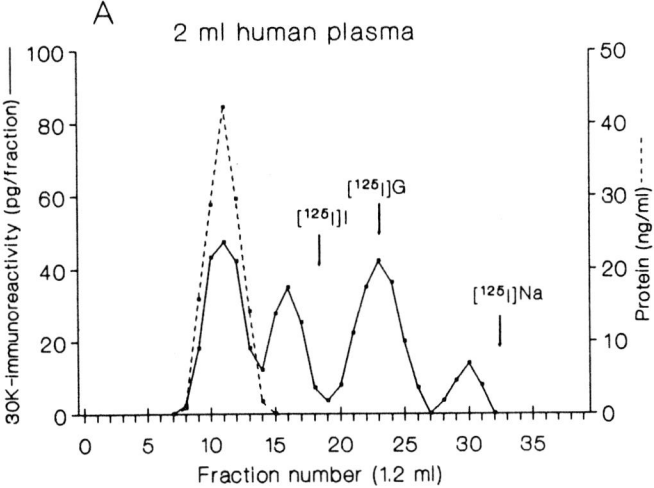

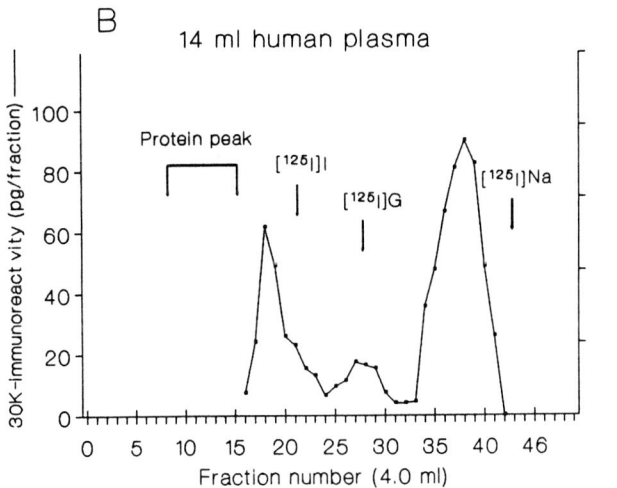

Figure 1 (A) Gel filtration and analysis of the C-terminal glucagon immunoreactivity in human plasma with high glucagon$_{3500MW}$ content, (total IRG: 245 pg/ml) by Biogel P-30 (40 × 1 cm) using RIA assay buffer as eluant (0.2 M glycine, pH 8.8, containing 0.25% HSA); in this case, 0.5 ml of each eluate was assayed in duplicate, and the resulting glucagon$_{3500MW}$ concentration was 90 pg/ml of plasma. (B) Gel filtration and analysis of C-terminal glucagon immunoreactivity in human plasma with low glucagon$_{3500MW}$ content (total IRG: 150 pg/ml) by Biogel P-30 (100 × 1.5 ml), using 10 mM CO$_3$HNH$_4$, pH 8.8, containing 0.125% HSA, as eluant. The eluates were lyophilized and redissolved in 0.5 ml of RIA assay buffer without albumin, and, 0.2 ml of each was assayed by duplicate. The concentration of glucagon$_{3500MW}$ was 7 pg/ml of plasma. Arrows indicate the elution position of the added radioactive markers: insulin, glucagon, and NaI.

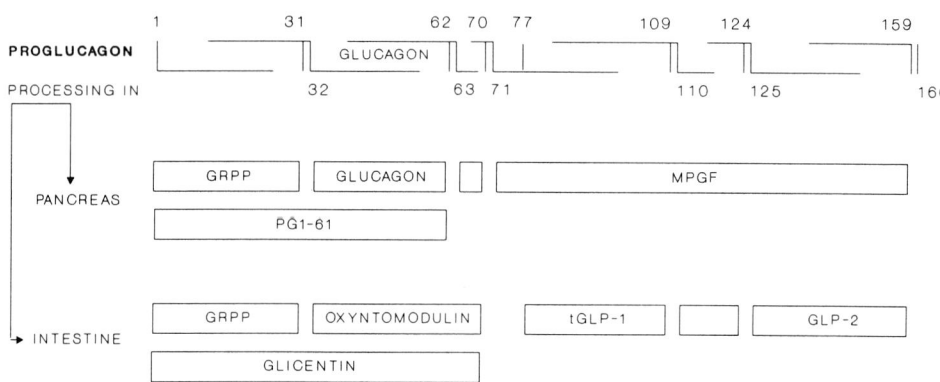

Figure 2 Mammalian proglucagon schema and the proglucagon-derived peptides originated by proglucagon processing in the pancreas and in the intestine.

sequence and thus can be detected by some antiglucagon sera, specifically, by those containing antibodies directed to the N-terminal and central part of the glucagon molecule. These antisera also react with glucagon and PG1–61, which have, respectively, molecular sizes similar to that of oxyntomodulin and glicentin; because of this, it is not possible to use only gel filtration to differentiate the pancreatic from the intestinal molecules and, consequently, two separate RIA assays of the eluates, with each of the two types of antiglucagon serum, are needed.

Attempts have been made to produce antisera which would react only with the intestinal forms. For this purpose, synthetic C-terminal fragments of oxyntomodulin, 30–37 and 19–37, were used as immunogens. These produced, respectively, antisera with non, FAN (Blache et al., 1988a), or low, R4804 (Yanaihara et al., 1984), cross-reactivity with glucagon. Even so, these antisera react with oxyntomodulin, glicentin, and PG64–69. An N-terminal antiglicentin serum, the R-64, was also produced using native glicentin as immunogen (Moody et al., 1981); this antiserum does not react with oxyntomodulin nor with glucagon, but it detects GRPP (PG1-30). Thus, to determine quantitatively any of these peptides, their physical separation by gel filtration is also required.

1. Gel Filtration

The gel filtration protocol is the same as that just described for glucagon measurement. In this case, the eluates must be assayed in separate RIAs with two types of antiglucagon sera and then the estimated levels of glicentin, in glucagon equivalents, will be the difference between the values obtained with

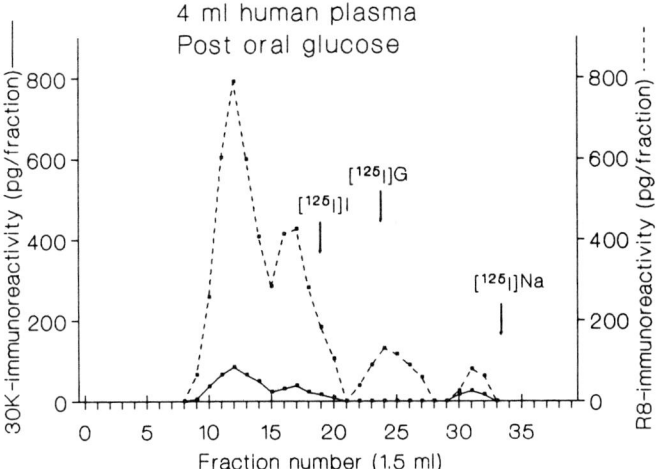

Figure 3 Gel filtration and analysis of the N- and C-terminal glucagon immunoreactivity in plasma from a gastrectomized patient, 60 min after oral glucose intake. Arrows indicate the elution position of the added radioactive markers: insulin, glucagon, and NaI.

the N- and the C-terminal assays of the peak appearing in the 9000-Da region. The level of oxyntomodulin, as with glucagon equivalents, will be again the difference between the N- and C-terminal values of the peak appearing in the 3500-Da region. Figure 3 shows an example of this.

2. Reversed-Phase HPLC

Bataille and co-workers standardized a HPLC (high-pressure liquid chromatography) method which separates glicentin, oxyntomodulin, glucagon, and other proglucagon-derived peptides (Kervran et al., 1987). It consists of one, or two in tandem, μBondapack C18 columns (0.39 × 30 cm i.d. Waters), a C18-sized precolumn (Waters), and a linear gradient of 20–40% acetonitrile in solvent A (1% trifluoroacetic acid, buffered to pH 2.5 with diethylamine), at a flow rate of 1.5 ml/min. During a first period (30 min), 1.5-ml fractions are collected and 0.5-ml fractions are collected thereafter; acetonitrile is evaporated in a Speed-Vac and, later, the samples are lyophilized prior to reconstitution in the appropriate buffer for RIA assay. In this system (Fig. 4), calibration with native or synthetic peptides shows glicentin to appear at 32 min, oxyntomodulin at 33 min, glucagon at 37.5 min, and PG1–61 at 35.5 min. Figure 4 shows the analysis of the N- and C-terminal glucagon-immunoreactive components in a dog ileum extract and in a human glucagonoma tumor (Alarcón et al., 1991). A better separation of glicentin from oxyntomo-

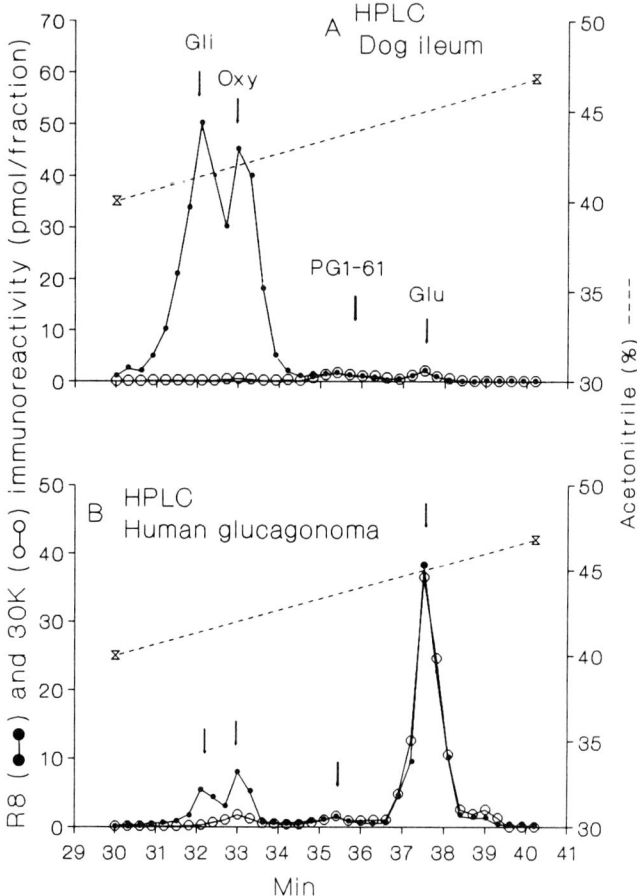

Figure 4 Reversed-phase HPLC of the N- and C-terminal glucagon immunoreactivity present in (A) a dog ileum extract and (B) a human glucagonoma tumor extract. Arrows indicate the elution position of the synthetic or native peptides used for calibration: glicentin, oxyntomodulin, proglucagon 1–61, and glucagon.

dulin can be achieved by a two-slope linear gradient: the first 30 min at 0–30% acetonitrile, followed by a second 30-min period at 30–50% acetonitrile in solvent A (Blache et al., 1988b). The two intestinal peptides can be equally measured with an N-terminal antiglucagon serum or with a C-terminal antiglicentin serum (FAN), using oxyntomodulin (19–37) as ^{125}I-Tyr tracer and standard (Blache et al., 1988b). This system allows the quantitation of the two intestinal peptides in tissue extracts and also in plasma previously deproteinated. For deproteination, 5 ml of plasma, diluted 1:3 with the described solvent A, is applied on a series of two Sep-Pak cartridges previously washed with 10 ml of acetonitrile and with 10 ml of solvent A; then, the cartridges are washed with 10 ml of solvent A and, after, the retained peptides, free of ~90% of the plasma proteins, are eluted with 4 × 1 ml of 40% acetonitrile in solvent A (Kervran et al., 1987). The recovery of added native or synthetic peptides always must be checked.

C. Truncated GLP-1 Measurement

C- and N-terminal-directed antisera for synthetic GLP-1 were produced by research groups, but some antisera of this type are available commercially; RIAs for this part of the proglucagon molecule have been standardized (Uttenthal et al., 1985; Orskov and Holst, 1987; Mojsov et al., 1986; Shima et al., 1987; Takahashi et al., 1988). The truncated forms of GLP-1 (PG 78–108 and PG 78–107 amide) have a number of biological activities, including highly insulinotropic activities (Kreymann et al., 1987; Holst et al., 1987; Mojsov et al., 1987). Only the C-terminal-directed antiserum reacts with the truncated GLP-1 forms. Using the two types of GLP-1 antiserum available, it has been possible to demonstrate the presence of a great immunoreactive heterogeneity in tissue and plasma, by gel filtration and/or HPLC. The need to determine concentrations and fluctuations of the biological active molecule (GLP-1 7-36 amide, tGLP-1) again requires the isolation, by gel filtration, of this peptide.

1. Gel Filtration

Orskov et al. (1991) have standardized a method using Sephadex G-50 SF columns (100 × 1.6 cm) and 20 mM veronal, pH 8.4, containing 1% bovine serum albumin and 0.6 mM thiomersal as eluant. In this system, they can distinguish, using a C-terminal GLP-1 antiserum for detection, between the peak of GLP-1 and that of tGLP-1. For plasma samples, they propose concentrating and partially purifying the small peptides, prior to gel filtration, passing 20 ml of the plasma sample trough a Sep-Pak cartridge, which is washed with 10 ml of 0.2% trifluoroacetic acid (TFA) and then with 5 ml of 20% acetonitrile in 0.2% TFA. The elution of the retained small peptides is achieved with 4 ml of 45% acetonitrile in 0.2% TFA (Orskov et al., 1991).

2. Reversed-Phase HPLC

Several groups have used various systems of reversed-phase HPLC to separate immunoreactive GLP-1 moieties (Mojsov et al., 1986; Shima et al., 1987; Orskov et al., 1989, 1991; Suda et al., 1988). We have been able to

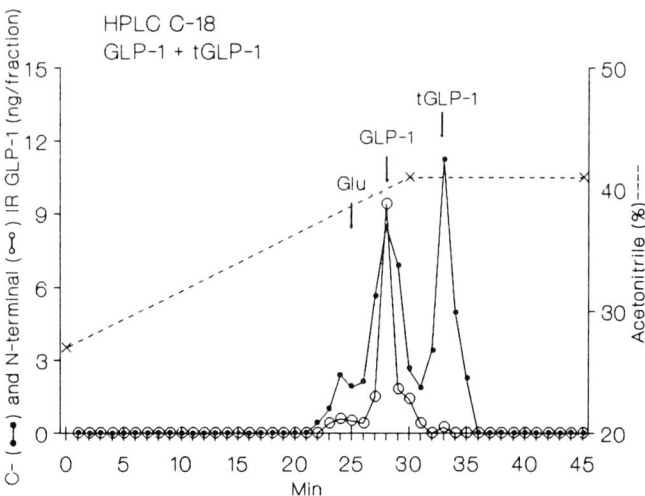

Figure 5 Reserved-phase HPLC of a mixture of synthetic amidated GLP-1 and tGLP-1 (20 ng, each), as determined with a C-terminal (Holst No. 2135) and an N-terminal (Peninsula No. 60027) anti-GLP-1 sera. Arrows indicate the elution position of the native or synthetic peptides individually used for calibration: glucagon, glucagonlike peptide-1, and truncated glucagonlike peptide-1.

achieve a good separation of GLP-1 from tGLP-1, both in their amidated forms (Fig. 5), using a system similar to that reported by Mosjov *et al.* (1990). It consists in a μBondapack C_{18} column (0.39 × 30 cm i.d. Waters) and a linear gradient of 27–41% acetonitrile in 0.1% phosphoric acid, buffered to pH 7.0 with triethylamine, at a flow rate of 1 ml/min. Prior to RIA, the 1-ml fractions collected are devoided of acetonitrile by a Speed-Vac, lyophilized to dryness, and then reconstituted in the convenient volume of assay buffer, containing 1% HSA to prevent the proven adsorption of the peptides by the tube material.

III. Conclusion

Most of the antisera raised against a particular peptide cross-react with other peptides of chemically related structure. To study the participation of a given peptide in a total immunoreactive heterogeneity, it is advisable, if not mandatory, to fractionate the sample into separate immunoreactive peptides which then can be analyzed individually. The method for fractionation of a sample (plasma, tissue extract, perfusion fluids, etc.) depends highly on its general composition and, mainly, on its total protein content. Samples with low immunoreactivity of peptide/s per milligram of protein require treatment to increase their specific activity; that can be the case of a plasma with low peptide contents or when an HPLC study is required.

Acknowledgments

We are grateful to Dr. R. H. Unger and to Dr. J. J. Holst for the generous supply of antisera, and we thank Ms. N. Martínez for her excellent secretarial work. We acknowledge financial support from FIS and DGICYT (Spanish Official Institution).

References

Alarcón, C., Kervran, A., Bataille, D., and Valverde, I. (1991). *Horm. Metab. Res.* **23**, 251–306.
Baldissera, F. G. A., and Holst, J. J. (1986). *Diabetologia* **29**, 462–467.
Bell, G. I. (1986). *Peptides* **7**, Suppl. 1, 27–36.
Blache, P., Kervran, A., Le-Nguyen, D., Laur, J., Cohen-Solal, A., Devilliers, G., Mangeat, P., Martinez, J., and Bataille, D. (1988a). *Biomed. Res.* **9**, suppl. 3, 19–28.
Blache, P., Kervran, A., Marliner, J., and Bataille, D. (1988b). *Anal. Biochem.* **173**, 171–179.
Cohen, R. M., Nakabayashi, T., Blix, P. M., Rue, P. A., Shoelson, S. E., Root, M. A., Frank, B. H., Revers, R. R., and Rubenstein, A. H. (1985). *Diabetes* **34**, 84–91.
Cohen, R. M., Given, B. D., Licinio-Paixao, J., Provow, S. A., Rue, P. A., Frank, B. H., Root, M. A., Polonsky, K. S., Tager, H. S., and Rubenstein, A. H. (1986). *Metab., Clin. Exp.* **35**, 1137–1146.
Deacon, C. F., and Conlon, J. M. (1985). *Diabetes* **34**, 491–497.
Eisentraut, A., Ohneda, A., Parada, E., and Unger, R. H. (1968). *Diabetes* **17**, 321.
Hampton, S. M., Beyzavi, K., Teale, D., and Marks, V. (1988). *Clin. Endocrinol.* **29**, 9–16.
Hartling, S. G., Dinesen, B., Kappelgaard, A. M., Faber, O. K., and Binder, C. (1986). *Clin. Chim. Acta* **156**, 289–298.
Heding, L. G. (1977). *Diabetologia* **13**, 467–474.
Holst, J. J., Orskov, C., Nielsen, O. V., and Schwartz, T. W. (1987). *FEBS Lett.* **211**, 169–174.
Kervran, A., Blache, P., and Bataille, D. (1987). *Endocrinology (Baltimore)* **121**, 704–713.
Kreymann, B., Williams, G., Ghatei, M. A., and Bloom, S. R. (1987). *Lancet* **ii**, 1300–1304.
Mojsov, S., Heinrich, G., Wilson, I. B., Ravazzola, M., Orci, L., and Habener, J. F. (1986). *J. Biol. Chem.* **261**, 11880–11889.
Mojsov, S., Weir, G. C., and Habener, J. F. (1987). *J. Clin. Invest.* **79**, 616–619.
Mojsov, S., Kopczynski, M. G., and Habener, J. F. (1990). *J. Biol. Chem.* **265**, 8001–8008.
Moody, A., Holst, J. J., Thim, L., and Lindkaer-Jensen, S. (1981). *Nature (London)* **289**, 514–516.
Novak, U., Wilks, A., Buell, G., and McEwen, R. (1987). *Eur. J. Biochem.* **164**, 553–558.
Orskov, C., and Holst, J. J. (1987). *Scand. J. Clin. Lab. Invest.* **47**, 165–174.
Orskov, C., Bersani, M., Jonsen, A. H., Hojrup, P., and Holst, J. J. (1989). *J. Biol. Chem.* **264**, 12826–12826.
Orskov, C., Jeppesen, J., Madsbad, S., and Holst, J. J. (1991). *J. Clin. Invest.* **87**, 415–423.
Rainbow, S. S., Woodhead, J. S., Yue, D. K., Luzio, S. D., and Hales, C. N. (1979). *Diabetologia* **17**, 229–234.
Rigopoulou, D., Valverde, I., Marco, J., Faloona, G., and Unger, R. H. (1970). *J. Biol. Chem.* **245**, 496–501.

Shima, K., Hirota, M., Ohboshi, C., Sato, M., and Nishino, T. (1987). *Acta Endocrinol. (Copenhagen)* **114**, 531–536.
Suda, K., Manaka, H., Takahashi, H., Tominaga, M., and Sasaki, H. (1988). *Biomed. Res.* **9**, Suppl. 3, 39–45.
Takahashi, H., Manaka, K., Suda, K., Fukase, K., Takahashi, M., Tominaga, M., and Sasaki, H. (1988). *Biomed. Res.* **9**, Suppl. 1, 90.
Unger, R. S., Eisentraut, A. M., McCall, M. S., and Madison, L. L. (1961). *J. Clin. Invest.* **40**, 1280–1289.
Uttenthal, L. O., Ghiglione, M., George, S. K., Bishop, A. E., Polak, J. M., and Bloom, S. R. (1985). *J. Clin. Endocrinol. Metab.* **61**, 472–479.
Valverde, I. (1983). *In* "Handbook of Experimental Pharmacology" (P. J. Lefèbvre, ed.), Vol. 66/I, pp. 223–244. Springer-Verlag, Berlin.
Valverde, I., and Rovira, A. (1983). *In* "Diabetes, Obesity and Hyperlipidemias" (G. Crepaldi, P. J. Lefebvre, and D. J. Galton, eds.), Vol. 2, pp. 289–296. Academic Press, London.
Valverde, I., Villanueva, M. L., Lozano, I., and Marco, J. (1974). *J. Clin. Endocrinol. Metab.* **39**, 1090–1098.
Weir, G. C., Turner, R. C., and Martin, D. B. (1975). *J. Clin. Endocrinol. Metab.* **40**, 296–302.
Yalow, R. S. (1974). *Recent Prog. Horm. Res.* **30**, 597–610.
Yalow, R. S., and Berson, S. A. (1960). *J. Clin. Invest.* **39**, 1157–1174.
Yanaihara, C., Matsumoto, T., Nishida, T., Uchida, T., Kobayashi, S., Moody, A. J., Orci, L., and Yanaihara, N. (1984). *Biomed. Res.* **5**, Suppl., 19–32.
Yoshioka, N., Kuzuya, T., Matsuda, A., Taniguchi, M., and Iwamoto, Y. (1988). *Diabetologia* **31**, 355–360.

9

Why Use a Flow Cytometer for Endocrine Cell Analysis?

Frank M. Perez, Daniel R. Deaver, and Wesley C. Hymer

I. Introduction
II. Capabilities of a Flow Cytometer
III. Identification of Pituitary Cells by Flow Cytometric Immunofluorescence
   A. Principles
   B. Flow Cytometric Analysis
   C. Data Analysis
   D. Experimental Considerations
IV. Sorting of Pituitary Cells by Light-Scatter Properties
   A. Principles
   B. Cell Processing
   C. Flow Cytometric Cell Analysis
   D. Cell Sorting by Light Scatter
   E. Data Analysis
   F. Experimental Considerations
V. Sorting of Pituitary Cells by Fluorescence
   A. Principles
   B. Fluorescent-Activated Cell Sorting
   C. Data Analysis
   D. Experimental Considerations
VI. Concluding Remarks
   References

I. Introduction

A chapter on flow cytometry may seem somewhat tangential in a handbook which focuses primarily on biochemical and molecular techniques applied to endocrine research. However, regardless of whether one measures hormone concentration, receptor number or ligand binding, "second messengers," mRNA expression, or gene regulation, each parameter relates ultimately to the cell. It is at the level of the cell where the flow cytometer shows its strength and versatility as an analytical research tool.

What has become increasingly apparent to those who study cells is the remarkable heterogeneity that exists within a given population. As an example, somatotrophs [growth hormone (GH)-producing cells] are composed of a variety of subpopulations that differ in both morphological and functional characteristics. These include buoyant density, ultrastructural appearance, light-scatter characteristics, rate of protein synthesis, response to secretagogues, ratios of bioactive to immunoreactive GH stored and released, relative amounts of stored and released GH variants, and the ability to support somatic growth upon implantation into hypophysectomized animals (Hopkins and Farquhar, 1973; Snyder *et al.*, 1977; Hymer and Wilbur, 1980; Frawley and Neill, 1984; Perez and Hymer, 1990; Tietien *et al.*, 1990; Farrington and Hymer, 1990). Similar heterogeneity has been found also for other adenohypophyseal cell types, as well as other endocrine cells such as the β-cells of the pancreas (Salomon and Meda, 1980; Bosco and Meda, 1991).

In considering cell heterogeneity as it relates to other physiological processes, we feel that the flow cytometer provides endocrinologists with an opportunity to integrate cell analysis with biochemical and molecular studies. For example, measurements of hormone concentration, receptor distribution, mRNA expression, or gene regulation provide data only on the total cell population. However, flow cytometric analysis gives additional information on the cells which constitute the population. This kind of analysis provides greater insight into the interrelationships of regulatory mechanisms which govern the physiological state.

Although flow cytometers have been used extensively in the field of cell biology, their full potential is far from being realized in the area of endocrinology. This underutilization of the flow cytometer may be attributed to practical considerations such as cost constraints and complexity of operation. In addition, it seems likely that the instruments' limited application stems from almost universal acceptance of endocrinologists to analyze cells by conventional microscopic methods. We hope that by presenting a brief description of the analytical features of a flow cytometer and selected examples of its application, our colleagues will be more inclined to use this apparatus in their research.

II. Capabilities of a Flow Cytometer

Equipped with laser light sources, appropriate detectors, and other instrument hardware, a flow cytometer is capable of simultaneously measuring and correlating up to six different cellular parameters. For example, the instrument can measure cell light-scatter properties, intracellular calcium and pH, membrane potential, DNA, RNA protein and lipid content, nuclear and cell surface antigens, enzyme activity, and organelle distribution (Crissman and Steinkamp, 1973; Shapiro, 1981a,b; Miller and Whitlock, 1981; Johnson et al., 1982; Grynkiewicz et al., 1985; Greenspan et al., 1985; Muscrove and Hedley, 1990). Such measurements are obviously useful for studying the cell cycle, cell proliferation, transformation and heterogeneity, signal transduction, and cytogenetics.

Flow cytometric cell analysis is impressively rapid (on the order of thousands of cells per minute) and precise. Data that are obtained from a flow cytometer are objective, quantitative, and reproducible. These characteristics offer obvious advantages over conventional microscopic methods of cell analysis which are often time consuming, labor intensive, and subject to investigator bias. The speed of a flow cytometer makes it ideally suited for studies that involve large numbers of cells. By analyzing thousands or even tens-of-thousands of cells, an investigator benefits from more reliable data and improved test statistics.

In addition to its analytical capabilities, a flow cytometer can sort cells on the basis of one or more of the parameters mentioned above. This feature permits isolation of specific cells contained in a mixed population. When used with selective staining procedures, flow cytometric cell sorting is capable of enriching specific, viable cell types to almost complete purity (St. John et al., 1986; Wynick et al., 1990).

We now describe how this instrument is used successfully to tackle the problems of identification, analysis, and isolation of specific cell types contained in a heterogeneous population.

III. Identification of Pituitary Cells by Flow Cytometric Immunofluorescence

A. Principles

Figure 1 presents a simplified diagram of how a flow cytometer analyzes cells. A suspension of individual cells is injected into a chamber (flow cell) in which fluid is flowing. This moving fluid sheath vertically aligns the cells as they pass through a small orifice which is located at the bottom of the flow cell. As the cells intersect the laser beam, they scatter light which is detected by photomultiplier tubes (PMTs) or photodiodes. When coupled with specific filters, these devices detect and amplify light emissions from cells which contain specific laser-excitable dyes. The PMTs transfer signals to a computer which is responsible for data analysis, display, and storage.

With conventional immunofluorescence-staining methods, specific cell

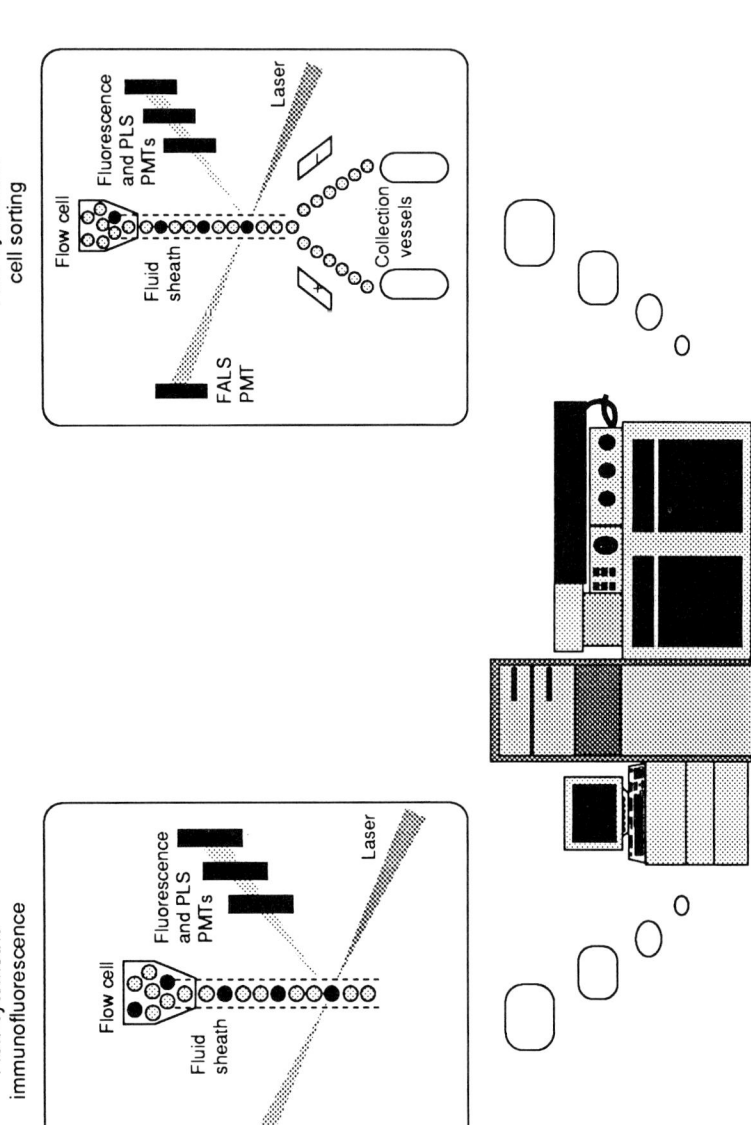

Figure 1 Diagram illustrating how a flow cytometer analyzes and sorts cells. PMT, photomultiplier tube; FALS, forward angle light scatter; PLS, perpendicular angle light scatter; solid circles represent cells containing a fluorophore.

types contained in a mixed population are easily identified and analyzed on a flow cytometer. The number of parameters which can be studied simultaneously will depend on the laser source (single or dual beam) and excitation wavelength of the fluorophores. For example, with a single beam argon–ion laser and appropriate mirrors and fluorophores, it is possible to monitor up to three distinct immunological parameters (Delia et al., 1991). When used in conjunction with light-scatter profiles, these measurements provide a wealth of information about cell-type distribution, size, staining intensity, and granularity. Some of the more popular fluorophores that are used for these purposes are shown in Table I. For a more comprehensive listing of fluorophores, refer to Waggoner (1990).

The protocols listed below permit identification and analysis of GH cells contained in the rat and bovine adenohypophyses. These procedures can be modified easily for other antigens.

CELL PROCESSING

This tissue-dissociation method (Hymer et al., 1972) produces a high cell yield (2.5×10^6 cells/pituitary gland from a male Sprague–Dawley rat aged 50 days) and excellent cell viability ($>95\%$).

1. Collect glands in Eagle's minimum essential medium with spinner culture salts (SMEM, pH 7.3).
2. Mince tissue into small blocks (1 mm^3).
3. Transfer tissue fragments into a spinner flask that contains SMEM with 0.3% bovine serum albumin (BSA) (w/v), 0.3% trypsin (w/v), and 100 µl of DNase (2 mg/ml), 110 U/ml penicillin, and 100 µg/ml streptomycin.
4. Stir tissues for two hours at 37°C with periodic mechanical trituration using a flame-tapered Pasteur pipette.
5. Wash cells once by centrifugation (100g for 5 min).
6. Resuspend cells in SMEM containing 0.1% BSA (w/v) and soybean trypsin inhibitor (0.1 mg/ml).
7. Filter cells through a 45-µm nylon mesh prior to use.

STAINING PROTOCOL

The following method (Hatfield and Hymer, 1985) is used for identifying rat or bovine GH cells in a mixed cell population:

C. Data Analysis

Figure 2 shows the typical appearance of bovine pituitary cells after staining for GH immunofluorescence; this is what a flow cytometer "sees" during cell analysis. Bright green (FITC) fluorescent staining (open arrows) is shown only in the cytosol of cells which contain GH. The red stain (closed arrows), which is due to DNA staining by propidium iodine, is found in the nuclei of both immunopositive and -negative cells. Only particles containing the red stain are counted; this criterion excludes debris and red blood cells

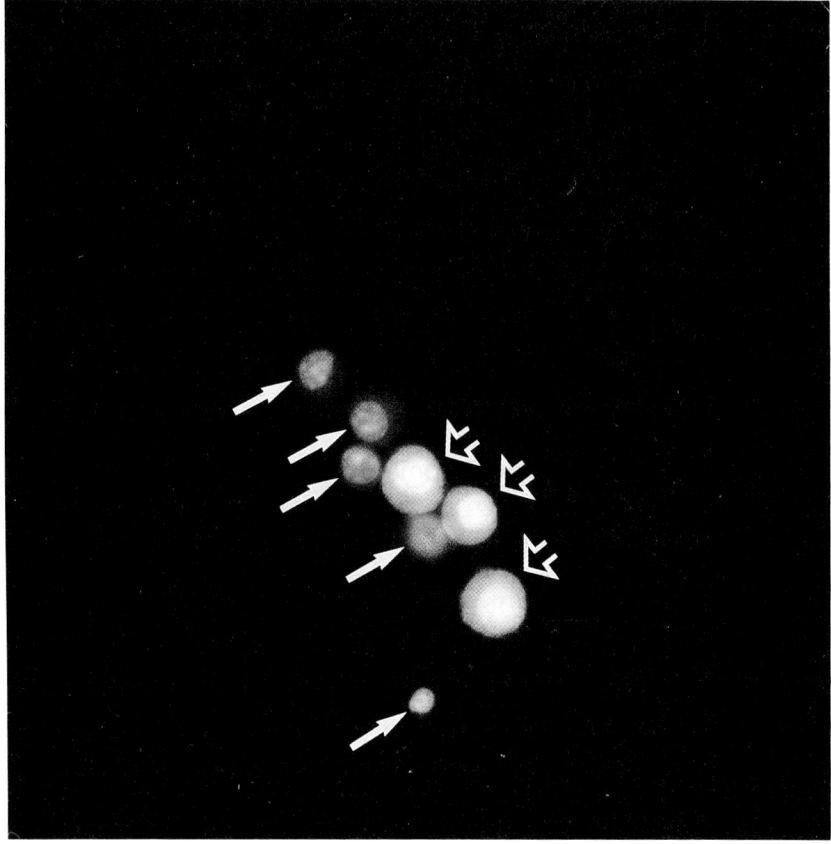

Figure 2 Light micrograph of bovine somatotrophs stained for GH immunofluorescence. A single cell suspension is obtained with the Cell Processing protocol described under Section III. Cells are incubated with rabbit anti-bovine GH (1:10,000) followed by staining with FITC and propidium iodine.

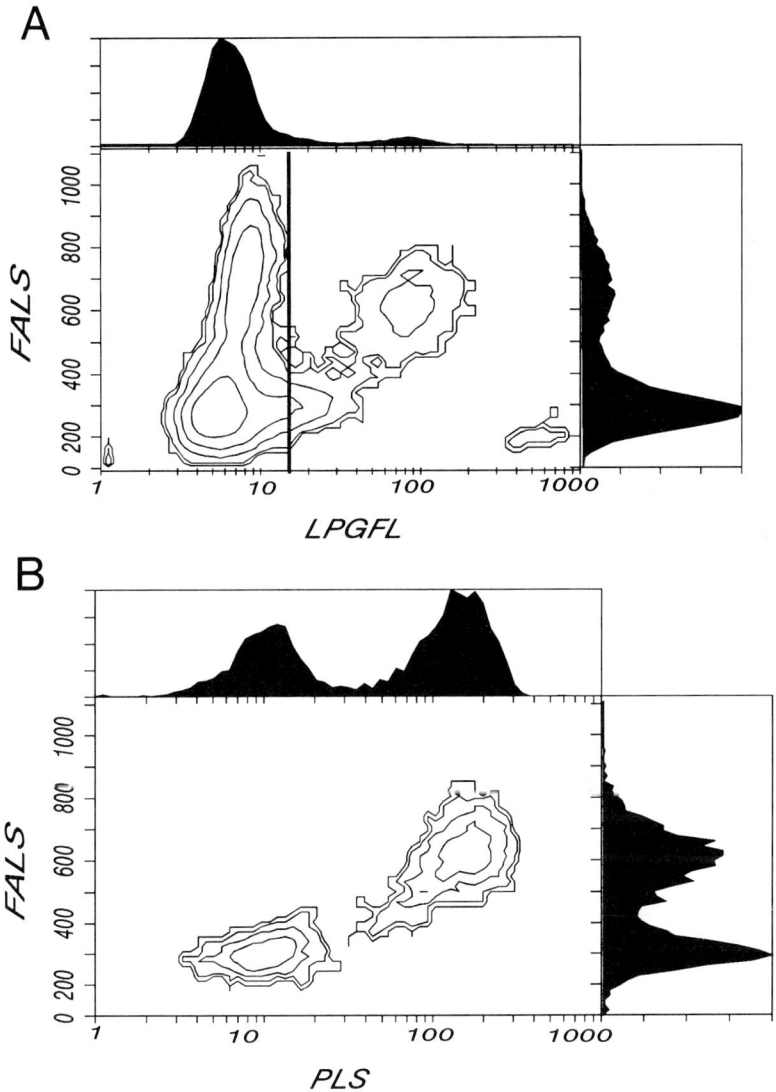

Figure 3 Bivariant frequency distribution of bovine adenohypophyseal cells. Following trypsinization, cells (4.0 × 10^5) are stained for GH immunofluorescence and then analyzed on the flow cytometer. (A) The distribution of all pituitary cells; (B) GH-positive cells only. All data are shown in the form of contour plots and two-dimensional histograms. Values for FALS signals are on a linear scale; those for perpendicular light scatter (PLS) are logarithmic.

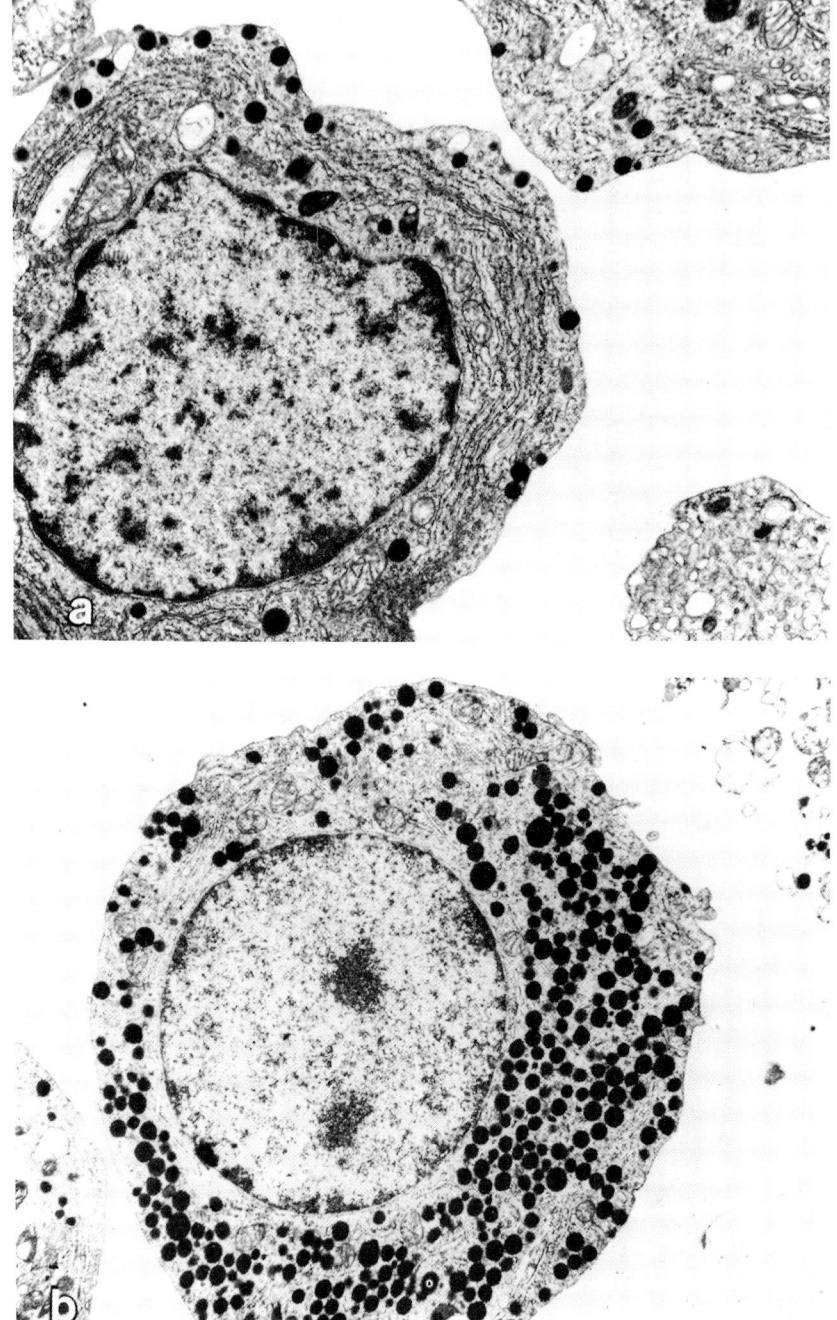

which commonly occur in sample preparations. Flow cytometric analysis of cells such as these reveals a frequency distribution which is shown in Fig. 3. An initial analysis of these data indicates that approximately 16% of the total cell population is immunopositive for GH. To further analyze this population, a gate (vertical line, Fig. 3A) is set so that only the immunopositive GH cells are identified. This procedure reveals two distinct subpopulations of somatotrophs (Fig. 3B) that differ on the basis of their size [forward angle light scatter (FALS)] and granularity [perpendicular light scatter (PLS)]. These subpopulations represent the Type I and II somatotrophs (Fig. 4) which have been reported for the rat (Snyder *et al.*, 1977).

D. Experimental Considerations

Flow cytometric immunofluorescence methods offer several advantages over conventional light-microscopic techniques in terms of cell identification. These include acquisition of objective, quantitative data, rapid cell analysis, and improvements in test statistics. The limitations of the techniques described above are that they are performed on fixed, dead cells and may fail to detect cells with limited intracellular stores of antigen. In addition, data on fluorescence-staining intensity are useful for identifying relative differences in the immunopositive cell population. However, caution should be used when interpreting these data; a decrease in staining intensity is suggestive of cell degranulation, but not proof.

When preparing cells for flow cytometry, it is important that the tissue-dispersion method produce individual cells; clusters of cells tend to clog the orifice. Several tissue-dissociation methods which use proteolytic enzymes such as trypsin or collagenase produce suitable cell suspensions. Cell yields from these methods should be considered since some cells (about 25%) are invariably lost during the staining procedure. To our knowledge there are no data indicating preferential cell loss during staining or analysis on the flow cytometer. We have found that a minimum of 1.0×10^5 cells in the initial suspension (prior to staining) are adequate for analysis on the flow cytometer.

The type of fixative and duration and temperature of fixation should be determined empirically since these factors affect antigenicity. After fixation, cells can be processed immediately or stored at 5°C. With adenohypophyseal

---

Figure 4 Subpopulations of somatotrophs obtained from the male rat adenohypophysis. Following trypsinization of tissue, the cell suspension is layered on a BSA density gradient and centrifuged. Type I somatotrophs (a) are found in the upper gradient region (1.065–1.070 g/cm^3); Type II cells (b) occur in a denser gradient region (1.072–1.085 g/cm^3).

cells, storage for up to 2 weeks has no detectable effect on immunofluorescent staining. The conditions for incubation of antisera are best determined experimentally. The duration of incubation of antisera will vary; increasing the temperature accelerates antibody–antigen interactions. Although pituitary cells are usually incubated with primary antiserum overnight at 5°C, a 3-hr incubation at room temperature produces adequate cell staining.

Supplements in the buffer have specific purposes and are used routinely. Sodium azide reduces cell loss and prevents cell clumping (Hatfield and Hymer, 1986a); Triton X-100 makes cells permeable to antibodies.

Background staining and instrument noise are problems which may be encountered when using immunofluorescence methods. Nonspecific binding of antibodies is the primary cause of background staining. This problem may be reduced as follows: (a) increase the concentration of BSA in the buffer, (b) increase the number of washes and/or extend the wash periods after incubation of antisera, (c) incubate cells overnight with nonimmune serum (5.0% in buffer) obtained by the host animal that supplies the secondary antibody, and (d) affinity purify antisera. Background noise results from cell autofluorescence which is attributed to the presence of flavins and other coenzymes found in cells (Benson et al., 1979). In the case of adenohypophyseal cells, autofluorescence is negligable (Hatfield and Hymer, 1986a). For other cells, this problem may be reduced by using appropriate filters that select for the emission maxima of a fluorophore.

In situations where the antigen concentration is low and a weak fluorescence signal is produced, the high-affinity binding properties of avidin–biotin reagents may improve signal amplification.

Controls for the staining procedure include replacement of the primary antibody with nonimmune serum, staining with propidium iodine only, and measurement of autofluorescence in cells that are not reacted with antibodies or the nuclear stain.

Each fluorophore has several spectral and physical characteristics that should be considered when used alone or in combination with other probes. First, the absorption wavelength of the fluorophore should match closely that of the light source. In the case of fluorescein, for example, its absorption maxima (495 nm) is suitable for an argon–ion laser line of 488 nm. Second, the selected fluorophore should have a high extinction coefficient. This refers to the capacity of the probe to absorb light. Third, the fluorophore should have a high quantum yield. This characteristic reflects the efficiency at which the probe emits light (i.e., its fluorescence). Fourth, differences between the absorption and emission wavelengths (referred to as the Stokes shift) of a probe should be large enough so that the weaker fluorescence light is discernible from the stronger excitation light. Finally, fluorophores may be sensitive to changes in pH and quenched by the presence of other fluorescent molecules (Loken, 1990).

Probes such as those listed in Table I can be used in a variety of combinations, thus permitting simultaneous measurement of several specific parameters. For example, fluorescein and phycoerythrin are ideally suited for detecting two distinct cellular antigens (Oi et al., 1982; Loken, 1986). Both probes are excited by an argon–ion laser line of 488 nm and each emits light of sufficiently separate wavelengths (520 and 576 nm, respectively). These fluorophores have been used with 7-aminocoumarin to obtain three-color immunofluorescence with a single argon–ion laser (Delia et al., 1991). Although multiple-label immunofluorescence can be a powerful research tool, it should be emphasized that the level of difficulty encountered during staining and subsequent cell analysis increases with each additional fluorophore.

IV. Sorting of Pituitary Cells by Light-Scatter Properties

A. Principles

Specific viable cells contained in a mixed population can be physically isolated with great precision by using the sort function of a flow cytometer (Fig. 1). This function is based on the principle that cells in a fluid stream, after passing through the laser beam, are captured in droplets which break off at the end of the stream. Each droplet containing a cell of interest (as determined by a previously set sort gate) is given an electrical charge. As the droplet passes between two electrically charged plates, it is deflected into a collection vessel. Three distinct cell types can be collected simultaneously depending upon whether the droplet is charged positively or negatively, or remains neutral. Each droplet can be delivered accurately on to the surface of a microscope slide or to a variety of collection vessels such as tissue culture tubes, flasks, petri dishes, or microwell plates.

Cells have been sorted on the basis of their fluorescence staining (see Section V, for an example), light absorption and intrinsic light scattering properties. This last application is based on the principal that as a cell passes through a laser beam, it scatters light which is detected and measured accurately. FALS (4° to 19°) is proportional to cell size (Mullaney et al., 1969); whereas, 70° to 110° scatter (referred to as perpendicular, orthogonal, or 90° light scatter) is related to intracellular structure (Salzman et al., 1975). This includes the ratio of nuclear to cytoplasmic diameters, nuclear shape, and the distribution of secretory granules and other organelles (Brunsting and Mullaney, 1974; Meyer and Brunsting, 1975; Salzman et al., 1975; Kerker et al.,

1979; Benson *et al.*, 1984). In the case of adenohypophyseal cells such as the somatotroph, secretory granules are the predominate intracellular structure affecting 90° light scatter.

Cell sorting on the basis of light scatter has been used for (a) separating blood cells into lymphocyte, monocyte, and granulocyte fractions, (b) isolating bone marrow neutrophils, (c) enriching mouse pancreatic B-cells, (d) monitoring rat granulosa cell size and granularity, (e) separating basal and secretory rat tracheal epithelial cells, and (f) enriching for rat GH, prolactin (PRL) and adrenocorticotrophic hormone (ACTH) cells (Salzman *et al.*, 1975; Watt *et al.*, 1980; Nielson *et al.*, 1982; Hatfield and Hymer, 1986a; Johnson *et al.*, 1990; Rao *et al.*, 1991). This last application is described below and has been used for monitoring pituitary cells under conditions which alter the endocrine status of female rats (Hatfield and Hymer, 1986b).

B. Cell Processing

A suspension of single adenohypophyseal cells is obtained according to the protocol, Cell Processing, described under Section III.

C. Flow Cytometric Cell Analysis

We developed the present application by using an EPICS V flow cytometer (Coulter Electronics, Hialeah, FL) equipped with an argon–ion laser beam. To prepare the instrument for cell analysis, the laser is tuned to 488 nm at 150 mW of power. A 1.5-A neutral-density filter is installed for detecting FALS. Since adenohypophyseal cell autofluorescence is negligible, no filters are required for measuring the 90° signal. A three-decade logarithmic amplifier is used to accommodate the large range of perpendicular light-scatter signals that are generated by these cells. A pulse width time of flight circuit (50–50% pulse width, 7 μsec delay, range setting of 3) can be used for particle size and gating of the FALS and orthogonal signals. Fluorospheres (10.3 μm) are used to standardize the instrument. Forward and perpendicular angle light scatter positions of these spheres are adjusted to channels 107 and 147, respectively.

D. Cell Sorting by Light Scatter

To prepare the flow cytometer for cell sorting, a quartz flow body with a 76-μm orifice is installed so that droplet formation occurs beyond the point where the laser beam intersects the flow stream. If droplets form prior to

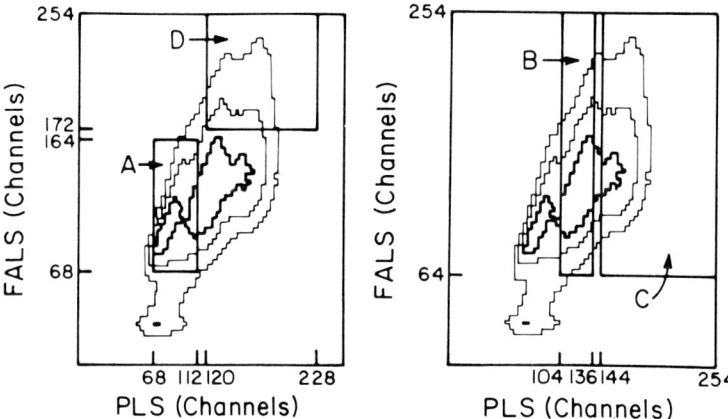

Figure 5 Selected regions of a light-scatter frequency distribution that are used for adenohypophyseal cell sorting. Cells contained in regions A and D (left side) or B and C (right side) are sorted concurrently. The solid line indicates the limits of each sort gate.

intersecting the beam, the light which is scattered by the droplet could overwhelm the signal produced by the cell. Establishing the proper position of droplet formation is discussed below under Section F, Experimental Considerations.

A bivariant frequency distribution showing the placement of sorting gates is illustrated in Fig. 5. All data are analyzed with a standard multiparametric data acquisition and display system (MDADS) supplied by Coulter. From these data, the operator selects the limits of a gate (solid lines) and uses the analytical feature of MDADS for determining the frequency of particles contained within each region (designated A, B, C, and D). Cells from regions A and D are sorted (2500 cells/sec) concurrently by using an exclusion sort logic for both FALS and perpendicular light scatter. This procedure ensures that no parameters overlap. Cells contained in regions B and C are then sorted concurrently by using an exclusion sort logic for perpendicular light scatter and an identical logic (overlapping parameters) for FALS.

E. Data Analysis

Figure 6 shows a bivariant frequency distribution of live male rat adenohypophyseal cells before and after cell sorting. Three distinct regions (designated a, b, and c) of adenohypophyseal cells are noted in the unfractionated sample. Sorting of these cells according to gates defined in Fig. 6 resolves each region into separate cell populations. Flow cytometric immunofluorescence

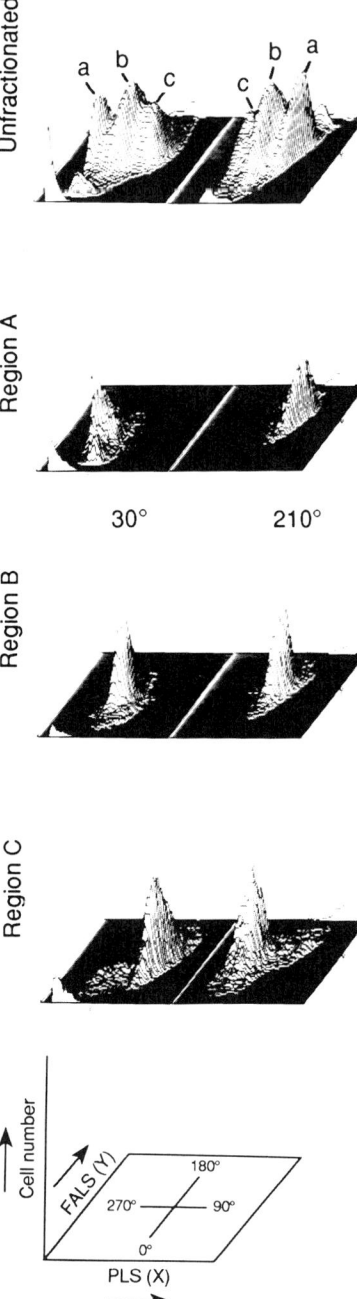

analysis of cells contained in each region reveals that GH cells are enriched (2.4-fold) in region C. Prolactin and ACTH cells are enriched (1.7- and 2.5-fold, respectively) in region B. Sort region A has over 65% nonstaining cells while region D (not shown) contains the highest concentration of luteinizing hormone (LH) and thyrotropin (TSH) cells (Hatfield and Hymer, 1986b).

F. Experimental Considerations

Sorting on the basis of light-scatter properties is a useful technique for isolating subpopulations of viable (>90%) cells that differ by as little as 5% in their FALS and orthogonal signals. These physical characteristics could be used for isolating specific cell types which have undergone hypertrophy and degranulation as the result of changes in their physiological states. In addition, because the cells in this procedure are viable, vital dyes can be used for investigating other parameters such as intracellular Ca^{2+}, pH, and membrane potential. Cell sorting by flow cytometry may be impractical in situations where the frequency of the desired cell is low (<20%) or where several millions of cells are required for experimentation.

After weighing the pros and cons and deciding in favor of cell sorting, the user needs to establish optimum operational conditions. Factors to consider include sort efficiency, cell recovery, and cell purity (these characteristics are discussed in detail elsewhere (Lindmo *et al.,* 1990). Briefly, sort efficiency refers to the frequency of occurrence of a selected cell in the sorted droplet. For example, efficiency decreases as the number of sorted droplets which contain unselected cells increase. Cell recovery refers to the ratio of cells that are sorted versus the number of sorted events detected by the instrument. Cell purity refers to the percentage of selected cells contained in the sort fraction. This characteristic is ascertained by sorting a small number of cells directly onto the surface of a microscope slide for examination.

Undoubtedly, the single most important factor which influences sort efficiency and cell recovery and purity is the position of the cell during droplet formation (i.e., droplet break-off point). For optimum sorting, the cell should enter the droplet at the break-off point. Adjustment of this point is accomplished by varying the frequency and amplitude of the vibration that is applied

Figure 6 Three-dimensional histograms of viable pituitary cells before and after sorting. Adenohypophyseal cells (7.0×10^5) are sorted according to the gates shown in Fig. 5. Each separated region (designated A, B, and C) corresponds to those (a, b, and c) that are shown in the unfractionated cell sample. Histograms are presented in two orientations; conventional points of reference are illustrated in the diagram.

to the flow cell by the piezoelectric transducer. Once the droplet break-off point is established, it should remain constant and be checked often during the sort. Alteration of the break-off point is caused most commonly by obstruction of cells in the flow cell orifice. The occurrence of this problem rises as the sort time increases because cells in suspension tend to form clumps. Although filtration through a nylon mesh removes these clusters, it is wise to have a second flow cell on hand.

V. Sorting of Pituitary Cells by Fluorescence

A. Principles

Specific cell types contained in a heterogeneous population can be enriched to purities of greater than 90% with fluorescent-activated cell sorting (FACS). Unlike the immunofluorescence-staining technique described under Section III, which uses fixed, dead cells, this technique usually involves applying fluorescent probes to living cells. These probes can be attached covalently to polynucleotides, lipids, and proteins such as ligands that bind to receptors or antibodies. In addition, fluorescent labels can associate noncovalently with molecules contained in cells. These vital dyes elucidate DNA, RNA, and lipid composition, changes in Ca^{2+}, pH, or membrane potential, and structural properties of organelles.

Flourescence-activated cell sorting has been used in the field of immunology for isolating cells which express specific surface antigens (Lanier et al., 1983). This application has proved rewarding in the area of endocrinology where viable PRL and GH cells have been enriched to almost complete purity on the basis of surface hormone staining (St. John et al., 1986; Wynick et al., 1990). A method for obtaining a highly purified population of rat PRL cells is presented below (St. John et al., 1986).

CELL PROCESSING

1. Collect glands in Ca^{2+}- and Mg^{2+}-free Hanks' balanced salt solution (CMF–HBSS, pH 7.3) containing 25 mM Hepes and BSA (4 mg/ml), and mince into small pieces.
2. Incubate (36°C with agitation) tissue fragments in CMF–HBSS containing collagenase (2 mg/ml) for an initial 30 min.
3. Incubate tissues in CMF–HBSS containing collagenase (2 mg/ml), trypsin (1 mg/ml), and deoxyribonuclease I (1 mg/ml) for an additional 30 min.

4. Rinse cells by centrifugation once in CMF–HBSS.
5. Resuspend cells in CMF–HBSS with Ca^{2+} and Mg^{2+}.
6. Triturate the partially dissociated tissue with a flame-tapered Pastuer pipette.
7. Rinse cells several times by centrifugation.
8. Resuspend cells in fresh HBSS.

To facilitate FACS analysis, red blood cells are removed from the suspension. This is accomplished by layering the cell suspension (10^7 cells/2.0 ml) over a CMF–HBSS/BSA (200 mg/4.0 ml) density gradient and centrifuging (50 g) for 10 min.

STAINING PROTOCOLS

Viable adenohypophyseal cells in suspension are stained on the basis of their surface PRL immunoreactivity according to the following protocol (St. John et al., 1986).

1. Place cells in culture medium and incubate for 2 hr at 37°C; subsequent steps are conducted at 4°C.
2. Centrifuge cells at 300g for 5 min.
3. Resuspend cells ($1-2 \times 10^6$ cells/ml) in buffer containing rabbit anti-rat PRL (1:1000); incubate for 60–120 min.
4. Rinse cells once by centrifugation.
5. Incubate cells with goat anti-rabbit IgG conjugated to FITC (1:50) for 60 min.
6. Rinse cells once by centrifugation.
7. Cells are ready for FACS or storage at 4°C for 4–6 hr.

FITC-stained cells obtained from FACS can be examined further for intracellular antigens by using a rhodamine-conjugated secondary antibody. The staining procedure is as follows.

1. Fix cells with 4% formaldehyde in 100 mM phosphate buffer (pH 7.2) for 30–60 min at 4°C.
2. Rinse cells three times by centrifugation with buffer only.
3. Permeabilize cells by resuspending in buffer that contains 0.5% saponin or 0.25% Triton X-100.
4. Rinse cells once before incubating with primary and secondary antibodies.

B. Fluorescent-Activated Cell Sorting

In the present application cells are analyzed and sorted on a FACS 440 (Becton-Dickinson, Mountain View, CA) equipped with 5-W argon-ion and 2-W krypton-ion lasers. The argon-ion laser is tuned to 488 nm at 400 mW and the krypton laser operates at 568 nm at 200 mV of power. In dual-labeling studies the argon-ion laser is used for FITC (identifying cell surface PRL) excitation; the krypton laser is used for rhodamine (localizing intracellular PRL). A 490-nm long-pass filter is installed for all fluorescence emissions. A half-height mirror is used to separate argon and krypton laser emissions from fluorophores. Fluorescein emissions excited by the argon laser are passed through a 530/30 bandpass filter; emissions from rhodamine excited by the krypton laser are passed through a 625/35 bandpass filter. Light emitted by each fluorophore is collected and measured in separate photomultiplier tubes. The absence of crossover of FITC and rhodamine signals is verified for double-label experiments.

Cells that are not sorted are analyzed on the flow cytometer at rates of 2000–6000 particles/sec; a rate of 2500 particles/sec is used for sorting. Sort gates, set on FALS and fluorescence intensity, include intact cells and separate positive from negative cells.

All data obtained from light scatter and fluorescence intensity profiles are analyzed on a PDP 11/23-based computer (Consort 40, Becton-Dickinson) and a VAX 750 (Digital Equipment Corp., Marlboro, MA).

C. Data Analysis

On the basis of FACS analysis, 29% of rat adenohypophyseal cells stain positive for cell surface PRL (Fig. 7). In this group 85–99% of the cells (viability >90%) stain positive for intracellular PRL while only 2% contain other hormones (GH, ACTH, LH, or TSH). Staining for surface PRL accounts for about half of all cells that stain positive for intracellular PRL (St. John *et al.*, 1986).

D. Experimental Considerations

The advantages and drawbacks of cell sorting that were discussed under Section IV,F, apply also for FACS. Factors which affect sort conditions such

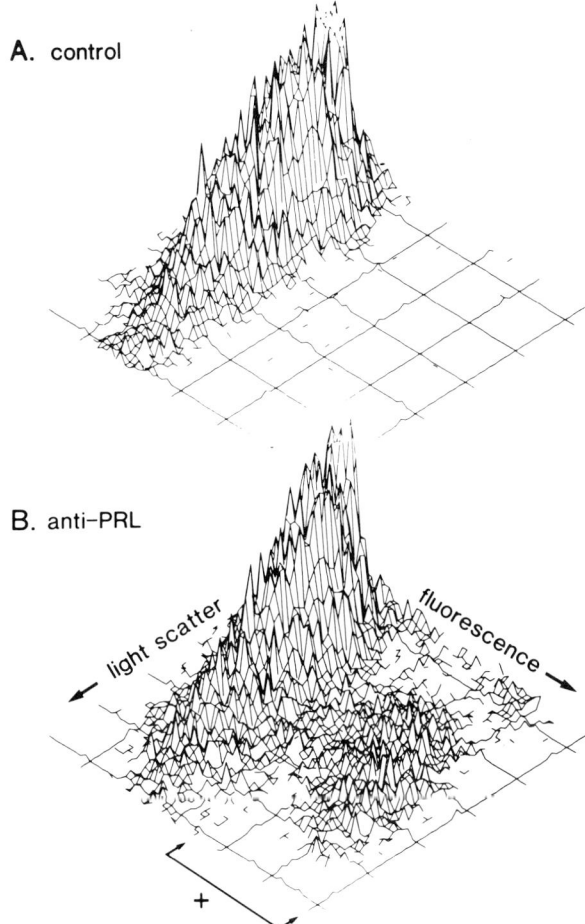

Figure 7 Fluorescent-activated cell sorting of live adenohypophyseal cells that are stained for cell surface PRL immunoreactivity. (A) Cells incubated with the fluorescent secondary antibody only (control). (B) Cells incubated with anti-PRL followed by the secondary antibody. Both figures are presented as hidden-line histograms that show FALS (linear scale) vs fluorescence-staining intensity (logarithmic scale). Arrows indicate increasing values for each signal. Cells contained in the region marked by + are positive for surface PRL immunoreactivity. (Courtesy of St. John et al., 1986.)

as droplet formation and obstruction of the flow cell orifice should be considered also for the present application.

Fluorescent-activated cell sorting on the basis of surface labeling is capable of isolating specific cells to almost complete purity (St. John *et al.*, 1986; Wynick *et al.*, 1990). However, this method has certain limitations. In some cases, antibodies that are used for immunocytochemistry will not recognize cell surface antigens. In other instances, antibodies may select only for a subpopulation of cells that contain immunoreactive surface antigens. Consequently, cells which do not express surface antigens will be missed in the analysis. Similarly, fluorescent-labeled ligands identify only cells that express the appropriate receptor, these cells do not necessarily constitute the entire population (Edwards *et al.*, 1983).

In addition to sorting cells on the basis of surface labeling, FACS can be used for isolating cells that contain vital dyes. Ideally, these probes should not interfere with the parameter of interest or disturb normal functions in cells. In practice, some fluorophores, by making cells photosensitive, may inflict damage during laser beam excitation. However, most vital stains are sufficiently nonintrusive so as to provide reliable quantitative data on a variety of cellular parameters.

VI. Concluding Remarks

Cell heterogeneity is one example of the complexity which constitutes the physiological state. If we as endocrinologists are to contribute to an understanding of this state, then it is important to integrate information obtained at the cellular, biochemical, and molecular levels. This can be accomplished, in part, by complementing biochemical and molecular studies with cell analysis by flow cytometry. Such interdisciplinary studies have the potential of broadening our perspectives on general physiology and of opening new avenues of investigation.

Acknowledgments

The authors thank Elaine Kunze for her expert technical assistance. This work was supported by the following grants: PHS CA-23248 (W.C.H.), NASA NCC 2-370 (W.C.H.), NASA NAG 8-807 (W.C.H.), NASA NAGW-1196 (W.C.H.), HATCH PROJECT 2961 (D.R.D.),

HATCH PROJECT 3113 (D.R.D.), USDA 88-37240-3912 (D.R.D.), NSF R11-9018700 (F.M.P.), PHS HD04358-3251 (F.M.P.).

References

Arndt-Jovin, D. J., and Jovin, T. M. (1977). *J. Histochem. Cytochem.* **25**, 585.
Benson, M. C., Meyer, R. A., Zaruba, M. E., and McKhann, G. M. J. (1979). *J. Histochem. Cytochem.* **27**, 44.
Benson, M. C., McDougal, D. C., and Coffey, D. S. (1984). *Cytometry* **5**, 515.
Bosco, D., and Meda, P. (1991). *Endocrinology (Baltimore)* **129**, 3157.
Brunsting, A., and Mullaney, P. F. (1974). *Biophys. J.* **14**, 439.
Crissman, H. A., and Steinkamp, J. A. (1973). *J. Cell Biol.* **59**, 766.
Crissman, H. A., Mullaney, P. F., and Steinkamp, J. A. (1975). *Methods Cell Biol.* **9**, 179.
Darzynkiewicz, Z., Kapuscinski, J., Traganos, F., and Crissman, H. A. (1987). *Cytometry* **8**, 138.
Delia, D., Martinez, E., Fontanella, E., and Aiello, A. (1991). *Cytometry* **12**, 537.
Edwards, S. A., Trotter, J., Rivier, J., and Vale, W. (1983). *Mol. Cell. Endocrinol.* **30**, 21.
Farrington, M., and Hymer, W. C. (1990). *Endocrinology (Baltimore)* **126**, 1630.
Frawley, L. S., and Neill, J. D. (1984). *Neuroendocrinology* **39**, 484.
Greenspan, P., Mayer, E. P., and Fowler, S. D. (1985). *J. Cell Biol.* **100**, 965.
Grynkiewicz, P., Poenie, M., and Tsien, R. Y. (1985). *J. Biol. Chem.* **260**, 3440.
Hatfield, J. M., and Hymer, W. C. (1985). *Cytometry* **6**, 137.
Hatfield, J. M., and Hymer, W. C. (1986a). *Endocrinology (Baltimore)* **119**, 2670.
Hatfield, J. M., and Hymer, W. C. (1986b). *Endocrinology (Baltimore)* **119**, 2683.
Haugland, R. P. (1983). In "Excited States of Biopolymers" (R. F. Steiner, ed.), p. 29. Plenum, New York.
Hopkins, C. R., and Farquhar, M. (1973). *J. Cell Biol.* **59**, 276.
Hymer, W. C., and Wilbur, D. L. (1980). In "Functional Correlates of Hormone Receptors in Reproduction" (V. B. Mahesh, T. G. Muldoon, B. B. Saxena, and W. A. Sadler, eds.), p. 13. Elsevier/North-Holland, Amsterdam.
Hymer, W. C., Kraicer, J., Bencosme, S. A., and Haskill, J. S. (1972). *Proc. Soc. Exp. Biol. Med.* **141**, 966.
Johnson, L. V., Walsh, M. L., and Chen, L. B. (1980). *Proc. Natl. Acad. Sci. U.S.A* **77**, 990.
Johnson, L. V., Walsh, M. L., Bockus, B. J., and Chen, L. B. (1982). *J. Cell Biol.* **88**, 526.
Johnson, N. S., Wilson, J. S., Habbersettm, R., Thomassen, D. G., Shopp, G. M., and Smith, D. M. (1990). *Cytometry* **11**, 395.
Kerker, M., Chew, H., McNutty, P. J., Kratohvil, J. P., Cooke, D. D., Sculley, M., and Lee, M. P. (1979). *J. Histochem. Cytochem.* **27**, 250.
Lanier, L. L., Engleman, E. G., Gatenby, P., Babcock, G. F., Warner, N. L., and Herenberg, L. A. (1983). *Immunol. Rev.* **74**, 143.
Lindmo, T., Peters, D. C., and Sweet, R. G. (1990). In "Flow Cytometry and Sorting" (M. R. Melamed, T. Lindmo, and M. L. Mendelsohn, eds.), p. 145. Wiley-Liss, New York.
Loken, M. R. (1986). In "Methods in Hematology" (P. Beverly, ed.), p. 132. Churchill-Livingstone, London.
Loken, M. R. (1990). In "Flow Cytometry and Sorting" (M. R. Melamed, T. Lindmo, and M. L. Mendelsohn, eds.), p. 341. Wiley-Liss, New York.
Meyer, R. A., and Brunsting, A. (1975). *Biophys. J.* **15**, 191.
Miller, A. G., and Whitlock, L. (1981). *J. Biol. Chem.* **256**, 2433.

Mullaney, F. P., Van Dilla, M. A., Coulter, J. R., and Dean, P. N. (1969). *Rev. Sci. Instrum.* **40**, 1029.
Muscrove, E. A., and Hedley, D. W. (1990). *Methods Cell Biol.* **33**, 59.
Nielson, O., Larsen, L. K., Christensen, I. J., and Lernmark, A. (1982). *Cytometry* **3**, 177.
Oi, V., Glazer, A. N., and Stryer, L. (1982). *J. Cell Biol.* **93**, 981.
Perez, F. M., and Hymer, W. C. (1990). *Endocrinology (Baltimore)* **127**, 1877.
Rao, I. M., Allsbrook, W. C., Conway, B. A., Martinez, J. E., Beck, J. R., Pantazis, C. G., Mills, T. M., Anderson, E., and Mahesh, V. B. (1991). *J. Reprod. Fertil.* **91**, 521.
Salomon, D., and Meda, P. (1980). *Exp. Cell Res.* **162**, 507.
Salzman, G. C., Crowell, J. M., Martin, J. C., Trujillo, B. A., Romereo, A., Mullaney, P. F., and LaBauve, P. M. (1975). *Acta Cytol.* **19**, 374.
Shapiro, H. M. (1981a). *Cytometry* **1**, 301.
Shapiro, H. M. (1981b). *Cytometry* **2**, 143.
Snyder, G., Hymer, W. C., and Snyder, J. (1977). *Endocrinology (Baltimore)* **101**, 788.
St. John, P. A., Dufy-Barbe, L., and Barker, J. L. (1986). *Endocrinology (Baltimore)* **119**, 2783.
Tietien, G. H., Grindeland, R. E., Hymer, W. C., and Vasques, M. (1990). *Program Annu. Meet. Endocr. Soc., 71st* Abstr. 785.
Titus, J. A., Haugland, R. P., Sharrow, S. O., and Segal, D. M. (1982). *J. Immunol. Methods* **50**, 193.
Waggoner, A. S. (1990). *In* "Flow Cytometry and Sorting" (M. R. Melamed, T. Lindmo, and M. L. Mendelsohn, eds.), p. 209. Wiley-Liss, New York.
Watt, S. M., Burgess, A. W., Metcalf, D., and Battye, F. L. (1980). *J. Histochem. Cytochem.* **28**, 934.
Wynick, D., Critchley, R., Ventikou, M. S., Burrin, J. M., and Bloom, S. R. (1990). *J. Endocrinol.* **126**, 269.

10

Identification and Characterization of Insulin-like Growth Factor-Binding Proteins

Yvonne W-H. Yang and Matthew M. Rechler

I. Introduction
II. Detection
   A. Binding of ^{125}I-IGF to IGFBPs in Solution
   B. Ligand Blotting
III. Identification of IGFBPs
   A. Size, Glycosylation, Chromatographic Properties, and Binding Specificity
   B. Immunological Identification
   C. Identification of IGFBP mRNAs
   D. Quantitation of IGFBPs
IV. Conclusions
   References

I. Introduction

The insulin-like growth factors, IGF-I and IGF-II, are homologous to insulin and have many properties in common (Rechler and Nissley, 1990). They bind to IGF-I receptors which, like insulin receptors, are heterotetramers that contain a cytoplasmic tyrosine kinase domain. Both IGFs and insulin exhibit insulin-like and mitogenic activity *in vitro* (Rechler and Nissley, 1990). One way that the IGFs differ from insulin, however, is that they occur in plasma, extravascular fluids, and cell culture media complexed to a family of proteins, the IGF-binding proteins (IGFBPs). The interaction of IGFs with IGFBPs accounts for the fact that IGFs are present at much higher concentrations than insulin in human plasma (~ 100 nM vs ~ 100 pM), that IGF biological activity and immunologic reactivity are associated with 150- and 50-kDa complexes in human plasma rather than the free 7.5 kDa peptides, and that the half-life of infused radiolabeled IGF-I in the 150 kDa complex is longer (12–15 hr) than that of free IGF-I (10–12 min) (Guler *et al.*, 1989). The

150-kDa complex is thought to prevent clinical hypoglycemia by restricting IGFs to the blood.

For many years, little attention was paid to the possible physiological significance of the IGFBPs. Instead, they received negative press because they interfered with the measurement of IGF-I and IGF-II by radioimmunoassays or radioreceptor assays (Nissley and Rechler, 1984; Daughaday and Rotwein, 1989). False negative results might arise from the inability of IGFs complexed to IGFBPs to compete for the binding of radiolabeled IGF with antibody or receptor. False positive results might arise from radioligand binding to unoccupied IGFBPs and partitioning together with unbound (free) radioligand. To eliminate possible interference by IGFBPs, prior to radioligand assay endogenous IGFs are dissociated from IGFBPs at acid pH and then separated by size or ethanol precipitation. Because of the ubiquitous expression of IGF-I and IGF-II in different tissues, similar strategies are necessary when measuring tissue IGFs.

IGF:IGFBP complexes were thought to be biologically inert because they did not bind to IGF receptors (Knauer and Smith, 1980). This concept was challenged by the provocative observation of Elgin et al. (1987) that an IGFBP purified from amniotic fluid potentiated the mitogenic effects of IGF-I *in vitro*. This report focused attention on the IGFBPs as potential modulators of IGF action.

Since 1988, six members of a family of IGFBPs, designated IGFBP 1–6, have been purified and cloned (Rechler and Nissley, 1990; Rechler and Brown, 1992). Their properties are summarized in Table I (Rechler, 1993). Human IGFBPs have been purified from amniotic fluid (IGFBP-1), cerebrospinal fluid (CSF; IGFBP-2 and IGFBP-6), serum or plasma (IGFBP-2, -3, -4, and -6), bone (IGFBP-5), and media conditioned by osteosarcoma (IGFBP-4), glioblastoma (IGFBP-5), or fibroblast (IGFBP-6) cultures. IGFBP-3 and IGFBP-4 occur in N-glycosylated forms. Human IGFBP-6 is O-glycosylated; it also contains a potential N-glycosylation site that is not used (Bach et al., 1992). IGFBPs 1–5 bind IGF-I and IGF-II with similar affinities. By contrast, IGFBP-6 has a 50-fold preferential affinity for IGF-II over IGF-I. None of the IGFBPs binds insulin.

IGFBP-3 is the N-glycosylated binding subunit of the 150-kDa IGF:IGFBP complex that is the predominant carrier of IGFs in adult plasma.[1] The 150-kDa complex consists of a binary complex of IGF-I (or IGF-II) and IGFBP-3 associated with an acid-labile nonbinding subunit (Rechler, 1993). The acid-labile subunit is present in excess in human plasma, so that most of the IGFBP-3 occurs in the 150-kDa complex. Steric or conformational factors may be important in this interaction, since binary complexes formed from

---

[1]The 150-kDa complex is only rarely found outside plasma, but has been identified in human and pig milk, pig colostrum, and rat lymph (Rechler, 1993).

Table I
Properties of Human IGFBPs

IGFBP	Source	Molecular mass[a]	Gel filtration	Molecular mass[b]	Glycosylation[c]	Binding specificity[d]	mRNA[e]
IGFBP-1	Amniotic fluid	25.3	50	30	None	I = II	1.55
IGFBP-2	CSF, serum	31.4	50	34	None	I = II	1.38
IGFBP-3	Plasma	28.7	150/50	38–42	N-linked	I = II	2.5
IGFBP-4	Serum, osteosarcoma cells	26.0	50	24	None	I = II	2.6
				28[f]	N-linked[f]		
IGFBP-5	Bone, glioblastoma cells	28.5	50	30–31	ND[g]	II > I	6
IGFBP-6	CSF, serum, human fibroblasts	22.8	50	28–32	O-linked	II ≫ I	1.3

[a] Molecular masses (in kDa) deduced from the cDNA sequences (Eechler, 1993).
[b] Apparent molecular masses (in kDa) calculated from ligand blots after SDS–polyacrylamide gel electrophoresis in the absence of reducing agents.
[c] Glycosylation determined by enzymatic deglycosylation. A potential N-glycosylation site in IGFBP-6 is not used.
[d] I = II, affinities for IGF-I and IGF-II within 1- to 3-fold; II > I, 5- to 10-fold higher affinity for IGF-II than IGF-I; II ≫ I, > 50-fold higher affinity for IGF-II than IGF-I.
[e] Size in kilobases.
[f] The predominant IGFBP-4 in osteosarcoma cell-conditioned medium is the nonglycosylated 24-kDa protein. The minor 28-kDa protein has been identified as the N-glycosylated form of the 24-kDa by ligand blotting.
[g] Not determined.

certain IGF molecules (such as partially processed IGF-II precursors in non-islet cell tumor hypoglycemia) (Baxter and Daughaday, 1991) bind poorly to the acid-labile subunit and do not form the 150-kDa complex.

The other IGFBPs form complexes with IGF-I (or IGF-II) but do not associate with other protein subunits. They appear in the 50-kDa region after gel filtration at neutral pH and predominate in all fluids except adult plasma. Of these, IGFBP-1 has been most extensively characterized (Rechler and Nissley, 1990; Rechler, 1993). It can be phosphorylated intracellularly on serine residues, increasing its affinity for IGF-I and possibly making the complex inhibitory; nonphosphorylated IGFBP-1 may potentiate IGF action by mechanisms that are presently poorly understood (Clemmons, 1992). IGFBP-1 is present at low concentrations in human plasma, but increases dramatically in pregnancy, fasting, and diabetes, and decreases rapidly after birth, feeding, and insulin treatment. The physiological significance of this regulation is unclear.

It is essential to define the role of the IGFBPs to understand how the IGF system functions in different physiological and pathological conditions. This chapter describes basic methodologies used (a) to detect the presence of IGFBPs, (b) to identify which IGFBPs are present, and (c) to quantitate their concentrations. Initially IGFBPs were recognized by their ability to bind IGF-I and IGF-II and by their molecular masses. Similar methods have been used to study other hormones and their binding proteins. For example, activated charcoal was used to separate free ligand from complexed ligand in steroid radioimmunoassays (Smith and Sestili, 1980), and growth hormone-binding proteins have been identified by gel filtration (Herington et al., 1986). IGFBPs also may be identified immunologically. Antibodies to IGFBP-1 and IGFBP-3 have been available for some time, and antibodies are now becoming available for the other IGFBPs. Hybridization probes also are available to identify which IGFBP mRNAs are present in tissues and cultured cells.

II. Detection

A. Binding of ^{125}I-IGF to IGFBPs in Solution

1. Overview

One of the first clues to the existence of IGFBPs occurred from the screening of culture media or biological fluids for IGF activity by radioimmuno- or radioreceptor assay following gel filtration at neutral pH. Inhibition of IGF tracer binding to antibody or receptor frequently was seen in fractions of higher molecular weight than those of IGF-I or IGF-II. These fractions contained IGFBPs that formed complexes with IGF tracer preventing it from binding to antibodies or receptors.

The presence of IGFBPs in column fractions or unknown samples was demonstrated directly by specific binding of ^{125}I-IGF-I or ^{125}I-IGF-II. In untreated samples, only those IGFBPs containing IGF-binding sites that were not occupied by endogenous IGF-I or IGF-II (or that readily exchanged with tracer) were identified. (For example, IGF-I tracer did not bind to the 150-kDa IGFBP complex in normal adult rat serum because it is saturated with endogenous IGF-I (Yang et al., 1989).) Total IGFBPs may be detected after dissociation of IGFs from IGF:IGFBP complexes at acid pH (see below).

Samples containing native or acid-stripped IGFBPs are incubated in solution with IGF tracer at pH 6–7.4, typically overnight at 4°C, to achieve steady-state binding. At the end of the incubation, unbound IGF tracer is separated from radiolabeled IGF:IGFBP complexes. This may be accomplished by gel permeation chromatography at neutral pH using Sephadex G-50 or Superdex G-75. To conveniently process multiple samples, unbound IGF may be adsorbed with activated charcoal, or IGF:IGFBP complexes precipitated using polyethylene glycol and γ-globulin carrier protein (Busby et al., 1988).[2] The charcoal separation method is routinely used in our laboratory and is described in the protocol, Charcoal-Separation Solution Assay, at the end of the section. In other laboratories, polyethylene glycol precipitation has been used with equivalent success. Charcoal or polyethylene glycol separations should be validated by comparison with column separations.

2. Dissociation of Endogenous IGFs from IGFBPs

To measure total IGFBPs by binding activity, samples may be treated by acidification to dissociate IGFs from IGFBPs and fractionated by size to obtain IGFBPs uncontaminated by IGFs. This may be accomplished by Sephadex

---

[2]Besides the charcoal-separation solution assay described in Protocol 1, alternative procedures also have been used to separate unbound and bound ^{125}I-IGF. (a) Protamine sulfate (0.02%, w/v) has been included with bovine serum albumin (0.25%, w/v) in the buffer used to suspend the charcoal (0.5%, w/v) (Martin and Baxter, 1986). Protamine sulfate had different effects on the adsorption of complexes formed from different IGFBPs (Conover et al., 1989). (b) Concanavalin A (0.2%, w/v) in combination with polyethylene glycol 6000 (8000) (17%, w/v) has been used to precipitate complexes of ^{125}I-IGF and glycosylated IGFBP-3 (Martin and Baxter, 1986). Wheat germ agglutinin was used in place of concanavalin A for O-glycosylated IGFBP-6. (c) ^{125}I-IGF:IGFBP complexes also have been precipitated using antisera raised against specific IGFBPs followed by second antibody and polyethylene glycol 6000 (8000) (6%, w/v) (Martin and Baxter, 1986).

G-50 gel filtration at acid pH (acetic acid, pH < 4).[3] Samples may be acidified by adding acetic acid (2 or 17.4 M) to a final concentration of 1 M before gel filtration. The elution volume of free tracer should be determined at the outset and confirmed after a series of column runs to make sure that the calibration has remained constant. Aliquots of fractions from the column void volume to the IGF tracer region are neutralized or spin-vacuum-dried[4] and resuspended in neutral buffer, then assayed for IGF-binding activity in a charcoal-separation assay. Since IGFs interfere with the measurement of IGFBPs in the solution assay, care should be taken not to include fractions containing IGFs in the IGFBP pool.

3. Separation by Column Chromatography at Neutral pH

Samples are incubated with IGF tracer (e.g., 100,000 cpm, 4°C overnight) and applied to a gel permeation column (e.g., Sephadex G-50) equilibrated with neutral buffer containing 0.15 M NaCl and supplemented with 1 mg/ml bovine serum albumin (BSA) (Romanus et al., 1986). If unoccupied IGF-binding sites are present, some of the IGF tracer radioactivity forms complexes and shifts from 7.5 kDa to a higher molecular mass (30–50 kDa). The specificity of this shift may be demonstrated by showing that it does not occur when excess unlabeled IGF is included in the incubation. Alternatively, IGF:IGFBP complexes may be recovered, made 1 M in acetic acid, and gel-filtered at acid pH (e.g., 1 M acetic acid); all of the tracer radioactivity should appear as free IGF-I or IGF-II.

4. Charcoal-Separation Solution Assay

This method was developed for the separation of bound and free radioligand in the radioimmunoassay of steroids and was adapted for the measurement of IGFBPs by Zapf et al. (1975) and modified by Moses et al. (1979). The separation is based on differential adsorption of molecules of different size. The protocol used in our laboratory is shown in Protocol 1. This method uses fatty acid-free bovine serum albumin in the buffers for binding incubation and charcoal separation since it increased the specific binding of IGF-II tracer to normal rat serum and decreased nonspecific binding (Moses et al., 1979). Unbound IGF is adsorbed to charcoal; IGF radioactivity remaining

---

[3] Alternatively, Ultrogel AcA 54 (IBF Biotechnics Inc., Columbia, MD) or FPLC/HPLC has been used for acid gel filtration. Rapid methods have been used to dissociate IGFs from IGFBPs such as Sep-Pak (C18 reverse phase, Millipore Corp., Bedford, MA) or acid-ethanol extraction. It has not been established whether these procedures yield IGFBP preparations that are uncontaminated with IGFs.

[4] Savant Speed-Vac Concentrator, Savant Instruments, Inc., Hicksville, NY.

in the charcoal supernate represents IGF:IGFBP complexes. Validity of the separation should be established for individual samples by comparison with column separations. Two problems are possible. First, IGFs may be stripped from the complex by charcoal treatment, as reported for diabetic rat serum by Unterman et al. (1990). Bach, however, has been unable to confirm this observation with human IGFBP-1 or other IGFBPs.[5] Second, complexes of IGFBPs (especially smaller proteins and IGFBP fragments) may adsorb to activated charcoal. Bach has estimated this loss as 10–20% for IGFBPs and other proteins of 20–100 kDa. Partial adsorption of IGFBPs to charcoal may limit the use of charcoal assays for quantitation of IGF-binding capacity (see below), but should not interfere with estimates of relative IGF-binding affinities.

Initial experiments to detect whether IGFBPs are present in a given sample are performed with fixed amounts of tracer radioactivity and different concentrations of sample. At low fractional occupancy, B/F increases linearly with the concentration of IGFBP: $(B/F) = [\text{complex}]/[\text{free IGF}] = K_a[\text{IGFBP}_{\text{total}}] - K_a[\text{complex}] \simeq K_a[\text{IGFBP}_{\text{total}}])$. At higher concentrations of complex, the dissociation reaction becomes significant, and binding approaches a steady state. The concentration of IGFBP that gives half-maximal binding is used in competitive binding studies (see below).

5. Affinity Crosslinking

Solution assays measure ^{125}I-IGF-binding activity without identifying the responsible IGF-binding species. Wilkins and D'Ercole (1985) used affinity crosslinking techniques developed to study ligand:receptor interactions to identify the sizes of IGF:IGFBP complexes. Disuccinimidyl suberate was used to covalently crosslink the ^{125}I-IGF:IGFBP complexes, with the size of the complexes determined after sodium dodecyl sulfate (SDS)–polyacrylamide gel electrophoresis.

Results of affinity crosslinking may be difficult to interpret. Multiple crosslinking events may occur, and the crosslinked products may not be completely linearized, so that their electrophoretic mobilities may not be directly related to molecular mass. Despite these limitations, the presence of 150-kDa complexes strongly suggests that ternary complexes of IGFBP-3:IGF:acid-labile subunit are present in the sample. Crosslinked complexes also may be immunoprecipitated and examined by gel electrophoresis.

---

[5]Bach (manuscript in preparation) compared radiolabeled noncovalent complexes with covalently crosslinked complexes (from which noncovalent complexes had been separated by acid gel filtration) using the charcoal assay. If stripping of the noncovalent complex had occurred, radioactivity in the charcoal supernate would have been lower for samples that had not been crosslinked than for the crosslinked samples. This was not seen.

PROTOCOL 1: CHARCOAL-SEPARATION SOLUTION ASSAY

Materials

1. *Tubes.* Conical polystyrene tubes (4.5 ml, 12 × 75 mm); Sarstedt, Inc. (Newton, NC; catalog No. 57.477).
2. *Bovine serum albumin (BSA).* Two preparations of BSA (essentially fatty acid free) from Sigma Chemical Co., Catalog Nos. A-7511 (crystallized and lyophilized) and A-6003, have been used. The quality of the fatty acid-free BSA may vary from lot to lot, so that pretesting before purchase is highly recommended.
3. *Tracers:* ^{125}I-*IGF-I* and ^{125}I-*IGF-II*. Specific activity ~267 Ci/g; Amersham Corp. (Arlington Heights, IL).
4. *Activated charcoal suspension.*
    a. Dulbecco's phosphate-buffered saline (PBS, pH 7.4, without Ca^{2+} or Mg^{2+})
    b. 20 mg/ml BSA, fatty acid free
    c. 5% (w/v) activated charcoal (untreated; Sigma, C-5260)
    d. 0.02% (w/v) sodium azide

    Charcoal suspension should be prepared at least 24 hr before use and can be stored at 4°C for at least several months. Charcoal is suspended at 4°C by gentle mixing using a magnetic stirrer; vigorous mixing should be avoided.
5. *Binding protein assay buffer.*
    a. Dulbecco's PBS (pH 7.4, without Ca^{2+} or Mg^{2+})
    b. 2 mg/ml BSA, fatty acid free
    c. 0.02% (w/v) sodium azide

    The buffer is filtered (0.45 μm) and stored at 4°C.

Procedures (Moses et al., 1979)

1. Samples containing IGFBPs are incubated with ^{125}I-IGF (~20,000 cpm per tube) in binding protein assay buffer (final volume 0.4 ml/tube) overnight at 4°C. Control samples that lack added IGFBPs are used to determine nonspecific binding. A positive control sample is included in each assay (e.g., 5 μl of adult rat serum with ^{125}I-IGF-II) to monitor day-to-day consistency.
2. Bound and unbound ^{125}I-IGF are separated by addition of a suspension of activated charcoal (0.5 ml/tube), vortex mixing, incubation on ice (10 min), and centrifugation (TJ-6, Beckman, Palo Alto, CA; 3000 rpm, ~1300g; 30 min; 4°C).
3. The charcoal supernate containing ^{125}I-IGF:IGFBP complexes is recovered, and radioactivity quantitated using a gamma counter.

Comment

Nonspecific binding is defined as the radioactivity present in the charcoal supernate in the presence of excess of unlabeled IGF (i.e., nonsaturable sites). Empirically we have observed that this value is the same as radioactivity in the charcoal supernate following incubation without added IGFBP (with or without unlabeled IGF). Nonspecific binding is typically $\leq 5-10\%$ of input radioactivity.

B. Ligand Blotting

Hossenlopp et al. (1986) described a powerful technique to identify denatured IGF-binding subunits according to their different molecular sizes. In this procedure, called ligand blotting, the IGFBPs are dissociated and separated from endogenous IGFs and the acid-labile subunit by denaturation with SDS under nonreducing conditions. (IGF-binding activity would be destroyed by disulfide reduction in the presence of denaturing agents). Proteins in the denatured samples are fractionated according to size by SDS–polyacrylamide gel electrophoresis and electrophoretically transferred onto nitrocellulose membranes. Among the proteins retained on the membrane, the IGFBPs can be identified by their ability to bind ^{125}I-IGF and visualized as specific protein bands after autoradiography. The ligand blotting procedure is present in Protocol 2 at the end of the section.

The ligand blotting technique has proven extremely powerful in identifying and characterizing IGFBPs. It obviates the need for acid dissociation of endogenous IGFs. Only small samples sizes are needed. By visualizing IGF-binding proteins without the complexities of association with ligand or the acid-labile subunit, their electrophoretic mobilities become a signature. For example, the two to four bands in the 38- to 42-kDa region seen with normal adult serum (Fig. 1, lanes 1 and 5) are larger than other IGFBPs. They represent IGFBP-3 N-glycosylation variants and are readily identified. Additional information may be obtained by enzymatically deglycosylating IGFBPs prior to ligand blotting. For example, treatment of rat serum with N-glycanase reduces the glycosylation variants of IGFBP-3 to a single protein of 34 kDa (Fig. 1, lanes 1 and 2 and lanes 5 and 6.). The protocols for enzymatic deglycosylation are given below (Protocols 3 and 4). In addition, IGFBPs may be immunoprecipitated with specific antibodies and the immunoprecipitates solubilized and examined by ligand blotting (see Section IIIB).

Despite its great utility, ligand blotting also has its limitations. First, because of its preferential affinity for IGF-II, IGFBP-6 might be missed if incubation were performed only with IGF-I tracer (Fig. 1). Second, partially proteolyzed IGFBP-3 in pregnancy serum remains immunoreactive and binds

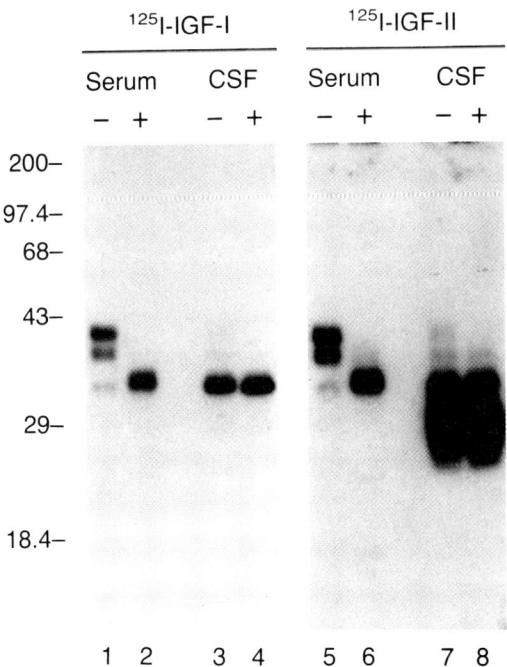

Figure 1 Ligand blot of human serum and cerebrospinal fluid (CSF). Human serum (1 μl) or cerebrospinal fluid (40 μl) was incubated without (−) or with (+) N-glycanase (2 Units/tube) before SDS–polyacrylamide gel electrophoresis. After electrotransfer of the fractionated proteins, the membrane was incubated with ^{125}I-IGF-I (lanes 1–4) or ^{125}I-IGF-II (lanes 5–8) and autoradiographed. The positions of ^{14}C-labeled molecular mass markers (in the presence of 0.1 M dithiothreitol) are indicated in kilodaltons on the left.

native IGF-I but does not bind iodo-IGF-I, so that it is not visualized by ligand blotting (Suikkari and Baxter, 1991; Suikkari and Baxter, 1992). Similarly, proteolytic fragments of IGFBP-2 (McCusker *et al.*, 1991), IGFBP-4 (Fowlkes and Freemark, 1992), and IGFBP-5 (Jones *et al.*, 1993) were detected by immunoblotting but not by ligand blotting.

PROTOCOL 2: LIGAND BLOTTING

Materials

1. *Apparatus for SDS–polyacrylamide gel electrophoresis.* Commercially available units of various sizes may be used. In our laboratory, we have used an Aquebogue Model 200 (gel size: 13.5 × 16.5 × .12 cm; Aquebogue Machine & Repair Shop Inc., Aquebogue, NY) and a Novex XCELL mini-cell and blot module (gel size: 6.5 × 8 × 0.15 cm; Novel Experimental Technology, San Diego, CA).

2. *Electrophoretic blotting apparatus.* Model TE52, Hoefer Scientific Instruments (San Francisco, CA).
3. *^{14}C-Labeled protein molecular mass markers.* (Bethesda Research Laboratories, Gaithersburg, MD).
4. *Electroblot transfer buffer.* Composed of 20 mM Tris base – 150 mM glycine – 20% (v/v) methanol.
5. *Bovine serum albumin.* RIA grade (Sigma Chemical Corp., St. Louis, MO).
6. *Nonidet P40 (NP-40).* (Fluka Chemical Corp., Ronkonkoma, NY.)
7. *Tween-20 (polyoxyethylenesorbitan monolaurate).* (Sigma Chemical Corp.)
8. *TS buffer.* Composed of 10 mM Tris – Cl, 150 mM NaCl, pH 7.4.
9. *Nitrocellulose membrane.* Pore size 0.2 μm. 15 × 15 cm, (BA83, Schleicher and Schuell, Inc., Keene, NH) or 8 × 7.5 cm (LC2001, Novel Experimental Technology, San Diego, CA).

Procedures

The ligand-blotting procedure was originally described by Hossenlopp *et al.* (1986) and modified by Yang *et al.* (1989).

1. Electrophoresis of *IGFBPs and other proteins on SDS – 10% polyacrylamide gels (Laemmli, 1970).* Conditions are chosen to resolve proteins of 10 – 50 kDa. The number of samples loaded onto each gel (and the width of each lane) may vary, being determined in part by volume of the sample (e.g., maximal loading volume for Aquebogue model, ≤ 80 μl for a 12-lane gel; Novex model, ≤ 50 μl for a 10-lane gel). When necessary, samples are concentrated by ultrafiltration using centricon 10 micro concentrators (Amicon, Beverly, MA). Samples should be prepared in the absence of reducing agent since disulfide reduction interferes with the ability of IGFBPs to bind IGFs. Molecular-mass markers are prepared in the presence of reducing agent (e.g., dithiothreitol) to ensure linearization of the protein. At least one blank lane should separate the marker from the closest sample to avoid diffusion of the reducing agent and skewing of the bands. Samples should be heated to 95° C for 5 min (or 60° C for 10 min) prior to loading. The approximate times for running these gels are 3.5 hr and 2 hr for the Aquebogue and Novex models, respectively.
2. *Electrophoretic transfer of proteins onto nitrocellulose membranes.* The nitrocellulose membrane and the gel are immersed in transfer buffer on a rocking platform for 5 – 10 min before assembly of the electrophoretic blotting apparatus. Electrotransfer is carried out at room temperature using the same transfer buffer. Electrophoresis is

at 70 V overnight, with cooling by circulating tap water. (Transfer is accomplished in 2 hr at 30 V for Minigels.) At the end of the transfer, the gel may be stained with Coomassie brilliant blue R-250 to detect any untransferred proteins.

3. *Washing.* Membranes are washed sequentially in the following buffers on a rocking platform at 4°C, 200–500 ml/wash (100 ml for Minigels).

   a. TS buffer–0.5 mg/ml sodium azide–3% (v/v) NP-40; 30 min, once.

   b. TS buffer–1% (w/v) BSA (RIA grade); 2 hr, once.

   c. TS buffer–0.1% (v/v) Tween 20; 20 min, twice.

4. *Incubation with ^{125}I-IGF.* Membranes are incubated overnight at 4°C with ^{125}I-IGF (400,000 cpm) in 20 ml of buffer [TS buffer supplemented with 1% (w/v) BSA (RIA grade) and 0.1% (v/v) Tween 20] per membrane (10 ml for Minigels). The signal is typically greater with ^{125}I-IGF-II than with ^{125}I-IGF-I.

5. *Washing.* Membranes are washed as described in Step 3.

   a. TS buffer–0.1% (v/v) Tween 20; 15 min, twice.

   b. TS buffer; 15 min, two to three times.

6. *Autoradiography.* The membrane is blotted dry with filter paper, sealed in a plastic bag, and exposed to Kodak X-Omatic AR film at −70°C in a Kodak X-Omatic cassette containing regular intensifying screens (Eastman–Kodak Co., Rochester, NY). Typical exposure times vary from 1 day to 1 week.

Comment

1. *Samples.* Each gel should include suitable standards with known IGFBPs (such as serum). It may be necessary to concentrate dilute samples (e.g., using Centricon ultrafiltration) before electrophoresis. Care should be taken to avoid protein overloading (resulting in distortion of the lane), especially with serum samples (e.g., use 1 μl human serum/lane of a large gel). Even though samples are not reduced, molecular-mass marker proteins should be reduced to give accurate calibration.

2. *Electroblotting.* After electroblotting SDS gels of fractionated rat serum, ≥95% of the stained proteins are transferred to nitrocellulose. The untransferred proteins are ≥200 kDa.

3. *Choice of tracer.* IGFBPs 1–5 bind IGF-I and IGF-II with similar affinity, so that either tracer may be used (although IGF-II tracer usually gives a stronger signal). Since IGFBP-6 has >50-fold higher

affinity for IGF-II than IGF-I, it may not be visualized using IGF-I tracer, but may only be seen with IGF-II tracer. For example, in the ligand blot of human cerebrospinal fluid shown in Fig. 1, a major broad band of 28–32 kDa IGFBP-6 was appreciated with IGF-II tracer (lanes 7 and 8) but was not seen when a parallel blot was incubated with IGF-I tracer (lanes 3 and 4).

III. Identification of IGFBPs

A. Size, Glycosylation, Chromatographic Properties, and Binding Specificity

Although definitive identification of IGFBPs requires recognition by specific antibodies, protein sequence, or mRNA hybridization, the IGFBPs may be distinguished and presumptively identified based on their physical, chemical, and binding properties (Table I).

1. Presence of 150-kDa IGF:IGFBP Complexes

Identification of 150-kDa IGF:IGFBP complexes in samples fractionated on gel permeation columns at neutral pH indicates that IGFBP-3 (or 29-kDa NH_2-terminal fragments of IGFBP-3) is present (Fig. 2). The other five IGFBPs do not bind to the acid-labile subunit and are not found in the 150-kDa region. Samples are fractionated on a suitable gel permeation column (e.g., Sephadex G-200, Sephacryl S-200 or S-300, Ultrogel AcA 44, Superose-12, or Superadex G-75) equilibrated at neutral pH. Columns may be calibrated with proteins of known ansd appropriate molecular weight or with serum.

The 150-kDa IGF:IGFBP complexes may be identified in several ways. First, samples may be incubated with IGF tracer before application to the column to identify the unoccupied IGF-binding sites: 150 kDa in normal adult rat serum (Fig. 2), 50 kDa in adult human serum (White et al., 1981). Second, aliquots of column fractions may be examined for IGF-binding activity in the charcoal-separation assay. With normal human serum, binding of IGF tracer to IGFBPs in 150-kDa fractions only is seen after separation from the excess 50-kDa unoccupied binding sites by gel filtration; this binding may require exchange of tracer with endogenous IGFs (White et al., 1981). In other cases, it may be necessary to dissociate endogenous IGFs (e.g., by acid gel filtration) before assay.[6] Third, fractions in the 150- and 50-kDa regions after neutral gel filtration also may be examined by ligand blotting (see below).

---

[6]After neutral gel filtration, fractions containing IGFBPs may be concentrated by Centricon ultrafiltrationor lyophilization prior to application to the acid gel filtration column. After acid gel filtration, IGF-binding activity may be lost after lyophilization, so that ultrafiltration is preferable (Yang et al., 1989).

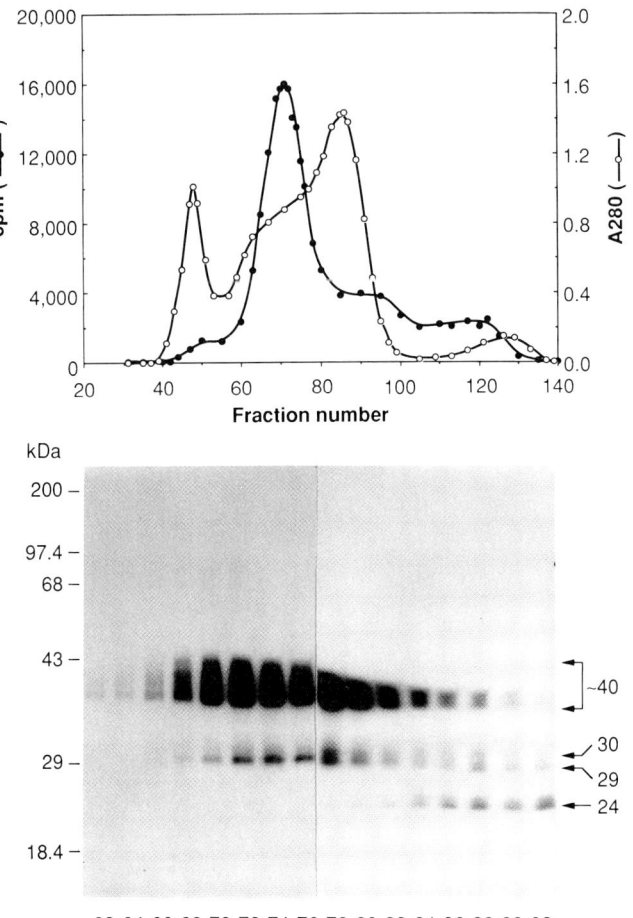

Figure 2 Ligand blot of IGFBPs in adult rat serum fractionated on Sephadex G-200 at neutral pH. (Top) Adult rat serum (1 ml) was incubated with ^{125}I-IGF-II (200,000 cpm) and applied to a Sephadex G-200 column (1.6 × 90 cm; bed volume, 130 ml) in 0.05 M boric acid–NaOH, 0.12 M NaCl, and 0.02% (w/v) sodium azide (pH 8.5) at 4°C. The profiles of ^{125}I-IGF-II binding (solid circle) and absorbance at 280 nm (open circle) are plotted. The absorbance peak at fractions 45–55 corresponds to macroglobulins, the shoulder at fractions 55–75 to γ-globulins, and the peak from fractions 80–95 to albumin. Fractions 60–80 correspond to the 150-kDa region, and fractions 82–100 to the 50-kDa region. (Bottom) Aliquots of alternate fractions were concentrated, using Centricon 30 Microconcentrators, and examined by ligand blotting. Aliquots of 100 μl were examined for fractions 62–76, and aliquots of 200 μl for fractions 78–92. The two blots were exposed at −70°C for the same length of time, and the photographs were placed side by side. Native IGFBP-3 (38–42 kDa) and truncated IGFBP-3 (29 kDa) are seen in the 150-kDa region. The 24-kDa IGFBP in the 50-kDa region probably represents IGFBP-4. The identity of the 29- to 30-kDa proteins in the 50-kDa region is not known. (From Yang et al., 1989, with permission.)

2. Identification of IGFBPs by Ligand Blotting, Enzymatic Deglycosylation, and Binding Specificity

The analytical power of ligand blotting may be enhanced (a) by analyzing samples that have been fractionated by neutral gel filtration into 150- and 50-kDa IGFBPs, (b) following enzymatic deglycosylation of N-linked and O-linked oligosaccharides, and (c) using IGF-I and IGF-II tracers.

Apparent molecular mass presumptively identifies N-glycosylated IGFBP-3 and nonglycosylated IGFBP-4. IGFBP-3 is the largest of the IGFBPs, occurring as two to four N-glycosylated forms of 38–42 kDa. Since these are the major IGFBPs in adult human or rat serum, identification of bands in unknown samples having a similar size provides strong preliminary identification. This identification is strengthened if these proteins are associated with IGFBP complexes in the 150-kDa region of a neutral gel permeation column (Fig. 2). The apparent molecular mass of native IGFBP-3 is greater than the 29-kDa deduced from nucleotide sequence because it contains N-linked oligosaccharides. Enzymatic removal of these carbohydrate chains using N-glycanase (see protocols at the end of this section) or endoglycosidase F significantly reduces the apparent molecular mass of IGFBP-3.[7] A well-resolved 24-kDa nonglycosylated IGFBP in serum probably represents IGFBP-4. IGFBP-4 also occurs as a less-abundant N-glycosylated 28-kDa form.

Resolution of IGFBPs by ligand blotting is most difficult in the ~30-kDa-size range. This may include nonglycosylated IGFBP-1, IGFBP-2[8] and IGFBP-5; an N-glycosylated NH$_2$-terminal 29- to 30-kDa fragment of IGFBP-3; N-glycosylated 28-kDa IGFBP-4; and O-glycosylated 28- to 32-kDa IGFBP-6. (a) Truncated IGFBP-3 may occur in the 150-kDa region of serum (Fig. 2), either because the binary complex of IGF and the 29-kDa IGFBP-3 fragment associates with the acid-labile subunit or because proteolysis of IGFBP-3 occurs within the 150-kDa ternary complex without dissociation of the product. (b) IGFBPs of ~30 kDa whose size is reduced by N-glycanase treatment might be a glycosylated 29-kDa fragment of IGFBP-3 or glycosylated 28-kDa IGFBP-4. These possibilities might be resolved for cultured cells that only express IGFBP-3 or IGFBP-4 mRNA. For example, bovine parathyroid endothelial cells express IGFBP-4 but not IGFBP-3 mRNA (Yang *et al.*, 1993).

IGFBP-6 is significantly reduced in size after O-glycanase treatment (see Protocol 4 at the end of this section (Bach *et al.*, 1992). O-linked oligosaccha-

---

[7] N-Glycanase cleaves the β-aspartylglucosylamine bond between the innermost GlcNAc and the asparagine residue; endoglycosidase F cleaves between the two GlcNAc residues within the oligosaccharide core. Some complex N-linked oligosaccharides may be removed by N-glycanase but not by endoglycosidase F.

[8] In human serum, IGFBP-2 is somewhat larger (34–36 kDa) than IGFBP-1, whereas rat IGFBP-1 and -2 are nearly identical in size (Yang *et al.*, 1989).

rides account for the difference in apparent molecular mass between the 28- to 32-kDa band seen on SDS–polyacrylamide gels and the 23-kDa molecular mass deduced from the nucleotide sequence. In human cerebrospinal fluid, IGFBP-6 appears to occur only as O-glycosylated species (Bach et al., 1992). IGFBP-6 also may be recognized by its strong preferential affinity for IGF-II vs IGF-I (Fig. 1).

IGFBPs also differ in their hydrophobicity, resulting in a distinctive elution profile on reversed-phase high-performance liquid chromatography (HPLC) or FPLC. For example, the IGFBPs in human serum are eluted from reverse-phase HPLC in the order: 30-kDa IGFBP-3 fragment, IGFBP-6, IGFBP-3, 28- and 24-kDa IGFBP-4, followed by IGFBP-2 (Kiefer et al., 1991). The greater hydrophobicity (and later elution) of IGFBP-2 compared to IGFBP-6 also was seen with human cerebrospinal fluid (Bach et al., 1992; Roghani et al., 1991).

PROTOCOL 3: ENZYMATIC DEGLYCOSYLATION OF IGFBPs: N-GLYCANASE

Materials

1. *N-glycanase.* Recombinant N-glycanase (Genzyme Corp., Boston, MA).
2. *1, 10-Phenanthroline hydrate.* (Genzyme Corp.)

Procedures

Adapted from procedures recommended by manufacturer (Yang et al., 1989). A positive control (e.g., human or rat serum) containing N-glycosylated IGFBP-3 should be included in the same experiment.

1. *N-glycanase reaction.* Resuspend samples in 0.55 M sodium phosphate buffer (pH 8.6). Add SDS (final concentration, ~0.17%), 1, 10-phenanthroline (protease inhibitor; final concentration, 10 mM), and NP-40 (final concentration, 1.25%). Samples are not boiled or incubated with reducing agent before incubation with N-glycanase in order to preserve IGF-binding activity. Start reaction by adding N-glycanase (1–2 U/μl serum), and incubate at 37°C for 20–24 hr.
2. *Identification of IGFBPs.* At the end of the incubation, dilute the reaction mixture with an equal volume of SDS–gel sample buffer (Laemmli, 2×), and boil for 5 min. IGFBPs are identified after fractionation by SDS–polyacrylamide gel electrophoresis and ligand blotting (see Protocol 2).

Comment

The concentration of N-glycanase used is higher than that recommended by the manufacturer since the proteins are not reduced and denatured.

PROTOCOL 4: ENZYMATIC DEGLYCOSYLATION OF IGFBPs: O-GLYCANASE

Materials

1. *O-Glycanase.* (Genzyme Corp., Boston, MA)
2. *Neuraminidase.* (Genzyme Corp.)
3. *Fucosidase.* (Boehringer-Mannheim, Indianapolis, IN)
4. *Incubation buffer* 0.2 M sodium phosphate buffer–2 mM calcium acetate, pH 6.6.

Procedures

Adapted from procedures recommended by the manufacturers for the exo- and endoglycosidases (Bach *et al.*, 1992). Suitable positive control (e.g., transferrin) should be included in the same experiment.

1. *Removal of sialic acid and fucose residues.* Resuspend samples (~1 µg) in incubation buffer. Incubate with neuraminidase (final concentration, 50 mU/50 µl) and fucosidase (final concentration, 40 mU/50 µl) for 1 hr at 37°C.
2. *O-Glycanase reaction.* Incubate with O-glycanase (final concentration, 5 mU/50 µl) at 37°C for 18–24 hr.
3. *Identification of IGFBPs.* At the end of the incubation, dilute the reaction mixture with an equal volume of SDS–gel sample buffer (Laemmli, 2×), and boil for 5 min. Treated and untreated IGFBPs are fractionated by SDS–PAGE and examined by ligand blotting (see Protocol 2).

Comment

The concentration of O-glycanase used is higher than that recommended by the manufacturer since the proteins are not reduced and denatured.

B. Immunological Identification

Antibodies raised against specific human IGFBPs allow the immunological identification of IGFBPs in an unknown sample. Sources of antibodies to particular IGFBPs are given in the following references: IGFBP-1 (Rutanen and Pekonen, 1991), IGFBP-2 (Romanus *et al.*, 1986; Clemmons *et al.*, 1991), IGFBP-3 (Baxter and Martin, 1986; Blum *et al.*, 1990), IGFBP-4 (Camacho-Hubner *et al.*, 1992), IGFBP-5 (Camacho-Hubner, *et al.*, 1992), and IGFBP-6 (Baxter and Saunders, 1992). The specific antibodies may be used in radioimmunoassays, immunoprecipitation, or immunoblotting. Protocols for immunoprecipitation and immunoblotting follow (Protocols 5 and 6, respectively). The validity of the interpretation of results using any of these antibodies critically depends on establishing the specificity of the antibody for a particular IGFBP. Antibody interactions are usually independent of ligand binding.

1. Immunoprecipitation

Samples for immunoprecipitation may be radiolabeled (prepared by biosynthetic labeling, *in vitro* translation, iodination, affinity crosslinking to radiolabeled IGFs) or unlabeled (in which case IGFBPs may be identified by ligand-blotting following immunoprecipitation (Lamson *et al.*, 1989)). When radiolabeled samples are used, one may determine radioactivity in the immunoprecipitates (Yang *et al.*, 1989), or the size of the immunoreactive species after SDS–polyacrylamide gel electrophoretic fractionation and autoradiography. Ligand blotting of unlabeled immunoprecipitates (Lamson *et al.*, 1989) is highly specific since it demonstrates that the immunoreactive protein also binds IGFs.

2. Immunoblotting

For immunoblotting, samples containing IGFBPs are fractionated by SDS–polyacrylamide gel electrophoresis and electroblotted to nitrocellulose membranes as described for ligand blotting, but IGFBPs are identified by binding of specific antibodies rather than binding of IGF tracer. Commercial kits are available to detect antibody bound to a protein band on nitrocellulose using coupled enzymes to enhance the sensitivity. The immunoblotting procedure currently used in our laboratory is described in Protocol 6. It identifies IGFBPs immobilized on nitrocellulose membranes using an enhanced chemiluminescence detection system. Membranes are incubated sequentially with (a) primary antibody to IGFBP (mouse monoclonal or rabbit polyclonal), (b) biotinylated second antibody raised in donkeys or goats against immunoglobulins from the same species as the primary antibody (mouse or rabbit), and (c) streptavidin-biotinylated horseradish peroxidase complexes. The specific binding between the primary and secondary antibodies, and between biotin and avidin, direct the horseradish peroxidase to the protein of interest. A peroxidase substrate (luminol) is added, which becomes oxidized by the bound peroxidase, resulting in a sustained and localized emission of light that can be detected on photographic films.

Immunoblotting allows one to identify the size of the immunoreactive proteins. In contrast to ligand blotting, proteins may be fractionated on SDS–polyacrylamide gels under reducing or nonreducing conditions (depending on which form of the protein is better recognized by the antibody). Nonreduced blots may be compared with ligand blots. In fact, the same nonreduced blot used for ligand blot (or immunoblot) may be stripped and reexamined by the other procedure. Proteins may be identified that are immunoreactive but do not bind IGF, or vice versa. For example, in fasted newborn pig serum, immunoblotting with antibodies to IGFBP-2 identified immunoreactive 14- and 22-kDa fragments that did not bind IGF-I; intact 34-kDa IGFBP-2 was not detected (McCusker *et al.*, 1991).

PROTOCOL 5: IMMUNOPRECIPITATION

Materials

1. *Protein A-containing* Staphylococcus aureus. Immunoprecipitin, 10% (w/v) (Bethesda Research Laboratories, Gaithersburg, MD).
2. *Primary antibody.* Polyclonal or monoclonal antibody raised against specific IGFBP. Control, preimmune or nonimmune antibody.
3. *Immunoprecipitation buffer stock solution.*
   a. 100 mM Tris–Cl (pH 7.4)
   b. 150 mM NaCl
   c. 5 mM EDTA (disodium)
   d. 1% (w/v) Sodium deoxycholate
   e. 1% (v/v) Triton X-100

Stock solution should be filtered (0.45 μm) after preparation and stored at 4°C in a sterile container.

4. *Protease inhibitors:* L-1-tosylamido-2-phenylethyl chloromethyl ketone (TPCK), Na-tosyl-L-lysyl chloromethyl ketone (TLCK), and phenylmethylsulfonyl fluoride (PMSF) (Sigma Chemical Co.). Stock solutions (10 mg/ml) of the protease inhibitors are prepared in ethanol and stored at -20°C.

Procedures

1. Add protease inhibitors (TPCK, TLCK, PMSF) to immunoprecipitation buffer [50 μg/ml (w/v), final concentration] immediately before use.
2. Prepare 10% (w/v) protein A-containing *S. aureus*. Wash protein A-containing *S. aureus* three times (500 μl per 1.5 ml microfuge tube) by alternately centrifuging (4°C, 2-3 min) and resuspending in immunoprecipitation buffer (1 ml). Resuspend the washed protein A-containing *S. aureus* in immunoprecipitation buffer as a 10% (w/v) suspension.
3. Preadsorb samples with protein A-containing *S. aureus*. Incubate samples with prewashed protein A-containing *S. aureus* (100-200 μl of 10% suspension/tube, 500 μl total volume in 1.5-ml microfuge tubes) for ≥ 30 min at 4°C. At the end of the incubation, recover the supernate after microcentrifugation (4°C, 2-3 min). Preadsorption should be performed at least once. In some instances, a larger volume of protein A-containing *S. aureus*, several rounds of preadsorption, or both may be required to reduce background. Extensive preadsorption, however, may decrease recovery of the immunoprecipitates.
4. Incubate preadsorbed sample with specific antibody (10 μl in 500 μl) overnight at 4°C.
5. Recover antigen-antibody complexes using protein A-containing *S. aureus*. Protein A-containing *S. aureus* should be freshly washed (as described in step 2), and added to each incubation mixture (20 μl 10% suspension/μl antibody used) for 30 min on ice. Pellets are recovered by microcentrifugation (4°C, 3 min) and washed three times with immunoprecipitation buffer, 0.5 ml/pellet/wash, by alternate resuspension and centrifugation.
6. Analyze immunoprecipitates by SDS-polyacrylamide gel electrophoresis (Laemmli, 1970). Antigen-antibody complexes are dissociated from the pellets by resuspension in Laemmli sample buffer (2×, 50 μl/pellet), boiling for 5 min, and centrifugation.

PROTOCOL 6: IMMUNOBLOTTING

Materials

1. *Nitrocellulose membrane*. Pore size 0.2 μm. BA83 (Schleicher and Schuell, Inc.).
2. *Bovine serum albumin*. RIA grade (Sigma Chemical Corp., St. Louis, MO).

3. *Enhanced chemiluminescence (ECL)*. *Western blotting kit* containing two detection solutions (Amersham Corp., Arlington Heights, IL).
4. *Primary antibody.* Polyclonal (rabbit) or monoclonal (mouse) antibody raised against specific IGFBP.
5. *Second antibody.* Biotinylated second antibody raised against the primary antibody; e.g., donkey biotinylated antibody to rabbit immunoglobulin (Amersham Corp., Cat. No. RPN.1004), or sheep biotinylated antibody to mouse immunoglobulin (Cat. No. RPN.1001).
6. *Streptavidin-biotinylated horseradish peroxidase.* (Amersham Corp., Cat. No. RPN.1051.)
7. *TS-T buffer.* Composed of 10 mM Tris–Cl, 150 mM NaCl, 0.1% (v/v) Tween 20, pH 7.4.

Procedures

All procedures are carried out at room temperature and on a rocking platform.

1. Following electrophoresis, transfer proteins onto nitrocellulose membrane by electroblotting.
2. Block the membrane with 2% (w/v) BSA (RIA grade, Sigma) in TS-T buffer, 1 hr.
3. Wash the membrane extensively with TS-T buffer (≥ 1 ml/cm^2, three to four times).
4. Incubate membrane with the primary antibody in TS-T buffer in a sealed bag (20 ml/blot, 1 hr).
5. Wash the membrane extensively with TS-T buffer.
6. Incubate the membrane with biotinylated second antibody (e.g., 1:10,000 dilution in TS-T buffer), 20–50 ml/blot, 20 min.
7. Wash the membrane extensively with TS-T buffer.
8. Incubate the membrane with streptavidin-biotinylated horseradish peroxidase complex, 1:5,000 dilution in TS-T buffer, 20–50 ml/blot, 20 min.
9. Wash the membrane extensively with TS-T buffer.
10. Mix equal volumes of the two detection solutions together in a clean container. (Use enough reagent to cover the whole blot, ~30–50 ml/blot.)
11. Incubate the membrane with the detection solution for exactly 1 min, drain excess reagent.

12. Cover the membrane with Saran Wrap, and in a dark room expose the membrane to Kodak X-Omatic AR film (Eastman–Kodak Co.) or Hyperfilm–ECL (Amersham Corp.) in a cassette. Exposure time may vary depending on the sample and antibody used.

Comment

1. The above immunoblotting procedure can be performed on a blot previously used for ligand blotting. Strip the ^{125}I-IGF using TS buffer–0.5 mg/ml sodium azide–3% (v/v) NP-40 (1–2 hr, 4°C). The blot should be reblocked before immunoblotting as described above (step 2).
2. The concentration of primary antibody determines the background of the final image and should be optimized in preliminary experiments.

C. Identification of IGFBP mRNAs

Complementary DNA clones of all six IGFBPs have been isolated (Rechler and Brown, 1992). This enables Northern blot analysis (or *in situ* hybridization) using hybridization probes based on the nucleotide sequence to identify which IGFBPs are expressed in a tissue or cell. The size of IGFBP mRNAs ranges from 1.3 to 6 kb, as summarized in Table I. Demonstration of IGFBP mRNA in tissues helps establish that the observed IGFBP is synthesized rather than stored or sequestered in the tissue.

D. Quantitation of IGFBPs

Radioimmunoassay and (with some limitations) solution-binding assay are the most suitable of the assays described for IGFBPs for quantitative analysis. Although ligand blotting and immunoblotting are potentially amenable to quantitation (for example, by verifying transfer recoveries and establishing standard curves with known proteins), this application has not been reported.

Radioimmunoassay is a straightforward approach to measure all immunoreactive species. As discussed above, antibodies that can identify each of the six IGFBPs are available. Radioimmunoassay does not provide any information about whether the immunoreactive species bind IGFs and whether they are intact molecules or fragments with potentially altered biological properties.

We have used the charcoal-separation solution-binding assay to quantitate IGF-binding capacity of an unknown sample. Different concentrations of IGF-I (or IGF-II) are incubated with a concentration of sample containing IGFBP that gives half-maximal binding (as determined in a preliminary experiment). Binding data are analyzed without transformation using the Ligand program (Munson and Rodbard, 1984). The validity of the method hinges on the quantitative separation of free IGF from IGF bound to IGFBPs. As discussed above, complete separation may not be strictly attainable, since a fraction of the 20- 30-kDa IGF : IGFBP complexes and unoccupied IGFBPs are adsorbed by activated charcoal. Nonetheless, this approach provides a useful basis for comparing the relative abundance of IGFBPs in different samples.

IV. Conclusions

Characterization of IGFBPs is an essential component of understanding the biological role of the IGF system in a given cell or tissue. Practical methods have been presented to detect IGFBPs (by solution-binding assay, gel permeation chromatography, and ligand blotting), to identify the six IGFBPs (by presence in 150-kDa complexes, N- or O-glycosylation, subunit molecular mass, relative binding affinity for IGF-I and IGF-II, immunoreactivity, and Northern blot hybridization), and to quantitate their relative abundance. Future challenges will be to quantitate individual IGFBPs in tissues as well as in plasma; to determine the localization of IGFBPs to media, extracellular matrix, and the cell surface; to identify modified forms of IGFBPs (proteolytic fragments, and phosphorylated and glycosylated species); and to understand the biological implications of these differences. Only then will we begin to understand how IGFBP diversity modulates IGF action.

References

Bach, L. A., Thotakura, N. R., and Rechler, M. M. (1992). *Biochem. Biophys. Res. Commun.* **186**, 301–7.
Baxter, R. C., and Daughaday, W. H. (1991). *J. Clin. Endocrinol. Metab.* **73**, 696–702.
Baxter, R. C., and Martin, J. L. (1986). *J. Clin. Invest.* **78**, 1504–1512.
Baxter, R. C., and Saunders, H. (1992). *J. Endocrinol.* **134**, 133–139.
Blum, W. F., Ranke, M. B., Kietzmann, K., Gauggel, E., Zeisel, H. J., and Bierich, J. R. (1990). *J. Clin. Endocrinol. Metab.* **70**, 1292–1298.
Busby, W. H., Jr., Klapper, D. G., and Clemmons, D. R. (1988). *J. Biol. Chem.* **263**, 14203–14210.
Camacho-Hubner, C., Busby, W. H., Jr., McCusker, R. H., Wright, G., and Clemmons, D. R. (1992). *J. Biol. Chem.* **267**, 11949–11956.
Clemmons, D. R. (1992). *Growth Regul.* **2**, 80–87.
Clemmons, D. R., Snyder, D. K., and Busby, W. H., Jr. (1991). *J. Clin. Endocrinol. Metab.* **73**, 727–733.

Conover, C. A., Liu, F., Powell, D., Rosenfeld, R. G., and Hintz, R. L. (1989). *J. Clin. Invest.* **83**, 852–859.
Daughaday, W. H., and Rotwein, P. (1989). *Endocr. Rev.* **10**, 68–91.
Elgin, G. R., Busby, W. H., Jr., and Clemmons, D. R. (1987). *Proc. Natl. Acad. Sci. U.S.A.* **84**, 3254–3258.
Fowlkes, J., and Freemark, M. (1992) *Endocrinology (Baltimore)* **131**, 2071–2076.
Guler, H.-P., Zapf, J., Schmid, C., and Froesch, E. R. (1989). *Acta Endocrinol. (Copenhagen)* **121**, 753–758.
Herington, A. C., Ymer, S., and Stevenson, J. (1986). *J. Clin. Invest.* **77**, 1817–1823.
Hossenlopp, P., Seurin, D., Segovia-Quinson, B., Hardouin, S., and Binoux, M. (1986). *Anal. Biochem.* **154**, 138–143.
Jones, J. I., Gockerman, A., Busby, W. H. Jr., Camacho-Hubner, C., and Clemmon, D. R. (1993). *J. Cell Biol.* **121** 679–687.
Kiefer, M. C., Masiarz, F. R., Bauer, D. M., and Zapf, J. (1991). *J. Biol. Chem.* **266**, 9043–9049.
Knauer, D. J., and Smith, G. L. (1980). *Proc. Natl. Acad. Sci. U.S.A.* **77**, 7252–7256.
Laemmli, U. K. (1970). *Nature (London)* **227**, 680–685.
Lamson, G., Oh, Y., Pham, H., Guidice, L. C., and Rosenfeld, R. G. (1989). *J. Clin. Endocrinol. Metab.* **69**, 852–859.
Martin, J. L., and Baxter, R. C. (1986). *J. Biol. Chem.* **261**, 8754–8760.
McCusker, R. H., Cohick, W. S., Busby, W. H., and Clemmons, D. R. (1991). *Endocrinology (Baltimore)* **129**, 2631–2638.
Moses, A. C., Nissley, S. P., Passamani, J., White, R. M., and Rechler, M. M. (1979). *Endocrinology (Baltimore)* **104**, 536–546.
Munson, P. J., and Rodbard, D. (1984). In "Computers in Endocrinology" (D. Rodbard and G. Forti, eds.), pp. 117–145. Raven, New York.
Nissley, S. P., and Rechler, M. M. (1984). In "Hormonal Proteins and Peptides" (C. H. Li, ed.), Vol. 12, pp. 127–203. Academic Press, New York.
Rechler, M. M. (1993). *Vitam. Horm. (N.Y.)* **47**: 1–114.
Rechler, M. M., and Brown, A. L. (1992). *Growth Regul.* **2**, 55–68.
Rechler, M. M., and Nissley, S. P. (1990). In "Peptide Growth Factors and Their Receptors, I" (M. B. Sporn and A. B. Roberts, eds.), pp. 263–367. Springer-Verlag, Berlin.
Roghani, M., Lassarre, C., Zapf, J., Povoa, G., and Binoux, M. (1991). *J. Clin. Endocrinol. Metab.* **73**, 658–666.
Romanus, J. A., Terrell, J. E., Yang, Y. W.-H., Nissley, S. P., and Rechler, M. M. (1986). *Endocrinology (Baltimore)* **118**, 1743–1758.
Rutanen, E.-M., and Pekonen, F. (1991). *Acta Endocrinol. (Copenhagen)* **124**, 70–73.
Smith, R. G., and Sestili, M. A. (1980). *Clin. Chem.* **26**, 543–550.
Suikkari, A.-M., and Baxter, R. C. (1991). *J. Clin. Endocrinol. Metab.* **72**, 1377–1379.
Suikkari, A.-M., and Baxter, R. C. (1992). *J. Clin. Endocrinol. Metab.* **74**, 177–183.
Unterman, T. G., Patel, K., Mahathre, V. K., Rajamohan, G., Oehler, D. T., and Becker, R. E. (1990). *Endocrinology (Baltimore)* **126**, 2614–2619.
White, R. M., Nissley, S. P., Moses, A. C., Rechler, M. M., and Johnsonbaugh, R. E. (1981). *J. Clin. Endocrinol. Metab.* **53**, 49–57.
Wilkins, J. R., and D'Ercole, A. J., (1985). *J. Clin. Invest.* **75**, 1350–1358.
Yang, Y. W.-H., Wang, J.-F., Orlowski, C. C., Nissley, S. P., and Rechler, M. M. (1989). *Endocrinology (Baltimore)* **125**, 1540–1555.
Yang, Y. W.-H., Pioli, P., Fiorelli, G., Brandi, M. L., and Rechler, M. M. (1993). *Endocrinology* **133**, 343–351.
Zapf, J., Waldvogel, M., and Froesch, E. R. (1975). *Arch. Biochem. Biophys.* **168**, 638–645.

PART II

Histological and *in Situ* Approaches

11

Localization of Peptides: Double Labeling Immunohistochemistry

José E. García-Arrarás

I. Preparation of Tissues
   A. Fixation
   B. Sectioning
II. Single Labeling
   A. Technique
   B. Controls
III. Double Labeling
   A. Simultaneous Addition of Sera
   B. Direct and Indirect Immunofluorescence Combined
   C. Serial Sections
   D. Sequential Single Labeling
   E. Dual-Color Peroxidase Reaction
   F. Other Methods
IV. Triple Labeling
   References

Localization of hormones or other antigens to specific tissues or cells by immunochemical methods has been performed successfully for the past 3 decades. At present it still remains a powerful tool for the detection, localization, and description of different subpopulations of cells within endocrine and nervous tissues (Polak and Van Noorden, 1987). Moreover, with the explosive discovery of neuropeptides that has taken place within the past 15 years and the possibility of expressing peptide genes in transgenic cells or animals, the use of immunohistochemistry has become essential to most studies in cellular and molecular biology.

The discovery of large numbers of peptides led to the subsequent realization that, in many tissues, multiple peptides and/or hormones are co-localized within the same endocrine or nerve cells (Hökfelt *et al.*, 1980, 1987; Krieger, 1983; Sherman *et al.*, 1989). Double labeling techniques have been developed in order to reveal co-localization of hormones or peptides and thus provide an accurate description of the hormonal or peptidergic output produced by each cell (Hökfelt *et al.,* 1980, 1987; Furness *et al.,* 1987; Gibbins, 1989; Lindh *et al.,* 1989; Kupfermann, 1991). Procedures used for double labeling are fundamentally the same as those used for labeling single compounds. Therefore, we need to consider the problems and limitations inherent to single labeling in our discussion of double labeling. Two main techniques have been described for double labeling: immunofluorescent and enzymatic techniques. Both have their unique advantages and disadvantages (Bronckers *et al.,* 1987; De Mey, 1983; Gillitzer *et al.,* 1990; Starkweather, 1990; Valnes and Brandtzaeg, 1982). However, if a microscope equipped with a UV light source and the appropriate filters is available, immunofluorescence should be the method of choice because of the ease of performing the experiments. I focus on the immunofluorescent technique as it can be applied to the double labeling of peptides at the light-microscope level, but also discuss some aspects of the enzymatic technique for comparison. Although the main focus of the techniques described here is on determining coexistence of two peptides, the same techniques can be used to determine the coexistence of any two antigens that can be clearly recognized by two different antibodies.

I. Preparation of Tissues

A. Fixation

The first step, and one of the most important, is to choose the right fixative for your preparation. The fixative should be capable of retaining most of the antigenicity recognized by the antibody and at the same time maintain tissue and cellular morphology (Vielkind and Swierenga, 1989). Although no fixative can be considered to be universally adequate for preservation of antigens, 4% paraformaldehyde comes closest to this description. Other formaldehyde- and/or picric acid-based fixatives such as Zamboni and Bouins are also commonly used. Glutaraldehyde-based fixation generally results in poor immunocytochemical labeling and, whenever possible, should be avoided.

1. Cryostat Sectioning

Paraformaldehyde and Zamboni are the two most commonly used fixatives for cryostat sectioning.

a. Paraformaldehyde One of the most commonly used fixatives, 4% paraformaldehyde, is also one of the simplest to make, although it should be prepared fresh prior to use. To prepare 100 ml of fixative, 4 g of paraformaldehyde are added to 50 ml of distilled water and heated with constant stirring to 60°C. In order to dissolve the paraformaldehyde, about 1.5 ml of 0.1 M NaOH is added to the heated solution, drop by drop, until it clears. An equal volume of 0.2 M phosphate buffer is added, bringing the molarity of the solution to 0.1 M, and then it is cooled on ice. The pH of the solution is checked; some researchers have obtained better results when the pH is maintained near neutral (7-7.4) while others claim better results at basic pH (Berod *et al.*, 1981; Forsgren and Soderberg, 1987). We have found no obvious differences between pHs in our attempts to detect a variety of neuropeptides. Once the solution is at 4°C, it can be used to fix the tissue.

Tissue can be placed in the fixative for 1-2 hr at 4°C, depending on the size of the pieces. For large tissues or tissues that are difficult to dissect from the organisms it might be necessary to anesthetize the animal and perfuse it transcardially with a saline solution, followed by ice-cold fixative, in order to prevent degradation of the peptides during a lengthy dissection (McGinty and Bloom, 1983; Simerly *et al.*, 1986).

After fixation, the tissue is rinsed several times (3 × 15 min rinses) in 0.1 M phosphate-buffered saline (PBS) and placed in a solution of 30% sucrose-PBS overnight. The tissue initially will float in the sucrose solution; when it sinks, it is ready to be sectioned. Incubating the tissue in a sucrose solution preserves the morphology as it decreases crystallization, which occurs during freezing of the tissue.

This fixation method can also be used for tissue or cell culture; however, the last step, incubation in the sucrose solution, is omitted (García-Arrarás *et al.*, 1986, 1987).

b. Zamboni   A second fixative that has given excellent results in some preparations is the picric acid-formaldehyde (Zamboni and De Martino, 1967). It consists of 2% formaldehyde and 0.2% picric acid (saturated solution, filtered) in 0.1 M phosphate buffer, pH 7-7.4. The protocol for fixation with this fixative takes a little extra time but it offers some advantages that make it ideal for certain preparations or experimental situations.

The tissue is placed in the fixative for at least 12 hr in the refrigerator (overnight fixation is usually done). The tissue can be left in this fixative for a longer period than that in the paraformaldahyde fixation; tissues that have remained in the fixative for 1 week and even longer have been used successfully in our laboratory. However, the length of time it can be left in the fixative will depend on the peptide or antigen that is being detected. This fixative can also be used for transcardial perfusion.

The tissue is processed with graded alcohols to eliminate the picric acid and eventually through a dehydration and rehydration procedure in order to clear the tissue and make the antibodies more accessible to the antigens within the cells (Costa and Furness, 1984; Furness *et al.*, 1984). The following protocol is generally used:

1. Wash in 80% ethanol until picric acid is removed. This will usually take 3 × 15 min washes.
2. Wash 2 × 15 min in 95% ethanol.
3. Wash 2 × 15 min in 100% ethanol.
4. Wash 2 × 15 min in xylene.
5. Wash 2 × 15 min in 100% ethanol.
6. Wash 1 × 15 min in 95% ethanol.
7. Wash 1 × 15 min in 80% ethanol.
8. Wash 1 × 15 min in 50% ethanol.
9. Wash 1 × 15 min in distilled water.
10. Wash 1 × 15 min in PBS.
11. Let stand overnight in 30% sucrose–PBS.

For pieces of tissue larger than 1 cm it might be necessary to increase the length of time in each rinse up to 30 min.

The procedure has been shortened by using dimethyl sulfoxide (DMSO) for the clearing steps (Gibbins *et al.*, 1987). DMSO eliminates the need for the lengthy alcohol dehydration procedure. However, DMSO is cancerogenic and should be handled with caution. The protocol for this modification follows:

1. Wash in 80% ethanol until picric acid is removed. This will usually take 3 × 15 min washes.
2. Wash 3 × 15 min in DMSO.
3. Wash 3 × 15 min in PBS.
4. Let stand overnight in 30% sucrose–PBS.

Following fixation and the clearing procedure of choice, the tissue is left overnight in 30% sucrose–PBS and is ready for sectioning the following day. An additional advantage of this type of fixation is that tissues can be left in the sucrose solution for several days before being sectioned, while in the paraformaldehyde protocol, tissues are usually sectioned the day following fixation.

Unfixed tissue that has been rapidly frozen at −80°C can also be used successfully for immunohistochemistry (Forsgren and Soderberg, 1987).

2. Paraffin Sectioning

Bouins fixative is usually used for paraffin sections, although it can also be used for frozen sections. Bouins is prepared by mixing together 75 ml saturated aqueous picric acid, 25 ml 40% formalin, and 5 ml of glacial acetic acid (Humanson, 1972). Paraffin sections provide excellent tissue morphology; however, the immunochemical results are usually less satisfactory than those obtained from frozen sections. This is probably due to loss of antigenicity during the tissue processing. Recent techniques might provide a way to retrieve antigens from paraffin-embedded tissues (Shi *et al.*, 1991).

3. Whole Mounts

In tissue that is relatively thin (micrometers) or can be obtained as thin layers, it is possible to do immunohistochemistry without sectioning. The tissue is stretched and held with pins on a wax- or sylgard-covered petri dish, or on a piece of balsa wood (see Costa and Furness, 1983, for a detailed procedure) and left in Zamboni fixative overnight as described above. The pins are removed and the tissue is then cleared as described previously, except that the sucrose solution step is omitted and the tissue is left in PBS with 0.01% sodium azide until used.

B. Sectioning

Tissue left in a 30% sucrose solution can be sectioned in a cryostat after freezing in an embedding medium. This embedding medium can be the same sucrose solution, a polyethylene glycol (PEG) solution, a commercial compound (OTC, Miles, Inc.), or a mixture of these (Elias 1990; Barthel and Raymond, 1990; Watson *et al.*, 1986). In our laboratory we prefer the OTC compound. The tissue should be frozen as quickly as possible, and most cryostats provide a rapid-freezing device for this purpose. If not available, the tissue can be frozen by being placed over dry ice or by being immersed in liquid nitrogen. Sections, usually 5–15 μm, are obtained and picked up on previously treated glass slides. Techniques where 3-μm-thick cryosections are obtained have been described recently (Elias, 1990; Barthel and Raymond, 1990). Two common methods of treating slides so that the tissue sections adhere to the surface are subbing with a chrome alum-gelatin solution or treating them with a polylysine solution (Sigma). The latter method has proven to be less time consuming and more effective in preventing detachment of sections from slides. The sections are then dried under a cold blower at room temperature for at least 1 hr. We have left the slides in a desiccator with calcium chloride for up to 1 week after sectioning without any major loss of immunoreactivity.

II. Single Labeling

Single-labeling techniques provide the fundamental steps also required for double-labeling experiments. In addition, in order to test whether artifacts are present in double labeling it is necessary to compare the results with those obtained in single labeling. The single-labeling method described here is the indirect method described by Coons and colleagues in 1955 which has been modified by many investigators (Polak and Van Noorden, 1987; Lindh et al., 1989; García-Arrarás et al., 1986, 1987; Hartman, 1973).

A. Technique

The sections are initially treated with some type of serum in order to minimize background staining due to unspecific antibody binding. Goat serum is commonly used; however, sera from other species can also be used. The best strategy is to use normal serum from the species in which the secondary antibody was raised, thus, minimizing the possibility for cross-reactivity.

After 15 min incubation in normal goat serum/PBS (diluted 1/20–1/50), sections are rinsed in 0.05% Triton X-100–PBS for an additional 15 min. (Not all tissues need to be treated with Triton. In cases where the antibody penetrates well, this step can be omitted.) This is followed by 3 × 10 min rinses in PBS before addition of the primary antibody. Several dilutions of the primary antibody should be tested to determine which one works best. We currently test our polyclonal sera at 1/100, 1/250, 1/500, 1/1000, and 1/2000 dilutions in PBS–0.01% Na azide. Dilutions of monoclonal antibodies also must be tested. Monoclonal culture media are usually used directly or slightly diluted, while ascites fluid should be treated like polyclonal serum. Once the procedure is started it is important that the sections are never allowed to dry, not even for a few seconds. In addition, all serum dilutions should be centrifuged at 3000g before being placed on the tissue sections; this will prevent precipitates from causing fluorescence artifacts.

The serum is placed as a drop that fully covers the tissue section. It is very important to wipe around the edges of the tissue sections before adding the diluted serum or monoclonal media to prevent drops of primary antibody from running off the slide. The slides are incubated with primary antibody in a sealed humid chamber and left for 12–24 hr at room temperature. The following day, three rinses with PBS (5 min each) are done before the addition of the secondary antibody. Secondary antibody should be directed against the donor species of the primary serum and should be conjugated with a fluorescent molecule, fluorescein isothiocianate (FITC) or tetramethylrhodamine isothiocyanate (TRITC) are the most common. This secondary antibody is usually used at a dilution of 1/40–1/60 in PBS–Na azide and is left for about 1 hr. Three more rinses with PBS are done before the slides are coverslipped. Several mounting media have been used for coverslipping; however, we prefer a mixture of glycerol/phosphate buffer, pH 8.6. In addition, several chemicals (e.g., 0.1–0.25 M n-propyl gallate) (Giloh and Sedat, 1982)) have been recommended to prevent the rapid disappearance of the fluorescence as one observes the slides under the microscope. In order to prevent drying of the sections or movement of the coverslip it is wise to seal the edges of the sections with Permount or some other solidifying compound. Commercial nail polish can be used for this purpose.

Many modifications of the above protocol can be found in the literature. In some cases researchers add Triton X-100 (0.05–0.1%) to the primary or secondary antibody dilutions in order to increase the penetration of the antibodies. Others use Nonidet P40 (NP-40) or saponin (Willingham, 1990). In some cases normal serum is also added with the primary or secondary antibodies to diminish nonspecific binding.

In single labeling, the primary antibody can also be visualized by enzymatic techniques where alkaline phosphatase- or peroxidase-coupled secondary antisera are used and the reaction is developed to produce a colored product. The advantage of this technique is that the reaction products are permanent and can be used for long-term documentation or retrospective evaluation of the results. In addition, the results can be documented using a light microscope without the need of special filters or an ultraviolet light source.

The indirect immunolabeling method has been modified to increase the sensitivity of the reaction. Of special interest is the use of compounds with a strong affinity, such as avidin–biotin (ABC) or streptavidin–biotin, together with peroxidase or fluorescent probes (Polak and Van Noorden, 1987; Buckland, 1986; Hsu et al., 1981). However, these modifications are generally not used for double labeling, probably due to the increased chance of cross-reaction that is provided by the additional reagents.

B. Controls

One of the main problems in immunochemistry is determining exactly the specificity of your antibody (Leeuwen, 1986; Van der Sluis and Boer, 1986). This is particularly true today as many commercially available antibodies are not well characterized (Herman and Elfont, 1991). As Berkenbosch and Tilders (1987) report, "even the most pure antibody populations (purified conventional sera or monoclonal) are epitope specific rather than substance specific, in other words they may interact with various substances sharing that particular epitope."

None of the controls used in immunohistochemistry is completely satisfactory for clearly establishing the presence of the antigen. This is one of the reasons why the term peptide-*like immunoreactivity* has come into common use, until the specific nature of the antigen is firmly establish by other methods such as HPLC coelution, sequencing of the peptide or of its DNA, NMR, etc.

Nevertheless, several controls will help to establish the validity of the results. Primary among these is treating some sections with normal serum as the primary antibody. The best control is the preimmune serum from the same animal that produced the antiserum, when available. A second control is to incubate the antiserum with the specific antigen for 12–24 hr, in order to neutralize the antibody, before addition to the sections. Concentrations of antigen used for these controls are usually on the order of 10^{-5}–10^{-6} M or 1 μg/ml. After such treatment no immunoreactivity should be found in the sections. If a reduction of the reaction is observed it might be argued that the antigen present in the tissue or cells is similar but not identical to the antigen used for the absorption test. Complete blockage of the reaction is by itself no definite proof that the molecules are identical (Holzwarth and Brownfield, 1985; García-Arrarás and Martínez, 1990).

In the case of small molecules that have to be chemically coupled to a carrier molecule in order to induce the immunoreaction in the animal the best control is not the antigen by itself but the antigen–carrier complex that was used to produce the antibody in the host species. For example, serotonin has to be coupled to a large protein for it to become immunogenic. The antibody obtained is usually neutralized by the serotonin–protein complex, but not by serotonin or the protein alone (Holzwarth and Brownfield, 1985; García-Arrarás and Martínez, 1990; Schipper and Tilders, 1983).

III. Double Labeling

Double-labeling techniques can be used at the cell level and at the tissue level. At the cell level they are mainly used to determine whether two antigens are co-localized within the same cell and to study the internal distribution of co-localized antigens. At the tissue level the method is used to study the anatomical relationship of different cellular components. For double labeling two antibodies recognizing two different antigens are used. The principal variation in methodology lies in the timing of treatment and on the use of either immunofluorescent or immunoenzimatic techniques to visualize the antibodies. The two antibodies can be added either simultaneously or sequentially to the same section or alternatively they can be added independently to serial sections. The methodology to be used depends on the species from which the sera being used were obtained.

A. Simultaneous Addition of Sera

The easiest method to double label, and one that gives superb results, is to use antibodies that have been raised in two different species (Costa and Furness, 1983, 1984; García-Arrarás and Martínez, 1990; Erichsen et al., 1982; Macrae et al., 1986). Antisera from any species can be used, as long as the appropriate secondary antibodies are available. The two species most commonly used are rabbit and mouse. The former because most antibodies are usually raised in rabbits and the latter because of the increasing availability of mice monoclonal antibodies against various peptides. Secondary antibodies against rat, guinea pig, sheep, and goat immunoglobulins are also available.

The protocol for simultaneous double labeling is essentially the same as that described above for single labeling. However, the primary antibody consists of a mixture or "cocktail" of two antisera, mixed at the dilution that has proven effective for each antibody in prior single-labeling experiments. If the best dilution for each antibody is 1/500, prepare 1/250 dilutions of each and mix 1:1. This mixture is placed on the sections and left overnight in a sealed humid chamber.

A mixture of two secondary antibodies is used following the same protocol as that for single labeling. These must be against the sera of the species in which the primary antibodies were obtained and each should be labeled with a different fluorochrome. For example, if the primary antibody cocktail consists of a galanin antiserum raised in rabbit and a mouse anti-somatostatin monoclonal antibody, the secondary antibodies could be goat anti-rabbit FITC (GAR-FITC) and goat anti-mouse TRITC (GAM-TRITC). The cocktail is placed on the sections for 1 hr and then the technique proceeds as for single labeling.

It is important to emphasize here the convenience of performing single-labeling experiments prior to the double labeling. First, it is easier to determine the concentration at which each antibody works best. Second, because one sees what and how the antibody labels, one can detect an artifact or any possibility of cross-reactivity in the double-labeling reaction. If the antibody does not work as a single label it will surely not work for double labeling either. It is evident that the structures that are labeled and the pattern of

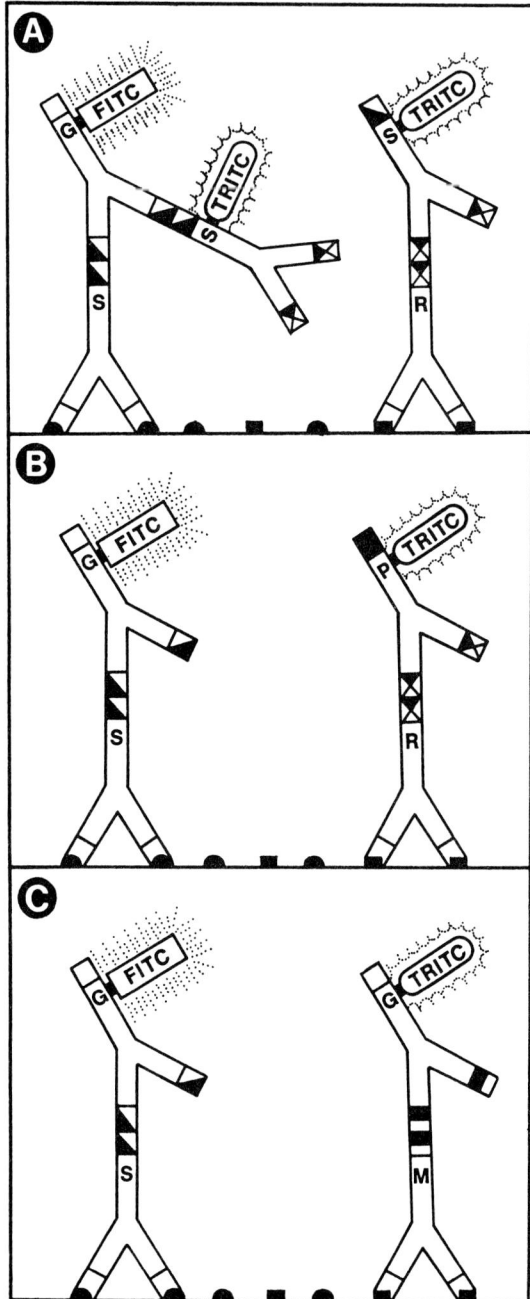

Figure 1 Double labeling for SS and galanin using antibodies that cause cross-reactivity (A) and alternate choices of adequate secondary (B) or primary (C) antisera.

labeling by a specific antibody should be the same whether the antibody is used in single- or in double-labeling. Any change in the pattern of labeling is indicative that cross-reaction is occurring or some artifact is present.

The main problem with double-labeling techniques, in general, is showing, using the right controls, that what one observes is real. In order to do this it is necessary to rule out all possibilities of cross-reaction between the antisera, in the case where they label the same cells, or the possibility of steric hindrance of one antibody over another, in the case where they are shown to label different cell populations.

Secondary antibodies for double labeling should be selected carefully. Ideally, both antisera should originate in the same species, decreasing the possibility of cross-reactivity. In addition, the sera used in the first step to reduce background staining should also, in the best of cases, originate from the same species as the secondary antisera. Figure 1 gives two examples of choosing secondary antibodies raised in different species for a double-labeling experiment. In Fig. 1A the primary antibodies were obtained from sheep (S) and rabbit (R); the secondary antibodies were a goat anti-sheep (GAS)-FITC and a sheep anti-rabbit (SAR)-TRITC. In such a scheme it is possible for the two secondary antibodies to bind to each other and show a false positive. The appropriate use of secondary antibodies is shown in Fig. 1B, where a pig anti-rabbit (PAR)-TRITC has been used, thus eliminating the possibility of cross-reactivity, or in Fig. 1C where the primary was changed to a mouse (M) serum and GAM-TRITC was used as a secondary.

Whichever antibodies are used, we have consistently performed what we refer to as the 3 by 3 experiment to rule out any possible cross-reactivity between the primary antibodies themselves, between the secondary antibodies themselves, or between inappropriate combinations of primary and secondary antibodies (Table I). The aim is to expose tissue sections to all the possible combinations of primary–secondary antibody interactions. At least nine

Table I
Cross-reactivity Test (3 × 3)

	GAR-FITC	GAM-TRITC	GAR-FITC GAM-TRITC
Anti-galanin (rabbit)	+	−	+FITC −TRITC
Anti-somatostatin (mouse)	−	+	−FITC +TRITC
Anti-galanin (rabbit) anti-somatostatin (mouse)	+FITC −TRITC	−FITC +TRITC	+FITC +TRITC

slides, containing experimental sections, are used. Three of them are treated with only one of the primary antibodies, three are treated with the other primary antibody, and the last three are treated with the mixture of primary antibodies. Every slide from each group receives a different secondary antibody treatment. One section is incubated with one type of secondary antibody, a second section is incubated with the other type of secondary antibody, while a third section is incubated with a mixture of both secondary antibodies. In this way, each of the nine slides represents a particular combination of one or both primary antibodies with one or both secondary antibodies. In the example presented in Table I, the + signs show where immunoreactivity should be expected while the − signs show where it should not occur. If sections treated with rabbit anti-galanin and GAM-TRITC show immunoreactivity it would indicate that the GAM antibody is binding the inappropriate primary antibody and therefore it should not be used for double-labeling experiments.

Paramount for the analysis of results obtained with the double-immunofluorescent method is the selectivity of the microscope filters (Haaijman, 1983; Wessendorf et al., 1990). The filters should be capable not only of transmitting the emission from a particular fluorochrome but also of blocking the emission of the other fluorochrome. Interference of the emission of one fluorochrome with the other can easily be checked by changing from one filter to the other. However, interference should also be tested at the photographic level, since sometimes a small amount of emission resulting from inappropriate fluorochrome excitation is enough to create an erroneous impression on the film.

Another source of error lies with the specificity of the antibodies used. If one of the primary antisera recognizes other antigens in addition to the one that the host was immunized against, inaccurate results will be obtained. Two examples of this phenomenon have been presented: Peptide histidine isoleucine amine (PHI) immunostaining in rat median eminence was found to arise from cross-reactivity of the antibody with corticotropin-releasing factor (CRF) (Berkenbosch et al., 1986). More surprising was the finding that antibodies directed to an amine (histamine) cross-react with luteinizing hormone-releasing hormone (LHRH) (Berkenbosch and Steinbusch, 1987).

Even when the antibodies are known to recognize different peptides it must be remembered that what they recognize are epitopes, which in some cases can be as small as one residue (Berkenbosch and Tilders, 1987). Thus, if the same epitopes are present in more than one molecule they can lead to erroneous conclusions. Otherwise, two antibodies can recognize different epitopes in the same molecule. This seems to be the case in the lobster, where antisera against FMRFamide and SCPb appear to recognize the molecule responsible for FMRFamide-like immunoreactivity (Arbiser and Beltz, 1991). Therefore, one needs to consider the homologies arising from peptides within the same family as well as those similarities that could be the product of convergent evolution.

When double labeling is performed and the population of cells appears to be different (i.e., they are labeled with one or another of the antibodies, but not with both) a second control should be performed to rule out steric hindrance of one of the antibodies over the other. In this case two antibodies to the same peptide (or antigen) raised in different species can be used as the primary antibody mixture. This secondary antibody mixture should be prepared as described previously. Results should show the same cells labeled by both antibodies.

The labeling of chromaffin cells in the adrenal medulla provides an excellent example of the use for simultaneous double-labeling immunofluorescence. More than a dozen different peptides have been found in cells of the adrenal medulla. In this gland, the expression and co-localization of the peptides have been shown to vary widely among species (Henion and Landis, 1990; Kong et al., 1989; Pelto-Huikko, 1989). In the avian adrenal gland different chromaffin cell populations can be described depending on the neuropeptides they express (I. Chévere-Colón, unpublished results). Thus, in the newly hatched chicken double-labeling immunocytochemistry detects a population of chromaffin cells that express neuropeptide tyrosine (NPY)-LI but not somatostatin (SS)-LI. A second population expresses SS-LI but not NPY-LI, while a third population express both peptides (Figs. 2A and 2B). On the other hand, most, if not all, cells that express SS-LI also express galanin-LI (Figs. 2C and 2D).

Double labeling with antibodies produced in two different species is always preferred since it produces better labeling, consumes less antisera, and implies less labor. However, in some cases, it is not possible to obtain the antibodies from two different species. Other alternatives are available: (1) one or both of the antibodies can be directly labeled, (2) a two-step washing off of the primary antibodies can be done or (3) serial sections can be stained.

B. Direct and Indirect Immunofluorescence Combined

This procedure is recommended only when one of the antibodies can be obtained in large amounts. For example, one of its main applications is when antibody-producing clones are available. The technique is to couple one of the antibodies directly to a fluorescent molecule (i.e., FITC) and visualize the other with a secondary antibody labeled with a different marker, such as TRITC. It is time consuming since one needs to partially purify the immunoglobulins, chemically couple them to the fluorescent marker, and dialyze and test them for immunoreactivity. During the process some of the antibody is lost and,

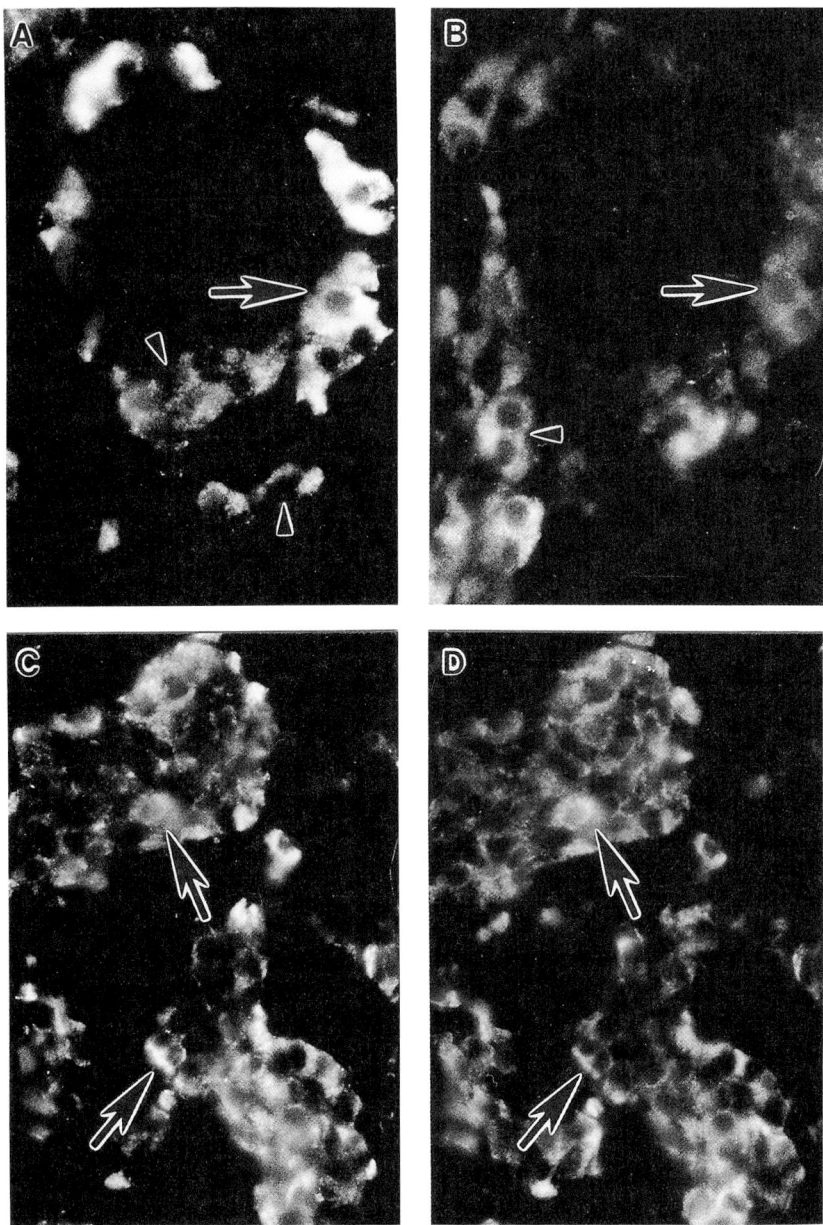

Figure 2 Double labeling for SS (A, C) with NPY (B) or with galanin (D) in the avian adrenal gland of the newly hatched chick (A, B) and the 10-day-old embryo (C, D). Co-localization can be seen in some cells (arrows) while others express only one peptide (arrowheads).

even when the final product is obtained, the direct immunofluorescent technique is never as sensitive as the indirect procedure. If the two antibodies have been coupled with different fluorochromes they could be added simultaneously to the sections. If only one of the antibodies is directly labeled then the double labeling has to be done in a sequential manner, where the first antibody application is followed by the appropriate secondary antibody (García-Arrarás *et al.,* 1986, 1987). The section is then incubated with a high concentration of normal serum to block any remaining binding sites in the secondary antibody. A fourth and last incubation is done with the directly coupled antibody. This is followed by several rinses before coverslipping. Modifications of this technique have also been used successfully for enzymatic labeling (Behringer *et al.,* 1991).

C. Serial Sections

This technique can be used when the cells that are being studied are large enough that two consecutive sections have portions of the same cell. The size limit will be directly associated with the embedding and sectioning technique that is being used (Karten and Brecha, 1980; Nitsch and Klauer, 1989). For example, paraffin embedding usually limits the thickness of the sections to 5–10 μm; thus, cells must need to be almost double this size to be recognized in adjacent sections. However, previously described embedding techniques for thinner cryosections or new ones using water-soluble plastic should permit sections where smaller cells can be visualized in serial sections. The sections are placed on two different slides, and each is treated with a different antibody as for single labeling. The same cell then has to be recognized in the adjacent sections to determine if it is double labeled or not. This technique has two main disadvantages: First, it is difficult to obtain serial sections. Second, the localization of the same cell in adjacent sections is extremely time consuming and its reliability depends on the availability of recognizable landmarks in the tissue sections.

D. Sequential Single Labeling

A third method for detecting antigens within the same cell, using antibodies made in the same species, relies on sequential single labelings (Tramu *et al.,* 1978; Vandesande and Dierickx, 1975). In this case, single labeling with antibody A is performed, as described above. The sections are mounted, observed in the microscope, and photographed. The coverslip is carefully removed and the sections are treated with a chemical (buffers with low pH, high ionic strength, dimethylformamide, or $KMnO_4$-H_2SO_4) to remove the primary antibody and its corresponding secondary antibody (Tramu et al., 1978; Nakane, 1968; Sternberger and Joseph, 1979). After several rinses single labeling is then performed with antibody B. Sections are mounted and the same areas photographed in order to determine if the same cells are labelled. Central to this technique is the issue of whether or not the first primary–secondary antibody complex is efficiently washed off from the tissue sections before addition of the second primary antibody. The disadvantages of the technique lie in the extra work and time necessary for its completion and on the difficulty of determining by photos whether the same cells are labeled. In addition, controls have to be done to determine unequivocally that the first antibody has been efficiently removed. Otherwise the possibility of cross-reaction cannot be discarded. An important control is to treat the sections with a secondary antibody after washing off the first antibody to see if only the secondary antibody has been removed and the primary is still bound. Additional difficulties are encountered by the series of rinses necessary for the addition and elimination of antibodies, which increase the possibility that the sections will detach from the slides.

E. Dual-Color Peroxidase Reaction

The peroxidase method has also been used for double labeling. Initially, the first antibody was removed leaving the reaction product, and the tissue was subsequently labeled with a second antibody and a different methodology was used to leave a second reaction product (Nakane, 1968). The method was later modified so that it is not necessary to remove the immunoreagents of the first staining sequence prior to applying the second antibody (Sternberger and Joseph, 1979). However, the use of immunoenzymatic double labeling has not become widely accepted probably due to problems of interference of the color reactions with each other (Gillitzer et al., 1990).

F. Other Methods

A new method for double labeling that combines an immunogold–silver-staining (streptavidin–biotin) method with an immunoenzymatic (PAP or alkaline phosphatase) method has recently been introduced (Gillitzer et al.,

1990). The method combines the advantages of the enzymatic reaction (use of bright-field lighting, reaction products that do not fade, and visualization of tissue morphology) with a clear separation of the two labels. However, until more investigators put this method to the test, its limitations, if any, will not be known.

Other methods which combine parts of the techniques described above have been used. For example, a combination of simultaneous immunofluorescent and immunoenzymatic staining has been done (Vandesande, 1983). Double labeling can also be done using enzymatic or fluorescent methods in combination with radiolabeling methods (Gasc et al., 1986; Pickel et al., 1986).

IV. Triple Labeling

New fluorescent markers are now available that permit the simultaneous labeling of three different antigens in the same tissue section (Wessendorf et al., 1990). These new markers fluoresce at wavelengths different from those of the other two commonly used, FITC and TRITC; therefore, a different filter has to be used to detect the fluorescence. The protocol for triple labeling is essentially the same as that for double labeling, except that the primary antibody mixture consists of three different sera at their previously tested working dilution. Similarly, the secondary antibody mixture will consist of three different antibodies aimed against three different species, one labeled with FITC, a second with TRITC, and a third with one of the new fluorescent markers. The main disadvantage of this technique, as should be evident from the above discussions, is that the primary antibodies must be made in different species, and the possibilities for cross-reactions are increased. Enzymatic triple labeling is also possible due to the availability of different chromogens; however, this combines direct, indirect, and ABC techniques, increasing the complexity of the technique (Van der Loos et al., 1987).

Acknowledgments

I express my appreciation to Mr. Iván Chévere-Colón for the use of his unpublished data; to Mr. Luis Jiménez for excellent technical assistance; and to Drs. N. Lugo, P. Orkand, and A. Segarra for helpful comments on the manuscript. The work on double labeling of the avian adrenal medulla was carried out at the University of Puerto Rico, Department of Biology, and was supported, in part, by NSF (BNS-8801538) and NIH-MBRS (GM-8102).

References

Arbiser, Z. K., and Beltz, B. S. (1991). *J. Comp. Neurol.* **306**, 417-424.
Barthel, L. K., and Raymond, P. A. (1990). *J. Histochem. Cytochem.* **38**, 1383-1388.
Behringer, D. M., Meyes, K.-M., and Veh, R. W. (1991). *J. Histochem. Cytochem.* **39**, 761-770.
Berkenbosch, F., and Steinbusch, H. W. M. (1987). *Brain Res.* **405**, 353-357.
Berkenbosch, F., and Tilders, F. J. H. (1987). *Neuroscience* **23**, 823-826.
Berkenbosch, F., Linton, E. A., and Tilders, F. J. H. (1986). *Neuroendocrinology* **44**, 338-346.
Berod, A., Hartman, B. K., and Pujol, J. F. (1981). *J. Histochem. Cytochem.* **29**, 844-850.
Bronckers, A. L. J. J., Gay, S., Finkelman, R. D., and Butler, W. T. (1987). *J. Histochem. Cytochem.* **35**, 825-830.
Buckland, R. M. (1986). *Nature (London)* **320**, 557-558.
Coons, A. J., Leduc, E., and Connolly, J. M. (1955). *J. Exp. Med.* **102**, 49-59.
Costa, M., and Furness, J. B. (1983). In "Immunohistochemistry" (A. C. Cuello, ed.), IBRO Handbook Series: Methods in the Neurosciences, pp. 373-397. Wiley, New York.
Costa, M., and Furness, J. B. (1984). *Neuroscience* **13**, 911-920.
De Mey, J. (1983). *J. Neurosci. Methods* **7**, 1-18.
Elias, J. M. (1990). *J. Histochem. Cytochem.* **39**, 549.
Erichsen, J. T., Reiner, A., and Karten, H. J. (1982). *Nature (London)* **295**, 407-410.
Forsgren, S., and Soderberg, L. (1987). *Histochemistry* **87**, 561-568.
Furness, J. B., Costa, M., and Keast, J. R. (1984). *Cell Tissue Res.* **237**, 328-336.
Furness, J. B., Costa, M., Morris, J. L., and Gibbins, I. L. (1987). In "Advances in Physiological Research" (H. McLennan, J. R. Ledsome, C. H. S. McIntosh, and D. R. Jones, eds.), pp. 143-165. Plenum, New York.
García-Arrarás, J. E., and Martínez, R. (1990). *Cell Tissue Res.* **262**, 363-372.
García-Arrarás, J., Fauquet, M., Chanconie, M., and Smith, J. (1986). *Dev. Biol.* **114**, 247-257.
García-Arrarás, J. E., Chanconie, M., Ziller, C., and Fauquet, M. (1987). *Dev. Brain Res.* **33**, 255-265.
Gasc, J.-M., Ennis, B. W., Baulieu, E.-M., and Stumpf, W. E. (1986). *J. Histochem. Cytochem.* **34**, 1505-1508.
Gibbins, I. L. (1989). In "Comparative Physiology of Regulatory Peptides" (S. Holmgren, ed.), pp. 308-343. Chapman & Hall, London.
Gibbins, I. L., Wattchow, D., and Coventry, B. (1987). *Brain Res.* **414**, 143-148.
Gillitzer, R., Berger, R., and Moll, H. (1990). *J. Histochem. Cytochem.* **38**, 307-313.
Giloh, H., and Sedat, J. W. (1982). *Science* **217**, 1252-1255.
Haaijman, J. J. (1983). In "Immunohistochemistry" (A. C. Cuello, ed.), IBRO Handbook Series: Methods in the Neurosciences, pp. 47-85. Wiley, New York.
Hartman, B. K. (1973). *J. Histochem. Cytochem.* **21**, 312-332.
Henion, P. D., and Landis, S. C. (1990). *J. Neurosci.* **10**, 2886-2896.
Herman, G. E., and Elfont, E. A. (1991). *Biotech. Histochem.* **66**, 194-199.
Hökfelt, T., Johansson, O., Ljungdahl, A., Lundberg, J. W., and Schultzberg, M. (1980). *Nature (London)* **284**, 515-521.
Hökfelt, T., Millhorn, D., Seroogy, K., Tsuruo, Y., Ceccatelli, S., Lindh, B., Meister, B., Melander, T., Schalling, M., Barfai, T., and Terenius, L. (1987). *Experientia* **43**, 768-780.
Holzwarth, M. A., and Brownfield, M. S. (1985). *Neuroendocrinology* **41**, 230-236.
Hsu, S.-M., Raine, L., and Fanger, H. (1981). *J. Histochem. Cytochem.* **29**, 577-580.
Humanson, G. L. (1972). "Animal Tissue Techniques." Freeman, San Francisco.
Karten, H. J., and Brecha, N. (1980). *Nature (London)* **283**, 87-88.
Kong, J. Y., Thureson-Klein, A., and Klein, R. L. (1989). *Neuroscience* **28**, 765-775.
Krieger, D. T. (1983). *Science* **222**, 975-985.

Kupfermann, I. (1991). *Physiol. Rev.* **71**, 683–732.
Leeuwen, F. V. (1986). *Am. J. Anat.* **175**, 363–377.
Lindh, B., Lundberg, J. M., and Hökfelt, T. (1989). *Cell Tissue Res.* **256**, 259–273.
Macrae, I. M., Furness, J. B., and Costa, M. (1986). *Cell Tissue Res.* **244**, 173–180.
McGinty, J. F., and Bloom, F. E. (1983). *Brain Res.* **278**, 145–153.
Nakane, P. K. (1968). *J. Histochem. Cytochem.* **16**, 557–560.
Nitsch, R., and Klauer, G. (1989). *Histochemistry* **92**, 459–465.
Pelto-Huikko, M. (1989). *J. Electron Microsc. Tech.* **12**, 364–379.
Pickel, V. M., Chan, J., and Milner, T. A. (1986). *J. Histochem. Cytochem.* **34**, 707–718.
Polak, J. M., and Van Noorden, S. (1987). "An Introduction to Immunocytochemistry: Current Techniques and Problems," Royal Microscopical Society, Microscopy Handbooks No. 11. Oxford Sci. Publ., Oxford.
Schipper, J., and Tilders, F. J. H. (1983). *J. Histochem. Cytochem.* **31**, 12–18.
Sherman, T. G., Akil, H., and Watson, S. (1989). *Discuss. Neurosci.* **6**, 1–58.
Shi, S.-R., Key, M. E., and Kalra, K. L. (1991). *J. Histochem. Cytochem.* **39**, 741–748.
Simerly, R. B., Gorski, R. A., and Swanson, L. W. (1986). *J. Comp. Neurol.* **246**, 343–363.
Starkweather, W. H. (1990). *KPL's Antibody Work* **Summer**, 11–12.
Sternberger, L. A., and Joseph, S. A. (1979). *J. Histochem. Cytochem.* **27**, 1424–1429.
Tramu, G., Pillez, A., and Leonardelli, J. (1978). *J. Histochem. Cytochem.* **26**, 322–324.
Valnes, K., and Brandtzaeg, P. (1982). *J. Histochem. Cytochem.* **30**, 518–524.
Van der Loos, C., Das, P. K., and Houthoff, H.-J. (1987). *J. Histochem. Cytochem.* **35**, 1199–1204.
Van der Sluis, P. J., and Boer, G. J. (1986). *Cell Biochem. Funct.* **4**, 1–17.
Vandesande, F. (1983). *In* "Immunohistochemistry" (A. C. Cuello, ed.), IBRO Handbook Series: Methods in the Neurosciences, pp. 257–272. Wiley, New York.
Vandesande, F., and Dierickx, K. (1975). *Cell Tissue Res.* **164**, 153–162.
Vielkind, U., and Swierenga, S. H. (1989). *Histochemistry* **91**, 81–88.
Watson, R. E., Jr., Wiegand, S. J., Clough, R. W., and Hoffman, G. E. (1986). *Peptides* **7**, 155–159.
Wessendorf, M. W., Appel, N. M., Molitor, T. W., and Elde, R. P. (1990). *J. Histochem. Cytochem.* **38**, 1859–1877.
Willingham, M. C. (1990). *Focus* **12**, 62–67.
Zamboni, L., and De Martino, C. (1967). *J. Cell Biol.* **35**, 148A.

12

Electron Microscopic Immunocytochemical Approaches to the Localization of Ligands, Receptors, Transducers, and Transporters

Robert M. Smith and Leonard Jarett

I. Introduction
II. Use of Nonimmunologic Electron-Dense Hormone Complexes to Localize Receptors and Ligands
III. General Considerations for Successful Immunocytochemical Electron Microscopy
    A. Preservation of Cellular Detail
    B. Preservation of Antigen Localization
    C. Preservation of Antigenicity
    D. Providing Antibody Access to Intracellular Antigens
    E. Effects of Epitope Masking/Unmasking
    F. Summary
IV. Localization of Hormones, Receptors, Transducers, and Transporters
    A. Specimen Preparation
    B. Immunolabeling Techniques
    C. Immunostaining Protocols
    D. Cellular Localization of Insulin, IGF-I, EGF, and TGFα
    E. Cellular Localization of Receptors for Insulin, IGF-I, and EGF
    F. Cellular Localization of $G_{i\alpha}$
    G. Cellular Localization of Glucose Transporters
V. Future Prospects for Electron Microscopy
    A. Immunocytochemical Electron Microscopy
    B. *In Situ* Hybridization Electron Microscopy
VI. Summary and Conclusion
    References

I. Introduction

Electron microscopic techniques are powerful tools, which, when combined with biochemical analysis, can show the relationship between the structural organization and biochemical function of cells and tissues. Cytochemical techniques were used to localize some enzymes and proteins over 40 years ago. Two major advances provided the techniques to study the cellular localization of receptors and their ligands. In 1959, Singer demonstrated that antibodies covalently attached to an electron-dense ferritin molecule could reveal the localization of cellular antigens. Faulk and Taylor (1971) are generally credited with providing the basis for using colloidal gold-complexed antigens in immunocytochemical electron microscopy. Refinements of their techniques have made immunocytochemical electron microscopy a valuable adjunct in the study of cell biology, endocrinology, and metabolism.

There are excellent reviews, which are referenced throughout this chapter, that cover every technical and theoretical aspect of immunocytochemical electron microscopy. This chapter is intended to introduce immunocytochemical electron microscopy to biologists and endocrinologists who are interested in the relationships between cell structure and function and, hopefully, have access to an electron microscopy laboratory. The methods described in this chapter are designed to be implemented in any electron microscopy laboratory.

II. Use of Nonimmunologic Electron-Dense Hormone Complexes to Localize Receptors and Ligands

Investigators have used ^{125}I-labeled insulin and light or electron microscopic autoradiography to demonstrate the binding of the hormone to cell surface receptors and the internalization of the ligand (Carpentier *et al.*, 1978; Maxfield *et al.*, 1978; Bergeron *et al.*, 1979). The major disadvantage of autoradiography is its limited spacial resolution. The potential distance between the autoradiographic grain and its source makes it difficult to determine the location of the iodinated hormone. To overcome this problem, our laboratory developed a ferritin-labeled insulin conjugate (Jarett and Smith, 1974; Smith and Jarett, 1982a). These studies showed there were cell-specific differences in the organization and distribution of insulin receptors and that the internalization and intracellular processing of the insulin–receptor complex was not the same as well-characterized nonhormonal ligands (reviewed in Smith and Jarett, 1988). These studies demonstrated that the electron-dense ligand needed to meet certain criteria to be a valid electron-dense label for the insulin receptor and later showed these same criteria were equally important for gold-labeled hormones (Smith *et al.*, 1988; Smith and Jarett, 1991). First, electron-dense hormone complexes must retain the full biological activity of the unlabeled hormone. Any reduction in biological activity suggests an impairment in receptor–ligand interaction or subsequent signal transduction processes. Second, the complex must bind to the receptor with the same affinity as the native, or ^{125}I-labeled, hormone. Increased affinity can be an indication of multivalency and could cause artifactual aggregation of receptors (Perelson and DeLisi, 1980) and abnormal ligand recycling and intracellular processing. Multivalency is generally the rule with gold-labeled proteins (see Section IV,B,2) but can be avoided (Smith and Jarett, 1991). Third, oligomers or aggregates of electron-dense particles must be eliminated in order to assess receptor clustering and ligand-induced receptor aggregation (Jarett and Smith, 1977). Although these criteria should be applied to electron-dense labeled ligands used to study receptor dynamics *in vivo* or *in vitro*, they are not essential for immunocytochemical studies performed on fixed tissues where the receptor is immobile.

Analysis of the intracellular processing of hormones and the hormone – receptor complex is difficult. Biochemical studies suggest that the processing of the hormone – receptor complex begins in endosomes (Hamel *et al.*, 1988) or acidic organelles (Yamashiro and Maxfield, 1984) and the fate of the ligand – receptor complex in these organelles plays a crucial role in determining the intracellular pathway of the ligand. Iodotyrosine, resulting from the degradation of ^{125}I-labeled insulin, is rapidly released from the cell, which makes it difficult to determine the site of insulin degradation using autoradiographic electron microscopy. Ferritin-labeled insulin particles were transported to lysosome-like structures and reached a steady state within 30-60 min in rat adipocytes (Smith and Jarett, 1982b) but gold-labeled insulin particles accumulated for many hours (Goldberg *et al.*, 1987). These differences are related to whether the electron-dense particle is degraded. Unfortunately, the presence of either electron-dense molecule in the lysosome does not necessarily reflect the processing pathway followed by the ligand. Little insulin is degraded in lysosomes (Ward, 1984). Insulin degradation is caused most likely by insulin-degrading enzyme in endosomes or cytoplasm (Hamel *et al.*, 1988; Goldstein and Livingston, 1980, 1981; Akiyama *et al.*, 1988). We have never observed gold particles in the cytoplasm of cells and the presence of endogenous ferritin made it impossible to conclude that ferritin – insulin was in the cytoplasm (Blackard *et al.*, 1986). Although ferritin – insulin (Smith and Jarett, 1987) and intact ^{125}I-insulin (Soler *et al.*, 1989; Thompson *et al.*, 1989) accumulated in the nucleus of intact cells, gold-labeled insulin has not. However, when gold-labeled insulin was prepared with ^{125}I-insulin, autoradiography revealed the ^{125}I-insulin was translocated to the nucleus while the gold particles accumulated in lysosomes (Smith and Jarett, 1991). The conflicts between biochemical and ultrastructural results demonstrate that one must be aware of potential artifacts and exercise constraint in interpreting results.

Despite the limitations in demonstrating intracellular processing of the hormone – receptor complex, studies using electron-dense ligands have provided a great deal of valuable information about the binding, aggregation, and internalization of receptors on the cell membrane (reviewed in Brown and Goldstein, 1979; Willingham and Pastan, 1981; Smith and Jarett, 1988). The use of electron-dense ligands should not be abandoned in favor of immunocytochemical electron microscopy; each approach should be used when appropriate.

III. General Considerations for Successful Immunocytochemical Electron Microscopy

The two types of immunocytochemical electron microscopy, "preembedding" or "postembedding," are based on when the immunolabeling is performed in relation to the fixation and preparation of the specimen for electron microscopy. Preembedding techniques are similar to the use of nonimmunologic electron-dense labels and are restricted to extracellular antigens or those intracellular antigens made accessible by permeabilizing cells or sectioning tissue. When the antibody:antigen complex is formed before the specimen is processed for electron microscopy, fixation and embedding have no effect on formation of the antibody:antigen. The following discussion pertains to postembedding immunocytochemical electron microscopy since tissue preparation techniques performed before the formation of the antibody:antigen complex have a direct bearing on the success of the investigation.

Successful localization of antigens by immunocytochemical electron microscopy requires several conditions be fulfilled simultaneously. Cellular detail must be preserved so structures within the cell can be identified. The antigen must not redistribute to artifactual locations. The antibody must be able to bind to an exposed epitope of the antigen (i.e., antigenicity must be preserved). The technique must provide the antibody access to the antigen. Lastly, the antigen must be present at sufficiently high concentrations to be detected above background labeling.

A. Preservation of Cellular Detail

Ultrastructure preservation and antigenicity preservation sometimes appear to be conflicting objectives. "Antigen inactivation" frequently results from conditions that are not related to specimen preparation techniques. Nevertheless, the choice of specimen preparation methodology may improve the chance for successful results. Immunocytochemical electron microscopy is one way to improve the spacial resolution obtained with immunofluorescent light microscopy and one may be tempted to use the same specimen preparation techniques. Unfortunately, permeabilization and fixation techniques that give good immunofluorescent results permeabilize the plasma membrane and result in the loss of soluble cytoplasmic proteins. These results are frequently unacceptable at the level of resolution obtained in electron microscopy.

Two primary types of tissue preparation techniques have evolved for immunocytochemical electron microscopy: cryoultramicrotomy (Tokuyasu, 1973) and resin embedding (Carlemalm et al., 1982). Cryoultramicrotomy was thought to increase antibody penetration and antigen preservation compared to resin-embedded material. Cryoultramicrotomy is a difficult technique to master and antibody penetration may not be improved (Stierhof et al., 1991). Some antigens may be preserved better by cryoultramicrotomy by avoiding the potentially deleterious effects of antigen extraction or inactivation during dehydration and infiltration with resin monomers (Kellenberger and Hayat, 1991). Our laboratory has used a variety of chemical fixation procedures coupled with resin-embedding techniques. The ultrastructural preservation of a wide variety of tissue and cell types has been adequate and antigenicity of receptors, hormones, growth factors, immunoglobulins, transporters, G-proteins, and other cellular proteins has been retained (see Sections IV, D–G). A major advantage of resin-embedding techniques is that they can be implemented in any electron microscopy laboratory at no additional expense for equipment and no retraining of technical personnel. These techniques are described in Section IV.

B. Preservation of Antigen Localization

There is little chance of artifactual redistribution of membrane-associated antigens caused by specimen preparation techniques as long as membranes remain intact. In contrast, it is difficult to prove that specimen preparation techniques have not allowed, or caused, a redistribution of soluble cytoplasmic antigens. It had been suggested that cryoultramicrotomy avoids these problems; it may not (Stierhof et al., 1991; Kellenberger and Hayat, 1991). Most cryoultramicrotomy is performed on "lightly fixed" or cryoprotected tissues. Fixation techniques prior to cryoultramicrotomy and resin embedding are almost identical. In both cases it is hoped that fixation suspends the soluble protein in the resulting cytoplasmic gel rather than crosslinks it in an artifactual manner to a cytoplasmic organelle. One hopes that epitopes on the antigen remain exposed after the protein is trapped in the cytoplasmic gel.

The difficulty in detecting soluble antigens also can be made clear by explaining the physical aspects of the sectioning and staining processes. Antibodies do not penetrate cryosections or resin-embedded sections (Stierhof et *al.,* 1991, and references therein). Immunostaining results only when the antigen is exposed on the surface of the section. Exposure or protrusion of proteins above the section surface is called "relief." Relief is affected by the physical and chemical properties of the specimen and embedding material and the interaction between the two at their interface (Kellenberger *et al.,* 1987). Large soluble antigens are more likely to have satisfactory relief than small soluble molecules. Dense protein containing granules, for instance insulin granules in pancreatic islet β-cells, show substantial relief when sectioned. This relief accounts for the ease with which insulin has been detected in granules in tissues subjected to fixation conditions that would be totally unsuitable for other antigens. Furthermore, differences in the protein or lipid composition of cellular membranes may cause some membranes to be revealed to a greater extent than others. These differences in relief make quantitative assessments of antigen distribution difficult and prone to error. For example, sectioning reveals more of a membrane's surface than it does small components in the cytoplasmic gel (Stierhof *et al.,* 1991). Therefore, even if an antigen is expressed at equal concentrations in the cytoplasm and on membranes, staining of the membrane-associated antigen will appear to be more intense. Quantifying a stimulus-induced translocation of cytoplasmic antigens to relief-rich structures, such as the plasma membrane, other organelle membranes, or the nucleus, is difficult. These analyses are valid only when the effects of relief have been considered. One might consider using other ultrastructural or biochemical techniques to perform more precise quantitative analysis.

C. Preservation of Antigenicity

From the earliest days of immunocytochemical electron microscopy investigators have struggled with the problem of antigen "inactivation." Inactivation is presumed to occur when a antibody that gave good results in immunoassays, Western blots, or immunoprecipitation fails to react with the antigen in immunocytochemistry. Solubilized, detergent-extracted, often reduced, proteins in Western blots have an entirely different conformation than the same proteins in the cell (Reynolds and Tanford, 1970). Antibodies that recognize solubilized proteins may be unable to bind to the protein denatured by organic solvents such as those used in electron microscopy, because, in most cases, antibodies are directed against conformation-dependent determinants (Sela, 1973). [Refer to Kellenberger and Hayat (1991) for a complete discussion of the effects of fixatives, organic solvents, and heat polymerization on antigen conformation.] To circumvent some of these problems, antibodies have been prepared using antigens exposed to fixatives (Harrach and Robenek, 1990). Although there is no doubt that some antigens are inactivated by

Table I
Effects of Specimen Preparation Techniques on Immunocytochemical Detection of Insulin or IGF-I Receptors

Antibody	4% Paraformaldehyde			1% Glutaraldehyde			4% Paraformaldehyde/ 0.5% glutaraldehyde			Freeze substitution
	Lowicryl	LR White	Spurr	Lowicryl	LR White	Spurr	Lowicryl	LR White	Spurr	
A410	++++	++++	++	++++	++++	++	++++	++++	++	+++
RPN	+++	+++	+	+++	++	+	+++	+++	+	ND
UBI	++	++	++	++	++	+	++	++	+	+++
BEST	+++	+++	++	+++	++	++	++	++	+	+++
AbP5	–	–	–	–	–	–	–	–	–	–
B-d	+	+	–	+	+	–	+	+	–	ND
B2	+	+	–	+	+	–	+	+	–	ND
B10	++	+	–	++	+	+	++	+	+	ND
IR3	+++	+++	++	+++	+++	++	+++	+++	++	+++
BERT	+++	+++	++	+++	+++	+	+++	+++	+	++
LAO	+	+	–	+	+	–	+	+	–	+

Antibodies studied. A410: rabbit polyclonal anti-rat insulin receptor (Jacobs *et al.*, 1978); RPN: Amersham RPN.538, monoclonal anti-human insulin receptor; UBI: Upstate Biotechnologies Inc., monoclonal anti-human insulin receptor; BEST: polyclonal anti-peptide (α subunit 657–670 residues, human insulin receptor) (Rosenzweig *et al.*, 1990); AbP5: polyclonal anti-peptide (β subunit 1328-1343 residues, human insul n receptor) (Herrera *et al.*, 1985); B-d, B2, and B10: autoantibodies to the insulin receptor from Type B patients (Kahn *et al.*, 1981); IR3: monoclonal anti-human IGF-I receptor (Kull *et al.*, 1983); BERT and LAO: polyclonal anti-peptide IGF-I receptor antibody (Rosenzweig *et al.*, 1990).

Note. Cell species types were used in which immunoprecipitation of the receptor had been demonstrated (see various references). Identical concentrations of purified IgG were used for each specimen preparation protocol. All sections were then stained with gold-labeled anti-species antibodies. Immunolabeling was assessed by the relative specific labeling of the plasma membrane to the nonspecific labeling in nuclei. Highly specific labeling is indicated by "++++," nonspecific labeling is indicated by "–," and not determined is indicated by ND.

fixation, dehydration, or resin components, systematic investigations in our laboratory suggest that these processes are not the only reasons, and sometimes not the primary reason, for failure to detect cellular antigens. Normal processes, e.g., change in phosphorylation state and dissociation of protein subunits, change the conformation of a proteins, which may mask or unmask antigenic epitopes (see Section III,E). In addition, antibodies prepared from solubilized proteins, or peptide sequences, may be directed against epitopes that are not exposed in the cell (transmembraneous domains) or are highly conformation dependent and therefore more susceptible to the effects of specimen preparation procedures.

We compared the efficacy of eight anti-insulin receptor and three anti-insulin-like growth factor-I (IGF-I) receptor antibodies to detect rat or human (IM-9 lymphocytes or MG63 osteosarcoma cells) insulin or IGF-I receptors by immunocytochemical electron microscopy. Each of the antibodies has been reported to immunoprecipitate their respective receptor; many, but not all, detected receptors in Western blots. Three aldehyde fixatives, 4% paraformaldehyde, 1% glutaraldehyde, and a combination of 4% formaldehyde and 0.5% glutaraldehyde, were tested. Three different resins, Lowicryl (Carlemalm et al., 1982), LR White (Newman, 1989), and Spurr (Spurr, 1969) were used. In addition, freeze substitution without fixation (Linner et al., 1986) was tested. Thin sections of these 10 conditions were all stained using identical techniques (described in Section IV,C, under the protocol Indirect Immunostaining Procedure). An antibody was considered to give a positive reaction when it detected receptors on the plasma membrane. As shown in Table I, the different fixatives had little or no effect on the efficacy of labeling. Samples embedded in Spurr resin had increased nonspecific labeling, probably absorption of the immunoglobulin, to the resin. There was no correlation in the ability of these antireceptor antibodies to react in immunoprecipitation, Western blot, and immunocytochemical electron microscopy systems. We have performed similar analyses of antibodies to a variety of membrane-associated and cytoplasmic antigens. Unfortunately, the ability of antibodies to react in Western blot analysis does not guarantee that they will react in immunocytochemical studies. These findings suggest that the lack of a reaction in immunocytochemical electron microscopy is more likely to be caused by using an antibody directed at a hidden or masked epitope or a conformation-dependent epitope than by a general inactivation of the antigen. Our experience, and that of others (Kellenberger and Hayat, 1991), suggests that polyclonal antibodies are frequently superior to monoclonals. If the monoclonal antibody is directed against a masked or conformation-dependent epitope it is more likely to be nonreactive than a polyclonal antibody with multiple epitopes. There may be advantages to using antipeptide antibodies (see Section V,A) in immunocytochemistry, particularly if they are directed against epitopes that are known to be accessible and not easily deformed by specimen preparation techniques. Peptides of 12–19 amino acids have produced antibodies that appear to specifically recognize glucose transporters in both Western blot and immunocytochemical electron microscopy (Slot *et al.*, 1991; Friedman *et al.*, 1991; Smith *et al.*, 1991). However, an antipeptide antibody directed against a 15-amino acid carboxyl-terminal sequence (1328-1343) on the β subunit of the insulin receptor (Herrera *et al.*, 1985), which demonstrated great specificity for the immuno-purified insulin receptor in Western blots, bound to virtually every cellular structure in several cell types tested (unpublished observations, R. M. Smith and L. Jarett). This localization, while immunologically "specific," was not thought to be specific for the insulin receptor. Techniques used to assess antigen specificity in immunocytochemical electron microscopy are discussed under Section IV,C,2.

D. Providing Antibody Access to Intracellular Antigens

Antibodies cannot penetrate cell membranes, although they can be internalized by receptor-mediated processes (Rodewald and Kraehenbuhl, 1984). Mechanical processes must be used to bring intracellular antigens and the antibody together. A number of different techniques exist; each has its benefits and problems. Some (e.g., freeze fracture, microinjection, mechanical shearing) may require specialized equipment or technical expertise. Permeabilization techniques similar to those used in light microscopy that result in permanent holes in or destruction of the plasma membrane have limited, though valuable, applications, if one accepts the loss of cytoplasmic proteins and generally poor ultrastructural preservation. Temporary permeabilization, e.g., hypotonic shock, electroporation, digitonin treatment, and microinjection, are good ways to get gold-labeled particles inside cells. These techniques generally have a high level of nonspecific labeling because there is no easy way to remove nonspecifically associated ligand except to repermeabilize the cell. However, temporary permeabilization is useful in studying the movement of labeled proteins in cells. For instance, microinjection of RNA- and nucleoplasmin-labeled gold particles has been used successfully to study nucleocytoplasmic translocation of cellular proteins in *Xenopus* oocytes (Feldherr *et al.*, 1984).

Sectioning tissue after cryofixation or resin embedding provides access to both intracellular and extracellular antigens, but only at the surface of the embedding material as explained above. Several laboratories have experimented with variations on the sectioning process. Fresh or lightly fixed tissue can be sliced ($\sim 20-300$ μm thick) using a cryomicrotome or vibratome. The tissue sections are then incubated with labeled antibodies or ligands. These thick sections are then washed to remove nonspecifically associated ligand, fixed, dehydrated, and embedded in resin, and thin (700 Å thick) sections are cut for observation in the electron microscope. Soluble antigens may redistribute in unfixed tissue after the thick sections are cut. Labeling of extracellular antigens is significantly improved in unfixed tissues, although care has to be taken to wash out nonspecifically trapped gold-labeled antibody complexes. Improved labeling of intracellular membrane-associated antigens occurs only in the cells that have been cut open by the microtome but, unless some form of fixation is used prior to the sectioning, this layer of cells is badly deformed (Priestly and Cuello, 1983). If the tissue is fixed, the antibodies penetrate no better in thick sections than in thin sections (Piekut and Casey, 1983).

Several studies have been performed on "resin-less" sections (Capco et al., 1984; Fey et al., 1986; Wang and Traub, 1991). These have limited applications because the antigen of interest must be localized to a cellular structure, e.g., filament, membrane, nuclear matrix, that remains intact during the harsh deplasticizing process. Others (Pinto da Silva et al., 1971; Robenek et al., 1982) have combined freeze-fracture/etching techniques and immunocytochemical electron microscopy. The "replica" technique and several variations that pop open cells have been used to localize antigens to one or both sides of the plasma membrane (Rutter and Hohenberg, 1991).

E. Effects of Epitope Masking/Unmasking

Recent studies in our laboratory suggest that epitopes of some cellular proteins can be masked or exposed as a consequence of cellular processes. For instance, carboxyl-terminal antipeptide antibodies to GLUT4 did not detect intracellular glucose transporters in rat adipocytes (Smith et al., 1991). Intracellular GLUT4 was readily detected with amino-terminal antibodies. However, both antibodies reacted with transporters translocated to the plasma membrane as a result of insulin treatment. These results suggest that the intracellular carboxyl-terminal epitopes of GLUT4 were masked. Epitope masking has been reported by other laboratories as well (Takamiya et al., 1980; Sheehan et al., 1991). Epitope masking presents some obstacles to immunoelectron microscopists at the present time, particularly when it is interpreted as antigen inactivation. Epitope masking can also complicate assessments of antigen translocation if the antigen is largely masked in one location and unmasked in another. However, future studies may be able to take advantage of epitope unmasking. If the masking or unmasking of specific epitopes in cellular proteins can be related to conformational changes resulting in (or from) activation or deactivation processes, and antipeptide antibodies can be prepared against those epitopes, biologists will have another tool to explore cell metabolism (see Sections IV,F and V,A).

F. Summary

Several specimen preparation techniques are available that preserve adequate cellular detail and antigenicity. Some of these techniques, particularly the ones presented below, can be implemented in any electron microscopy laboratory with little or no additional training or cost. Based on our experience, the most important factor in determining whether an antigen can be localized by immunocytochemical electron microscopy is the antibody. This might appear to be an absurdly obvious observation. But what it implies is that, with relatively limited and fundamentally similar specimen preparation techniques, an antibody is likely fail or succeed based solely on the epitope to which it is directed, irrespective of which immunocytochemical specimen preparation technique is used.

In addition, one should consider using alternative ultrastructural or biochemical techniques, rather than immunocytochemical electron microscopy, based on the objectives of the investigator. These alternatives may be particularly important when the biologist is interested in quantitative data rather than gaining high-resolution structural–functional information. It is easy to count gold particles on a micrograph. Therefore, it is tempting just to count the number of particles clustered over some cellular organelle, compare it to a control tissue, and plug the data into a paired t test to determine if the difference is significant. This type of analysis is a gross oversimplification and probably invalid more times than it is valid. Inaccuracies come from the problems specific to immunocytochemical electron microscopy we have already discussed, e.g., variations in relief and epitope masking. In addition, nonspecific absorption, differences in antibody affinity, the size of the gold particle, the molar ratio of ligand or antibody to gold particle, the number of gold particles per antigen, whether a direct or indirect labeling method was used, etc., also contribute to errors in quantitative analysis (Park *et al.,* 1989; Kellenberger and Hayat, 1991). These sources of error are in addition to the variability in specimen sampling common to all thin-section analysis (Hammel, 1986). In many cases, quantitative analysis in immunocytochemical electron microscopic is more suited to phrases like "apparently different" than "significantly different ($P < 0.05$)."

IV. Localization of Hormones, Receptors, Transducers, and Transporters

In the following pages methods are described that we have used to localize hormones, receptors, and other cellular proteins involved in cellular metabolism. In addition, brief examples of some of the results obtained in collaboration with investigators at the Diabetes and Endocrinology Research

Center at the University of Pennsylvania are presented as illustrations of the types of studies that can utilize immunocytochemical electron microscopy.

A. Specimen Preparation

As described above (Section III), specimen preparation methods must fulfill four criteria: the preservation of (1) cellular detail, (2) antigen location, and (3) antigenicity and (4) the provision of antibody access to the antigen. Specimen fixation and preparation techniques play a critical role in meeting these objectives.

1. Fixation Techniques

We initially used several standard types of aldehyde fixatives as described in Table 1. The contrast and cellular detail, particularly of cellular membranes, is less than that found with routine electron microscopy fixation techniques in which osmium tetroxide is used as a secondary fixative because of its ability to fix lipids. Many investigators have experienced decreased antigenicity or increased nonspecific reactions when osmium is used (Roth *et al.*, 1981), probably due to the deposition of heavy metals in lipids. In contrast, some studies have reported successful localization of hormones and, in one case, glucose transporters (Friedman *et al.*, 1991) in osmicated tissue. Methods have been suggested for reversing the deleterious effects of osmium (Baskin *et al.*, 1979; Bendayan and Zollinger, 1983). However, the efficacy of such procedures is controversial (Varndell and Polak, 1987). In our experience, membrane-associated antigens are almost always inactivated by osmium and we avoid its use. We have tried several different fixative recipes that avoid osmium but increase preservation of cell membranes and contrast of cellular proteins.

Studies have demonstrated that buffered aldehyde fixation retains immunoreactivity of a large number of cellular proteins (Van Ewijk *et al.*, 1984) including hormone receptors (Robenek *et al.*, 1982) (see Table I and Section IV,E). However, some investigators have reported that aldehyde fixation frequently results in decreased specific and increased nonspecific binding of *hormones* to receptors (Salih *et al.*, 1979; King and Baskin, 1991). In some cases, decreased specific binding may be caused by a lack of receptor internalization and recycling in fixed cells, thus reducing the total number of receptors exposed on the cell surface during the time course of an incubation and the total level of receptor occupancy. Other studies reported a decrease in receptor affinity (Salih *et al.*, 1979) that may result from restricted receptor mobility and an impairment of cooperative interactions between hormone-occupied receptors. These effects are important considerations if the investigator plans to incubate fixed cells with a hormone and then use anti-hormone antibodies and immunocytochemical techniques to localize the receptor. However, the availability of specific antibodies to the desired receptor would eliminate the need to use this indirect procedure.

Additions to the primary aldehyde fixative, such as picric acid, potassium ferricyanide, calcium, and sucrose, have been reported to improve preservation and/or cellular detail without affecting immunoreactivity. In the examples of mammalian tissues and cultured cells shown below we have used two fixatives: (PFeCN)—4% paraformaldehyde, 50 mM $K_3Fe(CN)_6$, 1 mM $CaCl_2$ in 0.1 M sodium cacodylate buffer, pH 7.4; or (GPA)—1% glutaraldehyde, 0.2% picric acid, 154 mM NaCl in 50 mM sodium phosphate buffer, pH 7.4.

Paraformaldehyde-potassium ferricyanide fixative

20 ml	0.2 M sodium cacodylate-HCl buffer, pH 7.4
10 ml	deionized water
10 ml	16% paraformaldehyde (EM grade, sealed under nitrogen in 10-ml ampules)
0.66 g	$K_3Fe(CN)_6$
4.4 mg	$CaCl_2$

This fixative is usually prepared just before use but may be stored for no longer than 24 hr at 4°C in the dark. During fixation the tissue should be light protected.

Glutaraldehyde-picric acid fixative (GPA):

17.7 ml	40 ml 0.1 M sodium phosphate buffer, pH 7.4
17.7 ml	deionized water
10 ml	8% glutaraldehyde (EM grade, sealed under nitrogen in 10-ml ampules)
0.72 g	NaCl
12.3 ml	1.3% (w/v) picric acid

This fixative is stable for at least 1 week at 4°C. (Caution should be exercised in storing the stock solution of picric acid to prevent its evaporation, and in its handling to prevent combustion. Read the cautions on the bottle label.)

Isolated or cultured cells and small (<3 mm thick) tissues are fixed by immersion in either fixative for 0.5-2 hr. We have observed no deleterious effects on antigenicity (compared to 1-hr fixations) in specimens immersed in fixative for up to 4 days (the longest time systematically investigated). This means that samples can be fixed in one laboratory and carried or shipped (properly packaged and labeled) to an electron microscopy laboratory elsewhere. Tissues are usually fixed at 4°C to reduce proteolysis. However, cultured cells attached to culture dishes or porous membrane supports frequently lift off of those surfaces when fixed at temperatures below 15-24°C. After trying numerous types of porous cell culture membranes, we found the Falcon Cell Culture Insert (catalog No. 353095) superior. Cells spread evenly and do not grow through the pores. The filter withstands dehydration and embedding, sections well, and is stabile under the electron beam. (See Fig. 3.) When possible, larger organs, e.g., liver, muscle, kidney, should be fixed by local or whole-body perfusion. Following the initial perfusion fixation, tissues should be dissected and carefully minced in the fixative into small (2-3 mm^3) pieces.

2. Dehydration and Embedding Procedures

Dehydrating biologic tissues with solvents, infiltrating the tissue with epoxide resins, and heat curing the resins are, at least theoretically, procedures that could denature antigens (Kellenberger and Hayat, 1991). Cryoultramicrotomy techniques avoid these problems but ultrastructural detail is frequently sacrificed. Furthermore, cryoultramicrotomy presents formidable barriers to many biologists, e.g., availability of equipment and technical expertise. Two resin-embedding techniques provide excellent ultrastructural detail and good antigen preservation. LR White is a water-miscible acrylic resin that avoids the use of high ethanol concentrations. It is usually polymerized by heating at 50-60°C, but can be polymerized by catalyst or UV radiation. The methacrylate-based Lowicryl resins (K4M, HM20, K11M, and HM23) require complete dehydration. Because they are low-viscosity resins, dehydration, infiltration, and curing can be performed at $-20°C$ or lower which should decrease any deleterious effects on antigenicity. LR White is usually easier to section than Lowicryl resins and may be the best multipurpose resin, especially when the laboratory does not have freezers set to -20 and $-35°C$, with UV lamps in the $-35°$ freezer as required for Lowicryl. Our laboratory has used LR White resin almost exclusively as described below. The use of Lowicryl resins has been described recently in great detail (Newman, 1989; Villiger, 1991) and vendors supply basic instructions for their use.

DEHYDRATION, INFILTRATION AND EMBEDDING USING LR WHITE RESIN

The following procedures are performed at room temperature. Specimens are usually processed in small (3–5 ml) glass vials with plastic caps filled to ~80% of capacity.

1. Wash PFeCN fixed specimens (for 5 min each) twice with 0.1 M sodium cacodylate buffer and twice with deionized water to remove unbound fixative and salts. Omit aqueous washes of tissue fixed with picric acid-containing fixatives and tissue goes immediately into 50 or 70% ethanol.
2. Dehydrate the tissue with three 10-min immersions in 70% ethanol.
3. Infiltrate the specimen with one part 70% ethanol:two parts LR White resin in two 30-min treatments. (If holes are observed in the sectioned specimen, dehydration was not sufficient. If antigenicity is adequate at 70% ethanol, change the dehydration procedure to one 70% and three 80%, or one each 70% and 80% and three 90%, ethanol treatments. Infiltrate with one part ethanol at the highest concentration used to two parts LR White resin). The specimen vials are attached to a rotating device (rotating in a nearly vertical plane at 2–5 rpm) to facilitate mixing and reduce infiltration time.
4. Continue the infiltration with two changes of LR White resin for 30–60 min.
5. Add fresh resin and leave overnight on the rotator.
6. The following morning, transfer the specimens to fresh resin for another 30–60 min before embedding in gelatin capsules. It is extremely important that the capsules be completely filled with resin and the capsules sealed.
7. Cure the resin at 50–60°C for 24–36 hr.

Various modifications of this procedure have been reported. For instance, final ethanol concentrations range from 30 to 100%. The lower concentrations, while risking poor infiltration of the resin, appear to enhance antigenicity of some cytosolic antigens. Concentrations higher than 70% may improve, or shorten the time required to obtain, infiltration of the resin in the tissue at the risk of decreasing antigenicity. LR White also can be polymerized at low temperatures using a catalyst or UV radiation.

Thin (~700 Å) sections are cut with a diamond knife and collected on uncoated nickel grids. The copper grids used in routine electron microscopy generally react with the staining solutions and should be avoided. Magnetized forceps must be demagnetized before being used with nickel grids. These grids cause astigmatism in the electron microscope (objective lens) which must be corrected frequently. Anecdotal evidence suggests that immunostaining of some membrane-associated antigens is better when specimens are stained within 4 hr, as compared to several days, after sectioning (Smith et al., 1991).

B. Immunolabeling Techniques

There are two general methods for detecting antigens exposed on the surface of thin sections: direct and indirect. Direct staining is accomplished by complexing the antibody to the gold particle and incubating the section with the complex. Indirect staining entails incubating the section with the primary antibody. The primary antibody is then detected with gold complexed to a second antibody or protein A or G or AG. Other differences between the two techniques are discussed below (Section IV,C). Irrespective of which technique you use, proper preparation of the electron-dense gold-labeled ligands is important. Fortunately, ligand preparation is easy and well within the abilities of any laboratory.

PREPARATION OF COLLOIDAL GOLD

There are several good methods of preparing colloidal gold particles of various sizes: 3–5 nm, Faulk and Taylor (1971); 5–15 nm, Slot and Geuze (1985); 16–150 nm, Frens (1973). The method described here was developed by Slot and Geuze (1985) to prepare 10-nm-diameter particles. Particle diameter is modified by changing the tannic acid concentration.

To produce 100 ml of a 10-nm-diameter colloid prepare two solutions in clean glassware as follows:

Solution A: 1 ml 1% (w/v) $HAuCl_4$ and 79 ml deionized water

Solution B: 4 ml 1% sodium citrate, 0.1 ml 1% tannic acid (Aleppo tannin, Mallinckrodt Chemical Company, St. Louis, MO), 0.1 ml 25 mM K_2CO_3, and 15.8 ml deionized water

All reagents are dissolved in deionized water; stock solutions of gold chloride, sodium citrate, and tannic acid can be refrigerated for 1 month. The 25 mM K_2CO_3 solution should be prepared on the day of use. Filter (0.22-μm filter) all solutions prior to use. Solution A should be in a 250-ml beaker with a magnetic stirring bar, solution B in a 50-ml beaker. Heat solutions A and B to exactly 60°C in a water bath (cover the beakers with watch glasses to reduce evaporation). Move solution A to hot plate stirrer (which has been preheated sufficiently to boil water) and begin mixing vigorously. Immediately add the entire contents of solution B *in one quick motion* to solution A. Mix the solution until it reaches 100°C. The solution will become wine or red colored. After the solution has turned red and been at 100°C for 2-5 min, remove and allow to cool. Check the volume and adjust to 100 ml with deionized water. Filter the colloid through a 0.45-μm sterile filter and store at 4°C in a clear glass bottle. No significant aggregation of the colloid has been observed during storage. The cost of making 100 ml colloidal gold is less than $10.00 (in 1992).

The size of the gold particles should be checked by drying down a 1/10 dilution of the colloid on a carbon-coated grid. Photograph the particles at high magnification and determine their size on suitably enlarged electron micrographs. The average size of particles in a single preparation should not vary more than ±10-15% from the desired diameter. The range of sizes can be appreciably worse if the mixing of solutions A and B is not done instantaneously. The molar concentration, i.e., number of particles per milliliter, should be estimated as described by Horisberger (1979) or Ackerman *et al.* (1983). (Our estimates of the particles per milliliter prepared with this method are 5 nm = 4×10^{13}; 10 nm = 5×10^{12}; 15 nm = 1.5×10^{12}.)

PREPARATION OF COLLOIDAL GOLD-IgG COMPLEXES

Gold particles in low ionic strength solutions are stable. However, the particles flocculate if exposed to electrolytes and the solution turns from red to blue-gray. When sufficient protein is added to the gold solution under appropriate conditions, the protein will absorb to the gold particles and protect them from salt-induced flocculation. This phenomena is used to determine the amount of protein needed to "stabilize" the colloid. Although the precise nature of the interaction (electrostatic or hydrophobic) between the gold and the added proteins is still debated, it is generally accepted that the pH of the gold and protein solutions should be adjusted to slightly above (+0.5 units) the isoelectric point of the protein (Geoghegan and Ackerman, 1977; Horisberger, 1979). The reaction between the gold particles and added proteins occurs rapidly, usually in minutes if not seconds. The number of protein molecules complexed to a gold particle depends on the relative sizes of both and, for a given gold particle size, is inversely proportional to the size of the protein. For instance, 4 immunoglobulin, 60 protein A, or >200 insulin molecules can absorb to a 15-nm gold particle (Baudhuin *et al.*, 1989). Although the multivalency resulting from these complexes is not important in postembedding immunocytochemical studies, its effects should be considered in preembedding procedures and steps taken to reduce the ratio of ligand to gold particle (Smith and Jarett, 1991).

The preparation of immunoglobulin gold complexes is quite simple and straightforward:

1. Immunoglobulin fractions should be prepared from sera by any acceptable technique, e.g., ammonium sulfate fractionation, Affigel or protein A affinity chromatography, or ion-exchange chromatography. The resulting Ig should be extensively dialyzed against 2 mM borax-HCl buffer, pH 9.0, to reduce ionic strength. If necessary, the protein should be concentrated to ≥ 1 mg/ml. Protein aggregates should be removed by centrifugation at 100,000 g immediately prior to use.
2. The pH of the gold solution should be adjusted to pH 9.0 with 0.2 M K_2CO_3 just before use.
3. Determine the concentration of immunoglobulin required to stabilize the colloid. Place 0.2-ml aliquots of the gold solution in 10 (1- to 2-ml) glass test tubes. Add 10 μl of the serially diluted protein solution (10-500 μg/ml) to each tube and mix. After 5 min add 50 μl 10% NaCl and mix. Tubes in which the gold solution turns blue-gray have insufficient protein; tubes that remain red are stabilized. Depending on the increments used in the serial dilutions, and the concentration difference between the lowest protein concentration stabilizing the colloid and the highest protein concentration permitting flocculation, the assay should be repeated over a narrower protein concentration range. [The color change can usually be perceived quite easily, especially with particles 10 nm or larger. However, for smaller particles or investigators with color perception problems, the assay can be scaled up to volumes suitable for spectrophotometric determinations. The optical density of the solution is then determined at 580 nm (Geoghegan and Ackerman, 1977).]
4. The immunoglobulin-gold complex is formed by mixing together 10 ml of gold colloid (pH adjusted) and protein 5-10% in excess of the concentration required to stabilize the colloid. We rapidly add the immunoglobulin to the mixing gold colloid. Some investigators suggest slowly adding protein to gold, others rapidly add gold to protein. Mix the solution and let it stand for 5 min.
5. Add 10% bovine serum albumin (dissolved in water and adjusted to pH 8.5-9.0 with 0.2 M K_2CO_3) to a final concentration of 1% to further stabilize the colloid.
6. The excess uncomplexed immunoglobulin must be removed from the solution. Several authors use ultracentrifugation (see Leunissen and de Mey, 1989, for appropriate centrifugation conditions for different sizes of particles). Centrifugation usually results in gold aggregates that must be discarded. We prefer to use exclusion column chromatography to separate the gold complex, in the void volume, from the

free ligand which is retained. After chromatography, we use Amicon ultrafiltration devices with high-molecular-weight cutoff filters (YM-100) to concentrate the gold complex solution. If small (1- to 2-ml) volumes of gold complex are prepared, ultrafiltration, repeated three to five times with 10-fold dilutions, adequately removes uncomplexed proteins and avoids both ultracentrifugation and column chromatography.

7. Sterile filter the complex (0.22-μm filter) and store in 500-μl aliquots at 4°C. Immunoglobulin-gold complexes are stable for many months as long as sterility is maintained.

1. Preparation of Colloidal Gold-Protein A, G, AG Complexes

The preparation of gold-protein A complexes has been described (Romano and Romano, 1977; Roth et al., 1978). In addition, gold complexes with protein G (Balslev and Hansen, 1989; Bendayan, 1989) and protein AG (Ghitescu et al., 1991) can be prepared. These complexes react with the Fc regions of a wide variety of species and classes of polyclonal and monoclonal immunoglobulins. The only practical difference between the preparation of immunoglobulin-gold complexes as described above and protein A/G/AG complexes is the pH used. The isoelectric points for proteins A, G, and AG are 4.75, 4.4, and 4.3, respectively (Ghitescu et al., 1991). Reactions performed at pH values of 5.5 to 6.0 provide satisfactory results. Flocculation assays should be performed as described above.

C. Immunostaining Protocols

There are two approaches for detecting antigens exposed on the surface of thin sections: direct and indirect immunostaining. Direct immunostaining is accomplished by incubating sections with a colloidal gold-antibody complex that directly binds to the antigen of interest. Indirect staining entails incubating the section with the primary antibody, which is then detected with gold complexed with a second antibody or protein A, G, or AG.

Indirect labeling frequently, if not always, results in the observation of multiple gold particles per cellular antigen. This is a disadvantage if one is interested in determining whether an antigen is aggregated because multiple gold particles may be clustered over a single antigenic site. The ratio of gold particles to antigenic sites is affected by the size of the gold particles and whether the primary and/or secondary antibodies are poly- or monoclonal. The ratio of gold particles to antigenic sites may be higher in areas where the antigen is dispersed (and there is little steric interference) than in areas where sites are actually clustered and there is appreciable steric interference. Therefore, assessing aggregation or dispersion resulting from some metabolic stimulus may be difficult using indirect techniques. On the other hand, because indirect labeling increases the number of antigen-associated gold particles it increases sensitivity and is valuable in combined light and electron microscopic studies of low-concentration antigens. Indirect labeling is also easier from a logistical viewpoint. A laboratory with gold-labeled protein A and G or AG or gold-labeled secondary antispecies antibodies to frequently used primary species, e.g., mouse and rabbit, can use indirect techniques with virtually any antibody to attempt to localize antigens.

In general, direct labeling methods result in less nonspecific background labeling because there is only the one reagent capable of nonspecific association. Spacial localization is more precise because the distance between the gold particle and the antigen is less than that in indirect methods (because of fewer intervening molecules between the gold particle and the antigen). There is also a better chance at performing semi-quantitative analysis than in indirect techniques because steric inhibition should be reduced, though not eliminated, in highly clustered antigens. [Steric inhibition is always greater with large (≥ 15 nm) particles than with small particles]. Furthermore, multiple labeling of antigens in a single section in one staining operation can be accomplished with different primary antibodies complexed to different-sized gold particles. Multiple staining using indirect techniques is possible but more cumbersome (see Section IV,C,3).

In practice, our laboratory uses indirect labeling techniques to determine whether or not an antigen can be detected with a particular antibody. If successful, and if the antibody is available in sufficient quantities, direct labeling is performed using primary antibody–gold complexes. As alternatives to the procedures described below, Varndell and Polak (1987) and Bendayan (1989) provide step-by-step instructions.

INDIRECT IMMUNOSTAINING PROCEDURE

Sections (~ 700 Å thick) are collected on nickel grids and allowed to dry in a dust-free area. Anecdotal evidence suggests that sections should be stained the same day they are cut. Airborne dust, if present, must be avoided. We use a 15-cm glass petri dish as an incubation chamber to protect the grids during staining procedures and cover the grids with an inverted Plexiglas box during washing. Adequate agitation of the grids during the washing steps reduces the amount of time and the number of washes needed. Our laboratory uses a Belly Dancer (Stovall Life Sciences, Inc) that provides an adjustable orbital motion that moves the grids around the top of the fluid. Others place the grids on waxed surfaces on magnetic stirrers and agitate the grids by turning on the stirrer at low speed. (This magnetizes the nickel grids and they should be demagnetized before being examined in the electron microscope.) If you cannot agitate the grids, increase the number of washes twofold:

1. Cut a piece of Whatman filter paper to fit the bottom of a 15-cm glass petri dish and moisten the filter paper with deionized water. Cut a piece of Parafilm or clean dental wax and place it on top of the filter paper. (The Parafilm should be large enough to allow the grids to be separated by $\frac{1}{2}$ inch, but smaller than the moistened filter paper.) Pipette onto the Parafilm drops ($\sim 20\ \mu l$) of a wetting (blocking) solution: 1% ovalbumin, 0.2% cold-water fish skin gelatin, in 10 mM phosphate-buffered normal saline, pH 7.4 (PBS). [If the primary antibody is detected with a gold-labeled anti-species antibody, Fc receptors, when present, can be blocked with a normal immunoglobulin from a third species (see Section IV,C,2).]

2. Float the grids, section-side down, on top of the drop of wetting solution, one grid per drop. Allow to stand with gentle agitation for 30–60 min at room temperature.

3. Pipette onto the Parafilm drops ($\sim 20\ \mu l$) of appropriately diluted (in PBS) primary immune antisera or control antisera (see Section IV,C,2).

4. Remove the grids from the wetting solution with fine forceps. Drain the wetting solution from the grids as completely as possible using fiber-free absorbant filter paper by drawing the liquid from between the forceps points. Avoid touching the grids with the filter paper.

5. Float the grids, section-side down, on top of the drops of antisera, one grid per drop. Take care to avoid cross-contaminating the antisera by washing the forceps with deionized water from a squirt bottle and drying the forceps between picking up the grids.

6. Incubate the sections for 1–2 hr at room temperature with gentle agitation or overnight in the covered petri dish at 4°C. The length of the incubation varies depending on the affinity of the antisera, its dilution, etc., and must be determined experimentally.

7. The wells of disposable microtiter plates (96 well) are filled with wash buffer [10 mM Tris–HCl-buffered normal saline (TBS)] so that a convex surface forms. Fill four times the number of wells as there are grids to be washed. In addition, if room permits, fill another set of wells (same number as there are grids) with 1% ovalbumin in PBS.

8. Remove the grids from the antisera with fine forceps. Drain the fluid from the grids with fiber-free absorbant filter paper as before. Float the grids, section-side down, on top of the wash buffer for 5 min with gentle agitation. Pick up the grids, drain the fluid with filter paper, and transfer to next wash solution. Repeat for a total of four washes.
9. Remove the grids from the washing solution and drain the excess fluid as before. Place the grids on a drop of 1% ovalbumin in PBS for 10 min with gentle agitation.
10. Remove the grids from the 1% ovalbumin in PBS and drain the excess fluid as before. Place the grids on a drop of appropriately diluted gold-labeled protein A, G, AG, or antispecies secondary IgG and incubate for 1 hr. (Our complexes are stored concentrated 10-fold compared to the original gold solutions. We use a 1:10 dilution which brings the incubation concentration back to the original concentration of the gold solution before complexing with antibody. The solution should be visibly pink.)
11. As in step 7 above, fill wells of a microtiter plate (four for each grid) with TBS. Fill an additional two wells per grid with deionized water. Remove the grids from the gold-labeled reagent and remove excess fluid as before. Float the grids, section-side down, on top of the wash buffer for 5 min with gentle agitation. Pick up the grids, drain the fluid with filter paper, and transfer to the next wash solution. Repeat for a total of four washes with wash buffer and two washes with deionized water. Remove the grids from the washing solution and drain the excess fluid as before.
12. Counterstain the sections with 2% neutralized aqueous uranyl acetate for 3–5 min. Additional staining with bismuth subnitrate or lead citrate (for 0.5–2 min) can be used if desired.
13. Wash grids with deionized water as usual after counterstaining and dry in a dust-free area.

DIRECT IMMUNOSTAINING PROCEDURE

1. Cut a piece of Whatman filter paper to fit the bottom of a 15-cm glass petri dish and moisten the filter paper with deionized water. Cut a piece of Parafilm or clean dental wax and place it on top of the filter paper. (The Parafilm should be large enough to allow the grids to be separated by $\frac{1}{2}$ inch, but smaller than the moistened filter paper.) Pipette onto the Parafilm drops (~ 20 μl) a wetting (blocking)

solution: 1% ovalbumin, 0.1% bovine serum albumin, in 10 mM PBS, pH 7.4. [Fc receptors, when present, can be blocked with normal immunoglobulin (see Section IV,C,2).]

2. Float the grids, section-side down, on top of the drop of wetting solution, one grid per drop. Allow to stand with gentle agitation for 30–60 min at room temperature.

3. Pipette onto the Parafilm drops (~20 μl) of appropriately diluted (in PBS) gold-labeled immune antisera or control antisera (see Section IV,C,2). (Our complexes are stored concentrated 10-fold compared to the original gold solutions. We use 1:10–50 dilutions of the stored complexes. The solution should be visibly pink.)

4. Remove the grids from the wetting solution with fine forceps. Drain the wetting solution from the grids as completely as possible using fiber-free absorbant filter paper by drawing the liquid from between the forceps points. Avoid touching the grids with the filter paper.

5. Float the grids, section-side down, on top of the drops of appropriately diluted gold-labeled antisera, one grid per drop. Take care to avoid cross-contaminating the antisera by washing the forceps with deionized water from a squirt bottle and drying the forceps between picking up the grids.

6. Incubate the sections for 1–2 hr at room temperature with gentle agitation or overnight in the covered petri dish at 4°C. The length of the incubation varies depending on the affinity of the antisera, its dilution, etc., and must be determined experimentally.

7. The wells of disposable microtiter plates (96 well) are filled with wash buffer (10 mM TBS) so that a convex surface forms. Fill four times the number of wells as there are grids to be washed. In addition, if room permits, fill additional wells (two for each grid) with deionized water.

8. Remove the grids from the gold-labeled antisera with fine forceps. Drain the fluid from the grids with fiber-free absorbant filter paper as before. Float the grids, section-side down, on top of the wash buffer for 5 min with gentle agitation. Pick up the grids, drain the fluid with filter paper, and transfer to next wash solution. Repeat for a total of four washes with wash buffer and two washes with deionized water.

9. Counterstain the sections with 2% neutralized uranyl acetate for 3–5 min. Additional staining with bismuth subnitrate or lead citrate (for 0.5–2 min) can be used if desired.

10. Wash grids as usual after counterstaining and dry in a dust-free area.

2. Detection of Multiple Antigens

Multiple antigens can be detected using either the indirect or direct labeling techniques. In indirect multiple labeling, the primary antisera must be raised in different species. The gold-labeled anti-primary species antibodies must be prepared with different-sized gold particles. When using direct labeling, the size of the gold particles linked to the primary antibodies must be different. In general, it is best to start with single-labeling procedures to determine the optimum concentration of reagents and labeling conditions before attempting multiple-labeling protocols. Then incubate the grids with a mixture of primary antibodies using the same final concentrations of each that worked in single-labeling experiments. Then incubate with a mixture of gold-labeled anti-primary species antibodies. Sequential incubations sometimes result in a reduction in the labeling of the first antigen. The sections should also be incubated simultaneously with a mixture of gold-labeled anti-primary species antibodies. In direct multiple-labeling protocols, the sections should also be incubated simultaneously with both gold-labeled primary antibodies. Because the extent of labeling is somewhat dependent of the size of the gold particle, it may be necessary to perform multiple-labeling studies reversing the size of the gold particles on primary or secondary anti-primary species antibodies to establish the best conditions for localizing both antigens. For more detailed discussions of multiple labeling see Varndell and Polak (1984) and Doerr-Schott (1989).

3. Controls to Test Specificity

There are four potential sources of nonspecific labeling: (1) absorption of one or more ligands to the tissue or resin, (2) the binding of a gold-labeled antibody to cellular Fc receptors, (3) the binding of protein A, G, or AG to endogenous immunoglobulins, and (4) the reaction of the primary antibody with something other than the intended antigen.

Absorption, or low-affinity binding, is usually caused by using ligands at too high a concentration. Serial dilutions of all reagents should be used to determine the concentrations that eliminate absorption.

Potential binding of the gold-labeled antibodies to Fc receptors is generally ignored because of the belief that such receptors are inactivated by fixation and dehydration. That cannot be assumed. If a primary antibody binds to an Fc receptor in addition to its antigen, and the primary antibody is detected with gold-labeled protein A, G, or AG, no artifact would be expected because the gold-labeled molecule binds to the Fc portion of the primary antibody attached to its antigen and cannot bind to antibodies attached to Fc receptors. However, if a gold-labeled secondary anti-primary species antibody is used to label the primary antibody, the complex can bind to primary antibody bound to the antigen and Fc receptors. Therefore, Fc receptors should be blocked with an immunoglobulin fraction from a third species prior to the addition of the primary antibody when a gold-labeled secondary anti-primary species antibody is used. Similarly, an immunoglobulin fraction should be used to block Fc receptors when a gold-labeled primary antibody is used in direct labeling studies.

Immunostaining procedures using gold-labeled protein A, G, or AG should include as controls sections in which the primary antibody is replaced with a similar concentration of ovalbumin. The absence of labeling in these controls demonstrates that protein A, G, or AG is not binding to endogenous immunoglobulins.

The most basic, and limited, test for specificity is to replace the immune antisera with "normal" antisera. If labeling is detected with a normal sera it is by definition nonspecific, but does not necessarily mean that the labeling seen with the immune antibody is also nonspecific. Frequently nonspecific results obtained with sera can be eliminated by using the purified immunoglobulin fraction. On the other hand, the absence of labeling with a normal antisera cannot be taken as evidence that the labeling seen in the immune antibody is antigen specific. Although rarely available, preimmune serum is a good control for nonspecific immunoglobulins present in the animal in which the immune antibody was produced. A negative reaction with a preimmune serum is supportive, but not conclusive, evidence of an antigen-specific reaction with the immune sera. A positive reaction of a preimmune serum will cast doubt on the reaction observed with the immune antibody. In one study, a preimmune serum gave a high level of labeling to nuclei and mitochondria that were not eliminated by dilution or purifying the immunoglobulin fraction. The immune antibody displayed the same labeling pattern in addition to what we believed to be the intended antigen on the plasma membrane. The preimmune IgG was used as a blocking agent for subsequent direct labeling with a gold-labeled immune IgG. The preimmune antibody eliminated an immunologically specific, but not antigen-specific, reaction (unpublished observations, R. M. Smith and L. Jarett).

Even in the absence of foregoing nonspecific reactions, proving that an observed reaction actually is the intended antigen, and only the intended antigen, is a problem. Given the difficulty in comparing antibody efficacy in Western blotting and immunocytochemistry (see Section III,C), if the antibody recognizes only one protein in a Western blot of a cell homogenate, immunocytochemical specificity is supported, but not proven. Demonstrating that the antibody recognizes the affinity-purified anti-

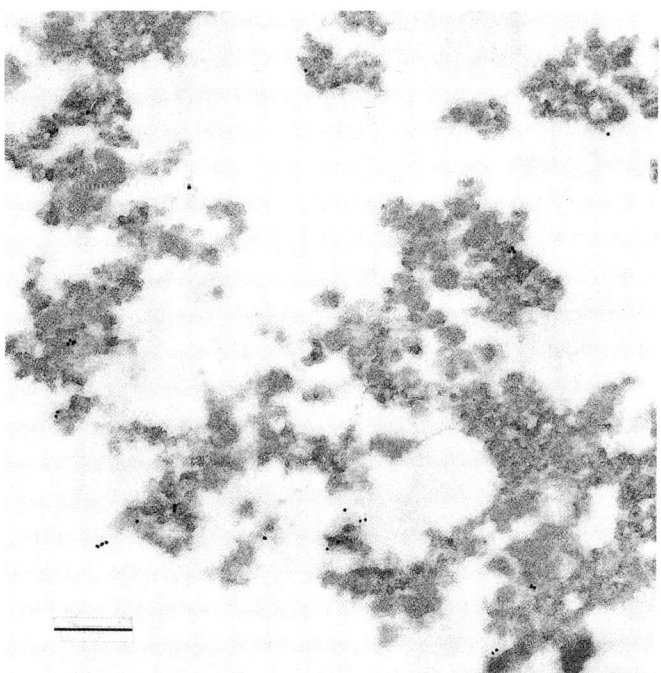

Figure 1 Immunocytochemical localization of insulin associated with nuclear matrix. Intact H35 hepatoma cells were incubated for 90 min at 37°C in the presence of 100 ng/ml insulin. Nuclear matrices were prepared using an agarose-encapsulating method (Thompson *et al.*, 1989). This material was fixed in 4% paraformaldehyde in 0.1 M sodium cacodylate buffer, pH 7.4, dehydrated, and embedded in LR White resin as described in Section IV,A. Thin sections were incubated with 8 nm colloidal gold-labeled polyclonal anti-insulin IgG as described in Section IV,C,2 under the protocol Direct Immunostaining Procedure. Scale bar: 0.25 μm.

gen in a Western blot does nothing to support an immunocytochemical observation. Blocking the antibody with purified antigen, or peptide sequence, is a frequently employed control, but suffers from a circular argument defect. For instance, the antibody could be recognizing an identical epitope in a protein other than the intended antigen. The purified antigen or peptide will block that reaction even though it is not the intended antigen. If one has multiple antibodies to different epitopes in the antigen, and all are positive and show the same localization, the chance that the labeling is antigen specific is certainly increased. However, if all of the epitopes are not exposed, the absence of a reaction with one or more antibodies would not necessarily prove a positive reaction with another antibody was not antigen specific. Proving antigen specificity is very difficult, but it rarely has been questioned when other routine controls prove negative. In some cases, e.g., the various isoforms of the glucose transporter, antigen specificity can be supported when no reaction is observed in cells that do not express the antigen.

D. Cellular Localization of Insulin, IGF-I, EGF, and TGFα

Immunocytochemical electron microscopy has been used to localized hormones and growth factors by so many laboratories that it would be impossible to review those here. Our laboratory has collaborated with other investigators in a number of studies that are summarized in this the next few sections to provide examples of the types of studies in which immunocytochemical electron microscopy can provide structural–functional information.

Insulin was localized in nuclei of insulin-treated cells (Soler et al., 1989) and was associated with the nuclear matrix as shown in Fig. 1 (Thompson et al., 1989). IGF-I was found in various intracellular organelles, e.g., endosomes, Golgi, lysosomes, and the nucleus of chicken lens epithelial cells (Soler et al., 1990) and toad bladder epithelia cells (Blazer-Yost et al., 1992). Insulin, IGF-I, and epidermal growth factor (EGF) have been detected in the capillaries, cells, and lumen of the mouse oviduct, in the mouse uterus, and in blastocyst-stage mouse embryos removed from the oviduct (Heyner et al., 1989; Smith et al., 1993b; Dardik et al., 1992). Immunocytochemistry demonstrated that transforming growth factor-α (TGFα) was endogenously produced by the cells of the inner cell mass and bound to TGFα/EGF receptors on the basolateral surface of polar trophectodermal cells (Dardik et al., 1992).

E. Cellular Localization of Receptors for Insulin, IGF-I, and EGF

Immunocytochemical localization of hormone receptors is frequently hindered by low receptor concentration and because only the receptors at the surface of the section can be stained. Therefore, postembedding immunocytochemical techniques reveal only a fraction of receptors that would be visualized by preembedding or nonimmunologic techniques. Nevertheless, immunocytochemical analysis can provide important information that nonimmunologic techniques cannot. In polarized cells, e.g., mouse blastocyst trophectoderm, toad bladder epithelium, gold-labeled insulin, IGF-I, and EGF, demonstrated receptors appear only on the surface exposed to the ligands. Immunocytochemical studies revealed that IGF-I and EGF/TGFα receptors were more concentrated on the basolateral surface than on the apical surface (Smith et al., 1993b; Dardik et al., 1992). The high concentration of IGF-I and IEGF/TGFα receptors on the bnasolateral, i.e., blastocoel surface, may be involved in the binding and internatlization of the IGF-I and TFGα produced in

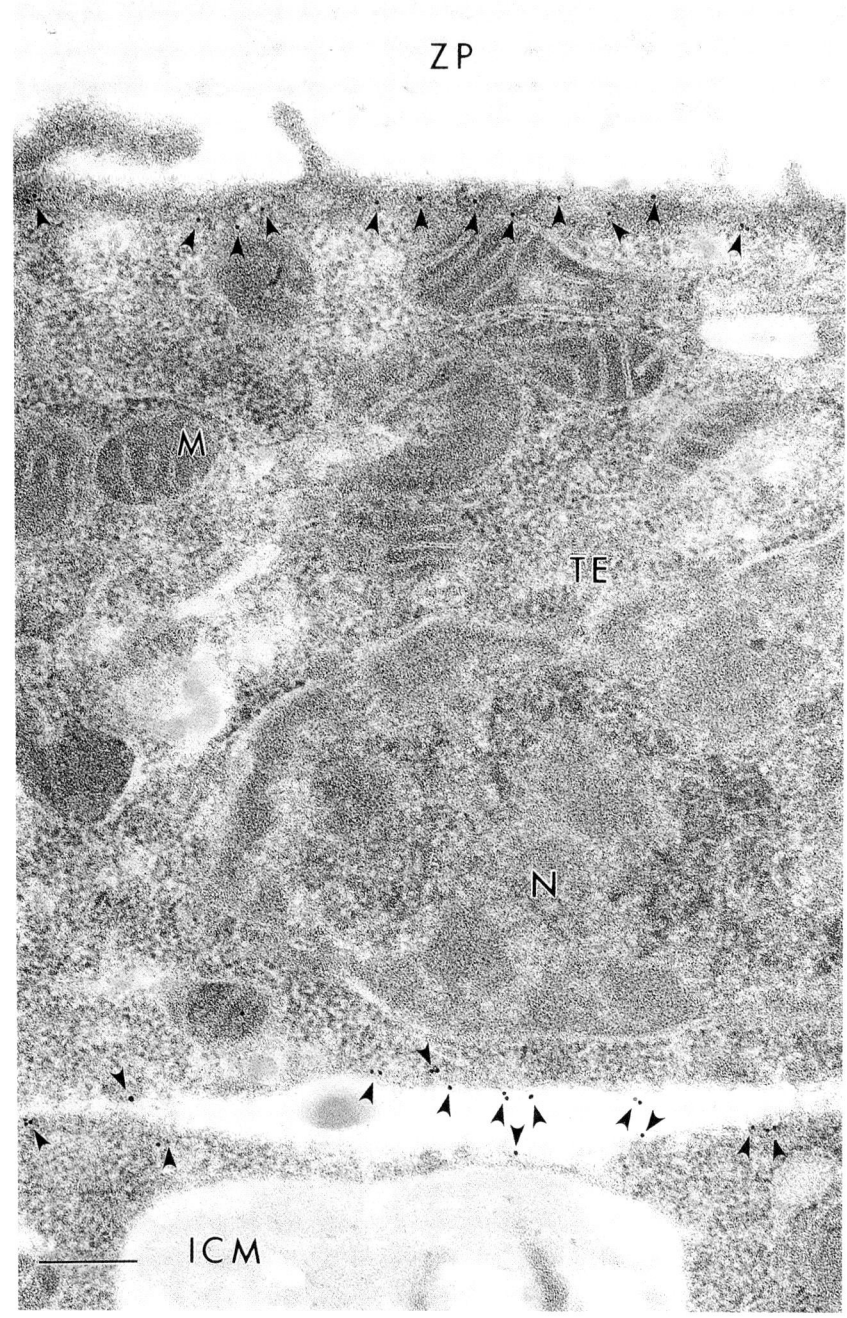

Figure 3 Immunocytochemical localization of $G_{i\alpha}$ in 3T3 fibroblasts. Cells were grown on microporous cell culture inserts (Falcon No. 353095) and fixed with GPA, dehydrated, and embedded in LR White resin. Thin sections were stained with 10 nm colloidal gold-labeled anti-$G_{i\alpha}$ polyclonal IgG (8729; Lewis et al., 1991). $G_{i\alpha}$ was detected distributed diffusely in the apical and basolateral plasma membranes. High concentrations of intracellular $G_{i\alpha}$ were detected in the electron-dense inclusions and membranes of tubular lysosomal structures (L). M, mitochondria; I, membrane of microporous cell culture insert. Scale bar: 0.5 µm.

the inner cell mass during early embryo development (Schultz et al., 1992). Insulin and receptors were detected on both apical and basolateral surfaces of the trophectoderm cells of the mouse blastocyst (Fig. 2) and toad bladder epithelium (Blazer-Yost et al., 1993).

Figure 2 Immunocytochemical localization of insulin receptors in mouse blastocyst. Mouse blastocysts were flushed from the oviduct and fixed with BPA. Blastocysts were attached to fibroblasts grown on Thermanox coverslips as described (Heyner et al., 1989). The cells were dehydrated and embedded in LR White resin. Sections were incubated with 8 nM colloidal gold-labeled anti-insulin receptor polyclonal IgG (A410; Jacobs et al., 1978). Insulin receptors (arrowheads) were detected in the plasma membrane and invaginations of the apical (facing the zona pellucida, ZP) and basolateral surfaces of trophectoderm cells (TE) and the inner cell mass (ICM). N, nucleus; M, mitochondria. Scale bar: 0.2 µm.

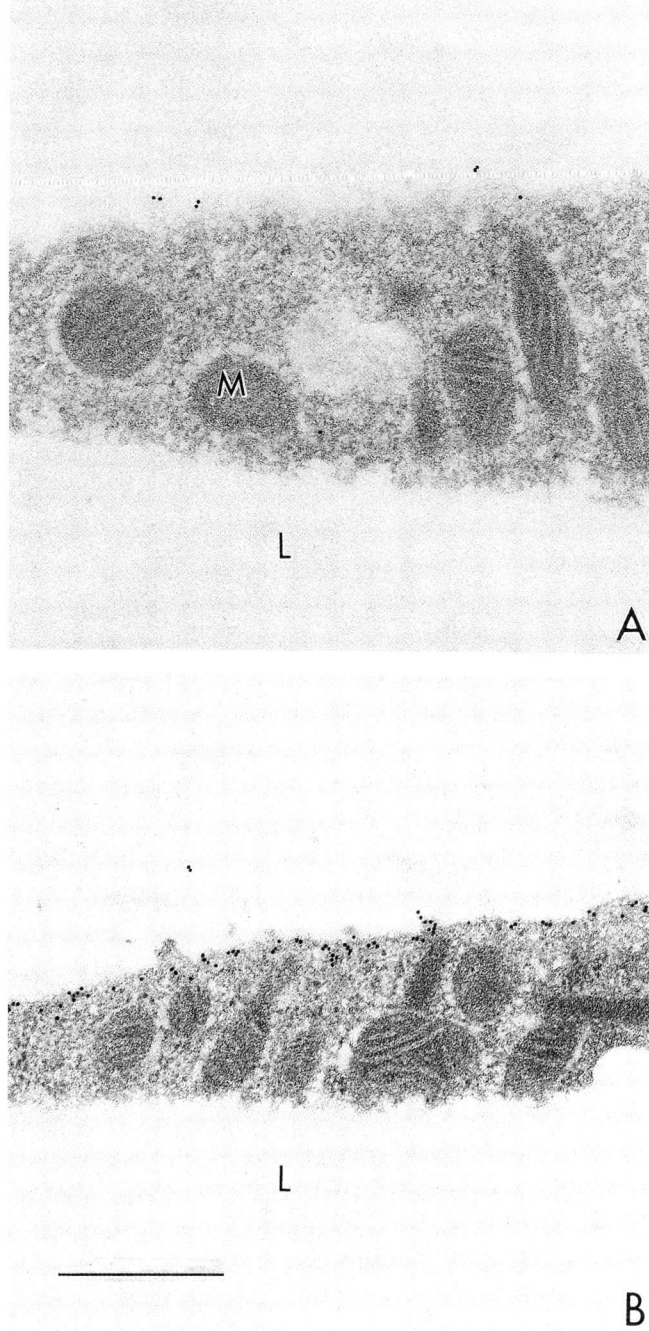

F. Cellular Localization of $G_{i\alpha}$

Guanine nucleotide-dependent regulatory proteins (G-proteins) are involved in the signal transduction process of a large number of receptor molecules (Gilman, 1987). Recent evidence suggests a role for G-proteins in metabolic effects of insulin and other growth factors (Gawler et al., 1987; O'Brien et al., 1987; Rothenberg and Kahn, 1988). A combined immunocytochemical, immunofluorescence study of $G_{i\alpha}$ distribution was done in collaboration with Manning and colleagues (Lewis et al., 1991). Immunofluorescence revealed two primary cellular locations of $G_{i\alpha}$; intense cytoplasmic labeling of thick tubule-like structures and diffuse plasma membrane labeling were seen. The labeled cytoplasmic organelle was thought to be mitochondria, although efforts to demonstrate $G_{i\alpha}$ in isolated mitochondria were unsuccessful. Immunocytochemical electron microscopy also revealed two major locations of $G_{i\alpha}$. In thin sections cut parallel to the substrate of attached cells, the most intense labeling was observed on the plasma membrane, particularly along the microvilli. Less intensely labeled were lysosomes [identified based on the positive staining by LAMP-1, a lysosomal membrane antibody (Chen et al., 1985)]. We have subsequently examined cells grown on a porous membrane support sectioned perpendicular to the substrate. As shown in Fig. 3, the apical and basolateral plasma membranes were diffusely labeled and there was dense labeling of the electron-dense inclusions and membranes of the tubular lysosomes. We believe that this pattern of immunocytochemical electron microscopic labeling is consistent with the immunofluorescent labeling originally reported (Lewis et al., 1991).

The effect of insulin on $G_{i\alpha}$ distribution in insulin-treated rat adipocytes was examined using an anti-peptide antibody directed to the α subunit of G_i, immunocytochemical electron microscopy, and Western blot analysis (Record et al., 1993). Immunocytochemical electron microscopy detected little or no $G_{i\alpha}$ in the plasma membrane of control cells (Fig. 4A). Insulin treatment greatly

Figure 4 Immunocytochemical localization of $G_{i\alpha}$ in rat adipocytes. Isolated adipocytes were prepared from epididymal fat pads and incubated in the absence (A) or presence (B) of 10 ng/ml insulin. The cells were fixed in PFeCN, dehydrated, and embedded in LR White resin. Thin sections were stained with 10 nm colloidal gold-labeled anti-$G_{i\alpha}$ polyclonal IgG. In control cells (A), little $G_{i\alpha}$ was detected. In insulin-treated cells (B), the plasma membrane and its invaginations were densely labeled. Although this increase in immunocytochemical labeling was observed, Western blot analysis demonstrated that insulin did not increase the amount of $G_{i\alpha}$ associated with the plasma membrane. These results suggest the antigenic epitope on the α subunit was masked in control cells and unmasked by insulin treatment, see Section III,E. L, central lipid droplet; M, mitochondria. Scale bar: 0.5 μm.

increased plasma membrane labeling as shown in Fig. 4B. However, Western blot analysis showed there was no increase in $G_{i\alpha}$ in the plasma membranes as a result of insulin treatment. Similar results were observed with other physiological stimuli. These results suggest that the α subunit of G_i is masked in control cells and its unmasking by insulin and other agents can be detected by appropriate antibodies.

G. Cellular Localization of Glucose Transporters

Several laboratories have recently used immunocytochemical techniques to demonstrate different isoforms of the glucose transporter in various tissues (Slot *et al.*, 1991; Friedman *et al.*, 1991; Takata *et al.*, 1991; Bornemann *et al.*, 1992; Aghayan *et al.*, 1992; Schultz *et al.*, 1992). Our study (Smith *et al.*, 1991) investigated the subcellular distribution and insulin-induced translocation of GLUT4 in isolated rat adipocytes under the same incubation conditions used in the studies by Cushman and Wardzala (1980) and Suzuki and Kono (1980). Two anti-peptide antibodies were used, one directed against the amino-terminus, the other against the carboxyl-terminus. In untreated cells, 95% of the labeling by the amino-terminal antibody was intracellular and on small cytoplasmic vesicles immediately beneath the plasma membrane or on plasma membrane invaginations. Insulin treatment increased plasma membrane labeling ~13-fold and caused a concomitant decrease in intracellular labeling of vesicles and invaginations. The labeling pattern observed with the carboxyl-terminal antibody was noticeably different. Labeling in control cells was very low. Insulin treatment resulted in a 20-fold increase in plasma membrane labeling. There was no change in intracellular labeling and total cell-associated labeling by the carboxyl-terminal antibody was increased 13-fold by insulin treatment. The results obtained with the amino-terminal antibody are consistent with the hypothesis of insulin-induced translocation of GLUT4 to the plasma membrane from submembranous invaginations or vesicles. However, the carboxyl-terminal antibody results suggest that the carboxyl-terminus may be masked in intracellular GLUT4 and is unmasked after translocation of the transporter to the plasma membrane. Genistein, a putative tyrosine kinase inhibitor (Akiyama *et al.*, 1987) did not affect insulin receptor autophosphorylation or kinase activity (Abler *et al.*, 1992), but inhibited basal and insulin-stimulated glucose transport by 50%. It had no effect on insulin-stimulated translocation of GLUT4 to the plasma membrane as detected by Western blot analysis (Smith *et al.*, 1993a). At the same time, genistein markedly reduced (~50%) the labeling of plasma membrane GLUT4 by the carboxyl-terminus antibody in control and insulin-treated cells. These data suggest that genistein inhibits the activity of glucose transporter in the plasma membrane of control or insulin-treated cells. The immunocytochemical data suggest that change in the carboxyl-terminus that results in the unmasking detected by the carboxyl-terminal antibody may be related to the activation of the transporter.

V. Future Prospects for Electron Microscopy

The combination of electron microscopic and biochemical analysis has and will continue to increase our understanding of complex cell biological systems. During the past 30 years the applications of electron microscopy have evolved to keep abreast of developments in other areas of scientific investigation. It is difficult to predict what new ultrastructural technologies will be developed as we continue to unravel the mysteries of how hormones and growth factors regulate cell metabolism and growth. Refinements in the areas of immunocytochemical and *in situ* hybridization electron microscopy hold particular promise.

A. Immunocytochemical Electron Microscopy

Epitope masking and unmasking, as we have discussed in reference to G proteins and glucose transporters, may prove to be an important area of future investigation. Our observations were clearly serendipitous. However, by carefully designing antipeptide antibodies to specific epitopes of cellular proteins, immunocytochemical electron microscopy may be able to demonstrate hormone-induced changes in protein structure directly related to changes in metabolic function. Currently available anti-phosphotyrosine antibodies (Rothenberg *et al.*, 1991; Unger *et al.*, 1991) are an example of this application.

B. *In Situ* Hybridization Electron Microscopy

Immunocytochemical electron microscopy provides a high-resolution method for detecting hormones, receptors, or other antigens in cells or subcellular compartments and satisfies the objectives of one group of investigators. However, it is well known that insulin and other hormones and growth factors regulate the expression of many cellular proteins by their effects on gene transcription (O'Brien and Granner, 1991). *In situ* hybridization techniques, which have been used extensively at the light-microscopic level, are being developed for electron microscopy. Troxler *et al.* (1990) recently published procedures utilizing colloidal gold-labeled streptavidin or anti-biotin antibodies to detect biotinylated DNA and RNA probes to localize poliovirus

RNA in thin sections of Lowicryl-embedded cells. Wachtler *et al.* (1992), using fixed, LR White-embedded Sertoli cells, localized the transcribed part of the human ribosomal RNA gene to the dense fibrillar component of the nucleolus.

These, and several other recent studies (Diamandis, 1990; Couwenhoven *et al.*, 1990; Detta and Hitchcock, 1991; Bugnon *et al.*, 1991), suggest that it is now possible to perform extensive studies on hormone action. Immunocytochemical techniques can be used to study the binding, internalization, and intracellular translocation of hormones and the effects of hormones on enzyme activity. Those studies can be augmented by *in situ* hybridization studies on similarly prepared tissues to localize specific cells or subcellular or subnuclear components involved in hormone-regulated gene transcription.

VI. Summary and Conclusion

In this chapter we have attempted to provide a general background for endocrinologists by discussing the potentials, problems, and procedures involved in immunocytochemical electron microscopy. While we strongly believe that electron microscopy provides a powerful tool to investigate hormonal regulation of cellular function, it is but one of many tools available to the cell biologist. We recognize the limitations of the techniques currently available and never hesitate to use a different structural or biochemical approach to attack our objectives. At the same time, electron microscopists are trying to develop methods to improve the reliability and results obtained with ultrastructural techniques. Hopefully, the discussion of the problems associated with immunocytochemical electron microscopy will not be a deterrent to pursuing appropriate studies.

Acknowledgments

We express our deepest gratitude to Mrs. Neelima Shah. Without her excellent technical skills, patience, contributions, and hard work over the past 10 years, the results described in this chapter would not have been possible. We thank our collaborators whose studies contributed to our understanding and development of immunocytochemical techniques or those who have generously provided antibodies: B. Blazer-Yost, G. Boden, J. Caro, M. J. Charron, S. Corvera, M. Cox, M. P. Czech, A. Dardik, R. I. Goldberg, I. D. Goldfine, M. Gouth, J. G. Haddad, S. Heyner, S. Jacobs, J. M. Lewis, H. F. Lodish, D. R. Manning, S. P. Nissley, J. M. Olefsky, F. de Pablo, R. D. Record, O. M. Rosen, S. A. Rosenzweig, P. L. Rothenberg, G. A. Schultz, R. M. Schultz, B. L. Seely, I. A. Simpson, A. P. Soler, S. I. Taylor, K. A. Thompson, and B. Thorens. This work has been supported in part by grants from the National Institutes of Health (DK28143 and DK19525) and the American Diabetes Association.

References

Abler, A. A., Smith, J. A., Randazzo, P. A., Rothenberg, P. L., and Jarett, L. (1992). *J. Biol. Chem.* **267**, 3946–3951.
Ackerman, G. A., Yang, J., and Wolken, K. W. (1983). *J. Histochem. Cytochem.* **31**, 433–440.
Aghayan, M., Rao, L. V., Smith, R. M., Jarett, L., Charron, M. J., Thorens, B., and Heyner, S. (1992). *Development* **115**, 305–312.
Akiyama, T., Ishica, J., Nakagawa, S., Ogawara, H., Watanabe, S., Itoh, N., Shibuya, M., and Fukami, Y. (1987). *J. Biol. Chem.* **262**, 5592–5595.
Akiyama, H., Shii, K., Yokono, K., Yonezawa, K., Sato, S., Watanabe, K., and Baba, S. (1988). *Biochem. Biophys. Res. Commun.* **155**, 914–922.
Balslev, Y., and Hansen, G. H. (1989). *Histochem. J.* **21**, 449–454.
Baskin, D. G., Erlandsen, S. L., and Parsons, J. A. (1979). *J. Histochem. Cytochem.* **27**, 1290–1292.
Baudhuin, P., Van der Smissen, P., Beauvois, S., and Courtoy, P. J. (1989). *In* "Colloidal Gold: Principles, Methods, and Applications" (M. A. Hayat, ed.), Vol. 2, pp. 1–17. Academic Press, San Diego.
Bendayan, M. (1989). *In* "Colloidal Gold: Principles, Methods, and Applications" (M. A. Hayat, ed.), Vol. 1, pp. 33–94. Academic Press, San Diego.
Bendayan, M., and Zollinger, M. (1983). *J. Histochem. Cytochem.* **31**, 101–109.
Bergeron, J. J. M., Sikstrom, R., Hand, A. R., and Posner, B. I. (1979). *J. Cell Biol.* **80**, 427–443.
Blackard, W. G., Smith, R. M., and Jarett, L. (1986). *Am. J. Physiol.* **250**, E148–E155.
Blazer-Yost, B. L., Shah, N., Jarett, L., Cox, M., and Smith, R. M. (1992). *Biochem. Intern* **28**, 143–153.
Bornemann, A., Ploug, T., and Schmalbruch, H. (1992). *Diabetes* **41**, 215–221.
Brown, M. S., and Goldstein, J. L. (1979). *Proc. Natl. Acad. Sci. U.S.A.* **76**, 3330–3337.
Bugnon, C., Bahjaoui, M., and Fellmann, D. (1991). *J. Histochem. Cytochem.* **39**, 859–862.
Capco, D. G., Krochmalnic, G., and Penman, S. (1984). *J. Cell Sci.* **98**, 1878–1885.
Carlemalm, E., Garavito, R. M., and Villiger, W. (1982). *J. Microsc. (Oxford)* **126**, 123–143.
Carpentier, J.-L., Gorden, P., Amherdt, M., van Obberghen, E., Kahn, R. C., and Orci, L. (1978). *J. Clin. Invest.* **61**, 1057–1070.
Chen, J. W., Pan, W., D'Souza, M. P., and August, J. T. (1985). *Arch. Biochem. Biophys.* **239**, 574–586.
Couwenhoven, R. I., Luo, W., and Snead, M. L. (1990). *J. Histochem. Cytochem.* **38**, 1853–1857.
Cushman, S. W., and Wardzala, L. J. (1980). *J. Biol. Chem.* **255**, 4758–4762.
Dardik, A., Smith, R. M., and Schultz, R. M. (1992). *Dev. Biol.* **154**, 396–409.
Detta, A., and Hitchcock, E. (1991). *Mol. Cell. Probes* **5**, 437–443.
Diamandis, E. P. (1990). *Clin. Chim. Acta* **194**, 19–50.
Doerr-Schott, J. (1989). *In* "Colloidal Gold: Principles, Methods, and Applications" (M. A. Hayat, ed.), Vol. 1, pp. 145–190. Academic Press, San Diego.
Faulk, W. P., and Taylor, G. M. (1971). *Immunocytochemistry* **8**, 1081–1083.
Feldherr, C. M., Kallenbach, E., and Schultz, N. (1984). *J. Cell Biol.* **99**, 2216–2222.
Fey, E. G., Krochmalnic, G., and Penman, S. (1986). *J. Cell Sci.* **102**, 1654–1665.
Frens, G. (1973). *Nature (London)* **241**, 20–22.

Friedman, J. E., Dudek, R. W., Whitehead, D. S., Downes, D. L., Frisell, W. R., Caro, J. F., and Dohm, G. L. (1991). *Diabetes* **40**, 150–154.
Gawler, D., Milligan, G., Spiegel, A. M., Unson, L. G., and Houslay, M. D. (1987). *Nature (London)* **327**, 229–232.
Geoghegan, W. D., and Ackerman, G. A. (1977). *J. Histochem. Cytochem.* **25**, 1187–1200.
Ghitescu, L., Galis, Z., and Bendayan, M. (1991). *J. Histochem. Cytochem.* **39**, 1057–1065.
Gilman, A. G. (1987). *Annu. Rev. Biochem.* **56**, 615–649.
Goldberg, R. I., Smith, R. M., and Jarett, L. (1987). *J. Cell. Physiol.* **133**, 213–218.
Goldstein, B. J., and Livingston, J. N. (1980). *Biochem. J.* **186**, 351–360.
Goldstein, B. J., and Livingston, J. N. (1981). *Endocrinology (Baltimore)* **108**, 953–961.
Hamel, F. G., Posner, B. I., Bergeron, J. J. M., Frank, B. H., and Duckworth, W. C. (1988). *J. Biol. Chem.* **263**, 6703–6708.
Hammel, I. (1986). *J. Histochem. Cytochem.* **34**, 941–944.
Harrach, B., and Robenek, H. (1990). *Arteriosclerosis* **10**, 564–576.
Herrera, R., Petruzzelli, L., Thomas, N., Bramson, H. N., Kaiser, E. T., and Rosen, O. M. (1985). *Proc. Natl. Acad. Sci. U.S.A.* **82**, 7899–7903.
Heyner, S., Rao, L. V., Jarett, L., and Smith, R. M. (1989). *Dev. Biol.* **134**, 48–58.
Horisberger, M. (1979). *Biol. Cell.* **36**, 253–258.
Jacobs, S., Chang, K.-J., and Cuatrecasas, P. (1978). *Science* **200**, 1283–1284.
Jarett, L., and Smith, R. M. (1974). *J. Biol. Chem.* **249**, 7024–7031.
Jarett, L., and Smith, R. M. (1977). *J. Supramol. Struct.* **6**, 45–59.
Kahn, C. R., Baird, K. L., Flier, J. S., Grunfeld, C., Harmon, J. T., Harrison, L. C., Karlsson, F. A., Kasuga, M., King, G. L., Lang, U., Podskalny, J. M., and Van Obberghen, E. (1981). *Recent Prog. Horm. Res.* **32**, 447–568.
Kellenberger, E., and Hayat, M. A. (1991). In "Colloidal Gold: Principles, Methods, and Applications" (M. A. Hayat, ed.), Vol. 3, pp. 2–30. Academic Press, San Diego.
Kellenberger, E., Dürrenberger, M., Villiger, W., Carlemalm, E., and Wurtz, M. (1987). *J. Histochem. Cytochem.* **35**, 959–969.
King, M. G., and Baskin, D. G. (1991). *Anat. Rec.* **231**, 467–472.
Kull, F. C., Jr., Jacobs, S., Su, Y.-F., Svoboda, M. E., Van Wyk, J. J., and Cuatrecasas, P. (1983). *J. Biol. Chem.* **258**, 6561–6566.
Leunissen, J. L. M., and de Mey, J. R. (1989). In "Immuno-Gold Labeling in Cell Biology" (A. J. Verkleij and J. L. M. Leunissen, eds.), pp. 3–16. CRC Press, Boca Raton, Florida.
Lewis, J. M., Woolkalis, M. J., Gerton, G. L., Smith, R. M., Jarett, L., and Manning, D. R. (1991). *Cell Regul.* **2**, 1097–1113.
Linner, J. G., Livesey, S. A., Harrison, D. S., and Steiner, A. L. (1986). *J. Histochem. Cytochem.* **34**, 1123–1135.
Maxfield, F. R., Schlessinger, J., Shechter, Y., Pastan, I., and Willingham, M. C. (1978). *Cell* **14**, 805–810.
Newman, G. R. (1989). In "Colloidal Gold: Principles, Methods, and Applications" (M. A. Hayat, ed.), Vol. 2, pp. 47–73. Academic Press, San Diego.
O'Brien, R. M., Houslay, M. D., Milligan, G., and Siddle, K. (1987). *FEBS Lett.* **212**, 281–288.
O'Brien, R. M., and Granner, D. K. (1991). *Biochem. J.* **278**, 609–619.
Park, K., Park, H., and Albrecht, R. M. (1989). In "Colloidal Gold: Principles, Methods, and Applications" (M. A. Hayat, ed.), Vol. 1, pp. 489–518. Academic Press, San Diego.
Perelson, A. S., and DeLisi, C. (1980). *Math. Biosci.* **48**, 71–110.
Piekut, D. T., and Casey, S. M. (1983). *J. Histochem. Cytochem.* **31**, 669–674.
Pinto da Silva, P., Douglas, S. D., and Branton, D. (1971). *Nature (London)* **232**, 194–196.
Priestly, J. V., and Cuello, A. C. (1983). In "Immunohistochemistry" (A. C. Cuello, ed.), pp. 273–321. Wiley, Chichester, England.

Record, R. D., Smith, R. M., and Jarett, L. (1993). *Exper. Cell Res.* **206**, 36–42.
Reynolds, A., and Tanford, C. (1970). *J. Biol. Chem.* **245**, 5161–5165.
Robenek, H., Rassat, J., Hesz, A., and Grunwald, J. (1982). *Eur. J. Cell Biol.* **27**, 242–250.
Rodewald, R., and Kraehenbuhl, J.-P. (1984). *J. Cell Biol.* **99**, 159s–164s.
Romano, E. L., and Romano, M. (1977). *Immunochemistry* **14**, 711–715.
Rosenzweig, S. A., Zetterström, C., and Benjamin, A. (1990). *J. Biol. Chem.* **265**, 18030–18034.
Roth, J., Bendayan, M., and Orci, L. (1978). *J. Histochem. Cytochem.* **26**, 1074–1081.
Roth, J., Bendayan, M., Carlemalm, E., Villiger, W., and Garavito, M. (1981). *J. Histochem. Cytochem.* **29**, 663–671.
Rothenberg, P. L., and Kahn, C. R. (1988). *J. Biol. Chem.* **263**, 15546–15552.
Rothenberg, P. L., Lane, W. S., Karasik, A., Backer, J., White, M., and Kahn, C. R. (1991). *J. Biol. Chem.* **266**, 8302–8311.
Rutter, G., and Hohenberg, H. (1991). In "Colloidal Gold: Principles, Methods, and Applications" (M. A. Hayat, ed.), Vol. 3, pp. 151–186. Academic Press, San Diego.
Salih, H., Murthy, G. S., and Friesen, H. G. (1979). *Endocrinology (Baltimore)* **105**, 21–26.
Schultz, G. A., Hogan, A., Watson, A. J., Smith, R. M., and Heyner, S. (1992). *Reprod. Fertil. Dev.* **4**, 361–371.
Sela, M. (1973). *Harvey Lect. Ser.* **67**, 214–246.
Sheehan, J. K., Boot-Handford, R. P., Chatler, E., Carlstedt, I., and Thornton, D. J. (1991). *Biochem. J.* **274**, 293–296.
Singer, S. J. (1959). *Nature (London)* **183**, 1523–1524.
Slot, J. W., and Geuze, H. J. (1985). *Eur. J. Cell Biol.* **38**, 87–93.
Slot, J. W., Geuze, H. J., Gigengack, S., Lienhard, G. E., and James, D. E. (1991). *J. Cell Biol.* **113**, 123–135.
Smith, R. M., and Jarett, L. (1982a). *J. Histochem. Cytochem.* **30**, 650–656.
Smith, R. M., and Jarett, L. (1982b). *Proc. Natl. Acad. Sci. U.S.A.* **79**, 7302–7306.
Smith, R. M., and Jarett, L. (1987). *Proc. Natl. Acad. Sci. U.S.A.* **84**, 459–463.
Smith, R. M., and Jarett, L. (1988). *Lab. Invest.* **58**, 613–629.
Smith, R. M., and Jarett, L. (1991). In "Colloidal Gold: Principles, Methods, and Applications" (M. A. Hayat, ed.), Vol. 3, pp. 243–263. Academic Press, San Diego.
Smith, R. M., Goldberg, R. I., and Jarett, L. (1988). *J. Histochem. Cytochem.* **36**, 359–365.
Smith, R. M., Charron, M. J., Shah, N., Lodish, H. F., and Jarett, L. (1991). *Proc. Natl. Acad. Sci. U.S.A.* **88**, 6893–6897.
Smith, R. M., Tiesinga, J. J., Shah, N., Smith, J. A., and Jarett, L. (1993a). *Arch. Biochem. Biophys.* **3001**, 238–246.
Smith, R. M., Garside, W. T., Aghayan, M., Shah, N., Jarett, L., and Heyner, S. (1993b). *Biol. Reprod. (in press)*.
Soler, A. P., Thompson, K. A., Smith, R. M., and Jarett, L. (1989). *Proc. Natl. Acad. Sci. U.S.A.* **86**, 6640–6644.
Soler, A. P., Alemany, J., Smith, R. M., de Pablo, F., and Jarett, L. (1990). *Endocrinology (Baltimore)* **127**, 595–603.
Spurr, A. R. (1969). *J. Ultrastruct. Res.* **26**, 31–43.
Stierhof, Y. D., Schwarz, H., Dürrenberger, M., Villiger, W., and Kellenberger, E. (1991). In "Colloidal Gold: Principles, Methods, and Applications" (M. A. Hayat, ed.), Vol. 3, pp. 88–115. Academic Press, San Diego.
Suzuki, K., and Kono, T. (1980). *Proc. Natl. Acad. Sci. U.S.A.* **77**, 2542–2545.
Takamiya, H., Batsford, S., and Vogt, A. (1980). *J. Histochem. Cytochem.* **28**, 1041–1049.
Takata, K., Kasahara, T., Kasahara, M., Ezaki, O., and Hirano, H. (1991). *Invest. Ophthalmol. Visual Sci.* **32**, 1569–1666.
Thompson, K. A., Soler, A. P., Smith, R. M., and Jarett, L. (1989). *Eur. J. Cell Biol.* **50**, 442–446.

Tokuyasu, K. T. (1973). *J. Cell Biol.* **57,** 551–565.
Troxler, M., Pasamontes, L., Egger, D., and Bienz, K. (1990). *J. Virol. Methods* **30,** 1–14.
Unger, J. W., Moss, A. M., and Livingston, J. N. (1991). *Neuroscience* **42,** 853–861.
Van Ewijk, W., Van Soest, P. L., Verkerk, A., and Jongkind, J. F. (1984). *Histochem. J.* **16,** 179–193.
Varndell, I. M., and Polak, J. M. (1984). *In* "Immunolabelling for Electron Microscopy" (J. M. Polak and I. M. Varndell, eds.), pp. 155–177. Elsevier, Amsterdam.
Varndell, I. M., and Polak, J. M. (1987). *In* "Electron Microscopy in Molecular Biology: A Practical Approach" (J. Sommerville and U. Scheer, eds.), pp. 179–200. IRL Press, Oxford.
Villiger, W. (1991). *In* "Colloidal Gold: Principles, Methods, and Applications" (M. A. Hayat, ed.), Vol. 3, pp. 59–71. Academic Press, San Diego.
Wächtler, F., Schöfer, C., Mosgöller, K., Weipoltshammer, K., Schwarzacher, H. G., Guichaoua, M., Hartung, M., Stahl, A., Bergé-Lefranc, J. L., Gonzalez, I., and Sylvester, J. (1992). *Exp. Cell Res.* **198,** 135–143.
Wang, X., and Traub, P. (1991). *J. Cell Sci.* **98,** 107–122.
Ward, W. F. (1984). *Horm. Metab. Res.* **16,** 509–512.
Willingham, M. C., and Pastan, I. H. (1981). *In* "Receptor-Mediated Binding and Internalization of Toxins and Hormones" (J. L. Middlebrook and L. D. Kohn, eds.), pp. 135–144. Academic Press, New York.
Yamashiro, D. J., and Maxfield, F. R. (1984). *J. Cell Biochem.* **26,** 231–246.

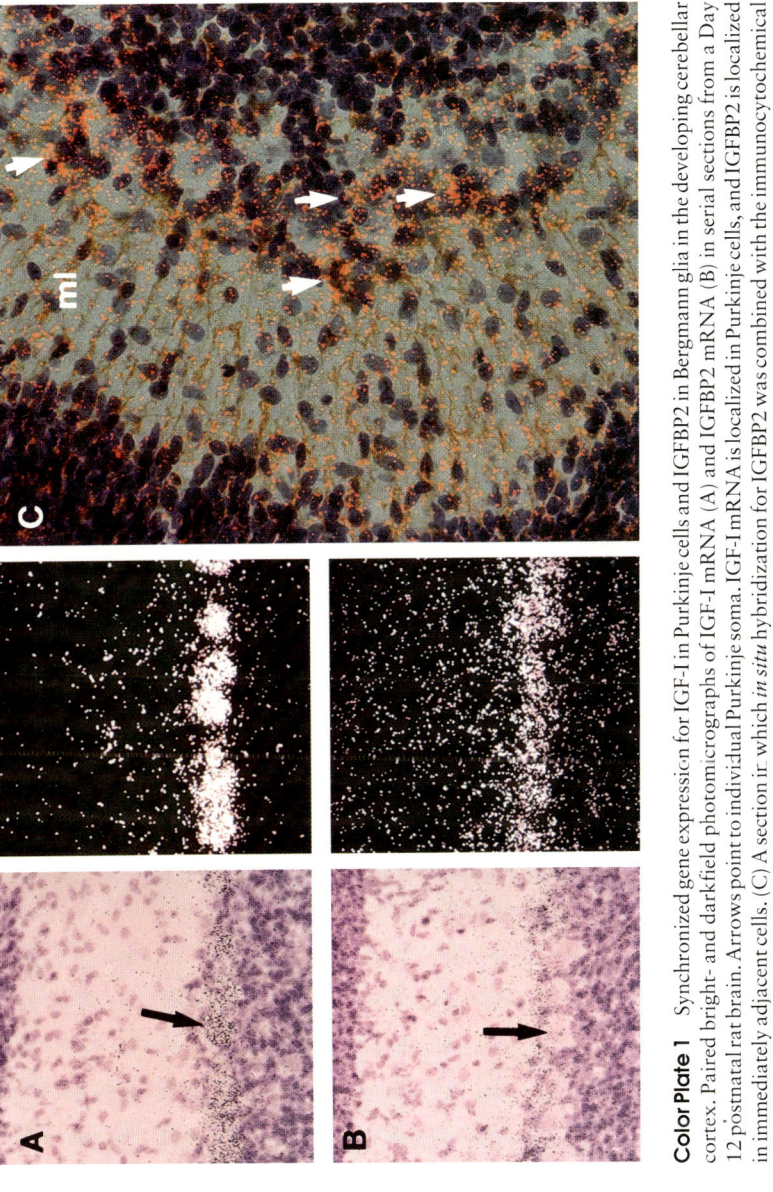

Color Plate 1 Synchronized gene expression for IGF-I in Purkinje cells and IGFBP2 in Bergmann glia in the developing cerebellar cortex. Paired bright- and darkfield photomicrographs of IGF-I mRNA (A) and IGFBP2 mRNA (B) in serial sections from a Day 12 postnatal rat brain. Arrows point to individual Purkinje soma. IGF-I mRNA is localized in Purkinje cells, and IGFBP2 is localized in immediately adjacent cells. (C) A section in which *in situ* hybridization for IGFBP2 was combined with the immunocytochemical detection of glial fibrillary acidic protein (GFAP)—an astrocyte-specific marker. The GFAP staining (brown) delineates Bergmann glial fibers extending from the level of the Purkinje cell layer outward through the molecular layer (ml). The IGFBP2 mRNA hybridization signal (red grains) is concentrated in cells adjacent to the Purkinje neurons from which these GFAP-positive fibers originate (arrowheads).

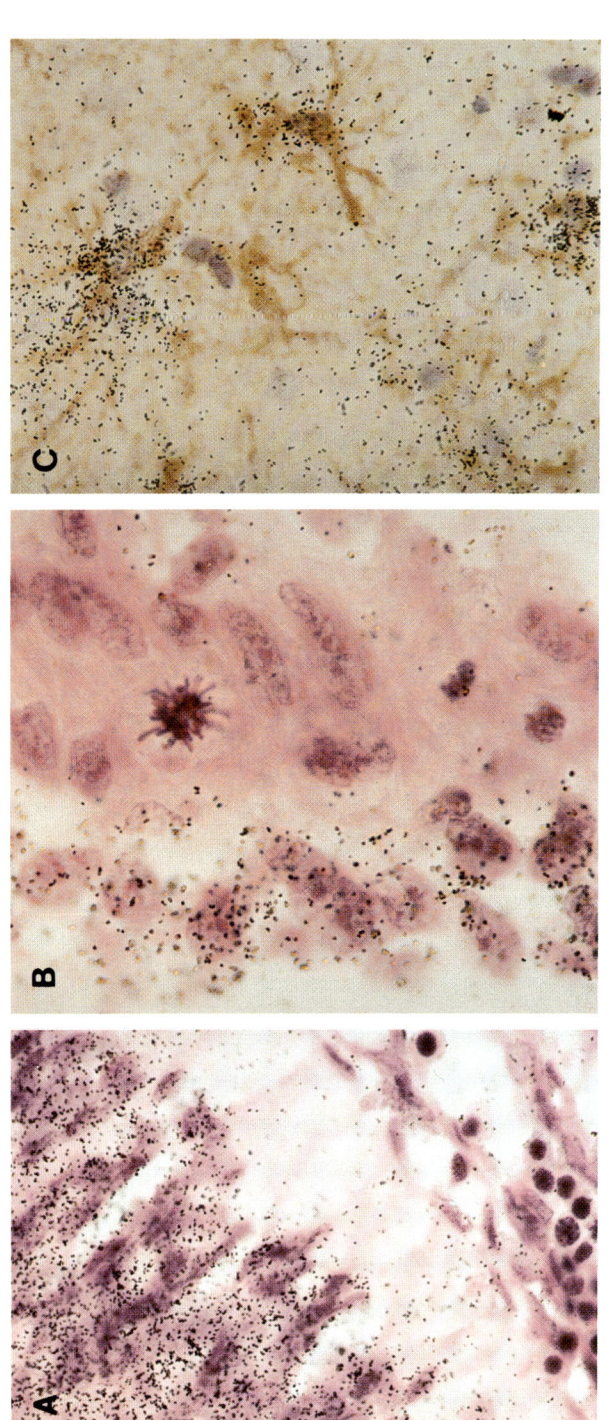

Color Plate 2 (A) IGF-1 receptor mRNA in the ventral floorplate of the hindbrain (from the region shown by arrows in Fig. 5). This midline structure consists of a pseudostratified columnar neuroepithelium containing cells whose processes extend from ventricular to pial surfaces of the brain and spinal cord. The pial surface, covered with blood vessels containing nucleated fetal red blood cells, is at the bottom of the picture. (B) IGF-1 mRNA hybridization in cells encircling the right atrium of the Embryonic Day 14 heart (see Fig. 5). Many mitotic figures are seen among the large myocytes which have foamy, pale pink (glycogen-containing) cytoplasm and multiple nucleoli. The cells containing IGF-1 mRNA have scant cytoplasm and do not show mitotic figures. The sections in both A and B are from embryos which were immersion fixed in Bouin's solution and paraffin embedded. The tissue was stained with hematoxylin and eosin following development of the emulsion coating. (C) An example of combination immunocytochemistry and *in situ* hybridization. The antibody was directed against glial fibrillary acidic protein (GFAP), a marker which is specific for astrocytes, and the brown immunostain outlines a typical astroglial morphology. The riboprobe was directed against glucose transporter 1 (Glut1), which is expressed in these reactive astrocytes in a section from a rat brain subjected to transient ischemia. The brain was fresh frozen and cut in 15-μm-thick sections on a cryostat; sections were fixed for 1 hr in Zamboni's fixative prior to immunocytochemical processing and counterstained lightly with cresyl violet following development of the emulsion coating.

13

In Situ Hybridization Histochemistry

Carolyn A. Bondy, Jian Zhou, and Wei-Hua Lee

I. Applications
   A. Complementary Uses of *in Situ* Hybridization, Immunocytochemistry, and Ligand-Binding Autoradiography
   B. The Use of *in Situ* Hybridization to Develop Functional Correlations
II. Methodological Issues
   A. Probe Choice
   B. Label Choice
   C. Control Procedures
III. *In Situ* Hybridization Protocols
   A. Synthesis of ^{35}S-Labeled Riboprobes
   B. Alkaline Hydrolysis
   C. Coating Slides
   D. Tissue Preparation
   E. Preparation of Sections for Hybridization
   F. Hybridization
   G. Stringency Washes
   H. Emulsion Coating
   I. Combined Immunocytochemistry and *in Situ* Hybridization
References

This chapter focuses on the use of *in situ* hybridization for the detection of peptide and receptor mRNAs in tissue sections. The first part discusses some important applications of the technique with special relevance for endocrinological investigations, the second part addresses major methodological issues, and the final portion provides detailed protocols for the synthesis and use of ^{35}S-labeled riboprobes for *in situ* hybridization alone and in combination with immunocytochemistry.

I. Applications

A. Complementary Uses of *in Situ* Hybridization, Immunocytochemistry, and Ligand-Binding Autoradiography

In situ hybridization is currently the method of choice for the identification of cells which produce peptide hormones or "growth" factors and cells which synthesize their receptors and binding proteins. The histological approach to the investigation of relationships between ligands and their receptors is based upon the idea that identification of these cells is an important prerequisite for the understanding of roles and modes of action of these factors *in vivo*. Complementary histochemical techniques such as immunocytochemistry and ligand-binding autoradiography provide confirmation that the mRNA of interest is translated, and also may supply important information about where the product is localized relative to its site of synthesis. In the case of a gene product like insulin, which is both synthesized and stored in the compact pancreatic β-cells, the two sites are identical, but in many instances, the situation is more complicated. For example, in the central nervous system, a given peptide or receptor may be localized in nerve terminal fields some distance away from nerve cell bodies, making it difficult to establish which neurons are expressing the gene product of interest. Sites of synthesis and accumulation may also diverge in the case of constitutively released peptides which are not stored inside the cell of origin, such as nerve growth factor (NGF) and insulin-like growth factors (IGFs). For example, NGF immunoreactivity is concentrated in sympathetic and sensory ganglia, but the peptide is not produced in these ganglia (Heumann *et al.*, 1984). Instead, it is synthesized in peripheral target zones innervated by ganglionic fibers, which internalize NGF and transport it back to nerve cell bodies located in the ganglia. Likewise, IGF-I immunoreactivity is concentrated in collecting ducts in the rat kidney (Hansson *et al.*, 1988; Kobayashi *et al.*, 1991), but its mRNA is localized upstream in the thick ascending limbs of Henle's loops (Chin *et al.*, 1992).

In situ hybridization is also useful for evaluation of the molecular specificity of ligand-binding sites. For example, IGF-I and -II may bind to three different membrane receptors (Rechler and Nissley, 1985) and as many as six different high-affinity binding proteins (Clemmons, 1990; Ooi, 1990). Prior to the cloning of these various components, this multiplicity in binding potential made the determination of molecular specificity in IGF binding and signal transduction very difficult. With the availability of specific nucleic acid probes, however, the ability to differentiate the different ligands, receptors, and binding proteins has become relatively straightforward. Cross-hybridization in the detection of mRNAs for different members of multigene families can usually be excluded by the synthesis of probes complementary to nonhomologous regions and by the use of stringent hybridization and wash conditions which eliminate mismatched hybrids.

B. The Use of *in Situ* Hybridization to Develop Functional Correlations

The use of *in situ* hybridization for the analysis of spatiotemporal patterns of gene expression during embryonic development is a powerful tool in the illumination of potential functional roles for regulatory peptides. While the mapping of cellular patterns of gene expression is simply descriptive, it provides an essential set of reference points for the experimental analysis of function. This has become a standard approach in the screening of novel genes for potential functions, with the emergence of PIT1 (GHF1) as a regulator of pituitary cell differentiation and hormone synthesis a prime example of its success (Dolle *et al.*, 1990). In addition, new roles for familiar old peptides have emerged from the study of their tissue-specific patterns of gene expression, with the expanded role of IGF-I (somatomedin C) as a paracrine/autocrine factor being a salient example (Underwood *et al.*, 1986).

The use of tissue homogenates for the examination of developmental changes in gene expression can be very misleading because the cellular composition of a given structure may change significantly as a function of development, and the contribution of contaminating membranes and connective tissue also changes during the course of organ growth. The pattern of IGF gene expression in the developing cerebellum provides a good illustration of these important considerations. IGF-I gene expression is abundant in the large projection neurons, or Purkinje cells (Fig. 1), of the cerebellar cortex during a brief window of time in their postnatal maturation (Bondy, 1991). This pattern of IGF expression is supported by evidence for IGF-I immunoreactivity in the developing cerebellar cortex (Andersson *et al.*, 1988). The specific timing of IGF-I expression correlates with the prolific growth of the Purkinje dendritic arbor and the reception of literally hundreds of thousands of synapses by each arbor, followed by myelination of the system. Thus, the specific cellular localization and timing of IGF-I expression in this system suggests that IGF-I may be involved in the growth of Purkinje dendrites, synapse formation, or myelination in the developing cerebellar cortex.

The analysis of IGF-I mRNA content in homogenates of the developing cerebellum results in quite a different picture, however. As shown in Fig. 1, the cellular composition of the cerebellar cortex changes dramatically during postnatal development. While the number of Purkinje cells is fixed at birth, the number of cerebellar granule cells increases astronomically in the first 3 postnatal weeks, hence the contribution of Purkinje cells to the total pool of cerebellar mRNA diminishes during this time. As a result of this change in

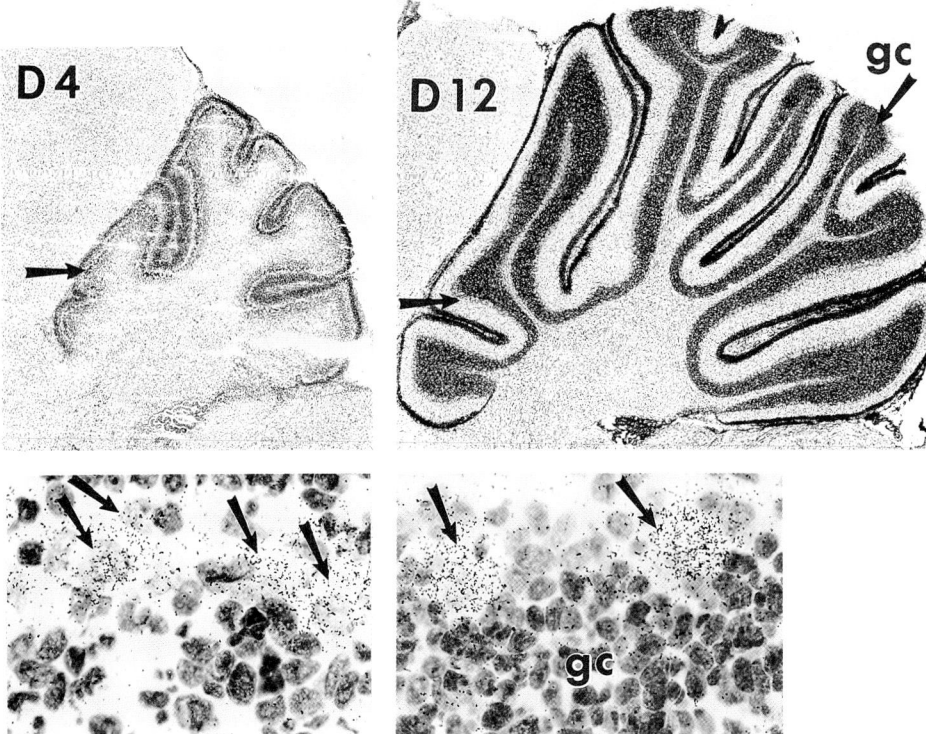

Figure 1 Postnatal development of the cerebellar cortex: IGF-I gene expression. (Top) Parasagittal sections from rat cerebella at 4 and 12 days after birth, a time of rapid cerebellar growth. Arrows point to the Purkinje cell layer. (Bottom) High-power micrographs of IGF-I mRNA hybridization in these same sections. Arrows point to individual Purkinje cells, which are jumbled closely together on Day 4 but are widely and evenly spaced by Day 12. The remarkable growth of the cerebellum over this time period is accounted for largely by the tremendous increase in granule cell (gc) numbers and by the elaboration of extensive processes by both granule cells and Purkinje cells. The Purkinje cells originate earlier than other constituents of the cerebellum, and their number is fixed before birth. They first demonstrate IGF-I gene expression just after birth and show increasing levels through the second week of life, during a time of prolific dendritic growth. As cerebellar maturation nears completion during the third and fourth postnatal weeks, Purkinje cell IGF-I gene expression diminishes and is nearly undetectable in the adult (Bondy, 1991).

cellular composition, IGF-I mRNA decreases as a percentage of total cerebellar mRNA from Postnatal Day 4 to Postnatal Day 12, when, in fact, as shown by *in situ* hybridization, the level of individual Purkinje cell IGF-I mRNA is increasing during this time period (Fig. 1).

Anatomical considerations also influence the interpretation of studies reporting the developmental regulation and differential regional distribution of IGF-II mRNA — based on the analysis of brain homogenates (Rotwein et al., 1988). In situ hybridization studies from a number of different laboratories have shown that IGF-II mRNA is localized in the brain's mesenchymal support structures, i.e., the choroid plexus, vasculature, and meninges (Hynes et al., 1988; Stylianopoulou et al., 1988; Bondy et al., 1990). During the course of brain development, its volume, composed of increasing numbers of neurons and glial cells and their processes, increases proportionately more than its surface area (and adhering meninges), and thus the measurement of IGF-II mRNA as a percentage of total brain mRNA shows a "developmental" decrease. However, evaluation of IGF-II gene expression in meninges and the other support structures by direct visualization does not support any developmental decrease; thus, we do not think that IGF-II has a specifically developmental role in brain function. Furthermore, the variation in IGF-II mRNA content in homogenates from different brain regions can be explained as a function of the degree of vascularization and meningeal contact of the different areas.

In situ hybridization may also be used to correlate cellular patterns of peptide or receptor gene expression with specific physiologic effects. For example, in order to evaluate the role of the IGF-I receptor in the regulation of cellular metabolism in ovarian tissue, we compared patterns of in vivo ^{14}C-labeled 2-deoxyglucose uptake with patterns of IGF-I and IGF-I receptor gene expression in serial sections through the rat ovary (Fig. 2). The animals were injected with this radiolabeled glucose analogue which competes with glucose for cellular uptake and 6-phosphorylation, but which is resistant to further metabolism and thus remains trapped inside metabolically active cells. The regional distribution of metabolic activity in tissue sections can then be analyzed autoradiographically, as shown in Fig. 2. These studies show that there is an impressive correlation between ovarian IGF-I receptor gene expression and metabolic activity.

Significant intermolecular relationships may emerge from the comparison of anatomical patterns of gene expression, such as the recent linkage of the NGF receptor and the cellular protooncogene product, trkA. The demonstration of similar patterns of neural expression for trkA (Martin-Zanaca et al., 1990) and NGF receptor mRNAs raised the possibility that these molecules might be linked in signal transduction, and subsequent studies have supported this view (Klein et al., 1991). We have observed a striking proximity in the timing and localization of gene expression for IGF-I and IGF-binding protein 2 (IGFBP2) in the developing brain (Lee et al., 1992, 1993). IGF-I mRNA is localized in projection neurons and IGFBP2 mRNA is located in immediately

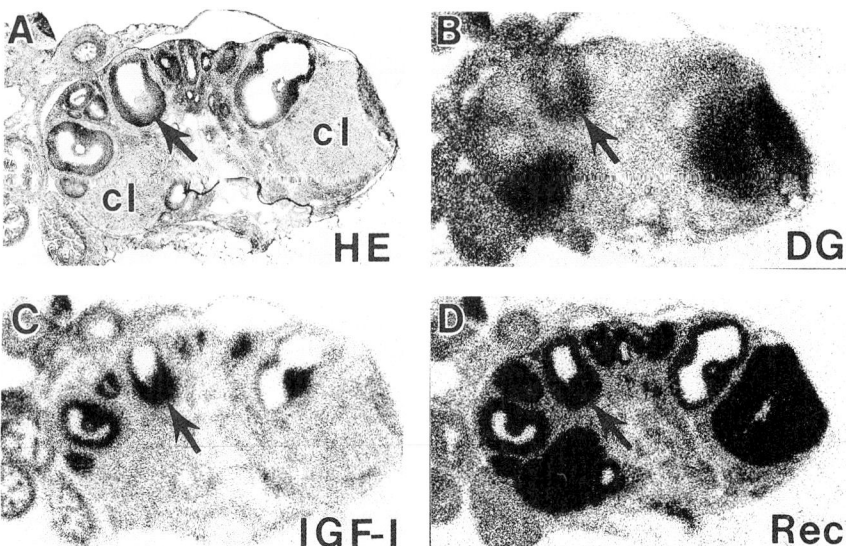

Figure 2 Comparison of patterns of glucose metabolism (B) and IGF-I (C) and IGF-I receptor (D) gene expression demonstrated by film autoradiography in the rat ovary. Glucose metabolism was measured by means of autoradiographic detection of ^{14}C-labeled 2-deoxyglucose (DG) uptake. The rat was injected with ^{14}C-labeled 2-deoxyglucose 30 min prior to sacrifice; some ovary sections were dried and exposed directly to film for determination of regional patterns of metabolic activity (B). Other sections were washed and delipidated, thus removing the intracellular [^{14}C]DG, and then hybridized to ^{35}S-labeled cRNA probes for IGF-I (C) or the IGF-I receptor (D). (A) A hematoxylin and eosin (HE)-stained section from the same ovary. Many large follicles are present in this ovary from a mature rat; an arrow points out the same follicle in the four serial sections. IGF-I receptor mRNA is abundant only in granulosa cells of growing follicles (see Zhou et al., 1991). IGF-I receptor mRNA is abundant in follicles of all stages and in corpora lutea (cl). DG localization parallels the pattern of IGF-I receptor gene expression, being most abundant in the corpora lutea. Oviducts are seen in the left-hand portion of the photographs.

adjacent astrocytes (Color Plate 1). This paralocalization of IGF-I and IGFBP2 has a number of significant functional implications. First, it suggests that IGFBP2 is specifically involved with IGF-I in its role in the process of neural differentiation. Second, it localizes the probable scene of IGF-I action to the area between the soma and *local* processes of the projection neurons (rather than distant axon terminal fields where the astrocytes do not project) and adjacent astrocytes. Finally, these observations provide evidence for a highly specific and novel form of communication between specific groups of neurons and selected local astrocytes during brain development.

II. Methodological Issues

A. Probe Choice

Three different types of probe are currently employed for *in situ* hybridization—biosynthetic cDNAs, synthetic oligodeoxynucleotides, and biosynthetic cRNAs (riboprobes). The pros and cons of each different type of probe have been compared in previous reviews (Tecott *et al.*, 1987; Lewis and Baldino, 1990). In brief, cDNA probes have a number of significant drawbacks and have been used much less frequently in recent years as riboprobe technology has become fairly standard. Nick translation or random priming of cDNA clones produces double-stranded cDNAs of variable length, with a significant percentage of extraneous vector sequences. Several factors militate against their optimal detection of tissue mRNA sequences. The double-stranded cDNA probe must be denatured prior to use, and the tendency for the complementary DNA strands to reanneal competes with their hybridization to target mRNA sequences (Cox *et al.*, 1984). Biosynthetically labelled cDNAs demonstrate poor tissue penetration, probably due to their relatively great length and to the formation of hyperpolymers between partially overlapping complementary sequences. The variability in cDNA probe length precludes the use of stringency conditions based on Tm calculations, and this factor, combined with the lack of "sense" controls for this type of probe, make the evaluation of nonspecific signal problematical.

Synthetic oligodeoxynucleotides have some advantages as probes for *in situ* hybridization. Large quantities of stable DNA probe which are tailor-made to sequences of special interest may be obtained from commercial or laboratory DNA synthesizers. Since published sequences may be more readily available than clones, synthetic oligonucleotides may be the most timely approach to the investigation of a new clone. The tailing reaction using terminal deoxynucleotidyl transferase for the 3' end labeling of oligonucleotides and the oligonucleotide hybridization and wash protocols are relatively simple (Lewis *et al.*, 1985), and results obtained for relatively abundant mRNAs are excellent. The major limitation in the use of oligonucleotides is their relative lack of potency in the detection of less-abundant mRNAs. This lack of sensitivity is due to the limited specific activity of labeled oligonucleotides and to the relative weakness of the short DNA:RNA hybrids.

The application of cRNA, or riboprobes, to *in situ* hybridization was pioneered by Angerer and colleagues (Cox *et al.*, 1984; Angerer *et al.*, 1987). The specific activity of enzymatically synthesized, single-stranded riboprobes is equal to that of cDNAs. Riboprobe length may be accurately adjusted by controlled alkaline hydrolysis to produce fragments of a size optimal for tissue

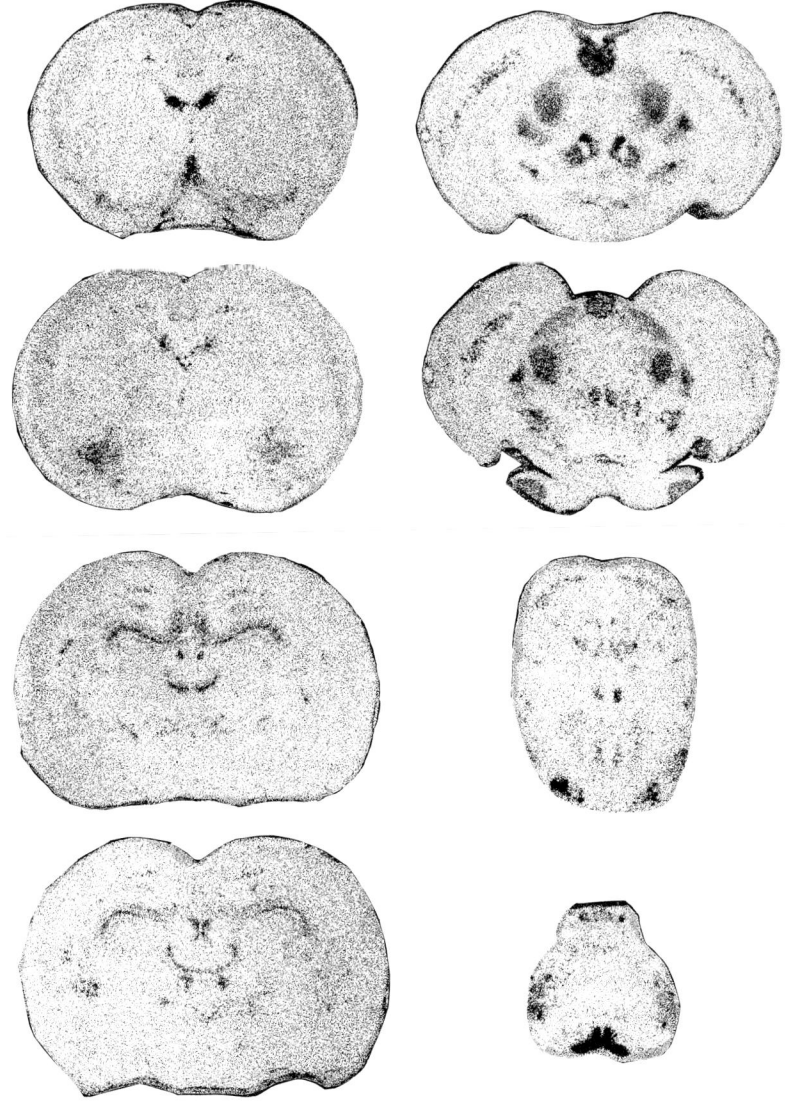

penetration. The intrinsically higher stability of RNA:RNA as compared with DNA:RNA hybrids and defined riboprobe length facilitates the use of high-stringency hybridization and wash conditions for the rigorous elimination of cross-hybridization. Also, riboprobe *in situ* hybridization facilitates the use of RNase to destroy nonhybridized probe and thus eliminate nonspecific signal caused by the sticking of probe to tissue. The result is the achievement of a high signal-to-noise ratio, or sensitivity, which is unsurpassed by any other current methodology. There are a number of commercially available transcription vectors which allow the introduction of cDNA sequences between dual opposed RNA polymerase sites, such that a single plasmid preparation may be used to make both "antisense" and "sense" probes of nearly equal length, with a minimum of extraneous vector sequence. The defined probe length and ready availability of an appropriate control are additional advantages of cRNA as opposed to cDNA probes.

B. Label Choice

Labels for nucleic acid probes falls into two broad categories: isotopic or nonisotopic. The range of nonisotopic labels currently available for nucleic acid probes has recently been reviewed in detail (Bloch, 1990; Lewis and Baldino, 1990). Aside from the practical convenience related to their stable and nonhazardous composition, nonisotopic labels have a theoretical advantage compared with isotopic labels: they provide subcellular-level resolution in mRNA detection. However this potential has not been fully realized. The relatively insensitive, nonisotopically labelled probes have been used successfully to detect abundant sequences such as viral mRNAs in infected cells and mRNAs for hormones and releasing factors concentrated in the hypothalamus and pituitary. If their sensitivity can be improved by amplification of colorimetric detection methods and by the efficient incorporation of the nonisotopic label into ribonucleotide probes, this technique should be useful in the mapping of subcellular pathways of RNA processing. Another limitation associated with nonisotopically labelled probes is that they are useful only for

---

Figure 3 Film autoradiographic survey of IGF-I mRNA localization in the neonatal rat brain. The exposure of hybridized tissue sections to high-performance autoradiography film such as Amersham's β-max facilitates the sensitive and high-resolution detection of target mRNAs and allows the screening of many sections with great ease. This figure shows film autoradiographs of coronal sections from the forebrain to the brain stem of a rat pup on the day of birth. The neuroanatomical loci demonstrating IGF-I gene expression have been detailed elsewhere (Bondy, 1991).

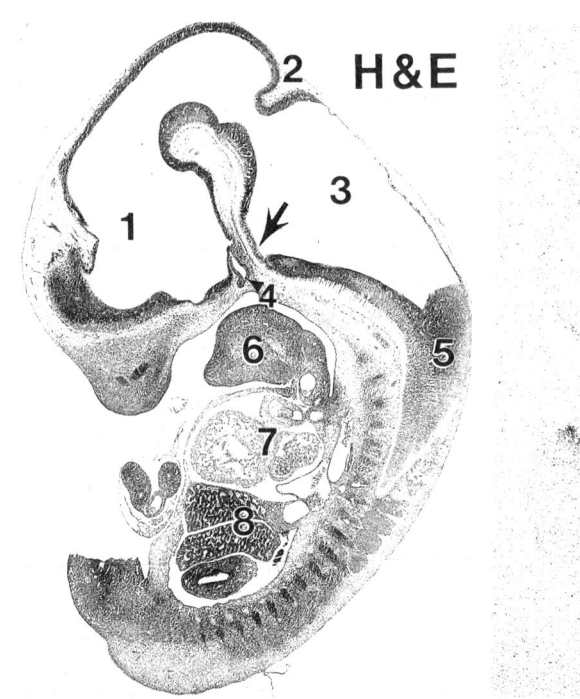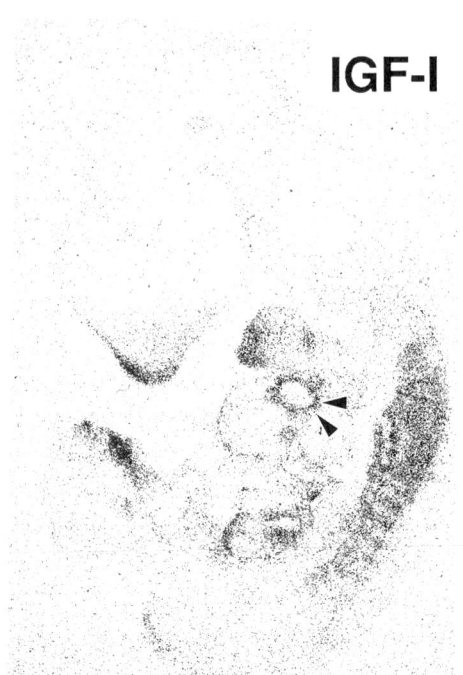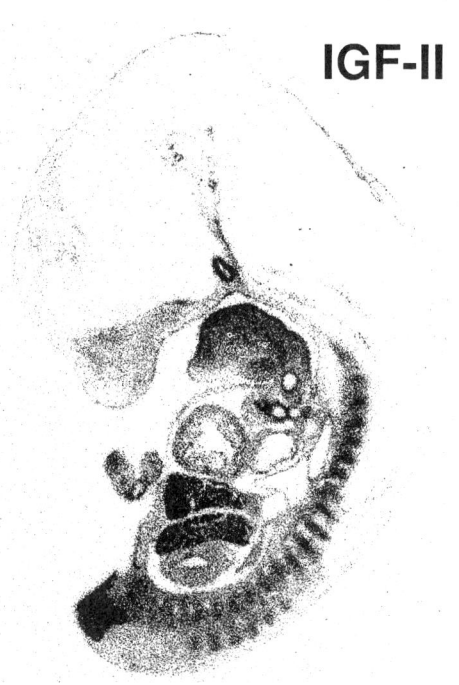

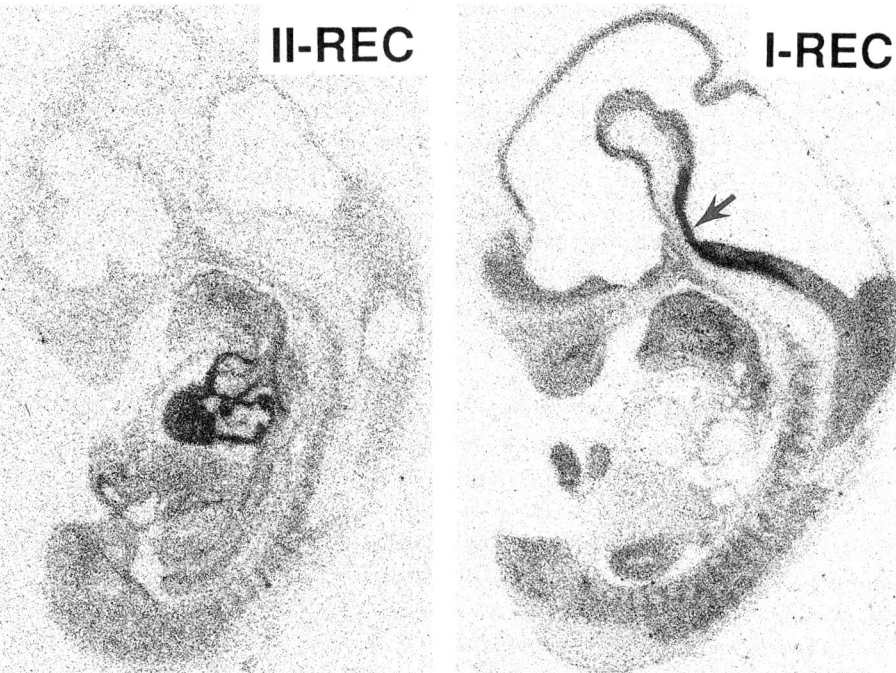

Figure 4 Patterns of IGF system gene expression in the rat embryo. Serial sections from an immersion-fixed (Bouin's solution), paraffin-embedded 14-day embryo were hybridized to the different cRNA probes and exposed to β-max film. The first panel shows an HE-stained section; the remaining panels show film autoradiographs. 1, 3rd ventricle in the forebrain; 2, cerebellar anlage; 3, 4th ventricle in the rhombencephalon; 4, Rathke's pouch or pituitary anlage; 5, spinal cord; 6, tongue/jaw; 7, heart; 8, liver. Arrowheads indicate IGF-I mRNA localized in a ring of cells surrounding the right atrium, a region shown in a high-power photomicrograph in Fig. 7A. Arrows point to the ventral floorplate of the hindbrain, where IGF-I receptor (I-Rec) mRNA is most abundant, and which is shown in a high-power micrograph in Fig. 7B. IGF-II or type-II receptor (II-Rec) mRNA is concentrated in the heart. (Some data from Bondy et al., 1990).

the qualitative detection of mRNAs, given the inherently nonquantitative nature of the detection methods. Finally, we rely heavily on the use of film autoradiography for screening and for the macroscopic view of patterns of gene expression throughout an entire tissue or a whole embryo (Figs. 3 and 4). With the use of nonisotopic probes, however, one is limited to the myopic range of the microscope.

The advantages and disadvantages of the different isotopic labels are considered briefly below and at length elsewhere (Angerer et al., 1987; Tecott *et al.*, 1987; Lewis and Baldino, 1990). ^{32}P provides probes of high specific activity, but the long track length of the high-energy β-emission almost always precludes a cellular level resolution of the signal, thus largely defeating the purpose of *in situ* hybridization. Radiolysis of the ^{32}P-labeled probe may also result in relatively high background levels. Tritium provides the best level of resolution, but the low-energy emission requires long periods of exposure. For most purposes, ^{35}S represents the best choice for an *in situ* probe label — affording good resolution, sensitivity, rapid detection, and probe stability.

In summary, the choice of probe and label for *in situ* hybridization depends on the specific application. For the *de novo* mapping of a specific mRNA, ^{35}S-labeled cRNA probes offer the best opportunity for high-sensitivity detection. A large number of sections from different tissues or developmental stages may be hybridized and screened using film autoradiography, followed by microscopic analysis of selected, emulsion-coated tissue sections. For quantitative studies of an mRNA of relative abundance in a well-described location, ^{35}S-labeled oligonucleotide probes are convenient and reliable. For *in situ* studies where high resolution is the paramount concern, tritium- or digoxigenin (Boehringer-Mannheim)-labeled cRNA probes offer the greatest potential.

C. Control Procedures

No single control procedure provides conclusive evidence of the specificity of a particular *in situ* hybridization pattern. A combination of two or more of the following criteria (adapted from Lewis and Baldino, 1990) will usually rule out any problems related to cross-hybridization or produced by probe/tissue artifacts.

1. A sense probe of similar length and G + C composition produces a negative pattern in the same tissue.
2. Pretreatment of tissue with RNase followed by proteinase K abolishes the hybridization signal.
3. Different probes with nonoverlapping sequences complementary to the same mRNA target produce identical hybridization patterns.
4. The hybridization signal disappears when the wash temperature exceeds the predicted T_m based on probe length, GC composition, formamide, and salt concentrations.
5. Excess unlabeled probe included in the prehybridization or hybridization buffers abolishes the hybridization signal.

6. The probe selectively hybridizes to a band of the predicted molecular weight on a Northern blot.
7. Co-localization of the encoded mRNA product can be accomplished by immunocytochemistry, enzyme activity, or other independent methods.

III. In Situ Hybridization Protocols

A. Synthesis of ^{35}S-Labeled Riboprobes

Single-stranded cRNA probes are produced by the *in vitro* transcription of DNA templates. The DNA to be transcribed is subcloned into the multiple-cloning site of a plasmid vector containing dual opposed RNA-polymerase promoter sites (see Fig. 5). The recombinant plasmid is grown and amplified in an appropriate bacterial strain, and 100–200 µg of supercoiled plasmid is purified using a Qiagen kit (Qiagen Inc., Chatsworth, CA). The plasmid is linearized (20 µg/ml) using a restriction endonuclease chosen to cut selectively adjacent to one promoter region thus opening the construct so as to allow transcription of the insert from the opposite promoter (Fig. 5). Sense transcripts provide a size and purine/pyrimidine content-matched control probe for the determination of nonspecific background, while antisense transcripts are used for the localization of target mRNA during *in situ* hybridization. It is essential to be certain that the supercoiled plasmid is completely linearized prior to transcription, since any supercoil remaining in the preparation will produce long vector sequence transcripts at the expense of insert transcription. The linearized fragments are purified using Geneclean II (Bio 101, Inc.).

Transcription is carried out with safeguards to protect the RNA product from degradation. Sterile plastic ware is used and reagents are prepared with diethylpyrocarbonate (DEPC)-treated water. DEPC (Sigma, D5758) is a chemical nuclease inactivator. A total of 0.1% DEPC is added to water contained in autoclavable bottles. The mixture is agitated vigorously until the DEPC is completely dispersed, after which the solution is allowed to stand overnight and autoclaved the next day. The following transcription protocol for the incorporation of two radiolabeled bases has been adapted from Zoeller *et al.* (1989):

1. DNA template (0.5–1.0 µg), [α-^{35}S]CTP, and [α-^{35}S]UTP (Amersham SJ 40382 and SJ 40383, 0.1 mCi each) are combined in a microfuge tube and dried in a speed vacuum. Into this same tube, the following are added.
    a. 2 µl NTP mix (500 µM each ATP and GTP and 25 µM each CTP and UTP)

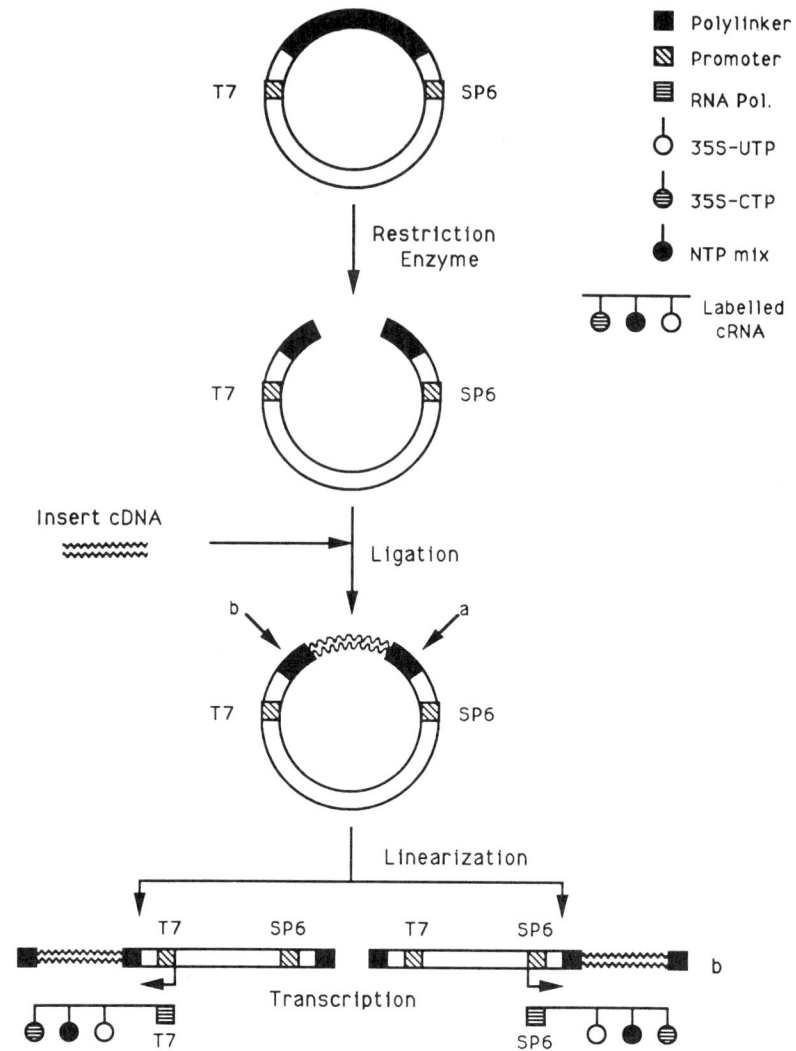

Figure 5 Radiolabeled riboprobe synthesis using a transcription vector. The cDNA to be transcribed is inserted into the "polylinker site" of a plasmid containing dual opposed promoter sites. The recombinant plasmid is then linearized with restriction enzymes chosen to give access to the insert from either the T7 (a) or SP6 (b) promoter sites. This allows transcription of the cDNA insert in either "sense" or "antisense" orientations.

b. 1 µl dithiothreitol (DTT, 100 mM)
c. 1 µl (50 U) appropriate RNA polymerase
d. 2 µl appropriate polymerase buffer (5X)
e. 1 µl RNasin (35 U/µl, Promega)
f. 3 µl DEPC-H$_2$O

The final volume will be 10 µl.

2. The tube is capped, vortexed, quick spun, and incubated for 60 min at 42°C.
3. An additional 1 µl RNA polymerase may be added at 30 min.
4. 1 µl RNase-free DNase I (1 U/µl, Promega) is added to digest the DNA template and the reaction incubated for 10 min at 37°C.
5. A sterile gel chromatography spin column (Bio-Spin 6, Bio-Rad) is used for separation of the radiolabeled cRNA probe from unincorporated nucleotides. After preparation of the column according to the manufacturer's directions using Tris (10mM)-EDTA (2mM) buffer, the same buffer is added to the reaction product to bring the final volume up to 75 µl. The sample is added to the column and then centrifuged at 1000 g for 2 min. The labeled probe in the collection tube is precipitated by the addition of 5 µl tRNA (10 µg/ml), 10 µl NaCl (5 M), 10 µl DEPC-H$_2$O, and 300 µl cold ethanol prior to freezing on dry ice for 30 min. The pellet is collected after centrifugation for 30 min (10,000 g) at 4°C and reconstituted as described below for alkaline hydrolysis.

B. Alkaline Hydrolysis

We make it a practice to reduce all our probes to ≈ 100 bases in length prior to hybridization using controlled alkaline hydrolysis, according to the protocol described by Angerer et al. (1987). We have found this length to be optimal in terms of mRNA detection (tissue penetration). It is also advantageous to have probes of a similar length for the standardization of hybridization and wash conditions, since hybrid length is one of the factors determining the Tm.

The pellet is resuspended in 50 µl H$_2$O and 50 µl carbonate buffer (80 mM NaHCO$_3$ and 120 mM Na$_2$CO$_3$ in DEPC-H$_2$O, pH 10.2) and incubated at 60°C for the time calculated by the formula (Cox et al., 1984) $t = (L_0-L_f)/kL_0L_f$, where t = time in minutes, L_o = initial probe length in kilobases, L_f = final probe length in kilobases, k = 0.11 scission/kb/min.

The hydrolysis is stopped by neutralization with 3 µl of 3 M sodium acetate (pH 6) and 5 µl 10% glacial acetic acid followed by ethanol precipitation. The dried pellet is reconstituted in 50 µl Tris buffer (10 mM Tris, 20 mM DTT) and 1 µl is counted. Counts in the range of $2-8 \times 10^6$ cpm/µl are expected. Analytical gels run after the alkaline hydrolysis step demonstrate that the yield of the ^{35}S-labeled RNA probe using this protocol is approximately 20-30 times the amount of starting DNA template, and the specific activity of the final product is $\approx 10^8$ dpm/µg. The ^{35}S-labeled cRNA probes (stored at $-70°C$) are good for at least 2 weeks.

C. Coating Slides

Glass slides must be treated with a coating to promote the adherence of tissue sections, and we have had great success with poly-L-lysine (Sigma P 1399) and the following simple protocol. Slides (50 per rack, with as many as 8 racks placed together in a plastic developing tray) are washed in detergent and hot tap water, rinsed, and soaked in 75% ethanol for 30 min. They are then rinsed with tap water and allowed to air dry. Next, they are soaked in poly-L-lysine (0.05 mg/ml in 10 mM Tris, pH 8.0; 500 ml in a staining dish which holds a rack of 50, is used to sequentially coat up to 8 racks or 400 slides) for 10 min, air-dried, and stored in covered slide boxes until use.

D. Tissue Preparation

Tissues destined for *in situ* hybridization may undergo a variety of specialized fixation and embedding protocols in order to optimize histological detail and the preservation of intact mRNA. In cases where the target mRNA is abundant and the cells are anatomically unmistakable (such as the vasopressin-containing magnocellular neurons of the hypothalamus, the POMC-containing cells of the pituitary or the IGF-I-containing Purkinje cells of the cerebellar cortex), little special treatment is necessary. In these situations, fresh frozen tissue treated only with a brief immersion in fixative immediately prior to hybridization is perfectly adequate, and this is certainly the easiest approach, particularly for laboratories without much investment in histological methods.

Retention of tissue mRNA and morphology are definitely improved, however, by more extensive fixation, i.e., perfusion *in situ* followed by immersion of the tissue in a crosslinking fixative. Fixation crosslinks tissue proteins and thus blocks the loss of mRNA by diffusion out of the tissue into aqueous buffers. Fixation may also compartmentalize the subcellular structure so that endogenous RNases are confined. We use more extensive fixation in order to improve morphology when cellular anatomy is an important consideration and when planning to combine *in situ* hybridization with immunocytochemistry, in which there is an anticipated loss of mRNA from the tissue during the processing involved in the immunocytochemical protocol. We use transcardial perfusion with 4% formaldehyde in phosphate-buffered saline (PBS) or Zamboni's fixative, followed by immersion of the tissue in the same fixative overnight. The tissue is equilibrated in sucrose gradients up to 30% prior to freezing for sectioning (15–20 μm) on a cryostat. For mRNA detection in young embryos where tissue morphology is critical, we use immersion–fixation in Bouin's solution (0.9% picric acid, 9% formaldehyde, and 5% acetic acid) for several hours to overnight, depending on specimen size. Samples are then washed in several changes of 50% ethanol, dehydrated through graded ethanols, and cleared in two changes of xylene prior to infiltration with paraffin as described in Angerer *et al.* (1987). Examples of embryonic tissue hybridized after this tissue preparation protocol are seen in Figs. 3 and Color Plate 2. When extensive fixation is used, tissue permeabilization with proteinase K treatment is required prior to hybridization. With fresh-frozen tissue, however, we do not find permeabilization treatments or prehybridization to be advantageous.

E. Preparation of Sections for Hybridization

The preparation of tissue sections for *in situ* hybridization is very simple. All solutions are made with DEPC-treated water and are stored in disposable plastic ware. Slides are brought to room temperature, placed in autoclaved containers, and immersed in fresh fixative (4% formaldehyde) in PBS for 5 min. The fix is drained and the slides are rinsed in PBS (1 min). Prefixed, proteinase K-treated tissue may enter the protocol at this point (see below). The next step, acetylation, is important for blocking the reaction of probe with tissue amino groups. Acetic anhydride (0.25%, v/v) is added to a container filled with the required amount of triethanolamine–HCl (0.1 M, pH 8)/NaCl (0.9 M) and vigorously mixed. The slides are immersed in this freshly prepared solution for 10 min. Slides are washed in 2× SSC (1 min) and then dehydrated in graded alcohols (70, 80, 95, and 100%, 1 min each) prior to immersion in chloroform for delipidation (5 min) and rehydrated (95% ethanol, 1 min). The delipidation step is beneficial for mRNA detection in brain and other lipid-laden tissues, but may not be necessary for all specimens (and is not applied to tissue which has come out of paraffin). The slides are allowed to air dry and then placed in covered plastic trays (Nunc square tissue culture dish, No. 166508) ready for hybridization. We do not find it advantageous to prehybridize tissue sections, except for those which have been processed for immunocytochemistry (see below).

F. Hybridization

Hybridization buffer composed of 50% formamide, 0.3 M NaCl, 20 mM Tris-HCl, pH 8, 5 mM EDTA, 500 µg tRNA/ml, 10% dextran sulfate, and 0.02% each of bovine serum albumin (BSA), ficoll, and polyvinylpyrollidone (1× Denhardt's solution) may be made in advance and stored at −20°C. The dextran, which must be of the highest grade (e.g., Sigma D6001), dissolves with some difficulty. This may be facilitated by warming the buffer up to 50°C briefly accompanied by gentle agitation. The ^{35}S-labeled probes (concentration of 10^7 cpm/ml or approximately 100 ng/ml) and fresh DTT (10 mM) are added to the hybridization buffer just prior to use. DTT is essential to prevent the oxidation of the ^{35}S and is not used for other isotopes or for nonisotopically labeled probes. After the ^{35}S-labeled probe in hybridization buffer is added to the sections (10 µl/cm²), the sections are gently covered with glass coverslips and placed in humidified chambers for overnight (14–16 hr) incubation at 55°C.

G. Stringency Washes

The principles governing nucleic acid hybridization *in situ* are essentially the same as those applied to hybridization in solution and on blots and have been discussed in cogent fashion in a number of previous reviews, including Angerer *et al.* (1987). The following stringency wash protocol for ^{35}S-labeled riboprobes has been adapted from Angerer *et al.* (1987). At this point, it is no longer necessary to use DEPC-treated solutions or sterile containers, as the mRNA of interest is now hybridized with the probe, and we will actually be treating the sections with exogenous RNase. It is important to take care in using the RNase so as not to contaminate benches or equipment with the highly concentrated enzyme. An inexpensive grade of DTT (≈ 2 mM) is added to all wash solutions.

1. Slides are placed in slide racks and immersed in containers of 4× SSC which are gently rocked to allow coverslips to fall off (≈ 30 min). Coverslips must not be forcibly removed or the tissue will be damaged.
2. Slides are washed three times (≈ 5 min) in fresh 4× SSC. Radioactivity in wash solutions after this point should be undetectable.

3. Sections are then dehydrated through a series of graded ethanols. Ethanols less than 95% are diluted with 2X SSC.
4. The slides are immersed in the first stringency wash solution (0.3 M NaCl, 50% formamide, 20 mM Tris–HCl, 1 mM EDTA) at 60°C and its temperature is monitored. When the temperature has returned to 60° (the addition of the glass slides will have cooled it) a 10-min incubation is begun.
5. The slides are immersed in room temperature 2X SSC to cool.
6. Sections are then treated with RNase A (Promega, 20 µg/ml in 0.5 M NaCl, 10 mm Tris, 1 mM EDTA) for 30 min at 37°C.
7. Slides are passed through graded salt solutions (2X, 1X, then 0.5X SSC) followed by a second stringency wash in 0.1X SSC at 55°C for 15 min.
8. Slides are immersed in room temperature 0.1X SSC to cool.
9. After a final dehydration, sections are air-dried and apposed to Hyperfilm β-max (Amersham).

H. Emulsion Coating

A darkroom equipped with a revolving or double door; sodium safelights with red and amber filters such as Thomas Duplex; a small, shallow water bath; and sufficient shelf or bench space to accommodate drying slides is required. The bath must maintain a stable temperature of 42°C. A tube filled halfway with distilled water is placed in the bath to warm. An equal volume of solid NTB2 (Kodak) emulsion is added to the container and gently pushed down and mixed with a glass rod. This is allowed to melt for 20 min after which the 1:1 diluted emulsion is stirred gently again. A clean glass slide is dipped in the mixture and held up to the red light to evaluate the emulsion for the presence of undissolved particles or bubbles. If the diluted emulsion is not smooth, then it should be stirred again and allowed rest for another 15 min. Prolonged heating of the emulsion will cause high levels of background. Slides are dipped into the emulsion for 3–5 sec using a steady motion to avoid creating bubbles and attain an even coating. The technique must be consistent from slide to slide to obtain a reproducible emulsion coating. Slides are dried in a vertical position in the darkroom for 2 hr and are then packed into clean, light-proof boxes containing dessicant capsules, which are sealed with black tape and wrapped in foil.

The boxes are stored at 4°C until slides are to be developed (exposure to emulsion is for two to three times the period required for optimal film exposure). Fresh Kodak D19 developer and fixer are brought to 16°C. Slide boxes are opened in the darkroom and slides are placed into racks and immersed in the developer for 3 min, then immersed in a container of water (16°) for 30 sec to stop, and finally immersed in fixer with gentle agitation for 4 min. Slides must then be washed in running tap water for 15-20 min to remove remaining emulsion prior to counterstaining with hematoxylin and eosin or a Nissl stain for microscopic evaluation. Nissl stains such as thionine or cresyl violet may provide less than optimal detail since RNase treatment destroys much of the Nissl substance normally taking up the stain.

I. Combined Immunocytochemistry and in Situ Hybridization

The combination of *in situ* hybridization and immunocytochemistry on the same tissue section has a number of important applications. Cell-type specific antibodies are used in order to identify the cell type in which a given gene is transcribed. An example of this type of application is shown in Color Plate 2C, which shows glucose transporter (Glut1) gene expression in reactive astrocytes. The success of this combination of techniques depends on both the preservation of the intensity of the immunostaining and morphological detail and the retention of maximal levels of endogenous mRNA. In our experience, best results are obtained when the immunostaining precedes the *in situ* hybridization, with the immunostaining protocol modified to safeguard tissue mRNA. We describe these modifications, assuming the investigators have already obtained specific immunostaining procedures. To protect mRNA from degradation by RNases which are present in the blocking serum and possibly in the primary and secondary antibodies, we use the following strategies:

1. RNase inhibitor (RNasin, Promega, 140 U/ml final concentration) and DEPC (0.04% final dilution) are included in the normal serum, primary and secondary antibodies, and avidin–biotin complex.
2. RNasin only is added to the peroxidase substrate mixture.
3. All solutions are prepared with DEPC-treated, autoclaved water.

Following development of the immunodetection color reaction, tissue sections must be treated with proteinase K and prehybridized in order to prevent high background levels. Prehybridization treatment for slides subjected to immunocytochemistry follows:

1. Treat with proteinase K (1 ng/ml in 100 mM Tris–HCl, pH 8.0, 50 mM EDTA) for 30 min at 37°C, followed by brief washes in PBS.
2. For acetylation, add 0.25% acetic anhydrate to triethanolamin–HCl (0.1 M, pH 8.0) immediately before use and incubate at room temperature for 10 min.

3. Dehydrate through an alcohol series.
4. Immerse in chloroform at room temperature for 5 min.
5. Rehydrate to 95% alcohol.
6. For prehybridization, add enough hybridization buffer to cover sections and place in covered humidified chamber at 50–60°C for a minimum of 2 hr.
7. Drain the prehybridization buffer before adding new hybridization buffer containing 50 mM DTT and 2×10^6 cpm radiolabeled probe/100 µl and coverslip.
8. Incubate overnight at 55°C.

Acknowledgments

We thank Ricardo Dreyfuss for expert photomicrography.

References

Andersson, I. K., Edwall, D., Norstedt, G., Rozell, B., Skottner, A., and Hansson, H.-A. (1988). *Acta Physiol. Scand.* **132**, 167–173.
Angerer, L. M., Stoler, M. H., and Angerer, R. C. (1987). In *"In situ* Hybridization; Applications to Neurobiology" (K. L. Valentino, J. H. Eberwine, and J. D. Barchas, eds.), pp. 42–70. Oxford Univ. Press, Oxford.
Bloch, B. (1990). In *"In situ* Hybridization Histochemistry" (M.-F. Chesselet, ed.), pp. 23–88. CRC Press, Boca Raton, Florida.
Bondy, C. A. (1991). *J. Neurosci.* **11**, 3442–3455.
Bondy, C. A., Werner, H., Roberts, C. T., and LeRoith, D. (1990). *Mol. Endocrinol.* **4**, 1386–1398.
Chin, E., Zhou, J., and Bondy, C. A. (1992). *Endocrinology* **130**, 3237–3245.
Clemmons, D. R. (1990). *Trends Endocrinol. Metab.* **1**, 412–417.
Cox, K. H., DeLeon, D. V., Angerer, L. M., and Angerer, R. C. (1984). *Dev. Biol.* **101**, 485–502.
Dolle, P., Castrillo, J. L., Theill, L. E., Deerinck, T., *et al.* (1990). *Cell* **60**, 809–820.
Hansson, H.-A., Nilsson, J., Isgaard, H., Billig, O., Isaksson, A., Skottner, I. K., and Rozell, B. (1988). *Histochemistry* **89**, 403–410.
Heumann, R., Korsching, S., Scott, J., and Thoenen, H. (1984). *EMBO J.* **3**, 3138.
Hynes, M. A., Brooks, P. J., Van Wyk, J. J., and Lund, P. K. (1988). *Mol. Endocrinol.* **2**, 47–54.
Klein, R., Jing, S., Nanduri, V., O'Rourke, E., and Barbacid, M. (1991). *Cell* **65**, 189–197.
Kobayashi, S., Clemmons, D. R., and Venkatachalam, (1991). *Am. J. Physiol.* **261**, F22–F28.
Lee, W.-H., Javedan, S., and Bondy, C. A. (1992). *J. Neurosci.* **12**, 4737–4744.
Lee, W.-H., Michaes, K., and Bondy, C. A. (1993). *Neuroscience* **53**, 251–265.
Lewis, M. E., and Baldino, F., Jr. (1990). In *"In situ* Hybridization Histochemistry" (M.-F. Chesselet, ed.), pp. 1–22. CRC Press, Boca Raton, Florida.
Lewis, M. E., Sherman, T. G., and Watson, S. J. (1985). *Peptides* **6**, Suppl. 2, 75–92.
Martin-Zanaca, D., Barbacid, M., and Parada, L. F. (1990). *Genes Dev.* **4**, 683–694.
Ooi, G. (1990). *Mol. Cell. Endocrinol.* **71**, C39–C43.
Rechler, M. M., and Nissley, S. P. (1985). *Annu. Rev. Physiol.* **47**, 425–442.

Rotwein, P., Burgess, S. K., Milbrandt, J. D., and Krause, J. E. (1988). *Proc. Natl. Acad. Sci. U.S.A.* **85,** 265-269.
Stylianopoulou, F., Herbert, J., Soares, M. B., and Efstratiadis, A. (1988). *Proc. Natl. Acad. Sci. U.S.A.* **85,** 141-145.
Tecott, L. H., Eberwine, J. H., Barchas, J. D., and Valentino, K. L. (1987). *In* "*In situ* Hybridization: Applications to Neurobiology" (K. L. Valentino, J. H. Eberwine, and J. D. Barchas, eds.), pp. 3-24. Oxford Univ. Press, Oxford.
Underwood, L. E., D'Ercole, A. J., Clemmons, D. R., and Van Wyk, J. J. (1986). *Clin. Endocrinol. Metab.* **15,** 59-78.
Zhou, J., Chin, E., and Bondy, C. A. (1991). *Endocrinology (Baltimore)* **129,** 3281-3288.
Zoeller, R. T., Lebacq-Verheyden, A. M., and Battey, J. F. (1989). *Peptides* **10,** 415-422.

PART III

Techniques for Receptors and Signal Transductors

14

Plasma Membrane Isolation Strategies for Cell Surface Receptors: Application for the Insulin Receptor

Maxine A. Lesniak, Joshua Shemer, and Phillip Gorden[1]

I. Introduction
II. Preparation of Plasma Membranes
   A. Cultured Lymphoid Cells Grown in Suspension
   B. Tissues from Vertebrate Species
III. General Considerations for Radioreceptor Assays
IV. Binding of ^{125}I-Insulin to Plasma Membranes
   A. Insulin Binding to Crude Plasma Membranes
   B. Insulin Binding to Solubilized Insulin Receptors
References

I. Introduction

Functional receptors for polypeptide hormones and growth factors, e.g., insulin, insulin-like growth factors (IGF-I, IGF-II), and human growth hormone are primarily located on the cell surface. This is in contrast to receptors for steroids and thyroxine which bind to intracellular receptors (Gorden and Weintraub, 1992). Although using target tissues for studying the specific receptor is optimal, it is not always practical. To study cell surface receptors such as those for insulin, we have prepared plasma membranes from several vertebrate tissues, circulating cells, and cultured cells. This report suggests strategies for preparing plasma membranes from various sources, describes general considerations for performing radioreceptor assays, and provides an example for using plasma membranes in the study of insulin receptors.

[1] We would like to dedicate this chapter in memory of J. A. Hedo, M.D.

II. Preparation of Plasma Membranes

Although purified plasma membranes are the most desirable, their preparation is tedious and time consuming and, in many instances, the conditions are organ specific (Neville, 1968). For most radioreceptor applications membranes that are crudely or partially purified have proved to be satisfactory. Various methods have been used to prepare plasma membranes depending upon the tissue source: cultured cells grown in suspension such as B lymphocytes (IM-9 or viral-transformed B lymphocytes from patients); cells grown as a monolayer such as human hepatocytes, Hep G2 cells; and, from vertebrate species, freshly dissected tissues such as liver and brain.

A. Cultured Lymphoid Cells Grown in Suspension

This strategy employs hypotonic lysis of cells after intracellular loading of glycerol to burst the cells. Plasma membranes and intact nuclei are obtained from this procedure. Other strategies use nonionic detergents like Triton X-100 for isolation of cell surface receptors from lymphoid cells (Cone, 1987).

Materials

1. Earle's balanced salts solution (EBSS)
2. Glycerol solution 90% (w:v) in EBSS—room temperature
3. Lysing buffer—ice cold

10 mM	Hepes
1 mM	$CaCl_2$
1 mM	$MgSO_4$

4. Sucrose, 40% (w:v)

PROCEDURE

1. Isolate the cells from the growth medium using low-speed centrifugation, 600g, for 10 min at room temperature. **NOTE** Do not use low temperatures at this step.

2. Wash the pelleted cells with EBSS (~1/10 original volume).

3. Once the cells are washed free of medium, resuspend the cells in EBSS (~1/100 of original volume), then incubate at 37°C for 30 min.

4. To load the cells with glycerol, use three additions of glycerol to achieve a final concentration of 30%. For example,
   a. Resuspend the cells in 12 ml of EBSS, add 2 ml of 90% glycerol, mix well, wait 5 min
   b. Repeat with two additions of 2 ml each of glycerol for a total of 6 ml of glycerol

   The cells containing glycerol are placed on ice for 10 min.
5. Pellet the cells by centrifugation at 600g for 10 min at 4°C, discard the supernatant, resuspend the cells in ice-cold hypotonic lysing buffer, then place the cells on ice for 5 min.
6. Subject the lysed cells to differential centrifugation with the initial centrifugation at 600g for 15 min at 4°C.
7. The pellet contains nuclei and cell debri. Transfer the supernatant to tubes suitable for ultracentrifugation with a swinging bucket rotor.
8. With a Pasteur pipette, underlay the supernatant with a cushion of 40% sucrose (0.5 – 1.0 ml).
9. Centrifuge at 20,000g for 90 min at 4°C. The plasma membranes are located at the interface (the lipid-phase floats on top and debri is in the pellet.)
10. Remove the supernatant by aspiration.
11. Recover the membranes with a minimal amount of supernatant and sucrose; transfer to another set of centrifuge tubes.
12. Dilute the suspended membranes with lysing buffer, at 10-fold volume.
13. Again underlay the solution with 40% sucrose.
14. Repeat the previous centrifugation step.

The resulting plasma membranes at the interface can be collected and aliquoted into separate tubes for storage at −70°C prior to their use in a radioreceptor assay. The membranes should be subjected to only one freeze–thaw (adapted from Jett *et al.*, 1977).

B. Tissues from Vertebrate Species

1. Crude Membrane Preparation

Cell surface receptors from various vertebrate species — rodents, chickens, frogs, lizards — have been studied. The plasma membranes from tissues, the brain and liver for example, are prepared using similar procedures. The tissues are freshly dissected from the desired vertebrate and are quickly frozen on dry ice. (Sacrifice of the animals requires adherence to recommended guidelines for animals.)

Materials

1. Tissue — 1.5 – 4.5 g
2. Glass/glass homogenizer
3. $NaHCO_3$, 1 mM, with protease inhibitors:
   phenylmethylsulfonyl fluoride (PMSF), 2 mM
   leupeptin, 10 μg/ml
   aprotinin at a final concentration of 1 trypsin-inhibitory unit/ml.
4. Krebs–Ringer phosphate (Ca^{2+}-free)-KRP

118 mM	NaCl
5 mM	KCl
1.2 mM	KH_2PO_4
10 mM	$NaHPO_4$
1.2 mM	$MgSO_4$

PROCEDURE

1. The tissues (1.5 – 4.5 mg) are placed in an all-glass homogenizer and processed using 20 strokes at 4°C in 15 volumes of 1 mM $NaHCO_3$ with protease inhibitors.
2. The homogenate is divided into tubes for centrifugation at 600g at 4°C for 10 min.
3. The pellets are discarded, and the supernatants are retained for centrifugation at 20,000g at 4°C for 30 min.
4. Now the pellets are retained and the supernatants are discarded.
5. The resulting pellets are resuspended in Krebs–Ringer phosphate buffer (Ca^{2+}-free) pH 7.8, at a protein concentration of 10 mg/ml.
6. The plasma membrane protein concentration is determined by the Lowry method.
7. Aliquots of the membranes are either prepared for storage at −70°C, — this is the "crude membrane preparation" — or further purified.
8. The "receptor" can be solubilized from the crude membrane preparation and purified by affinity chromatography using wheat germ agglutinin, i.e., "partially purified preparation." (Either membrane

preparation can be used in a radioreceptor assay.) When using crude membrane preparations such as liver in the radioreceptor assay (RRA) the addition of protease inhibitors is required.

NOTE The concentration of membrane proteins should be measured before the addition of the protease inhibitors which in turn should be added after the membranes are thawed just prior to their use in the RRA.

2. Preparation of Partially Purified Membranes: Solubilization of Plasma Membranes

PROCEDURE

1. The crude plasma membrane pellet is resuspended in 50 mM Hepes with 1% Triton X-100 and PMSF 2 mM, pH 7.6.
2. The undissolved material is sedimented by centrifugation at 40,000g for 45 min at 4°C.
3. The supernatant with solubilized receptors is retained for further purification (Shemer *et al.*, 1986).

3. Preparation of Partially Purified Membranes: Affinity Chromatography–Wheat Germ Agglutinin

Cell surface receptors that are glycoproteins, such as the insulin receptor, can be solubilized and purified using a variety of immobilized lectins. In practice, wheat germ agglutinin (WGA) is the most efficient lectin in terms of insulin receptor purification and recovery (Hedo *et al.*, 1981). (Any solubilized preparation, e.g., plasma membranes, microsomal membranes, or whole cells, can be used.) Note also that the source of WGA is important. Some WGA have a greater capacity for absorbing than others (Shemer and LeRoith, 1987). One of the advantages of using lectin columns compared to other affinity columns is that they can be eluted with simple saccharides without changes in pH or ionic strength and without the use of denaturants.

Materials

1. Wheat germ agglutinin immobilized on agarose
2. Small plastic columns (15 × 0.9 cm)
3. Buffer I.
   0.15 M NaCl
   50 mM Hepes
   0.1% Triton X-100

0.01% sodium dodecyl sulfate, pH 7.6 plus 2 mM PMSF (PMSF optional for lymphocyte preparations)
4. Buffer II.
   0.15 M NaCl
   50 mM Hepes
   0.1% Triton X-100
   0.3 M N-acetyl-D-glucosamine, pH 7.6
5. Buffer III.
   0.15 M NaCl
   50 mM Hepes
   0.1% Triton X-100, pH 7.6

PROCEDURE

1. Columns with WGA-agarose are first prepared. The amount of lectin-agarose depends upon both the capacity of each batch of lectin-agarose and the concentration of glycoprotein in the preparation. [As an index 2.0 ml of settled gel, ~1.0 mg of lectin-agarose, has a capacity to bind all the insulin receptor activity present in 5 mg of crude microsomal membranes from human placenta, one of the tissues with the highest insulin receptor concentrations (Hedo *et al.*, 1981).] Prior to addition of solubilized membrane, the lectin columns are prepared by washing with each of the buffers as follows: wash the column with 50 bed-volumes of Buffer I, then wash with 10 bed-volumes of Buffer II, and, finally, wash with 100 bed-volumes of Buffer III.
2. Apply the sample to a column and recycle the eluate through the column at least three times. For example, solubilized liver membrane at a concentration of 30 mg/ml is resuspended in a total volume of 7.0 ml of Buffer I and is applied to 2.0-ml WGA-agarose column. Wash the column with 50–100 bed-volumes of Buffer III.
3. Elute the receptors with Buffer II—at least 2 bed volumes—by collecting 1.0-ml fractions. Determine the protein concentration in each fraction and test each for binding activity (Fig. 1). The fractions with peak activity are divided into aliquots and stored at $-70°C$; we have used membranes that have been stored for up to 1 year. (The eluting monosaccharide can be removed by dialysis in Buffer III.) Recovery of insulin receptors is between 70–95% of the total activity applied. The degree of purification depends on the specific activity of the initial preparation. Table I shows examples of the expected fold purification of insulin receptors prepared from various tissues.

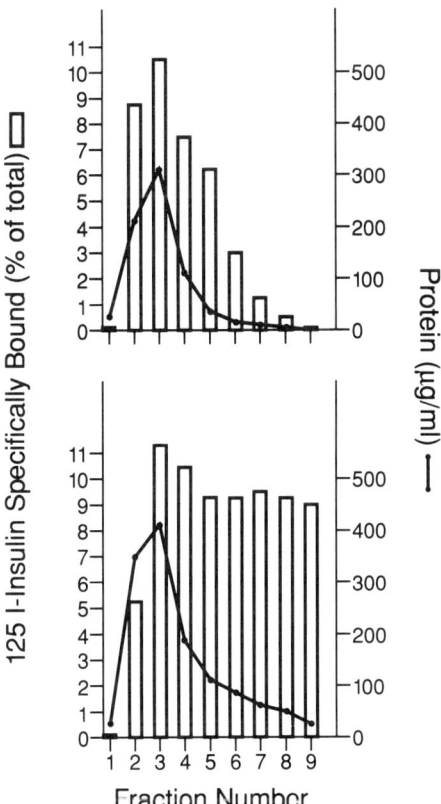

Figure 1 Wheat germ agglutinin chromatography elution profiles of ^{125}I-insulin binding (specific binding) to and protein concentration (mg/ml) of insulin receptors purified from membrane preparations of lizard livers (top) and brains (bottom) (From Shemer et al., 1986.)

NOTE To increase binding capacity one needs to increase receptor concentration. This can be achieved by increasing the protein concentration for tissues in which receptor concentration is expected to be low. In order to observe specific insulin binding activity of 6–8% in lizard tissues the initial protein concentration is ~30 mg for liver and ~240 mg for brain (Fig. 1) (Shemer et al., 1986).

4. In order to prevent receptor degradation, the chromatographic procedure is usually performed at 4°C. However, the WGA-agarose columns are equally efficient at room temperature. The columns can be reused, multiple times, for up to 1 year without apparent loss of

Table I
Purification of Insulin Receptor by Wheat Germ Agglutinin Chromatography

	Fold increase/mg protein (eluate/crude extract) R_0
Liver (rat)	11
Brain (rat)	14
Cultured lymphoid cells IM-9 (human)	21
Placenta (human)	22

Note. ^{125}I-insulin binding to crude extracts and eluates from what germ agglutinin chromatography; R_0-maximal binding capacityt. After Hedo *et al.* (1981) and Hendricks *et al.* (1984).

activity. Store columns in Buffer II with sodium azide 0.01% at 4°C and reactivate by following the initial washing steps.

III. General Considerations for Radioreceptor Assays

Many polypeptides labeled with ^{125}I and graded for use in an RRA are commercially available. [An investigator can radiolabel polypeptide ligands that retain biological activity using methods previously described (Roth, 1975).] Also polypeptides and proteins such as insulin, growth factors (IGF-I, IGF-II), and growth hormones are commercially available as recombinant products. The conditions for the binding assay vary with the interaction of the ligand with its receptor. Binding is highly dependent on the concentrations of reactant as well as temperature, i.e., the lower the temperature the slower the association/dissociation formation of $H + R \rightleftarrows HR$. Usually, steady state is reached within 1–2 hr at room temperature, but 18–24 hr may be required at low temperatures (4°C). One of the characteristics of the insulin receptor, regardless of the source, is that there is a sharp optimal pH requirement for binding (pH 7.6 or 7.8 for membranes or whole lymphoid cells, respectively). Other proteins, e.g., IGF-I, IGF-II, and human growth hormone, show little or no pH dependence but usually pH 7.4 is still selected. In addition to selecting a buffer to maintain pH throughout the incubation period, the binding buffer requires physiologic salts so that the integrity of the tissue is maintained during the incubation period (Gorden and Weintraub, 1992). Further, an unrelated protein, e.g., bovine serum albumin (BSA) or ovalbumin, is usually included in the buffer. When using crude membranes or partially purified membranes from tissues such as liver as the source for measuring the receptor, proteases (even in the presence of protease inhibitors) can degrade labeled and unlabeled ligands when they are present in low concentrations (<1 mg/ml). The unrelated proteins complete for the proteases as well as coat the surfaces of plastics used for the assay. The nonspecific binding of ligand to surfaces is especially true for insulin. (One must also be aware that ligands such as insulin may be present in commercial preparations of BSA. We recommend that if BSA is selected it should be tested for its ability to not degrade the polypeptide or otherwise interfere in the assay.)

IV. Binding of ^{125}I-Insulin to Plasma Membranes

A. Insulin Binding to Crude Plasma Membranes

The radioreceptor assay using crude membrane preparations is carried out in polyethylene microfuge tubes (400 μl) for convenience. The assay is as follows.

1. 50 μl of membrane preparation (~50 μg protein). The 20,000g pellet resuspended in buffer without BSA plus bacitracin 1.0 mg/ml
2. 50 μl of ^{125}I-insulin (5×10^{-10} M) in buffer with 3% BSA
3. 50 μl of unlabeled polypeptide in buffer with 0.1% BSA (varying concentrations of unlabeled ligand, e.g., $0 - 1.6 \times 10^{-7}$ M insulin and/or analog)

The buffer for the labeled and unlabeled ligands is Krebs–Ringer phosphate (liver, brain; Shemer *et al.*, 1986) or 50 mM Hepes (brain, lymphocytes; Lesniak *et al.*, 1987) plus 1% BSA, pH 7.8.

The incubation can be carried out either for 1.5 hr at 15°C when using brain or lymphoid cell preparations or 4.5 hr at 4°C when using crude liver membranes (Havrankova *et al.*, 1978). Overnight at 4°C has also been used (Shemer *et al.*, 1986).

B. Insulin Binding to Solubilized Insulin Receptors

The RRA using partially purified plasma membranes is performed in a total volume of 200 μl in 1.5-ml microfuge tubes. The assay is as follows:

1. 25 μl WGA-purified receptors (the aim is to have 20 μg/ml as a final protein concentration, thus the initial concentration should be ~160 μg/ml)

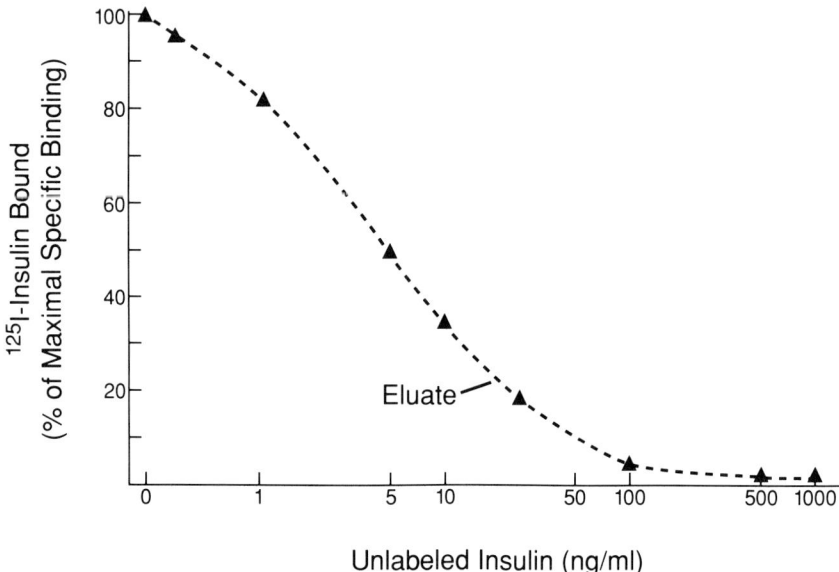

Figure 2 Competition-inhibition curve of insulin binding to insulin receptors from IM-9 lymphoid cells purified by wheat germ agglutinin chromatography. (After Hedo et al., 1981.)

2. 125 µl of ^{125}I-insulin (5×10^{-10} M) in buffer with bacitracin 1 mg/ml
3. 50 µl of buffer with unlabeled ligand or no unlabeled ligand

The buffer for labeled and unlabeled ligand is 50 mM Hepes, 150 mM NaCl, and 0.1% BSA, pH 7.8. For convenience the protease inhibitor, bacitracin, is added to the labeled ligand. The incubation is for 4 hr at room temperature. To terminate the reaction 100 µl of 0.3% bovine γ-globulin and 300 µl of 25% polyethylene glycol (PEG) are added to the tubes which are then immediately chilled on ice. The precipitates are sedimented by centrifugation at 2500 g for 15 min at 4°C and washed once with 300 µl of 12.5% PEG (Shemer et al., 1986).

To determine the amount of ^{125}I-insulin that is membrane bound, the tubes are centrifuged in a microfuge (bench top) at maximum speed for 3–5 min at 4°C. The supernatant is aspirated and then the tip of the tube which contains the pellet is excised and placed in a container suitable for counting the radioactivity in a gamma counter (Fig. 2). The specific activity is determined by subtracting the low-affinity binding, i.e., nonspecific binding, from the high-affinity binding, i.e., total binding. Nonspecific binding is the binding of labeled ligand in the presence of a high concentration of unlabeled ligand, e.g., for insulin ~5×10^{-7} M; total binding is the binding of labeled ligand in the absence of added unlabeled ligand. (Routinely the supernatant from the sample with no added unlabeled ligand is saved in order to test the integrity of the labeled peptide, i.e., to show that the membrane preparation did not degrade or otherwise interfere with ligand binding. To test the integrity of the labeled ligand, rebinding to a fresh aliquot of membrane is performed. Specific binding for the separate incubations should vary only a few percent. A quick, but less-precise method, is to determine the solubility of labeled insulin in 5% trichloroacetic acid; it should be negligible (5–10%).

Strategies for preparing and using whole-cell preparations for studying polypeptide hormone cell surface receptors in the RRA have been reported in detail elsewhere (Lesniak *et al.*, 1987). With the cloning of cDNA for peptide receptors as well as the elucidation of genomic DNA for several of the cell surface receptors investigators can now prepare receptors with mutations, e.g., using site-directed mutagenesis which mimics the natural occurring mutated receptors. These "genes" can be transfected into several different cultured cells as well as introduced into embryos to make transgenic mice. The effect of the mutated receptor's ability to bind ligand, to be internalized, to be phosphorylated, etc., can be examined in the target (and other) tissues using the methods described here and in the cited references.

References

Cone, R. E. (1987). *In* "Immunochemical Techniques, Part K: *In vivo* Models of B and T Cell Function and Lymphoid Cell Receptors" (G. Di Sabato, ed.), Methods in Enzymology, Vol. 150, p. 388. Academic Press, San Diego.
Gorden, P., and Weintraub, B. D. (1992). *In* "Williams' Textbook of Endocrinology" (J. D. Wilson and D. W. Foster, eds.), 8th Ed., p. 1647. W. B. Saunders, Philadelphia.
Havrankova, J., Roth, J., and Brownstein, M. (1978). *Nature (London)* 272, 827.
Hedo, J. A., Harrison, L. C., and Roth, J. (1981). *Biochemistry* 20, 3385.
Hendricks, S. A., Agardh, C.-D., Taylor, S. I., and Roth, J. (1984). *J. Neurochem.* 43, 1302.
Jett, M., Seed, T. M., and Jamieson, G. A. (1977). *J. Biol. Chem.* 252, 2142.
Lesniak, M. A., Hedo, J. A., Grunberger, G., Marcus-Samuels, B., Roth, J., and Gorden, P. (1987). *In* "Immunochemical Techniques, Part K: *In vivo* Models of B and T Cell Function and Lymphoid Cell Receptors" (G. Di Sabato, ed.), Methods in Enzymology, Vol. 150, p. 701. Academic Press, San Diego.
Neville, D. M., Jr. (1968). *Biochim. Biophys. Acta* 154, 540.
Roth, J. (1975). *In* "Hormone Action, Part B: Peptide Hormones" (B. W. O'Malley and J. D. Hardman, eds.), Methods in Enzymology, Vol. 37, p. 223. Academic Press, New York.
Shemer, J., and LeRoith, D. (1987). *Neuropeptides* 9, 1.
Shemer, J., Penhos, J. C., and LeRoith, D. (1986). *Diabetologia* 29, 321.

15

Analysis of Autophosphorylation and Substrate Phosphorylation by Receptor Tyrosine Kinases

Robert S. Garofalo

I. Introduction
II. Methods
   A. In Vitro Autophosphorylation
   B. Autophosphorylation and Substrate Phosphorylation in Cultured Cells
   C. Exogenous Substrate Phosphorylation
References

I. Introduction

Receptor tyrosine kinases (RTKs) comprise a class of receptors for hormones and growth factors which are distinguished by the presence of ligand-activated tyrosine kinases in their cytoplasmic domains. While the receptors can be grouped into several classes based on their structure (reviewed in Ullrich and Schlessinger, 1990), their activation mechanisms appear to share certain common features. The binding of ligand is followed by autophosphorylation of the receptor cytoplasmic domains on tyrosine residues. Autophosphorylation requires the interaction of two kinase domains which either associate following ligand binding in the case of monomeric RTKs (e.g., the epidermal growth factor (EGF) or platelet-derived growth factor (PDGF) receptors; Yarden and Schlessinger, 1987; Hammacher et al., 1989) or are preassociated into a disulfide-linked $\alpha_2\beta_2$ tetrameric complex (e.g., the insulin and insulin-like growth factor-I (IGF-I) receptors; Sweet et al., 1987; Boni-Schnetzler et al., 1988). Autophosphorylation serves at least two purposes: (a) to activate the tyrosine kinase of the receptor toward exogenous substrates (Rosen et al., 1983; White et al., 1988) and (b) to create phosphotyrosine-binding sites for signal transduction proteins with so-called SH2 (src homology 2) domains (Koch et al., 1991). Proteins such as the regulatory subunit of phosphatidylinositol-3 kinase, phospholipase C-γ, and the GTP-ase-activating protein of ras are examples of signal transduction proteins with SH2 domains which associate with RTKs following autophosphorylation (reviewed in Cantley et al., 1991). These interactions are receptor specific, with the amino acid sequence surrounding receptor phosphotyrosine residues influencing the relative affinity for different SH2-bearing proteins (Hu et al., 1992). For example, in vivo, the insulin receptor seems to interact only weakly with PI-3 kinase while the EGF or PDGF receptors exhibit more stable associations (Hu et al., 1992; Endemann et al., 1990; Kazlauskas and Cooper, 1990). In many cases, proteins which bind to receptors via their SH2 domains are themselves phosphorylated on tyrosine by the receptors. Thus, RTKs are ligand-activated transmembrane enzymes which phosphorylate themselves and other proteins as part of their signal transduction mechanism. In this chapter, I describe methods to study this kinase activity, their applications, and their limitations. Many of the examples utilized here involve the insulin and IGF-I receptors, although much of what is described pertains to RTKs in general.

Kinase activity can be assessed using two different end points: receptor autophosphorylation or exogenous substrate phosphorylation. Receptor autophosphorylation studies have helped to determine which tyrosines are phosphorylated, in what order, and the effects of these phosphorylations on receptor conformation, kinase activity, and signaling (O'Hare and Pilch, 1990; Rosen, 1987; Williams, 1989). For example, in the insulin receptor, the phosphorylation of three tyrosines in the kinase domain is associated with kinase activation toward exogenous substrates (Herrera and Rosen, 1986; Murakami and Rosen, 1991; White et al., 1988; Wilden et al., 1992) and induces a conformational change which can be detected by an antipeptide antibody (Herrera and Rosen, 1986). In the EGF and PDGF receptors, tyrosine residues have been identified which, upon phosphorylation, act as binding sites for SH2-bearing proteins. The importance of these interactions is demonstrated by impairment of receptor signaling after site-directed mutagenesis of these critical residues (Kazlauskas and Cooper, 1990; Seedorf et al., 1992; Vega et al., 1992). Autophosphorylation can also serve as a convenient means of labeling receptors or receptor subunits. Indeed, ligand-stimulated autophosphorylation of receptors can be used as a means of detecting their presence in tissues, measuring their relative abundance in comparison to other tissues or developmental periods, and identifying the size and number of their subunits. For example, autophosphorylation of insulin and IGF-I receptors in developing rat brain was used to demonstrate that the abundance of these receptors in fetal brain was much higher than that in adult brain and liver (Garofalo and Rosen, 1989). The level of receptors detected by autophosphorylation correlated very well with that found by exogenous substrate phosphorylation. In addition, structural information provided by sodium dodecyl sulfate–polyacrylamide gel electrophoresis (SDS–PAGE) analysis indicated that fetal brain IGF-I receptors contained two distinct β subunits associated together in a single tetrameric complex (Garofalo and Rosen, 1989). Similar hybrid receptors have also been seen in other cells and tissues (Moxham *et al.*, 1989; Garofalo and Barenton, 1992; Soos and Siddle, 1989). These are most likely composed of an insulin receptor $\alpha\beta$ dimer and an IGF-I receptor $\alpha\beta$ dimer and are now thought to comprise a significant portion of the total IGF-I receptor population in some tissues (Moxham and Jacobs, 1992). Autophosphorylation has been used to demonstrate multiple forms of other RTKs, such as the *met* (Rodrigues *et al.*, 1991), *trk* (Martin-Zanca *et al.*, 1989), and fibroblast growth factor (FGF) receptors (Bottaro *et al.*, 1990). Interactions between RTKs in cells can be studied by examining their phosphorylation state in response to heterologous ligands. Such transphosphorylation, whereby one receptor is phosphorylated following activation of another, has been shown to occur in the case of EGF-induced phosphorylation of the erbB2 *(neu)* protooncogene (Connelly and Stern, 1990; Stern and Kamps, 1988; King *et al.*, 1988).

Autophosphorylation assays have the advantage that the substrate for the kinase, the receptor itself, is present in the appropriate stoichiometry and contains the preferred phosphorylation sites for the kinase. In contrast, most exogenous substrate kinase assays utilize model substrates which may not closely resemble the substrates found *in vivo* and, as a result, exhibit fairly high K_m's for phosphorylation (0.1 to 8 mM; Kemp and Pearson, 1991). Few naturally occurring substrates for RTKs have been defined (Kasuga *et al.*, 1990; Roth *et al.*, 1992), and even fewer are available for use as substrates in kinase assays. The most widely used, commercially available substrates for exogenous kinase assays of RTKs are random amino acid copolymers containing tyrosine as the only phosphorylatable amino acid (Braun *et al.*, 1984) and short peptides, such as that derived from the autophosphorylation site of the src protein (Hunter, 1982). These have been very useful for comparisons of receptor activity under different conditions and in determining parameters such as divalent cation requirements, pH optima, and ligand specificity of activation. In comparative receptor studies, differences in kinase activity observed using a variety of model substrates has been taken to suggest distinct substrate specificities for the respective enzymes (Braun *et al.*, 1984; Zick *et al.*, 1985). It is difficult, however, to know what the affinities for these model substrates mean with regard to the true substrate specificities of the RTKs.

As more RTK substrates are identified by molecular cloning, the use of synthetic peptides derived from their deduced amino acid sequences is providing insight into the determinants of substrate specificity. Recently, a major endogenous substrate of the insulin receptor was identified and termed IRS-1 (insulin receptor substrate-1; Sun *et al.*, 1991). A number of tyrosine-containing synthetic peptides derived from the IRS-1 sequence were used in substrate phosphorylation experiments, and, notably, several of them exhibited K_m's in the range of 24 to 90 µM, which is quite low compared to most model substrates, indicating that these are excellent substrates for the insulin receptor (Shoelson et al., 1992). Synthesizing peptides with alterations in the amino acids surrounding the tyrosine residue revealed the importance of nearby methionines in a motif, YMXM, as well as the influence of an acidic amino acid N-terminal to the tyrosine (Shoelson et al., 1992). The K_m's for phosphorylation of these peptides were lower than that reported for a peptide corresponding to one of the major autophosphorylation sites of the insulin receptor (240 µM; Stadtmauer and Rosen, 1986). This suggests that, while primary sequence is a major determinant of specificity, interactions with other residues in the vicinity of the phosphorylation site are likely to influence the efficiency of phosphorylation. With the increasing availability of automated peptide synthesis, this approach is likely to become more widespread as more substrates are identified.

In most tissues and cell lines, some purification or enrichment is needed to detect autophosphorylated RTKs from crude lysates. The most common methods are (a) preparation of a glycoprotein fraction from solubilized membranes by lectin affinity chromatography using, for example, wheat germ agglutinin (WGA; Hedo et al., 1981) and (b) immunoprecipitation with receptor-specific or antiphosphotyrosine antibodies. Glycoprotein fractions from some tissues, such as rat liver and human placenta, contain abundant insulin receptors, and receptors autophosphorylated in the presence of [γ-^{32}P]ATP can be readily detected following SDS–PAGE and autoradiography. However, most tissues or cell lines contain fewer receptors and require immunoprecipitation to achieve sufficient concentration for visualization above background. If immunoprecipitation is employed, then autophosphorylation assays can also be carried out using detergent-solubilized membranes which have not been subject to chromatography (see below). Eliminating the lectin chromatography step is sometimes advantageous because this has been shown to lead to alterations in ligand affinities when compared to measurements carried out on membranes (Hedo et al., 1981). In addition, lectin-purified receptors sometimes exhibit increased basal autophosphorylation in the absence of hormones, although this seems receptor specific and the basis of this change is not known. For example, in vitro autophosphorylation and immunoprecipitation of WGA-purified insulin and IGF-I receptors from an epidermal cell line (KB) with an antibody which recognizes a common domain (AbP2; Herrera and Rosen, 1988; Garofalo and Barenton, 1992) recovers receptors with a high basal level of autophosphorylation (Fig. 1A, lane 1) and a small degree of stimulation by IGF-I (Fig. 1A, lane 3). When subsets of the total receptor population are recovered by receptor-specific antibodies, it is seen that insulin and hybrid receptors recognized by an insulin receptor-specific antibody (AbP5; Herrera et al., 1986) exhibit low basal activity (lane 5) and autophosphorylation is stimulated by insulin (lane 6) and most strongly by IGF-I (lane 7).

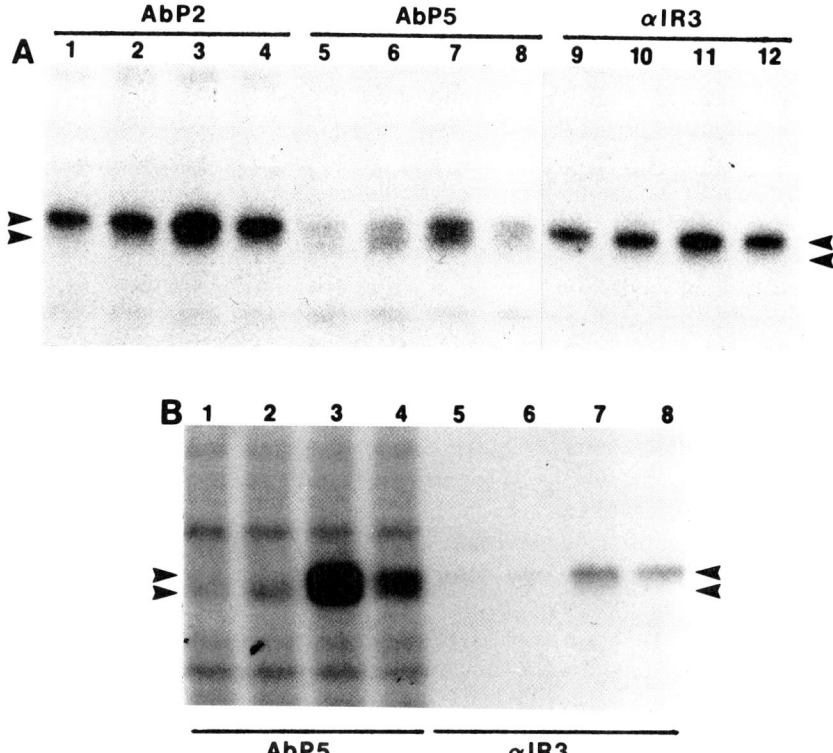

Figure 1 *In vitro* autophosphorylation and immunoprecipitation of insulin and IGF-I receptors from WGA (A) and solubilized membrane (B) preparations. (A) WGA-purified receptors from epidermoid cells (KB) were incubated in the absence of hormones (lanes 1,5,9), or in the presence of 1 nM insulin (lanes 2,6,10), IGF-I (lanes 3,7,11), or IGF-II (lanes 4,8,12) and autophosphorylation initiated by the addition of [γ-^{32}P]ATP (20 μM). Immunoprecipitation was carried out with an antibody that recognizes a well-conserved domain in both insulin and IGF-I receptors (AbP2, lanes 1–4), an insulin receptor-specific antibody (AbP5, lanes 5–8), and an IGF-I receptor-specific antibody (αIR-3, lanes 9–12). (B) Solubilized membrane preparations from KB cells were incubated in the absence of hormones (lanes 1,5) or in the presence of 1 mM insulin (lanes 2,6), IGF-I (lanes 3,7), or IGF-II (lanes 4,8) and autophosphorylation was initiated as described in A. Immunoprecipitation was carried out with AbP5 (lanes 1–4) and αIR-3 (lanes 5–8). The 102- and 95-kDa β subunits are indicated by the upper and lower arrowheads, respectively. Autoradiographic exposure was for 40 (A) or 20 (B) hr at $-70°$C with an intensifying screen. Neither gel was subjected to base treatment.

In contrast, immunoprecipitation with an IGF-I receptor-specific antibody (αIR3; Kull et al., 1983) recovers homotypic IGF-I receptors which exhibit a high basal activity (lane 9) and little response to any ligand (lanes 10–12). Therefore, within a single preparation of WGA-purified receptors subject to in vitro autophosphorylation under identical conditions, specific subpopulations of insulin and IGF-I receptors display quite different levels of basal and ligand-stimulated autophosphorylation. In contrast, when solubilized membrane preparations, not subject to lectin chromatography, are used, both receptor populations consistently exhibit low basal activity and clear ligand-stimulated autophosphorylation (Fig. 1B, lanes 1–8). Thus, it seems that the degree of purification affects the extent of ligand-independent autophosphorylation of the IGF-I receptor, but not the insulin receptor. The factors which influence autophosphorylation under these conditions are poorly understood, but may include changes intrinsic to the receptor which occur during purification, e.g., conformational changes induced by the lectin binding to α subunits, or interactions between receptors while bound to the affinity matrix, or, conversely, the loss of extrinsic components present in solubilized membrane preparations which modulate the basal level of autophosphorylation, e.g., phosphatases. Therefore, whenever possible, autophosphorylation assays using solubilized membrane preparations and immunoprecipitation appear preferable to using lectin-purified receptors.

II. Methods

A. In Vitro Autophosphorylation

Assays of in vitro autophosphorylation utilize receptors either from solubilized membrane preparations or from membrane preparations partially purified by lectin chromatography. Membranes can be prepared from cultured cells or tissues according to standard protocols. In general, homogenization is carried out in ice-cold buffer (e.g., 10 mM Tris, pH 7.5, 2 mM $MgCl_2$, 0.2 M sucrose) containing protease inhibitors (10 μg/ml each of leupeptin, aprotinin, and soybean trypsin inhibitor, 1 mM phenylmethylsulfonyl fluoride (PMSF), and 0.5 mM EDTA; Garofalo and Rosen, 1989). The homogenate is centrifuged at low speed (1000g, 15 min) to remove unbroken cells and large organelles such as nuclei and mitochondria. A membrane fraction is then prepared from the low-speed supernatant by centrifugation at 100,000g for 45–60 min at 4°C. In some cases, this crude microsomal membrane pellet is subjected to low osmotic strength lysis by resuspension and incubation (60 min, 4°C) in a large excess of 6 mM Tris, pH 8.0, 1 mM EDTA, and protease inhibitors as above. This ensures that the membrane vesicles formed upon homogenization are open and do not contain trapped cytoplasmic material. After collection of the lysed membranes by centrifugation at 100,000g, they are resuspended in 20 mM Tris, pH 7.5, 1 mM EDTA, and protease inhibitors (excluding PMSF) and stored in aliquots at $-70°C$ until use. We have found that membranes prepared in this way will exhibit reliable ligand-stimulated kinase activity even after several years of storage at $-70°C$.

Preparation of a glycoprotein fraction from these membranes is carried out essentially according to Hedo et al. (1981; Lesniak et al., Chapter 14, this volume). However, if solubilized membranes are to be used directly for autophosphorylation assays, Triton X-100 is added to the membrane suspension to a final concentration of 1–2% and mixed by vortexing, and membranes are allowed to solubilize for 30 min at 4°C on a shaker. Triton-insoluble material is then removed by centrifugation at 100,000g (15 min, 4°C, TLA 45 rotor, Beckman TL-100 Ultracentrifuge) and the supernatant recovered for use.

Reactions are carried out in a final volume of 50 μl and include 50 mM Hepes, pH 7.8, 2.5 mM $MnCl_2$, and 0.1–5 μg of WGA-purified receptors or 10–50 μg of membrane protein (as determined prior to solubilization). $MnCl_2$ can be included as the only divalent cation, as most tyrosine kinases exhibit a preference for $MnCl_2$ over $MgCl_2$ (Ek and Heldin, 1982; Zick et al., 1983). Some investigators include both $MgCl_2$ and $MnCl_2$, although the presence of Mg^{2+} may increase the background level of phosphorylation due to serine/threonine kinases and stimulate the activity of phosphotyrosine phosphatases (PTPases; Casnellie et al., 1982). Incubation with the ligand at an appropriate concentration(s) is carried out for 60 min at 4°C and the autophosphorylation reaction is initiated by the addition of ATP and 100 μM sodium orthovanadate, a potent inhibitor of PTPases (Gordon, 1991). The ATP concentration will be dependent on the K_m of the particular RTK for ATP, but generally is in the range of 10–100 μM. If $[\gamma\text{-}^{32}P]ATP$ is used for labeling the receptors, the concentration of unlabeled ATP should be kept as low as possible in order to achieve high specific activity. A concentration of 20–50 μM ATP (3–20 μCi/nmol) is sufficient to label insulin and IGF-I receptors from a variety of sources to a level which is usually detectable after 15–30 hr of autoradiography with an intensifying screen.

The autophosphorylation reaction is allowed to proceed for 1–20 min at 23°C and terminated by the addition of SDS–PAGE sample buffer (Laemmli, 1970) and boiling if samples are to be subjected directly to SDS–PAGE or by addition of a stop buffer if immunoprecipitation is to be carried out. In the latter case, the reaction is stopped with 200 μl of ice-cold buffer containing 10 mM Hepes, pH 7.8, 0.1% Triton X-100, 150 mM NaCl, 1 mM sodium orthovanadate, 4 mM EDTA, 4 mM EGTA, and protease inhibitors (0.4 mM PMSF, 10 μg/ml each of leupeptin, aprotinin, and soybean trypsin inhibitor). Antibody is then added at the appropriate concentration and allowed to react with receptor proteins for 3–15 hr at 4°C on a shaker. Protein A Sepharose (Pharmacia) is added (approximately 5 μl of a 50% slurry per μl of polyclonal antisera) and incubation is continued for an additional 60 min at 4°C. Immunoprecipitates are recovered by centrifugation in a microfuge and washed extensively: twice in 50 mM Hepes, pH 7.8, 150 mM NaCl, 0.1% Triton X-100 (Buffer A), once in 50 mM Hepes, pH 7.8, 500 mM NaCl, once again in Buffer A, and once in 5 mM Tris, pH 7.5, 75 mM NaCl (Garofalo and Rosen, 1989). All buffers contain 1 mM sodium orthovanadate. After aspiration of the final wash, receptors are released from immune complexes by addition of SDS–PAGE sample buffer and boiling. Depending on the IgG subclass, immunoprecipitation with monoclonal antibodies may require the use of protein G Sepharose or a rabbit anti-mouse IgG secondary antibody with protein A. However, the reaction of protein A with mouse antibodies is dependent on pH (Ey *et al.*, 1978), and immunoprecipitation at a pH near 8.0 may allow the use of protein A with mouse monoclonal antibodies, although the efficiency of immunoprecipitation should be verified empirically in such a case.

Immunoprecipitates are analyzed by SDS–PAGE and ^{32}P-labeled receptors are visualized following autoradiography of dried gels. For quantitation, an autoradiogram placed over the dried gel is used as a guide for the excision of the receptor bands which are subjected to Cerenkov or scintillation counting. Alternatively, autoradiograms on preflashed X-ray film can be quantitated by laser densitometry. A high background of phosphoproteins is sometimes seen after autoradiography which may make visualization of receptors difficult. The level of background depends on several factors including the source of receptors (WGA-purified or solubilized membranes) and the type of antibody (polyclonal antisera, affinity-purified polyclonal or monoclonal). In many cases, the background can be significantly reduced by base treatment of gels which selectively hydrolyzes phosphoserine residues, enriching the signal for phosphotyrosine (Cooper *et al.*, 1983). As seen in Fig. 2, the effect of base treatment can often be dramatic because only a small percentage of total cellular phosphoproteins are phosphorylated on tyrosine (Sefton *et al.*, 1980). After staining and destaining, or even after drying and initial autoradiographic exposure, gels are treated with 1 N NaOH for 90 min at 55°C, reequilibrated with destain (10% methanol, 7% acetic acid), dried, and subject to autoradiography.

Autophosphorylation of receptors can also be analyzed using anti-phosphotyrosine (anti-P-Tyr) antibodies, many of which are commercially available. Production and use of anti-P-Tyr antibodies is detailed in several recent reviews (Kamps, 1991; White and Backer, 1991). These antibodies can be used in a variety of strategies involving immunoprecipitation and immunoblotting.

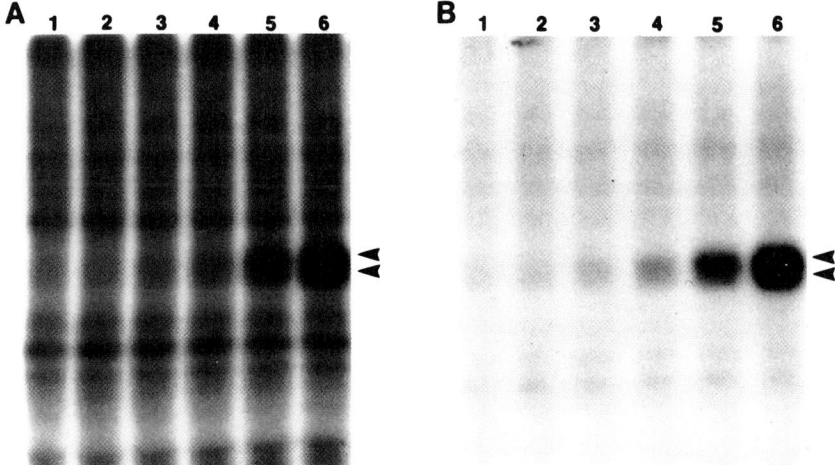

Figure 2 Base treatment of SDS–PAGE gels to enrich the signal for phosphotyrosine. Solubilized membrane preparations (KB cells) were incubated in the absence of hormones (lane 1) or in the presence of 0.1, 0.3, 1, 3, and 10 nM IGF-I (lanes 2–6, respectively) and autophosphorylation was initiated by the addition of [γ-^{32}P]ATP (20 μM). Immunoprecipitation of insulin and hybrid IGF-I receptors was carried out with AbP5. The autoradiogram of the gel in A was obtained prior to base treatment whereas that in B shows the same gel after incubation in 1 N NaOH at 55°C for 90 min, redrying, and reexposure. Autoradiographic exposure was nearly equivalent (80 hr for A, 70 hr for B, −70°C with an intensifying screen), as indicated by the similar intensity of the tyrosine-phosphorylated β subunits (arrowheads).

Immunoprecipitation from *in vitro* autophosphorylation reactions (carried out essentially as described above for receptor-specific antibodies) will recover more receptor after incubation with ligand (Fig. 3, arrow) although, in this case, little change can be detected in other major phosphotyrosine proteins. This approach is generally more informative when ligand-induced phosphorylation is examined in whole cells (see below) where the signal transduction pathways are intact and cytoplasmic substrates located downstream from the receptor are present. But if an immunoprecipitating, receptor-specific antibody is not available, anti-P-Tyr immunoprecipitation may be useful even in *in vitro* assays. A distinct advantage of anti-P-tyr antibodies for studies of RTK phosphorylation is that assays can be carried out in the absence of radiolabeled [γ-^{32}P]ATP. A combined immunoprecipitation–immunoblot approach employing either immunoprecipitation with anti-P-tyr and immunoblot with anti-receptor antibody or the converse can be used to measure increases in receptor autophosphorylation. In the first scheme, more receptor will be recovered by anti-P-tyr immunoprecipitation after autophosphorylation and the increase detected by anti-receptor immunoblotting. The converse strategy,

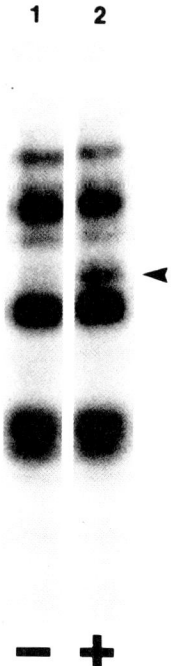

Figure 3 *In vitro* autophosphorylation and immunoprecipitation with antiphosphotyrosine (anti-P-Tyr) antibodies. Solubilized membrane preparations from fetal rat brain were incubated in the absence (lane 1) or presence (lane 2) of 100 nM insulin, allowed to undergo autophosphorylation (2 min) after addition of [γ-^{32}P]ATP, and immunoprecipitated with anti-P-Tyr antibodies. Increased recovery of the insulin receptor β subunit (arrowhead) is seen when autophosphorylation is carried out in the presence of insulin (lane 2). The specificity of anti-P-Tyr immunoprecipitation was confirmed by inclusion of p-nitrophenyl phosphate (2 mM) in the immunoprecipitation reaction. Under those conditions, none of the major tyrosine-phosphorylated proteins were recovered in the immunoprecipitates (data not shown). Autoradiographic exposure was for 15 hr at $-70°$C with an intensifying screen.

in which receptor-specific antibodies are used for immunoprecipitation, has the advantage that the subsequent immunoblot can be probed with both receptor-specific antibodies to control for variation in immunoprecipitation and anti-P-Tyr antibody to detect increases in autophosphorylation. In addition, any phosphotyrosine proteins which stably associate with the receptor after autophosphorylation will be detected in the subsequent anti-P-Tyr immunoblot.

B. Autophosphorylation and Substrate Phosphorylation in Cultured Cells

In vitro analysis of receptor autophosphorylation as described above can provide information on receptor abundance, structure, and intrinsic mechanism. However, it will provide only limited insight into questions regarding cellular actions of the kinase, *in vivo* half-life of the autophosphorylated receptors, or interactions with other cellular proteins. To address such issues, receptor and substrate phosphorylation can be studied in cultured cells. Detection of autophosphorylated receptors can be accomplished by either metabolic labeling of cellular ATP pools with $^{32}P_i$ or by anti-P-Tyr immunoblotting. The latter procedure seems more sensitive, providing higher a signal-to-noise ratio and in addition does not require the use of large amounts of radiolabel. However, some caution must be exercised in interpreting the results of anti-P-Tyr immunoblotting because different antibodies may not be equivalent in their ability to recognize all phosphotyrosine proteins (Kamps, 1991; Kozma *et al.*, 1991). Incubation of immunoblots with nonfat dry milk to block nonspecific binding of antibodies should also be avoided because anti-P-Tyr antibodies bind very strongly to components in the milk (Kamps, 1991).

Analysis of ligand-stimulated phosphorylation in intact cells requires the use of conditions which will maximize the signal due to receptor activation and minimize the effects of cellular PTPases. This is accomplished by (a) preincubation of cells in low serum (0.1–1%) which will lead to upregulation of receptors and/or a decrease in their steady-state level of phosphorylation and (b) lysis of cells under conditions designed to inhibit PTPases and preserve phosphotyrosine residues (Kamps and Sefton, 1988). Preincubation in low serum is generally carried out overnight (15–18 hr) but incubations as short as 1.5 hr in Hepes-buffered saline have been reported (Kozma *et al.*, 1991). This will vary depending on the cell type, abundance of the protein(s) to be detected, and the magnitude of the change in phosphorylation expected. Preincubation of cells in PTPase inhibitors has been used to increase the phosphotyrosine signal. For example, preincubation of cells transfected with normal and mutated insulin receptors for 4 hr in 0.5 mM vanadate followed by insulin treatment led to an increase in tyrosine phosphorylation of the insulin receptor, as well as several other cellular proteins, over that seen without vanadate treatment (Yonezawa and Roth, 1991). However, extended incubations in PTPase inhibitors may be toxic or elicit other effects not clearly related to inhibition of PTPase activity (Gordon, 1991). Vanadate toxicity has been reported to occur with incubation beyond 6 hr at a concentration of 100 μM (Gordon, 1991).

After preincubation in low serum, cells are stimulated by the addition of culture medium containing ligand at the appropriate concentration. Incubation with the ligand usually ranges from 5–20 min at 37°C and reactions are terminated by cell lysis. If total cellular phosphoproteins are to be analyzed by anti-P-Tyr immunoblotting, cells are lysed in hot SDS–PAGE sample buffer and boiled immediately (Kamps and Sefton, 1988). The viscosity of the cell lysate must be reduced by either sonication or shearing by passage several times through a 22-gauge needle. The samples are then stored at $-70°C$ until use. If immunoprecipitation is to be carried out, cells are lysed in Tris–RIPA buffer (50 mM Tris, pH 7.2, 150 mM NaCl, 1% Nonidet P-40, 0.1% sodium deoxycholate, 0.1% SDS, and 1% Trasylol; Kamps and Sefton, 1988) including 1 mM sodium orthovanadate and 2 mM EDTA. RIPA buffer, with the inclusion of sodium orthovanadate as PTPase inhibitor, has been shown to provide recovery of phosphotyrosine proteins from transformed fibroblasts equivalent to that obtained after lysis of cells directly in SDS–PAGE sample buffer (Kamps and Sefton, 1988). Cells are lysed for 20 min at 4°C, lysates are clarified by centrifugation (40 min, 20,000g), and the resulting supernatant is used for immunoprecipitation. Immunoprecipitates are then resolved by SDS–PAGE, transferred to nitrocellulose, and probed with anti-P-Tyr antibodies in order to detect changes in phosphorylation of receptors or other proteins.

Conclusive demonstration that a receptor or other associated protein is phosphorylated on tyrosine requires confirmation of positive anti-P-Tyr immunoblotting by phosphoamino acid analysis. This is accomplished by recovery of tyrosine-phosphorylated proteins from ^{32}P-labeled cells, acid hydrolysis, and resolution of phosphoamino acids by electrophoresis or chromatography (Hunter and Sefton, 1980; Cooper *et al.*, 1983; Boyle *et al.*, 1991; Neufeld *et al.*, 1989). General guidelines for *in vivo* labeling of cells with ^{32}P are discussed by Garrison (1983), Kasuga *et al.* (1985), and Cooper *et al.* (1983) but there are many variations for individual cell and receptor systems. Labeling can be carried out to achieve complete equilibration of ^{32}P$_i$ with intracellular ATP pools (e.g., 16–18 hr with ^{32}P$_i$ at a concentration of 1 mCi/ml in phosphate-free Dulbecco's modified Eagles medium (DMEM) supplemented with 4% fetal calf serum; Cooper *et al.*, 1983). Alternatively, labeling for shorter periods (e.g., 2 hr at 37°C in phosphate-free Krebs–Ringer bicarbonate with 0.1% dialyzed bovine serum albumin and ^{32}P$_i$ at a concentration of 0.125 mCi/ml; Garrison, 1983; Kasuga *et al.*, 1985), although not quantitative, may favor the visualization of hormone-responsive phosphotyrosine proteins because tyrosine phosphates in proteins appear to turn over more rapidly than serine or threonine phosphates (Cooper *et al.*, 1983). In addition, the use of serum-free medium will reduce the proportion of label incorporated into nucleic acid.

Labeling is followed by ligand stimulation, termination of reactions by cell lysis as described above, immunoprecipitation, and resolution of phosphoproteins by SDS–PAGE. In order to determine the phosphoamino acid content of the protein(s) of interest, ^{32}P-labeled proteins are recovered from dried SDS–PAGE gels (Cooper *et al.*, 1983; Boyle *et al.*, 1991) or after transfer to a polyvinylidene difluoride (PVDF) membrane (Immobilon, Millipore; Kamps and Sefton, 1989) and subjected to acid hydrolysis. The Immobilon procedure facilitates the analysis of phosphotyrosine proteins by eliminating several time-consuming steps which decrease the recovery of phosphoamino acids. Rather than elution of proteins from gel slices by homogenization or tryptic digestion and subsequent concentration by trichloroacetic acid (TCA) precipitation or lyophilization (Cooper *et al.*, 1983; Boyle *et al.*, 1991), proteins are electrophoretically transferred to Immobilon and localized by autoradiography of the membrane, the corresponding portion of the membrane is excised, and the protein is subjected directly to acid hydrolysis while still bound to the membrane (Kamps and Sefton, 1989). In either case, partial hydrolysis of phosphoproteins is carried out in 6 N HCl at 110°C for 60 min and the resulting phosphoamino acids are separated by thin-layer electrophoresis or chromatography on cellulose thin-layer plates (Cooper *et al.*, 1983; Boyle *et al.*, 1991; Neufeld *et al.*, 1989). One-dimensional separation by electrophoresis or chromatography at pH 3.5 is often adequate when the products of *in vitro* kinase reactions or immunoprecipitations are analyzed (Cooper *et al.*, 1983; Neufeld *et al.*, 1989). If the resolution of phosphotyrosine from phosphothreonine is not sufficient in the one-dimensional separation, then two-dimensional electrophoresis at pH 1.9 and 3.5 can be carried out (Cooper *et al.*, 1983; Boyle *et al.*, 1991). Detailed descriptions of the procedure for two-dimensional phosphoamino acid analysis can be found in Boyle *et al.* (1991) and Cooper *et al.* (1983). An apparatus which has been optimized for performing two-dimensional phosphoamino acid analysis and peptide mapping is available commercially (C.B.S. Scientific, Del Mar, CA).

C. Exogenous Substrate Phosphorylation

The ability of RTKs to phosphorylate exogenous substrates can be used to (a) detect and quantitate ligand-stimulated tyrosine kinase activity in cell and tissue extracts, (b) examine the ligand and substrate specificity of RTKs, (c) examine the actions of activators or inhibitors of RTK kinase activity, and (d) determine kinetic properties and cofactor requirements of the RTK enzyme. The receptor preparation utilized in exogenous substrate kinase assays can be fairly crude (Casnellie *et al.*, 1982), although a lower background level of phosphorylation is obtained with more purified preparations. Crude preparations will also contain PTPases, which may be inhibited by the inclusion of sodium orthovanadate and the use of $MnCl_2$ rather than $MgCl_2$ as divalent cation (Casnellie et al., 1982). Membranes, either intact or solubilized, generally have a background level of tyrosine kinase activity which is too high for detection of a ligand-stimulated RTK activity, unless the concentration of the RTK is very high (e.g., EGF receptors in A431 cell membranes; Braun et al., 1984). However, glycoprotein fractions obtained after lectin chromatography are often used. The relative level of tyrosine kinase activity can vary dramatically depending on the cell type or developmental stage. Embryonic tissues (Maher, 1991) and some transformed cell lines (Sefton et al., 1980) have high levels of tyrosine kinase activity and may require more purification to analyze substrate phosphorylation by RTKs *in vitro*. Very low background can be achieved by performing immune complex kinase assays, in which receptors are immunoprecipitated from crude extracts with antibodies that do not inhibit kinase activity and the immunoprecipitates used directly to phosphorylate substrates. The most widely used substrates are random amino acid copolymers, such as poly(Glu/Tyr)(4:1) (Braun et al., 1984) and short peptides, e.g., that derived from the tyrosine phosphorylation site of the src protein (Hunter, 1982). Poly(Glu/Ala/Tyr)(6:3:1) seems to be a preferred substrate for the EGF receptor (Braun et al., 1984; Zick et al., 1985). These substrates are available commercially (Sigma, St. Louis, MO). Other proteins such as histone (Stadtmauer and Rosen, 1983; Garofalo and Rosen, 1989), peptides such as angiotensin II (Wong and Goldberg, 1983), and, as mentioned above, synthetic peptides based on specific receptor or substrate sequences (Hunter, 1982; Shoelson et al., 1992; Stadtmauer and Rosen, 1986; Kemp and Pearson, 1991) have also been used. The tyrosine-containing polymers and short peptides have an advantage for analysis of RTKs in crude preparations because they contain tyrosine as the only phosphorylatable amino acid and thus eliminate any contribution from serine/threonine kinases.

Exogenous substrate kinase assays include two basic steps: (a) incubation of the RTK preparation with substrates in the presence of $[\gamma\text{-}^{32}P]ATP$ and (b) separation of the phosphorylated substrate from the unincorporated radiolabel. Separation is most commonly achieved by precipitation of high-molecular-weight amino acid copolymers onto filter paper discs in TCA (Braun et al., 1984; Racker, 1991) or by binding of peptide substrates onto phosphocellulose paper (Glass et al., 1978; Casnellie, 1991). Other methods, including dot blot (Glover and Allis, 1991), solid phase (Sahal et al., 1991), and a nonradioactive method using anti-P-Tyr antibodies to quantitate substrate phosphorylation (Rijksen et al., 1989, 1991), have been reported but have yet to see wide application in studies of RTKs. In all the methods described below, specific incorporation into the substrate is determined by subtracting the incorporation measured in the absence of substrate from that in its presence.

1. Assays Using Amino Acid Copolymer Substrates

Racker (1991) has recently reviewed the principles and techniques for kinase assays using amino acid polymers. Reaction conditions must be optimized for particular RTKs, but a typical protocol (25–50 μl final reaction volume) involves preincubation of WGA-purified RTKs in phosphorylation buffer (30 mM Hepes, pH 7.4, 2.5 mM $MnCl_2$) including polymer substrate (0.1–2 mg/ml) in the presence or absence of ligand at 4°C for 60 min to allow ligand binding, followed by initiation of the kinase reaction by the addition of [γ-^{32}P]ATP (0.5–2 μCi/nmol) and sodium orthovanadate (100 μM). After incubation for 5–60 min at 23°C the reaction is terminated by spotting an aliquot (20–30 μl) of the reaction mixture onto 2-cm squares of 3MM filter paper (Whatman) and immersion into ice-cold 10% TCA and 10 mM sodium pyrophosphate. Sodium pyrophosphate is included to decrease the nonspecific binding of ^{32}P to the filter. The filters are washed in this solution with changes every 15 min for 1–2 hr. Filters are then rinsed in 95% ethanol, air- or oven-dried, and counted in a scintillation counter. To achieve a constant rate of substrate phosphorylation without an initial lag (as seen for the insulin receptor; Rosen *et al.*, 1983), receptors can be preactivated by autophosphorylation prior to addition of substrate. In this case, the ligand-binding step is performed in phosphorylation buffer in the absence of substrate and followed by incubation with [γ-^{32}P]ATP, 100 μM vanadate for 30 min at 23°C. Substrate is then added and phosphorylation allowed to proceed for the appropriate incubation time. Preactivation of receptors also eliminates any inhibitory effect of substrates on activation which is a concern in kinetic studies (White *et al.*, 1988). A modification of this technique, the use of phosphocellulose paper (Whatman P81) for TCA precipitation (Sahal and Fujita-Yamaguchi, 1987), reportedly provides a higher signal-to-noise ratio than Whatman 3MM paper due to its lower level of nonspecific binding of [^{32}P]ATP. Polymer phosphorylation can also be analyzed by gel electrophoresis in polyacrylamide gels containing 9.4 M urea, in the absence of SDS (Schieven *et al.*, 1986), although this is less convenient for assaying large numbers of samples.

2. Assays Using Peptide Substrates

Kinase reactions are carried out essentially as described above for polymer substrates, except that the phosphorylated product is recovered by binding to phosphocellulose (P81) paper. A feature which is essential for quantitative retention of the peptides on the phosphocellulose paper is the presence of at least two basic residues (Casnellie, 1991). In cases where sufficient basic residues are not present in the sequence of interest they can be added at the amino- or carboxy-terminus of the synthetic peptide (Casnellie *et al.*, 1982; Shoelson *et al.*, 1992; Stadtmauer and Rosen, 1986). The washing of filters under acidic conditions promotes peptide binding to P81 by suppressing the negative charges of both the phosphate groups and the acidic amino acids which are usually present near tyrosine residues at RTK phosphorylation sites (Casnellie, 1991). The use and design of *in vitro* peptide substrates has been recently reviewed (Casnellie, 1991; Kemp and Pearson, 1991).

The method of terminating the phosphorylation reaction depends on the purity of the sample being assayed. For relatively purified preparations with a low background level of kinase activity, reactions are terminated either by spotting aliquots of the reaction mix onto P81 filter paper squares and immersion in 75 mM phosphoric acid or by addition of 30% acetic acid followed by spotting onto phosphocellulose (Casnellie, 1991). For cruder preparations (e.g., glycoprotein fractions, cell lysates) proteins are first precipitated by addition of an equal volume of 5–10% TCA and 10 μl of bovine serum albumin as carrier protein. The low-molecular-weight peptides remain in solution under these conditions. After incubation on ice (10 min), precipitated proteins are removed by centrifugation and aliquots of the supernatant containing phosphorylated peptides spotted onto phosphocellulose and washed in 75 mM phosphoric acid (Casnellie, 1991). Five washes of at least 5 min each are carried out in a beaker or similar container on a shaker at room temperature, followed by a 5-min wash in acetone. Filters are dried in a 50°C oven and quantitated by scintillation counting.

Depending on the purity of the RTK preparation being assayed and the level of tyrosine kinase activity in the tissue or cells being studied, the increase in substrate phosphorylation in response to ligand stimulation in these assays may range from 2- to 20-fold. If the source of the RTK has a relatively high level of tyrosine kinase activity and the abundance of the RTK is low, it may be difficult to detect any ligand-stimulated increase in phosphorylation over the high background level. In this case, immune complex kinase assays can be performed if an immunoprecipitating, receptor-specific antibody which does not interfere with tyrosine kinase activity is available. Such antibodies have been used in studies of the insulin receptor (Ganguly *et al.*, 1985; Garofalo and Rosen, 1989; Herrera *et al.*, 1985; Stadtmauer and Rosen, 1986), and at least one (monoclonal antibody CII25.3; Ganguly *et al.*, 1985) is commercially available (Oncogene Science, New York). Antibodies against other RTKs have been used in immune complex autophosphorylation assays (Rodrigues *et al.*, 1991; Martin-Zanca *et al.*, 1989) and presumably would also be useful in exogenous substrate phosphorylation assays. Immunoprecipitation of RTKs from membranes diluted 1 : 1 with 2× TX-100/saline (1× is 5 mM Hepes, pH 7.8, 2% Triton X-100, 150 mM NaCl, 10 μg/ml each of leupeptin and aprotinin, and 1 mM PMSF; Garofalo and Rosen, 1989) is carried out for

3–15 hr at 4°C and the immunoprecipitates are washed three times in Buffer A (see Section II,A, *In Vitro* Autophosphorylation). The washed beads are resuspended in phosphorylation buffer containing substrate with or without ligand and incubated further at 4°C (45–60 min) to allow ligand binding and the kinase reaction is initiated by the addition of [γ-^{32}P]ATP and sodium orthovanadate. After incubation for the desired time the immune complexes are recovered by centrifugation and the supernatant containing the phosphorylated substrate is withdrawn for quantitation as described above for peptide or polymer substrates. An advantage of the immune complex assay is that the phosphorylation state of receptors in the immune complex can also be determined (Garofalo and Rosen, 1989). After the kinase reaction, the immune complexes are washed as described above for autophosphorylation (Section II,A) and subject to SDS–PAGE and autoradiography to correlate autophosphorylation with substrate phosphorylation.

In conclusion, the dramatic increase in the number of RTKs recently identified by molecular cloning leaves us with the daunting task of determining their function and mode of action. The regulation by multiple RTKs, often expressed concurrently in the same cell, of complex processes such as proliferation and differentiation suggests that cellular response systems must interpret an intricate web of signals emanating from the activated receptor complexes. The nature and strength of these signals will be determined by the specific interactions of the receptors with other proteins (e.g., substrates) and the half-life of the activated complexes. Detailed analysis of autophosphorylation, substrate phosphorylation, and complex formation using the techniques described above (and references contained therein) will provide fundamental information regarding RTK mechanisms. RTK action must then be considered in different cellular contexts to understand how signals from multiple RTKs are integrated to drive cells toward their distinctive fates.

References

Boni-Schnetzler, M., Kaligian, A., DelVecchio, R., and Pilch, P. F. (1988). *J. Biol. Chem.* **263**, 6822–6828.

Bottaro, D. P., Rubin, J. S., Ron, D., Finch, P. W., Florio, C., and Aaronson, S. A. (1990). *J. Biol. Chem.* **265**, 12767–12770.

Boyle, W. J., Van Der Geer, P., and Hunter, T. (1991). *In* "Protein Phosphorylation," Part B (T. Hunter and B. M. Sefton, eds.), Methods in Enzymology, Vol. 201, pp. 110–149. Academic Press, Orlando, Florida.

Braun, S., Raymond, W. E., and Racker, E. (1984). *J. Biol. Chem.* **259**, 2051–2054.

Cantley, L. C., Auger, K. R., Carpenter, C., Duckworth, B., Graziani, A., Kapeller, R., and Soltoff, S. (1991). *Cell* **64**, 281–302.

Casnellie, J. E. (1991). *In* "Protein Phosphorylation," Part A (T. Hunter and B. M. Sefton, eds.), Methods in Enzymology, Vol. 200, pp. 115–120. Academic Press, San Diego.

Casnellie, J. E., Harrison, M. L., Pike, L. J., Hellstrom, K. E., and Krebs, E. G. (1982). *Proc. Natl. Acad. Sci. U.S.A.* **79**, 282–286.
Connelly, P. A., and Stern, D. F. (1990). *Proc. Natl. Acad. Sci. U.S.A.* **87**, 6054–6057.
Cooper, J. A., Sefton, B. M., and Hunter, T. (1983). In "Hormone Action," Part F (J. Corbin and J. Hardman, eds.), Methods in Enzymology, Vol. 99, pp. 387–405. Academic Press, Orlando, Florida.
Ek, B., and Heldin, C.-H. (1982). *J. Biol. Chem.* **257**, 10486–10492.
Endemann, G., Yonezawa, K., and Roth, R. A. (1990). *J. Biol. Chem.* **265**, 396–400.
Ey, P. L., Prowse, S. J., and Jenkin, C. R. (1978). *Immunochemistry* **15**, 429–436.
Ganguly, S., Petruzzelli, L. M., Herrera, R., Stadtmauer, L., and Rosen, O. M. (1985). *Curr. Top. Cell. Regul.* **27**, 83–94.
Garofalo, R. S., and Barenton, B. (1992). *J. Biol. Chem.* **267**, 11470–11475.
Garofalo, R. S., and Rosen, O. M. (1989). *Mol. Cell. Biol.* **9**, 2806–2817.
Garrison, J. C. (1983). In "Hormone Action," Part F (J. Corbin and J. Hardman, eds.), Methods in Enzymology, Vol. 99, pp. 20–36. Academic Press, Orlando, Florida.
Glass, D. B., Masaracchia, R. A., Feramisco, J. R., and Kemp, B. E. (1978). *Anal. Biochem.* **87**, 566–575.
Glover, C. V. C., and Allis, C. D. (1991). In "Protein Phosphorylation," Part A (T. Hunter and B. M. Sefton, eds.), Methods in Enzymology, Vol. 200, pp. 85–90. Academic Press, San Diego.
Gordon, J. A. (1991). In "Protein Phosphorylation," Part B (T. Hunter and B. M. Sefton, eds.), Methods in Enzymology, Vol. 201, pp. 477–482. Academic Press, San Diego.
Hammacher, A., Mellstrom, K., Heldin, C.-H., and Westermark, B. (1989). *EMBO J.* **8**, 2489–2495.
Hedo, J., Harrison, L. C., and Roth, J. (1981). *Biochemistry* **20**, 3385–3393.
Herrera, R., and Rosen, O. M. (1986). *J. Biol. Chem.* **261**, 11980–11985.
Herrera, R., and Rosen, O. M. (1988). In "Insulin Receptors, Part A: Methods for the Study of Structure and Function" (C. R. Kahn and L. C. Harrison, eds.), pp. 181–188. Alan R. Liss, New York.
Herrera, R., Petruzzelli, L., Thomas, N., Bramson, H. N., Kaiser, E. T., and Rosen, O. M. (1985). *J. Biol. Chem.* **82**, 7899–7903.
Herrera, R., Petruzzelli, L. M., and Rosen, O. M. (1986). *J. Biol. Chem.* **261**, 2489–2491.
Hu, P., Margolis, B., Skolnik, E. Y., Lammers, R., Ullrich, A., and Schlessinger, J. (1992). *Mol. Cell Biol.* **12**, 981–990.
Hunter, T. (1982). *J. Biol. Chem.* **257**, 4843–4848.
Hunter, T., and Sefton, B. M. (1980). *Proc. Natl. Acad. Sci. U.S.A.* **77**, 1311–1315.
Kamps, M. P. (1991). In "Protein Phosphorylation," Part B (T. Hunter and B. M. Sefton, eds.), Methods in Enzymology, Vol. 201, pp. 101–110. Academic Press, San Diego.
Kamps, M. P., and Sefton, B. M. (1988). *Oncogene* **2**, 305–315.
Kamps, M. P., and Sefton, B. M. (1989). *Anal. Biochem.* **176**, 22–27.
Kasuga, M., White, M. F., and Kahn, C. R. (1985). In "Hormone Action," Part I (L. Birnbaumer and B. O'Malley, eds.), Methods in Enzymology, Vol. 109, pp. 609–621. Academic Press, Orlando, Florida.
Kasuga, M., Izumi, T., Tobe, K., Shiba, T., Momomura, K., Tashiro-Hashimoto, Y., and Kadowaki, T. (1990). *Diabetes Care* **13**, 317–326.
Kazlauskas, A., and Cooper, J. A. (1990). *EMBO J.* **9**, 3279–3286.
Kemp, B. E., and Pearson, R. B. (1991). In "Protein Phosphorylation," Part A (T. Hunter and B. M. Sefton, eds.), Methods in Enzymology, Vol. 200, pp. 121–134. Academic Press, San Diego.
King, C. R., Borrello, I., Bellot, F., Comiglio, P., and Schlessinger, J. (1988). *EMBO J.* **7**, 1647–1651.

Koch, C. A., Anderson, D., Moran, M. F., Ellis, C., and Pawson, T. (1991). *Science* **252**, 668–674.
Kozma, L. M., Rossomando, A. J., and Weber, M. J. (1991). In "Protein Phosphorylation," Part B (T. Hunter and B. M. Sefton, eds.), Methods in Enzymology, Vol. 201, pp. 28–43. Academic Press, San Diego.
Kull, F. C., Jr., Jacobs, S., Su, Y. F., Svoboda, M. E., Van Wyck, J. J., and Cuatrecasas, P. (1983). *J. Biol. Chem.* **258**, 6561–6566.
Laemmli, U. K. (1970). *Nature (London)* **227**, 680–685.
Maher, P. A. (1991). *J. Cell Biol.* **112**, 955–963.
Martin-Zanca, D., Oskam, R., Mitra, G., Copeland, T., and Barbacid, M. (1989). *Mol. Cell. Biol.* **9**, 24–33.
Moxham, C. P., and Jacobs, S. (1992). *J. Cell. Biochem.* **48**, 136–140.
Moxham, C. P., Duronio, V., and Jacobs, S. (1989). *J. Biol. Chem.* **264**, 13238–13244.
Murakami, M. S., and Rosen, O. M. (1991). *J. Biol. Chem.* **266**, 22653–22690.
Neufeld, E., Goren, H. J., and Boland, D. (1989). *Anal. Biochem.* **177**, 138–143.
O'Hare, T., and Pilch, P. F. (1990). *Int. J. Biochem.* **22**, 315–324.
Racker, E. (1991). In "Protein Phosphorylation," Part A (T. Hunter and B. M. Sefton, eds.), Methods in Enzymology, Vol. 200, pp. 107–111. Academic Press, San Diego.
Rijksen, G., van Oirschot, B. A., and Staal, G. E. J. (1989). *Anal. Biochem.* **182**, 98–102.
Rijksen, G., van Oirschot, B. A., and Staal, G. E. J. (1991). In "Protein Phosphorylation," Part A (T. Hunter and B. M. Sefton, eds.), Methods in Enzymology, Vol. 200, pp. 98–107. Academic Press, San Diego.
Rodrigues, G. A., Maujokas, M. A., and Park, M. (1991). *Mol. Cell. Biol.* **11**, 2962–2970.
Rosen, O. M. (1987). *Science* **237**, 1452–1458.
Rosen, O. M., Herrera, R., Olowe, Y., Petruzzelli, L. M., and Cobb, M. H. (1983). *Proc. Natl. Acad. Sci. U.S.A.* **80**, 3237–3240.
Roth, R. A., Zhang, B., Chin, J. E., and Kovacina, K. (1992). *J. Cell Biochem.* **48**, 12–18.
Sahal, D., and Fujita-Yamaguchi, Y. (1987). *Anal. Biochem.* **167**, 23–30.
Sahal, D., Li, S.-L., and Fujita-Yamaguchi, Y. (1991). In "Protein Phosphorylation," Part A (T. Hunter and B. M. Sefton, eds.), Methods in Enzymology, Vol. 200, pp. 90–98. Academic Press, San Diego.
Schieven, G., Thorner, J., and Martin, G. S. (1986). *Science* **231**, 390–393.
Seedorf, K., Millauer, B., Kostka, G., Schlessinger, J., and Ullrich, A. (1992). *Mol. Cell. Biol.* **12**, 4347–4356.
Sefton, B. M., Hunter, T., Beemon, K., and Eckhart, W. (1980). *Cell* **20**, 807–816.
Shoelson, S. E., Chatterjee, S., Chaudhuri, M., and White, M. F. (1992). *Proc. Natl. Acad. Sci. U.S.A.* **89**, 2027–2031.
Soos, M. A., and Siddle, K. (1989). *Biochem. J.* **263**, 553–563.
Stadtmauer, L. A., and Rosen, O. M. (1983). *J. Biol. Chem.* **258**, 6682–6685.
Stadtmauer, L. A., and Rosen, O. M. (1986). *J. Biol. Chem.* **261**, 10000–10005.
Stern, D. F., and Kamps, M. P. (1988). *EMBO J.* **7**, 995–1001.
Sun, X. J., Rothenberg, P., Kahn, C. R., Backer, J. M., Araki, E., Wilden, P. A., Cahill, D. A., Goldstein, B. J., and White, M. F. (1991). *Nature (London)* **352**, 73–77.
Sweet, L. J., Morrison, B. D., Wilden, P. A., and Pessin, J. E. (1987). *J. Biol. Chem.* **262**, 16730–16738.
Ullrich, A., and Schlessinger, J. (1990). *Cell* **61**, 203–212.
Vega, Q. C., Cochet, C., Filhol, O., Chang, C.-P., Rhee, S. G., and Gill, G. N. (1992). *Mol. Cell. Biol.* **12**, 128–135.
White, M. F., and Backer, J. M. (1991). In "Protein Phosphorylation," Part B (T. Hunter and B. M. Sefton, eds.), Methods in Enzymology, Vol. 201, pp. 65–79. Academic Press, San Diego.

White, M. F., Shoelson, S. E., Keutmann, H., and Kahn, C. R. (1988). *J. Biol. Chem.* **263**, 2969–2980.
Wilden, P. A., Kahn, C. R., Siddle, K., and White, M. F. (1992). *J. Biol. Chem.* **267**, 16660–16668.
Williams, L. T. (1989). *Science* **243**, 1564–1570.
Wong, T. W., and Goldberg, A. R. (1983). *J. Biol. Chem.* **258**, 1022–1025.
Yarden, Y., and Schlessinger, J. (1987). *Biochemistry* **26**, 1443–1451.
Yonezawa, K., and Roth, R. A. (1991). *Mol. Endocrinol.* **5**, 194–200.
Zick, Y., Kasuga, M., Kahn, C. R., and Roth, J. (1983). *J. Biol. Chem.* **258**, 75–80.
Zick, Y., Grunberger, G., Rees-Jones, R. W., and Comi, R. J. (1985). *Eur. J. Biochem.* **148**, 177–182.

16

Involvement of Protein Tyrosine Phosphatase Activity in the Regulation of Insulin's Signal Transduction

Dalit Hecht and Yehiel Zick

I. Introduction
II. Regulation of Protein Tyrosine Phosphatase Activity
III. Cellular Function Mediated by Protein Tyrosine Phosphatases
IV. Protein Tyrosine Phosphatases as Mediators of Insulin Action
V. Alterations in Insulin Receptor-Protein Tyrosine Phosphatase Activity in Insulin-Resistant States
VI. Protein Substates for Protein Tyrosine Phosphatases
VII. Assay of Protein Tyrosine Phosphatase Activity
    A. Preparation of Lectin-Purified Insulin Receptor
    B. Assay of IRK Activity
    C. Autophosphorylation of Insulin Receptors
    D. Immunoprecipitation of ^{32}P-Labeled IRK
    E. Autophosphorylation of Immunoprecipitated Insulin Receptor
    F. Western Blotting with Anti-P-Tyr Antibodies
    G. Assay of PTP Activity Using Tyrosine-Phosphorylated IRK as a Substrate
References

I. Introduction

Protein tyrosine phosphorylation is involved in signaling cell growth, metabolism, differentiation, and transformation (Cantley *et al.*, 1991). The level of tyrosine phosphorylation at any instance reflects the relative activities of protein tyrosine kinases (PTKs) and protein tyrosine phosphatases (PTPs) that catalyze the interconversion processes. Members of the PTP family are divided into two major subclasses; receptor-like (RPTP) and nonreceptor-like PTPs (NRPTP) (Fischer *et al.*, 1991). The prototype of nonreceptor PTPs, PTP 1B, is a 50-kDa protein whose C-terminus end is hydrophobic and enables binding of PTP 1B to the endoplasmic reticulum (ER), with the phosphatase domain oriented toward the cytoplasm (Frangioni et al., 1992). Other members of this family include the T-cell PTP (Cool et al., 1989) and PTP 1C (Shen et al., 1991) which has two *src* homology (SH2) domains.

Most RPTPs contain a tandem repeat of the catalytic domain, separated by a stretch of nonconserved amino acids. All are transmembrane proteins, with a large extracellular region. Based on the characteristics of their extracellular domains these enzymes are divided into four subclasses (Fischer et al., 1991): Type I RPTPs, represented by CD45, have a heavily glycosylated external segment and a conserved cysteine-rich region; this varies in length from 400 to 550 residues, depending upon the pattern of expression of three exons which encode sequences at the N-terminus of the molecule (Trowbridge, 1991). Type II RPTPs contain several immunoglobulin G-like domains linked to two to nine fibronectin (Type III) repeats. The leukocyte common antigen (LCA)-related protein (LAR) (Streuli et al., 1990), LAR-related Drosophila proteins DLAR, and DPTP (Streuli et al., 1989) are representatives of this group. Type III RPTPs, like HPTPβ (Krueger et al., 1990), contain a single catalytic domain and only fibronectin (Type III) repeats, while Type IV RPTPs, represented by HPTPα (Sap et al., 1990) and HPTPϵ (Krueger et al., 1990), have relatively short extracellular regions.

II. Regulation of Protein Tyrosine Phosphatase Activity

Little data is available regarding the mechanisms that control and regulate the activity of PTPs *in vivo*. Alterations in RPTPs activity occur following crosslinking of their extracellular domains by appropriate antibodies, but no ligand has been described to date that binds and modulates RPTP function. PTPs serve as substrates for protein kinases. However, there are no changes in PTP activity following phosphorylation with a variety of Ser/Thr or Tyr kinases (Stover et al., 1991; Tonks et al., 1990b). PTPs activity is also modulated by inhibitory proteins. Two such proteins, inhibitor L (38 kDa) and inhibitor H (> 500 kDa) were purified from bovine brain (Ingebristen, 1989), but their physiological role is presently unknown.

An array of effectors modulate PTPs activity *in vitro* (Tonks et al., 1990b). All PTP activities are inhibited by vanadate (Swarup et al., 1982), molybdate, and Zn^{2+}. NRPTPs are partially inhibited by Mn^{2+} and pyrophosphate, while they are activated by NaF and EDTA (Tonks et al., 1988). Since PTP catalysis proceeds via a cysteine-phosphate intermediate (Guan and Dixson, 1991), they have an absolute requirement for sulfhydryl-containing compounds for activity. When assayed in buffers in the absence of reducing agents, PTP activity is largely diminished. However, the inhibition is reversible and the enzymes are reactivated by adding sulfhydryl reagents (Tonks et al., 1988). Consistent with these observations, oxidizing agents such H_2O_2, $KMnO_4$, phenylarsine oxide, and vitamin K_3 reversibly inhibit PTP activity *in vitro* (Liao et al., 1991; Hecht and Zick, 1992).

III. Cellular Functions Mediated by Protein Tyrosine Phosphatases

Control over the level of phosphorylation, particularly of tyrosine residues, seems to be necessary for normal cell growth and differentiation as excessive, unregulated PTK activity often leads to transformation of the affected cells (Cantley et al., 1991). Dephosphorylation by PTPs provides one potential mechanism; still, the role of PTPs in mediating and regulating cellular functions is only beginning to emerge.

One of the best-studied examples is the RPTP CD45 which regulates leukocyte growth by stimulating the kinase activity of $pp56^{lck}$ (Ostergaard et al., 1989), whose major putative substrate is the ζ chain of the T-cell receptor (June et al., 1991). CD45 has also been implicated in the effective coupling of the T-cell receptor to the phosphatidylinositol (PI) pathway (Koretzky et al., 1990), while stimulation of T lymphocytes by way of the accessory molecule CD2 requires the presence of a functional CD45 (Koretzky et al., 1991).

PTPs control cell cycle and cell growth. About a 10-fold increase in RPTP activity occurs in density-dependent growth-arrested fibroblasts, implicating PTPs as regulators of cell growth in high-density cultures (Pallen and Tong, 1991). $p34^{cdc2}$, which controls the transition from the G_2 phase of the cell cycle into mitosis, is activated by dephosphorylation of a tyrosine residue in its ATP binding site (Gould et al., 1990). The PTP that acts upon $p34^{cdc2}$ is presumably $p80^{cdc25}$ (Gautier et al., 1991; Kumagi and Dunphy, 1991), which shares structural homology with a PTP (VH1) from vaccinia virus (Moreno and Nurse, 1991).

PTPs are also involved in modeling of cell structure and in the regulation of cell–cell and cell–substrate interactions. Both intercellular and adherence junctions are major sites of action of PTKs (Comoglio et al., 1984; David-Pfeuty and Singer, 1980) and PTPs (Volberg et al., 1991, 1992).

IV. Protein Tyrosine Phosphatases as Mediators of Insulin Action

An important example for a role of tyrosine phosphorylation dephosphorylation reactions in signal transduction, is the response of cells to insulin. The insulin receptor (IR) is a transmembrane glycoprotein comprising two α subunits (135 kDa) and two β subunits (95 kDa), linked by disulfide bonds (reviewed in Zick, 1989). Insulin binding to the α subunits leads to autophosphorylation of the tyrosine kinase which is part of the receptor β subunits (insulin receptor kinase; IRK). Autophosphorylation takes place on five tyrosine residues, three of which (positions 1146, 1150, 1151) are localized within the kinase region (White et al., 1988). Insulin-stimulated autophosphorylation activates IRK and enables it to phosphorylate endogenous substrates. This property of the receptor set the stage for the hypothesis that insulin signaling is mediated through tyrosine phosphorylation of cellular proteins. Indeed, several such proteins were identified (White et al., 1985; Rees-Jones and Taylor, 1985; Bernier et al., 1987; Heffetz et al., 1990; Sun et al., 1991).

Insulin-stimulated autophosphorylation of IRK *in vivo* is immediately followed by rapid dephosphorylation once insulin is removed (Haring et al., 1984; Mooney and Anderson, 1989). This is associated with concomitant deactivation of IRK with the tris-phosphorylated form of the receptor being the preferred substrate (King and Sale, 1990). Similarly, insulin-dependent phosphorylation of its substrate pp180 is regulated by PTPs that rapidly dephosphorylate this protein despite the presence of insulin and an activated IRK (White et al., 1985; Heffetz and Zick, 1989).

PTP activities toward the autophosphorylated IR-β subunit (IR-PTPs) are present in rat liver (Strout et al., 1988), rat adipocytes (Mooney and Anderson, 1989), and human placenta (Roome et al., 1988). IR-PTPs are present both in soluble and particulate fractions and copurify with the receptor following lectin affinity chromatography (King and Sale, 1990; Goldstein et al., 1991; Hecht and Zick, 1992). Hepatic membranal PTPs are more effective than CD45 or LAR in dephosphorylating the insulin and epidermal growth factor (EGF) receptors (Goldstein et al., 1991), suggesting that PTPs with selected specificity toward these receptors might exist in liver.

Interestingly, insulinomimetic agents such as vanadate and H_2O_2 (Czech et al., 1974; Tamura et al., 1984; Kadota et al., 1987b) are potent inhibitors of PTPs (Swarup et al., 1982; Heffetz et al., 1990). Furthermore, the synergism between these agents in promoting insulin-like biological effects (Kadota et al., 1987a) correlates with their capacity to synergize in inhibiting PTP activity *in vivo* (Heffetz et al., 1990).

V. Alterations in Insulin Receptor–Protein Tyrosine Phosphatase Activity in Insulin-Resistant States

PTP activity is both regulating and being regulated by the insulin receptor. Insulin induces a rapid cytosol to membrane translocation of IR-PTPs in Fao cells (Goldstein et al., 1991), while injection of PTP 1B into Xenopus oocytes inhibits autophosphorylation of IRK, insulin-stimulated S6 kinase activity (Cicirelli et al., 1990), and insulin-induced maturation of these cells

(Tonks *et al.*, 1990a). Defects in IRK activity are associated with insulin resistance in patients (Grunberger *et al.*, 1984; Caro *et al.*, 1986; Odawara *et al.*, 1989) and animal models (Le Marchand-Brustel *et al.*, 1985); however, the relation between these defects and altered IR-PTP activity remains unclear. We have recently demonstrated (Nadiv *et al.*, 1993) a significant 30% elevation in cytosolic IR-PTP activity in livers of insulin-resistant old rats. Similarly, a twofold increase in cytosolic IR-PTP activity occurs in streptozotocin-induced diabetic rats (Meyrovitch *et al.*, 1989; Goldstein *et al.*, 1991). However, not all insulin-resistant states involve an increase in IR-PTP activity. For example, insulin resistance induced by fasting is associated with inhibition rather than stimulation of IR-PTP activity (Begum *et al.*, 1992).

VI. Protein Substrates for Protein Tyrosine Phosphatases

Unlike PTKs, which express a significant degree of substrate specificity (Ullrich and Schlessinger, 1990), the substrate specificity of PTPs remains elusive. No consensus recognition sequences for PTPs has emerged, and there is no evidence, yet, to support a model where a given tyrosine-phosphorylated protein serves as a specific substrate for a given PTP. Reduced carboxymethylated lysozyme (RCM-lysozyme); reduced carboxymethylated and -succinilated bovine serum albumin (RCS-BSA), and Myosin P-light chain are all dephosphorylated by CD45 at comparable rates while dephosphorylation of myelin basic protein (MBP) occurs 10-fold more effectively (Tonks *et al.*, 1990b). These findings implicate a certain degree of substrate specificity for CD45.

Both PTP 1B and CD45 are also capable of dephosphorylating proteins of more physiological relevance including tyrosine-phosphorylated PTKs. pp56^{lck} is an *in vitro* substrate for both phosphatases, as are the autophosphorylated insulin and EGF receptors (Tonks *et al.*, 1990b). In contrast, p80^{cdc25} specifically dephosphorylates p34^{cdc2} but no other tyrosine-phosphorylated proteins (Gautier *et al.*, 1991). Similarly, two membranal PTPs purified from 3T3-L1 adipocytes (Liao *et al.*, 1991) dephosphorylate pp15 (Bernier *et al.*, 1987), a target protein for IRK, more effectively than the autophosphorylated IRK or RCM-lysozyme. Hence, several PTPs do exhibit some aspects of substrate specificity.

VII. Assay of Protein Tyrosine Phosphatase Activity

Two types of substrates are currently being used when PTP activity is assayed *in vitro*. One is a group of proteins that have been chemically modified and subjected to *in vitro* phosphorylation by a variety of PTKs. These are excellent substrates for a quick screening for the presence of PTP activity in various samples, or during the process of PTP purification. Three such substrates are most commonly used. One is RCM-lysozyme, which is the preferred substrate for PTP 1B (Tonks *et al.*, 1991b); another is MBP, which is the preferred substrate for RPTPs like CD45 (Tonks *et al.*, 1991a). Casein is an efficient substrate for a variety of other PTPs (Ingebristen, 1991). Preparation of these tyrosine-phosphorylated proteins, as well as detailed protocols for the assay of various PTP activities using these proteins as substrates has been recently described (Ingebristen, 1991; Tonks *et al.*, 1991a,b).

Other model substrates for PTPs are synthetic peptides corresponding to the major autophosphorylation site of the insulin receptor kinase (Meyrovitch *et al.*, 1989). However, while the peptide substrates represent the primary structure of the original tyrosine-phosphorylated protein, they fail to mimic its secondary and tertiary structure. Since substrate specificity is determined, at least in part, by the conformation of the native protein and the availability of particular residues for modification, issues such as steric hinderance or the native environment of each P-Tyr residue are often overlooked when dealing with peptide substrates. For these reasons using whole proteins is expected to increase the specificity and physiological significance of the dephosphorylation reaction. As a model system we therefore describe assay of PTP activity using a tyrosine-phosphorylated IRK as a substrate.

Buffers

Buffer A. 50 mM Hepes, pH 7.6; 2 mM EDTA; 0.25 M sucrose; 1 mM phenylmethylsulfonyl fluoride (PMSF); 10 μg/ml aprotinin; 10 μg/ml leupeptin; 5 μg/ml soybean trypsin inhibitor (STI).

Buffer B. 50 mM Hepes; pH 7.6; 150 mM NaCl; 0.1% Triton X-100; 10% glycerol; 1 mM PMSF.

Buffer C. 50 mM Hepes, pH 7.6; 150 mM NaCl; 0.1% Triton X-100; 10% glycerol; 0.5 M N-acetyl-D-glucoseamine; 1 mM PMSF; 10 μg/ml aprotinin; 10 μg/ml leupeptin; 5 μg/ml STI.

Buffer D. 50 mM Hepes, pH 7.6; 0.1% BSA; 0.1% Triton X-100.

Buffer E. 0.1% Triton X-100; 0.1 M NaF; 80 mM NaCl; 20 mM sodium pyrophosphate; 10 mM EGTA; 20 mM EDTA; 50 mM Hepes, pH 7.6.

Buffer F. 25 mM imidazole–HCl; 1 mM EDTA; 1 mM EGTA; 10% glycerol; 0.5 mM digitonin; pH 7.2.

Buffer G. 0.25 M sucrose; 5 mM EDTA; 0.5 mM EGTA; 1 mM dithiothreitol (DTT); 2 mM benzamidine; 0.2 mM PMSF; 50 mM Hepes, pH 7.4.

A. Preparation of Lectin-Purified Insulin Receptor

Insulin receptors are purified from rat livers as previously described (Zick *et al.*, 1983). Unless indicated otherwise all subsequent steps are carried out at

4°C. Twenty grams of livers from 6-week-old Sprague-Dawley rats are cut into small pieces and homogenized at high speed in Waring blender in 3 vol of Buffer A. After centrifugation at 3000g for 30 min, the pellet is discarded and the supernatant is centrifuged at 100,000g for 60 min. The pellet is resuspended in 60 ml of Buffer A supplemented with 1% Triton X-100 and is left to agitate for 15 min. The suspension is centrifuged again at 100,000g for 60 min and the supernatant is collected. The supernatant is mixed with 5 ml of wheat germ agglutinin (WGA) coupled to Sepharose (e.g., Glycaminocylex, Bio-Makor, Rehovot, Israel), which is preequilibrated in Buffer B. The suspension is rotated end over end overnight. The slurry is poured into a 1-cm-diameter column and the column is washed with 100 ml of Buffer B. The bound insulin receptor kinase is eluted with 15 ml of Buffer C. One-milliliter fractions are collected and frozen at $-70°C$. This preparation is referred to as WGA eluate.

B. Assay of IRK Activity

IRK activity is assayed employing the tyrosine-containing polymer poly-(Glu,Tyr) 4:1 as a substrate. The reaction is carried out essentially as described (Heffetz and Zick, 1986), with some modifications. Briefly, 20-μl aliquots from each tube of WGA eluate are incubated with 10 μl of 6×10^{-7} M insulin in Buffer D. Phosphorylation of poly-(Glu,Tyr) 4:1 is initiated with 30 μl of a reaction mixture containing 100 μM [γ-^{32}P]ATP (400 μCi/ml), 80 mM magnesium acetate, 8 mM manganese acetate, 2 mM CTP, 0.1% Triton X-100, 4 mg/ml poly-(Glu,Tyr) 4:1, and 50 mM Hepes, pH 7.6. Reactions are allowed to proceed for 10 min at 22°C and are terminated by applying 40-μl aliquots onto Whatman No. 3 MM filter papers that are extensively washed in 10% trichloroacetic acid and 10 mM sodium pyrophosphate, rinsed with ethanol, dried, and counted by liquid scintillation in Betamatic counter. One unit of kinase activity is defined as the amount of enzyme required to incorporate 1 pmol ^{32}P into poly-(Glu-Tyr) 4:1 during 10 min. The active fractions of the WGA eluate (> 500 units/ml) are stored at $-70°C$ and serve as the source for the autophosphorylated receptors. When the autophosphorylated receptor requires further purification (see below) the active fractions are pooled and extensively dialyzed against Buffer B (to remove the N-acetyl-D-glucoseamine).

C. Autophosphorylation of Insulin Receptors

When the IRK is subjected to autophosphorylation *in vitro*, in the presence of insulin, the phosphate is incorporated exclusively into tyrosine residues (Kasuga *et al.*, 1982). This method is therefore being used to generate

Tyr-phosphorylated IRK. Autophosphorylation assays are carried out as previously described (Zick et al., 1983). WGA eluates (about 200 μg in 400 μl of Buffer C) are incubated for 30 min at 22°C with 50 μl of 10^{-6} M insulin in Buffer D. Autophosphorylation is initiated by adding 50 μl of a reaction mixture containing 250 μM [γ-^{32}P]ATP (1 mCi/ml), 50 mM magnesium acetate, 20 mM manganese acetate, 0.1% Triton X-100, and 50 mM Hepes, pH 7.6. Reactions are allowed to proceed for 10–20 min at 22°C and are terminated by the addition of 100 μl of 10 mM ATP, 0.1 M EDTA, and 50 mM Hepes, pH 7.6. Samples are then subjected to heat inactivation for 10 min at 56°C. The autophosphorylated receptors are kept at $-70°C$ until further use. A 0.5-ml preparation is usually used as a substrate source to assay 10 samples of PTP. Autophosphorylation can be carried out with unlabeled ATP. Under these conditions the ATP concentration during the reaction is increased to 0.5 mM. The extent of tyrosine phosphorylation of IRK is monitored by Western blotting with antibodies directed against P-Tyr residues (anti-P-Tyr antibodies, see below). Of importance is the fact that the WGA eluates from rat liver also contain significant amounts of EGF receptors, which are readily phosphorylated on tyrosine residues under these conditions. When the ^{32}P-labeled WGA is used as the substrate source, the dephosphorylation of the EGF receptor can be monitored as well. This is a useful control in experiments aimed at defining the substrate specificity of an insulin receptor PTP.

It should be noted that during termination of the reaction the specific activity of the [γ-^{32}P]ATP is reduced 20-fold, the magnesium and manganase are chelated by EDTA, and the proteins are heat inactivated. Still these samples contain enough [γ-^{32}P]ATP that could serve as a substrate for protein kinases that contaminate crude preparations of PTPs. Therefore, it is recommended that this preparation be applied as a substrate only for initial screenings or when kinase-free preparations of PTPs are available. In other cases, further purification of the ^{32}P-Tyr-labeled IRK is necessary (see below).

When further purification is required the autophosphorylation is carried out with a dialyzed receptor (see above). Following the kinase assay and termination of the reaction the ^{32}P-Tyr-labeled IRK is reapplied to a 0.5-ml WGA column. Excess salts, insulin, and [γ-^{32}P]ATP are removed in the "wash trough" fraction while the labeled receptor is then eluted in Buffer C. Fractions (300 μl) are collected and the presence of the ^{32}P-Tyr-labeled 95-kDa β subunits of IRK is detected by mixing a 30-μl aliquot from each tube with ($\times$2.5 concentrated) "stopping solution" containing 5% (w/v) SDS, 0.06 M Tris, 5 mM ATP, 25% (v/v) glycerol, 1.82 M mercaptoethanol, and 0.02% (w/v) bromophenol blue, pH 6.8. After being heated for 10 min at 95°C, the samples are analyzed by 7.5% SDS–polyacrylamide gel electrophoresis and autoradiography. Fractions containing the labeled IRK are kept at $-70°C$ and serve as substrate for assays of PTP activity.

D. Immunoprecipitation of ^{32}P-Labeled IRK

For most preliminary studies, when screening for the presence of IRK-specific PTP (IRK-PTP) activity, or when the IRK-PTP activity is assayed during enzyme purification, the above autophosphorylated IRK preparations serve as most suitable substrates. However, since IRK constitutes only ~1% of the total proteins present in the WGA eluate, it is often desirable to use a highly purified ^{32}P-labeled IRK. For that purpose the ^{32}P-labeled IRK is enriched by immunoprecipitation. In these experiments the autophosphorylation (see above) is terminated with 400 μl of Buffer E and no heat inactivation is done.

Various kinds of anti-receptor antibodies that could be used for immunoprecipitation are available now. Two different antibodies are being used in the experiments described below: One is sera from a patient with autoantibodies to the insulin receptor (B6) (a generous gift of Dr. P. Gorden, Diabetes Branch, NIH, Bethesda, MD); the second are rabbit polyclonal antibodies (ST-50) raised against a synthetic peptide corresponding to positions 1309–1324 of the human insulin receptor. The polyclonal antibodies are raised in New Zealand White rabbits according to standard immunization protocol (Heffetz et al., 1991), and the serum is further purified by affinity chromatography on immobilized peptide. To prepare a sample sufficient to serve as a substrate for a single PTP reaction, 7.5 μl of affinity-purified ST-50 antibodies are mixed with a 30-μl suspension of protein A Sepharose (50% v/v) and 100 μl of 0.1 M Tris/HCl buffer, pH 8.5. The mixture is suspended for 1 hr at 4°C. The precipitate is collected by centrifugation for 3 min at 12000g at 4°C and washed three times with 0.1 M Tris/HCl buffer, pH 8.5. The washed complex is resuspended with 100 μl of ^{32}P-labeled IRK and the mixture is suspended overnight at 4°C in an Eppendorf shaker. The immune complex is washed twice with 50 mM Hepes, 150 mM NaCl, 1% (v/v) Triton X-100, 0.1% SDS, pH 7.6, and once with 50 mM Hepes, 150 mM NaCl, and 0.1% (v/v) Triton X-100, pH 7.6. The immune complex is immediately subjected to dephosphorylation by the PTP under study.

E. Autophosphorylation of Immunoprecipitated Insulin Receptor

In an alternative approach, the IRK is directly immunoprecipitated from the WGA eluate, and the autophosphorylation is carried out in the immune complexes. For that purpose immune complexes of ST-50 antibodies are prepared as described above. The washed complexes are resuspended with 100 μl (~100 μg) of WGA eluate and the mixture is suspended overnight at 4°C in an Eppendorf shaker. The immune complexes are washed three times in

Buffer B (less glycerol), suspended in 20 μl of Buffer B (containing 10 μg/ml aprotinin), and incubated with 10 μl insulin (final concentration 10^{-7} M) for 30 min at 22°C. Phosphorylation is initiated by addition of 20 μl reaction mixture to yield the following final concentrations: 10 μM [γ-^{32}P]ATP (200 μCi/ml), 20 mM manganese chloride, 0.1% Triton X-100, and 50 mM Hepes, pH 7.6. The reaction is allowed to proceed for 10 min at 30°C, after which the immune complexes are extensively washed, five to seven times in Buffer B containing 2 mM EDTA, in order to remove all free radioactive material. The pellets are immediately subjected to PTP assay.

Autophosphorylation in the immune complexes can also be carried out with unlabeled ATP. Under these conditions the ATP concentration during the reaction is increased to 0.5 mM. Following extensive washings, the pellets are subjected to PTP assay as described below. The extent of dephosphorylation is then determined by immunoblotting the phosphorylated receptor (before and after incubation with the appropriate PTP) with anti-P-Tyr antibodies.

F. Western Blotting with Anti-P-Tyr Antibodies

Affinity-purified anti-P-Tyr antibodies are generated as previously described (Heffetz and Zick, 1989). Proteins separated by 7.5% SDS–polyacrylamide gel electrophoresis are electrophoretically transferred to nitrocellulose papers, incubated with anti-P-Tyr antibodies, and decorated with goat anti-rabbit antibodies as previously described (Heffetz et al., 1991). The latter antibodies can be either ^{125}I-labeled or conjugated to alkaline phosphatase or to horseredish peroxidase. This allows detection of the tyrosine-phosphorylated receptors by means of autoradiography, colorimetric reaction, or enhanced chemiluminescence.

G. Assay of PTP Activity Using Tyrosine-Phosphorlated IRK as a Substrate

1. Principle of Assay

The assay is based on the conversion of tyrosine-phosphorylated IRK into its dephosphorylated form. The content of the P-Tyr IRK is determined before and after incubation with the PTP and the extent of its reduction is taken as a measure of the PTP activity.

2. Source of PTP

As mentioned, purified PTPs such as LAR and CD45 dephosphorylate the IRK (Tonks *et al.*, 1990b). We (Hecht and Zick, 1992), like others (Goldstein *et al.*, 1991; Sale, 1991), have recently demonstrated the presence of liver PTPs that dephosphorylate the insulin receptor. Rat liver and rat hepatoma Fao cells are the sources of the PTP activity to be described. Other cell lines and tissues can be used, applying the same methodology.

3. Extraction of PTP Activity from Fao Cells

Monolayer cultures of rat hepatoma (Fao) cells are grown in 100-mm Nunc tissue culture dishes at 37°C in a humidified atmosphere composed of 95% air and 5% CO_2, in RPMI 1640 medium supplemented with 10% fetal calf serum. Soluble (cytosolic) and particulate extracts are prepared essentially as described in Relech *et al.* (1986), with the following modifications. The cells are washed with ice-cold phosphate-buffered saline (PBS) and frozen on liquid nitrogen. Solubilization is performed in 600 µl of Buffer F, and the extracts are centrifuged for 10 min at 4°C at 400g. The supernatants are collected and recentrifuged for 15 min at 4°C at 12,000g. These supernatants are defined as cytosolic extracts. The 12000g pellets are suspended in the same buffer and defined as the particulate extracts. The two fractions are assayed for PTP activity immediately after extraction.

4. Extraction of PTP Activity from Rat Liver

Male Sprague–Dawley rats (200–300 g) are anesthetized over dry ice. Livers are excised rapidly and frozen in liquid nitrogen. Tissue extracts are prepared by adding 3 ml/g of Buffer G. The homogenate is centrifuged for 15 min at 4°C at 500g, and the supernatant (s1) is further centrifuged for 60 min at 4°C at 100,000g. This supernatant (s2) is the cytosolic fraction. The pellet (p1) is suspended in Buffer G and is the particulate fraction. In some cases the pellet (p1) is suspended in 1% Triton X-100 in 50 mM Hepes, pH 7.4, and recentrifuged for 60 min at 4°C at 100,000g. The supernatant (s3) is collected and is the membrane-Triton extract.

5. Procedure

The assay is done in 1.5 ml plastic microfuge tubes. Depending on the substrate source, the following procedures are carried out.

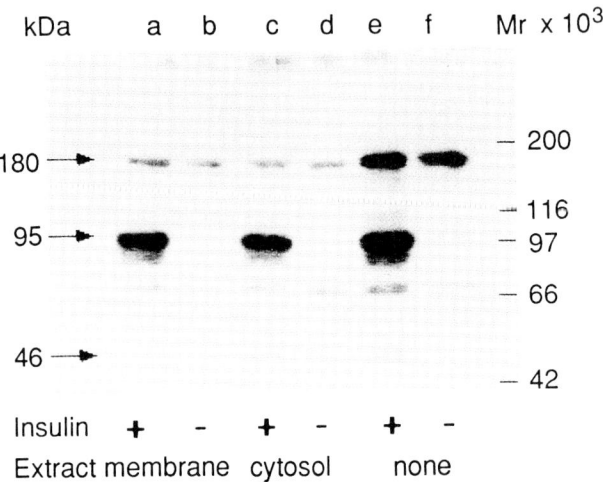

Figure 1 Assay of PTP activity using ^{32}P-labeled WGA eluates as substrates. ^{32}P-Labeled WGA eluates, which contain tyrosine-phosphorylated IRK are prepared as described (Section VII,A). The PTP source is soluble or particulate PTPs present in extracts of Fao cells. The assay is performed as detailed in the text. The intensity of the 95-kDa bands, corresponding to the β subunit of the insulin receptor, is quantitated by densitometry and the extent of the dephosphorylation is calculated. Reaction times and PTP concentrations are adjusted so as to yield linear dephosphorylation rates. Since the ^{32}P-labeled WGA eluate also contains substantial amounts of autophosphorylated EGF receptors, the extent of its dephosphorylation is readily quantitated by densitometric analysis of the 180-kDa band which corresponds to the EGF receptor.

a. Assay of PTP Activity Using ^{32}P-Labeled WGA Eluates as Substrates
^{32}P-Labeled WGA eluates, which contain tyrosine-phosphorylated IRK, are prepared as described above (Section VII,A). The enzyme source is soluble or particulate PTPs present in extracts of Fao cells. Fifty-microliter eluates are mixed with 5 μl of 16 mM DTT. The reaction is initiated by adding 25 μl of either cytosolic ($\sim$2.5 mg protein/ml) or particulate (1.5 mg protein/ml) extracts of Fao cells, or 25 μl of Buffer F as control. Dephosphorylation is allowed to proceed for 5–60 min at 30°C and is terminated by adding ($\times$5 concentrated) "sample buffer" (Laemmli, 1970) containing 5% (w/v) SDS, 0.06 M Tris, 25% (v/v) glycerol, 1.82 M mercaptoethanol, and 0.02% (w/v) bromophenol blue, pH 6.8. After heating for 10 min at 95°C, the samples are analyzed by 7.5% SDS–polyacrylamide gel electrophoresis and autoradiography. The intensity of the 95-kDa bands, corresponding to the β subunit of the insulin receptor, is quantitated by densitometry and the extent of the dephosphorylation is calculated. Reaction times and PTP concentrations are adjusted to yield linear dephosphorylation rates. Example of such an assay is presented

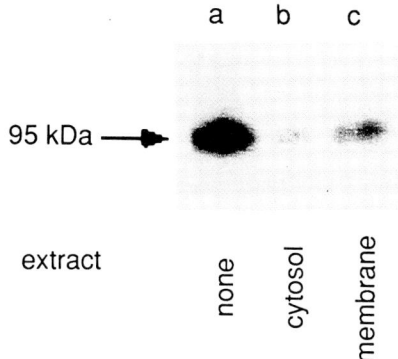

Figure 2 Assay of PTP activity using ^{32}P-labeled immunoprecipitated IRK as substrate. WGA eluates are immunoprecipitated and the IRK present in the immune complexes is subjected to autophosphorylation as described in the text. The dephosphorylation reaction is initiated by incubating the immune complexes with cytosolic or particulate fractions of rat liver. The samples are analyzed by 7.5% SDS-polyacrylamide gel electrophoresis, autoradiography, and densitometry as described in the text. The position of 95-kDa bands, corresponding to the β subunit of the insulin receptor, is marked with an arrow.

in Fig. 1. As mentioned, the ^{32}P-labeled WGA eluate also contains substantial amounts of autophosphorylated EGF receptors. The extent of their dephosphorylation is readily quantitated by densitometric analysis of the 180-kDa band which corresponds to the EGF receptor.

b. Assay of PTP Activity Using ^{32}P-Labeled Immunoprecipitated IRK as Substrates WGA eluates are first immunoprecipitated (with ST-50 antibodies) and the IRK present in the immune complexes is subjected to autophosphorylation as described above (Section VII,E). After extensive washes the phosphorylated immune complexes are subjected to dephosphorylation. The reaction is initiated by suspending the immune complexes in 30 μl of Buffer F, containing 1 mM DTT, or in 30 μl of cytosolic (~2 mg protein/ml) or particulate fractions (~2 mg protein/ml) of rat liver that serve as the PTP source. The reaction is allowed to proceed for 5-45 min at 30°C and is terminated by adding ($\times$5 concentrated) sample buffer (Laemmli, 1970) and heating for 10 min at 95°C. The samples are analyzed by 7.5% SDS-polyacrylamide gel electrophoresis, autoradiography, and densitometry as described above. Example of such an assay is presented in Fig. 2.

c. Assay of PTP Activity Using Unlabeled Immunoprecipitated and Autophosphorylated IRK as Substrates WGA eluates are subjected to autophosphorylation with unlabeled ATP as described above. The autophospho-

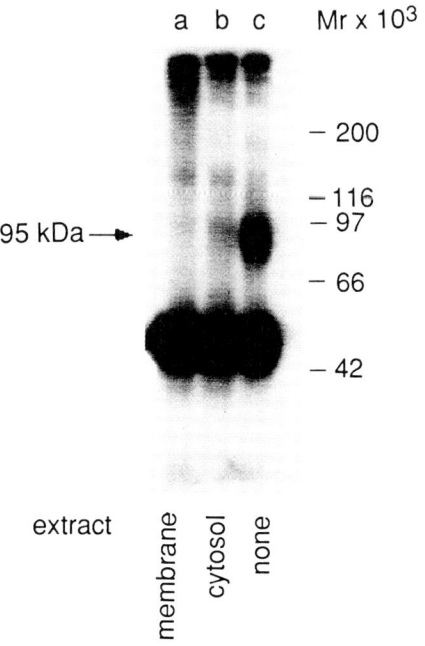

Figure 3 Assay of PTP activity using unlabeled immunoprecipitated and autophosphorylated IRK as substrates. WGA eluates are subjected to autophosphorylation with unlabeled ATP as described in the text. The autophosphorylated insulin receptors are then immunoprecipitated and the dephosphorylation reaction is initiated by adding liver extracts (cytosolic or particulate fraction) which serve as the PTP source. The samples are analyzed by 7.5% SDS–polyacrylamide gel electrophoresis, followed by Western blotting with anti-P-Tyr antibodies. The 95-kDa band corresponding to the insulin receptor β subunit is marked with an arrow. The labeled band of 50 kDa corresponds to the heavy chain of the anti-P-Tyr antibody that reacts with the second antibody (in this case, ^{125}I-goat anti-rabbit antibody).

rylated insulin receptors are then immunoprecipitated as described above with anti-receptor antibodies. The immune precipitates are suspended in 60 μl of Buffer B (without glycerol and PMSF). The reaction is initiated by adding 20 μl of liver extracts (cytosolic or particulate fraction, ~6 mg protein/ml each) which serve as the PTP source, 20 μl of Buffer F is added to the control tubes. The dephosphorylation is allowed to proceed for 5–45 min at 30°C and is terminated by adding (×5 concentrated) Laemmli sample buffer and heating for 10 min at 95°C. The samples are analyzed by 7.5% SDS–polyacrylamide gel electrophoresis, followed by Western blotting with anti-P-Tyr antibodies. The 95-kDa band corresponding to the insulin receptor β

subunit is further analyzed by densitometry. Example of such an assay is presented in Fig. 3.

References

Begum, N., Graham, A., Sussman, K., and Draznin, B. (1992). *Am. J. Physiol.* **262**, E142–E149.
Bernier, M., Laird, D. M., and Lane, M. D. (1987). *Proc. Natl. Acad. Sci. U.S.A.* **84**, 1844–1848.
Cantley, L. C., Auger, K. R., Carpenter, C., Duckworth, B., Graziani, A., Kapeller, R., and Soltoff, S. (1991). *Cell* **64**, 281–302.
Caro, J. F., Ittoop, O., Pories, W. J., Meelheim, D., Flickinger, E. G., Thomas, F., Jenquin, M., Silverman, J. F., Khazamie, P. G., and Sinha, M. K. (1986). *J. Clin. Invest.* **78**, 249–258.
Cicirelli, M. F., Tonks, N. K., Diltz, C. D., Weiel, J. E., Fischer, E. H., and Krebs, E. G. (1990). *Proc. Natl. Acad. Sci. U.S.A.* **87**, 5514–5518.
Comoglio, P. M., DiRenzo, M. F., Tarone, G., Giancotti, F. G., Naldini, L., and Marchisio, P. C. (1984). *EMBO J.* **3**, 483–489.
Cool, D. E., Tonks, N. K., Charbonneau, H., Walsh, K. A., Fischer, E. H., and Krebs, E. G. (1989). *Proc. Natl. Acad. Sci. U.S.A.* **86**, 5257–5261.
Czech, M. P., Lawrence, J. C., Jr., and Lyne, W. S. (1974). *Proc. Natl. Acad. Sci. U.S.A.* **71**, 4173–4177.
David-Pfeuty, T., and Singer, S. J. (1980). *Proc. Natl. Acad. Sci. U.S.A.* **77**, 6687–6691.
Fischer, E. H., Charbonneau, H., and Tonks, N. K. (1991). *Science* **235**, 401–406.
Frangioni, J. V., Beahm, P. H., Shifrin, V., Jost, C. A., and Neel, B. G. (1992). *Cell* **68**, 545–560.
Gautier, J., Solomon, M. J., Booher, J. F., Bazan, J. F., and Kirschner, M. W. (1991). *Cell* **67**, 197–211.
Goldstein, B. J., Meyrovitch, J., Zhang, W. R., Backer, J. M., Csermely, P., Hashimoto, N., and Kahn, C. R. (1991). *Adv. Protein Phosphatases* **6**, 1–17.
Gould, K. L., Moreno, S., Tonks, N. K., and Nurse, P. (1990). *Science* **250**, 1573–1576.
Grunberger, G., Zick, Y., and Gorden, P. (1984). *Science* **223**, 932–934.
Guan, K., and Dixson, J. E. (1991). *J. Biol. Chem.* **266**, 17026–17030.
Haring, H. U., Kasuga, M., and White, M. F. (1984). *Biochemistry* **23**, 3298–3306.
Hecht, D., and Zick, Y. (1992). *Biochem. Biophys. Res. Commun.* **188**, 773–779.
Heffetz, D., and Zick, Y. (1986). *J. Biol. Chem.* **261**, 889–894.
Heffetz, D., and Zick, Y. (1989). *J. Biol. Chem.* **264**, 10126–10132.
Heffetz, D., Bushkin, I., Dror, R., and Zick, Y. (1990). *J. Biol. Chem.* **265**, 2896–2902.
Heffetz, D., Fridkin, M., and Zick, Y. (1991). *In* "Protein Phosphorylation," Part B (T. Hunter and B. M. Sefton, eds.), Methods in Enzymology, Vol. 201, pp. 44–53. Academic Press, San Diego.
Ingebristen, T. S. (1989). *J. Biol. Chem.* **264**, 7754–7759.
Ingebristen, T. S. (1991). *In* "Protein Phosphorylation," Part B (T. Hunter and B. M. Sefton, eds.), Methods in Enzymology, Vol. 201, pp. 451–465. Academic Press, San Diego.
June, C. H., Fletcher, M. C., Ledbetter, J. A., Schieven, G. L., Siegel, J. N., Phillips, A. F., and Samelson, L. E. (1991). *Proc. Natl. Acad. Sci. U.S.A.* **88**, 7704–7707.
Kadota, S., Fantus, G. I., Deragon, G., Guyda, H. J., and Posner, B. I. (1987a). *J. Biol. Chem.* **262**, 8252–8256.
Kadota, S., Fantus, G. I., Deragon, G., Guyda, H. J., Hersh, B., and Posner, B. I. (1987b). *Biochem. Biophys. Res. Commun.* **147**, 259–266.

Kasuga, M., Zick, Y., Blithe, D., Crettaz, M., and Kahn, C. R. (1982). *Nature (London)* **298**, 667–669.
King, M. J., and Sale, G. J. (1990). *Biochem. J.* **266**, 251–259.
Koretzky, G. A., Picus, J., Thomas, M. L., and Weiss, A. (1990). *Nature (London)* **346**, 66–68.
Koretzky, G. A., Picus, J., Schultz, T., and Weiss, A. (1991). *Proc. Natl. Acad. Sci. U.S.A.* **88**, 2037–2041.
Krueger, N. X., Streuli, M., and Saito, H. (1990). *EMBO J.* **9**, 3241–3252.
Kumagi, A., and Dunphy, W. G. (1991). *Cell* **64**, 903–914.
Laemmli, U. K. (1970). *Nature (London)* **227**, 680–684.
Le Marchand-Brustel, Y., Gremeaux, T., Ballotti, R., and Van-Obberghen, E. (1985). *Nature (London)* **315**, 676–679.
Liao, K., Hoffman, R. D., and Lane, D. M. (1991). *J. Biol. Chem.* **266**, 6544–6553.
Meyrovitch, J., Farfel, Z., Suck, J., and Sechter, Y. (1987). *J. Biol. Chem.* **262**, 6658–6662.
Meyrovitch, J., Backer, J., and Kahn, C. R. (1989). *J. Clin. Invest.* **84**, 976–983.
Mooney, R. A., and Anderson, D. L. (1989). *J. Biol. Chem.* **264**, 6850–6857.
Moreno, S., and Nurse, P. (1991). *Nature (London)* **351**, 194–196.
Nadiv, O., Shinitzky, M., Manu, H., Hecht, D., Roberts, Jr., C. T., LeRoith, D., and Zick, Y. (1993). Abstracts of the V. International Symposium on Insulin Receptors and Insulin Action. Munich May 4–7.
Odawara, M., Kadowaki, T., Yamamoto, R., Shibasaki, Y., Tobe, K., Accili, D., Bevins, C., Mikami, Y., Matsuura, N., Akanuma, Y., Takaku, F., Taylor, S. I., and Kasuga, M. (1989). *Science* **245**, 66–68.
Ostergaard, H. L., Shackelford, D. A., Hurley, T. R., Johnson, P., Hyman, R., Sefton, B. M., and Trowbridge, I. S. (1989). *Proc. Natl. Acad. Sci. U.S.A.* **86**, 8959–8963.
Pallen, C. J., and Tong, P. H. (1991). *Proc. Natl. Acad. Sci. U.S.A.* **88**, 6996–7000.
Rees-Jones, R. W., and Taylor, S. I. (1985). *J. Biol. Chem.* **260**, 4461–4467.
Relech, S. L., Meier, K. E., and Krebs, E. G. (1986). *Biochemistry* **25**, 8348–8353.
Roome, J., O'Hare, T., Pilch, P. F., and Brautigen, D. L. (1988). *Biochem. J.* **256**, 493–500.
Sale, G. J. (1991). *Adv. Protein Phosphatases* **6**, 159–186.
Sap, J., D'Eustachio, P., Givol, D., and Schlessinger, J. (1990). *Proc. Natl. Acad. Sci. U.S.A.* **87**, 6112–6116.
Shen, S.-H., Bastien, L., Posner, B. I., and Chretein, P. (1991). *Nature (London)* **352**, 736–739.
Stover, D. R., Charbonneau, H., Tonks, N. K., and Walsh, D. A. (1991). *Proc. Natl. Acad. Sci. U.S.A.* **88**, 7704–7707.
Streuli, M., Krueger, N. X., Tsai, A. Y. M., and Saito, H. (1989). *Proc. Natl. Acad. Sci. U.S.A.* **86**, 8698–8702.
Streuli, M., Krueger, N. X., Thai, T., Tang, M., and Saito, H. (1990). *EMBO J.* **9**, 2399–2407.
Strout, H. V., Vicario, P. P., Saperstein, R., and Slater, E. E. (1988). *Biochem. Biophys. Res. Commun.* **151**, 633–640.
Sun, X. J., Rothenberg, P., Kahn, C. R., Backer, J. M., Araki, E., Wilden, P. A., Cahill, D. A., Goldstein, B. J., and White, M. F. (1991). *Nature (London)* **352**, 73–77.
Swarup, G., Cohen, S., and Garbers, D. L. (1982). *Biochem. Biophys. Res. Commun.* **107**, 1104–1109.
Tamura, S., Brown, T. A., Whipple, J. H., Fugita-Yamaguchi, Y., Dulber, R. E., Cheng, K., and Larner, J. (1984). *J. Biol. Chem.* **259**, 6650–6658.
Tonks, N. K., Diltz, C. D., and Fischer, E. H. (1988). *J. Biol. Chem.* **263**, 6731–6737.
Tonks, N. K., Cicirelli, M. F., Diltz, C. D., Krebs, E. G., and Fischer, E. H. (1990a). *Mol. Cell. Biol.* **10**, 458–463.
Tonks, N. K., Diltz, C. D., and Fischer, E. H. (1990b). *J. Biol. Chem.* **265**, 10674–10680.
Tonks, N. K., Diltz, C. D., and Fischer, E. H. (1991a). *In* "Protein Phosphorylation," Part B

(T. Hunter and B. M. Sefton, eds.), Methods in Enzymology, Vol. 201, pp. 442-451. Academic Press, San Diego.
Tonks, N. K., Diltz, C. D., and Fischer, E. H. (1991b). *In* "Protein Phosphorylation," Part B (T. Hunter and B. M. Sefton, eds.), Methods in Enzymology, Vol. 201, pp. 427-442. Academic Press, San Diego.
Trowbridge, I. S. (1991). *J. Biol. Chem.* **266**, 23517-23520.
Ullrich, A., and Schlessinger, J. (1990). *Cell* **61**, 203-212.
Volberg, T., Geiger, B., Dror, R., and Zick, Y. (1991). *Cell Regul.* **2**, 105-120.
Volberg, T., Zick, Y., Dror, R., Sabanay, I., Gilon, C., Levitzki, A., and Geiger, B. (1992). *EMBO J.* **11**, 1733-1742.
White, M. F., Maron, R., and Kahn, C. R. (1985). *Nature (London)* **318**, 183-186.
White, M. F., Shoelson, S. E., Keutmann, H., and Kahn, C. R. (1988). *J. Biol. Chem.* **263**, 2969-2980.
Zick, Y. (1989). *Crit. Rev. Biochem. Mol. Biol.* **24**, 217-269.
Zick, Y., Kasuga, M., Kahn, C. R., and Roth, J. (1983). *J. Biol. Chem.* **258**, 75-80.

one isoform does not appear to be activated by diacylglycerols or phorbol esters (Nishizuka, 1988; Bell and Burns, 1991; Stabel and Parker, 1991). Other differences exist among the various isoforms particularly with respect to tissue-specific expression, but also with respect to substrate specificity, catalytic behavior, response to autophosphorylation, response to downregulation by phorbol esters, and presumably other responses. This makes evaluation of PKC activation in cells an extremely tenuous business at the moment; it is made even more problematical by the fact that new isoforms are being discovered all the time with some differences in the properties alluded to above. Nevertheless, given these complexities and uncertainties, it is still possible to draw some conclusions about PKC involvement in signaling processes using a variety of experimental techniques. In this chapter, I describe some of the techniques in current use in our laboratory for evaluating PKC involvement in cellular processes; however, it should be borne in mind that these techniques may be superseded by others as new isozymes and their properties are discovered.

I. Measuring Activation of Protein Kinase C in Intact Cells

With many protein/serine/threonine kinases, it is relatively easy to measure the activation state in intact cells after stimulation with agonists. A commonly employed method is to expose cells to a hormone for various lengths of time, break open the cells, and then measure the increase in kinase activity resulting from the hormone stimulation; this is possible because hormone stimulation leads to the covalent modification of the specific kinase under study, and, in many cases, specific peptide substrates are also available. Examples of kinases for which this approach is useful are the mitogen-activated protein (MAP) or extracellular-signal-regulated kinases (ERK), the 70 and 90K ribosomal protein S6 kinases, and many others. In these cases, the addition of hormones to cells leads to the phosphorylation and stable activation of these kinases, which is maintained once the cells are broken and can be assayed by traditional phosphotransferase reactions using relatively specific substrates. Another method for assaying kinase activity is used in the setting of hormone-induced dissociation of a protein kinase complex, as is seen with the cyclic AMP-dependent protein kinase; in this case, activation of the kinase is the result of dissociation of the regulatory subunits from the catalytic subunits, and these free catalytic subunits can be measured in appropriate cell preparations following stimulation of the cells. A third method in frequent use for other serine/threonine kinases is to use electrophoretic analysis in combination with Western blotting to measure covalent modification of the kinases, i.e., autophosphorylation, leading to an electrophoretic mobility shift in the kinase itself. This has been particularly useful in the case of the MAP kinases alluded to above and the *raf*-1 protein kinase; in each case, a decrease in electrophoretic mobility occurs in response to activation of the kinase, which in turn is due to covalent modification by multiple-site phosphorylation.

None of these methods is particularly useful in the case of the various isozymes of PKC. For example, although at least some isoforms undergo autophosphorylation upon activation, there is not generally a mobility shift or a stable increase in activity that can be measured as an index of activation. There is no dissociation from regulatory subunits, as occurs in the case of the cyclic AMP-dependent protein kinase; the generation of a constitutively activated catalytic subunit, the so-called protein kinase M, from intact PKC has been said to occur in certain cell types, such as neutrophils, but it appears that this is not a common or general mechanism by which PKC is activated in other cell types.

Therefore, other more indirect measures must be used for PKC activation in intact cells. One of these involves the measurement of the so-called translocation of the kinase, in which specific isoforms of the kinase change from a predominantly cytosolic form to a predominantly membrane-associated form and can be assayed in these locations either immunologically or by measuring phosphotransferase activities. This technique has a number of limitations, including the fact that a typical cell contains four or more PKC isoenzymes, and thus translocation of enzyme activity may or may not be an expression of all of the isoforms in a given cell, given differences in substrate specificity of the different isoenzymes. A limitation of the immunological techniques is that, in the case of the typical cell with four or five different isoenzymes, antibodies to each isoenzyme must be used to document the translocation of one or more isoenzymes in response to the agonist. A final problem is that translocation of PKC could conceivably occur in the absence of activation; conversely, activation of at least some isoforms of PKC can occur in the absence of demonstrable translocation of various immunoreactive species. Therefore, although the technique of assessing PKC translocation is of some use, particularly with respect to measuring phorbol ester-induced translocation, it is limited by some of the problems listed above.

A second technique that we and others have used is the evaluation of specific PKC substrates in intact cells after exposure of the cells to hormones. This has a number of advantages over the translocation assay including sensitivity, at least in certain cell types; relative convenience, particularly when immunoprecipitation of ^{32}P-labeled proteins is carried out; and the fact that what is being measured is the ultimate result of protein kinase activation, i.e., phosphorylation of specific substrates. However, this technique is subject to one of the same major limitations as translocation; that is, the behavior of most known PKC substrates has not yet been studied in detail for all 10 or so isoenzymes, and it is possible that certain generally used PKC substrates will not be good substrates for individual isoforms and therefore will not prove useful as indices of activation of these isoforms.

A. Measuring PKC Translocation by Enzyme Activity

We routinely use a technique originally described by Glynn et al. (1985) and modified by us (Halsey et al., 1987) to measure the subcellular localization of PKC and its changes in response to agonists. This assay relies on the PKC-mediated phosphorylation of an N-bromosuccinamide (NBS) fragment of histone IIIS, from which has been removed the major phosphorylation sites for other protein kinases in crude cell extracts. This allows PKC activity to be measured in crude cell extracts without anion exchange column fractionation of the extracts. However, it relies on the fact that histone IIIS is a substrate for the PKC in the reaction mixture. This protein does not appear to be a good substrate for some of the calcium-independent PKCs. In addition, we generally assay the phosphotransferase activity of a cell extract in the presence or absence of Ca^{2+} and phosphatidylserine, thus failing to differentiate among the Ca^{2+}-dependent and independent isozymes. Therefore, what this assay is probably measuring is predominantly PKC-α, since β and γ isoforms are lacking in most of the cultured cells that we study routinely.

In a typical experiment, cells are deprived of serum for approximately 16 to 18 hr by incubating them in Dulbecco's modified Eagle's medium (DMEM) plus 1% bovine serum albumin (BSA; crystallized and lyophilized, Sigma Chemical Co., St. Louis, MO). We routinely check each lot of BSA used in our serum-deprivation medium for growth factor activity, since this can vary from lot to lot; obviously the presence of growth factor or insulin activity in the BSA could lead to lack of quiescence. In general, fibroblasts in a confluent state and deprived of serum overnight will be at or near G_o, with very low basal levels of PKC activity; however, this technique also seems to work for a variety of cells, including transformed cells, in which true quiescence is not achieved, but in which PKC seems to decrease in activity in response to the serum deprivation. Following this period of serum deprivation, the cells are exposed to hormones or buffer for variable lengths of time. It is always advisable to include in the experiment a positive control in the form of an active phorbol ester (e.g., 1.6 μM phorbol-12-myristate-13-acetate or PMA). After hormone exposure, the cells are washed rapidly three times with 4 ml of ice-cold phosphate-buffered saline (PBS) and then scraped into a small volume, usually less than 1 ml, of an ice-cold homogenization buffer: 100 mM β-glycerol phosphate, pH 7.5, 0.25 M sucrose, 2 mM EGTA, 2 mM EDTA, 2 mM dithiothreitol (DTT). The cells are homogenized using a Dounce homogenizer with 10–30 strokes, depending on tightness of fit, and then diluted rapidly with extra homogenization buffer as needed to maintain a constant protein concentration between 0.05 and 0.1 mg/ml in the resulting supernatant fraction. If cytosol vs membrane ratios are to be compared among samples, then it is very important to establish equal protein concentrations in the crude cellular homogenate at this point; similarly, care must be taken to wash out most of the BSA in the incubation medium so that it does not influence the protein concentration. Following equalization of the protein content of the crude homogenates, the samples are centrifuged at 200,000g for 1 hr at 4°C, and the supernatant fractions are then used directly for the assay of PKC and protein concentration. The membrane pellets are resuspended in the original volume of homogenization buffer containing 0.3% (v/v) Triton X-100; this can be done by gentle sonication or by passing the samples 10 times through a 25-gauge needle. This suspension is then incubated on ice for 30 min with frequent mixing, followed by centrifugation at 12,000g for 15 min at 4°C. This supernatant is also then used for the measurement of protein concentration and PKC activity.

We routinely measure PKC activity in these extracts without column chromatography, since many of the nonspecific or calcium and phospholipid-independent kinases do not phosphorylate the histone NBS peptide (details of peptide preparation and purification are given in Halsey et al., 1987). This can be seen by the extremely low reaction rates seen in the absence of added calcium and phosphatidylserine added to the reaction. A typical PKC assay is performed at 30°C in a final reaction volume of 40 μl containing final concentrations: 100 μM [γ-^{32}P]ATP (specific activity between 1 and 3000 cpm/pmol), 10 mM $MgCl_2$, 0.5 mM EDTA, 0.5 mM EGTA, 0.5 mg/ml histone NBS in 20 mM Hepes HCl (pH 7.5). The assays are performed in the presence or absence of 100 μM phosphatidylserine (PS) plus 1.5 mM $CaCl_2$, or in the absence of these two compounds; usually duplicate assays for the $+PS + CaCl_2$ and $-PS - CaCl_2$ conditions are used and averaged. If the protein content of the cell extract is well below 1 mg/ml, the reactions are usually linear well beyond 15 min, but this should be checked in each case since some cells contain much more PKC than others. In general, incubations are carried out for 15 min, at which point 20 μl of the reaction mixture is spotted onto a square of P81 paper (Whatman) and the protein is precipitated by immersing the papers in 25% (w/v) trichloroacetic acid (TCA) at 4°C. The papers are then washed for 1 hr at 4°C in the same solution, followed by four, 30-min washes in 20% TCA at 4°C. The papers are then washed sequentially in acetone and petroleum ether for 5 min each at room temperature, then dried under a lamp and subjected to scintillation counting in water or in fluor, depending on the number of counts per minute involved. PKC activity is then defined as the kinase activity present when both phospholipid and calcium are present in the reaction mixture minus the kinase activity present when these factors are absent. Using the concentrations of phosphatidylserine mentioned above, the addition of diolein leads to only a small increase in the protein kinase C activities and is not routinely used in these assays. However, when assays involve both the supernatant and the extracted particulate fraction, much higher lipid concentrations are generally used (600 μM phosphatidylserine plus 80 μM diolein); these high concentrations of lipids were found to be necessary to obtain optimal PKC reactivity in the presence of 0.3% Triton X-100. At least in our hands, Nonidet P-40 at a variety of concentrations was inhibitory to PKC activity, and its use is best avoided.

The substrate for these assays, histone NBS, can be prepared easily using commercially available histone IIIS (Sigma Chemical Co.) as described previously (Glynn *et al.*, 1985; Halsey *et al.*, 1987). Alternatively, if the activity of specific PKC isoforms is desired then specific peptide substrates for these isoforms can be synthesized and used in similar P81 filter paper assays. In some cases, using the smaller substrates, phosphoric acid precipitation and/or washes can be used in place of TCA precipitation on the P81 papers. Using these techniques, we have studied a wide variety of cells and shown that in the cytosolic fraction virtually all of the histone NBS kinase activity is calcium and phospholipid-dependent (this specific activity minus the calcium- and phospholipid-independent activity is referred to as PKC). In the particulate fraction, there is somewhat more histone NBS kinase activity in the calcium- and phospholipid-independent form; however, the majority of this activity is still calcium and phospholipid dependent (Halsey *et al.*, 1987). As a positive control, virtually all of the PKC activity in the cytosol should be rapidly displaced to the particulate fraction by phorbol ester treatment (Halsey *et al.*, 1987). This occurs within a minute or two of addition of active phorbol esters such as PMA to most cell types and results in quantitative removal of the enzyme activity from the cytosolic to the membrane fractions. If desired, the EGTA and EDTA can be left out of the original homogenization buffer and a small amount of calcium (1 μM) included; this will permit the analysis of the larger quantity of PKC that exists in a calcium-dependent, loosely membrane-associated form that can be seen to change with less radical stimuli than with phorbol esters. An example of this is shown in Halsey *et al.* (1987), in which the cell-permeable diacylglycerol diC8 was used to cause modest changes in cytosol:membrane partitioning of PKC activity in NIH 3T3 fibroblasts. The most useful comparison is the volumetric comparison between the cytosol and the particulate fractions, assuming that the samples were normalized for protein concentration before centrifugation; however, if desired, specific enzyme activity can also be determined in the individual fractions.

It is often quite useful to assess total cellular PKC activity directed at histone NBS, particularly when downregulation of PKC is being assessed. For this purpose, we often used a modification of this assay in which the original homogenization buffer described above is used, but which also contains 0.3% Triton X-100 (Halsey *et al.,* 1987). In this situation, the samples are homogenized in this buffer by douncing, then allowed to cool on ice for 30–45 min with frequent mixing in the presence of 0.3% Triton X-100 to extract the membrane-associated PKC activity. Samples are then centrifuged at 200,000g for 1 hr at 4°C, and the resulting supernatants are used in the assay of histone NBS kinase activity, using the high lipid concentrations described above. The particulate fraction has been shown to contain little or no detectable PKC activity or immunoreactivity, although these experiments were performed before many of the current antibodies to isoenzymes were available. Nonetheless, we believe that the Triton-extracted supernatant fraction contains most, if not all, of the PKC immunoreactivity and enzyme activity and can be used to assess total cellular PKC activity. This is most useful for documentation of PKC downregulation (see below); however, it has been useful in our hands for comparing PKC contents of different cell types, in various tissues in transgenic animals, and in other situations in which total cellular PKC activity is desired.

B. Measuring Translocation by Immunoassay

Measurement of PKC enzyme activity, even with a variety of peptide substrates, is still somewhat compromised by the fact that many cells express four or more PKC isozymes, each with different calcium concentration requirements, substrate specificities, etc. For this reason, translocation assays involving specific measurement of PKC isozymes by their immunoreactivity have several advantages over measurements of enzyme activity. One difficulty with this type of assay is that specific antibodies are required for each PKC isozyme under study, and species differences among animals sometimes requires the development of species-specific anti-peptide antibodies for specific cell types. Nonetheless, various investigators have developed panels of antibodies to nine or more PKC isotypes, and these are widely available for use in such experiments (Borner *et al.,* 1992; Wetsel *et al.,* 1992); some are also available commercially from suppliers such as GIBCO/BRL (Gaithersburg, MD). Similar techniques can be used to assess the effect of phorbol ester "downregulation" on the cellular concentration of PKC isozymes. In general, detection of PKC isoenzymes by Western blot is less sensitive than the measurement of enzyme activity described above, but it obviously lends an element of specificity to the assay. In practice, it is easiest to first determine the isozymes that are expressed in an individual cell type, so that the number of Western blots can be kept to a minimum in an individual experiment. For example, in a recent study (Borner *et al.,* 1992), the rat embryo fibroblasts used in their studies appeared to express α, δ, ϵ, and ζ but not β or γ. Therefore, they were able to limit their subsequent studies to measurement of mRNA and protein levels to these four isozymes by Northern and Western blotting analysis, respectively.

For measurement of PKC immunoreactivities, we generally use the homogenization buffer alluded to above for PKC assays, match the cellular homogenates for protein concentration prior to centrifugation for the reasons alluded to above, and then, immediately after centrifugation, prepare these supernatant fractions for electrophoresis by adding at least 2% final concentration SDS, dithiothreitol, sucrose, and pyronin Y and then boiling the samples for 3–5 min in preparation for electrophoresis. The particulate fractions can then be resuspended directly in SDS sample buffer of the same volume, or resuspended and extracted in Triton-containing buffers, centrifuged as described above, and then prepared for SDS electrophoresis. Equal volumes can then be loaded on SDS gels, transferred to nitrocellulose by routine techniques, and then immunoblotted with the specific antiserum under question; to determine specificity of the reaction, nonimmune or preimmune serum should be used in parallel reactions, and blocking with the specific peptides against which the antibodies were raised also supports the specificity of the immunological interactions. Once again, a positive control can be used in the form of PKC-α translocation in response to phorbol esters, which is virtually complete after 5 min exposure to 1.6 μM PMA. The above homogenization technique has an inherent disadvantage that relatively long periods time, i.e., 1 hr, are required for the initial centrifugation. This is particularly important since some of the PKC isozymes, notably ϵ and ζ seem to be susceptible to proteolysis even at 4°C. For this reason, and in certain circumstances, it is desirable to include panels of protease inhibitors in the homogenization buffer; in other circumstances, such as the assessment of downregulation of these isotypes, it is a good idea to harvest the cells directly in hot SDS sample buffer to minimize the possibility of subsequent proteolysis.

To date, no listing of commonly used cell types and their expressed PKC isozymes has been published. However, within a reasonable period of time, the cellular content and subcellular distribution of all known PKC isozymes should be published for many commonly used cell types, and investigators wishing to assess translocation or downregulation using isotype-specific antibodies will only need to investigate the isozymes present in that particular cell type.

C. Evaluation of PKC Activation by Measuring Substrate Phosphorylation

The measurement of PKC substrate phosphorylation after stimulation of intact cells with a given agonist has a number of advantages over the translocation assays described above. It does not involve any assumptions concerning the role of cytosolic vs chelator-displaceable vs detergent-displaceable PKC activity; it can be seen in circumstances in which translocation of PKC isozymes is not seen; it does not require the availability of antibodies to 9 or 10 PKC isozymes; and, given an appropriate substrate, it has the potential for much greater sensitivity than the translocation assays. On the other hand, disadvantages to this approach include the fact that there are only a few well-characterized PKC substrates that will serve this purpose well; some cell types do not express the well-characterized substrates at all or to any significant extent; either two-dimensional gels or antibodies are usually required for assessment of phosphorylation status of these substrates; the phosphorylated substrates themselves are sensitive to phosphatase activity, so that inhibition of phosphatases could conceivably result in elevated levels of phosphorylation rather than activation of PKC; and finally and most significantly, the specificity of substrate phosphorylation by all of the available PKC isozymes has not yet been determined. Nonetheless, within the limits of these caveats, evaluation of substrate phosphorylation is, in my view, the best means of assessing whether or not a given agonist activates the diacylglycerol and Ca^{2+}-dependent forms of PKC in a given cell type. In particular, a positive result with a specific substrate leads directly to the conclusion that PKC has been activated, and taken together with data on inositol phospholipid turnover and generation of DAG and/or IP3, leads to firm conclusions about the activation of the kinase. A negative result, however, does not necessarily exclude protein kinase activation, given the possibility that some subspecies of PKC could be activated that do not phosphorylate an individual cellular substrate. As in the case of the translocation assay described above, it is hoped that the substrate specificities of the various proteins for different PKC isozymes will soon be determined.

In brief, the approach involves incubating cells overnight in serum-free medium as described above, then incubating them with ^{32}P for 2–4 hr to label the cellular ATP pools to near constant specific activity, stimulating the cells for a relatively short period with the agonist under study, and then measuring the phosphorylation state of specific PKC substrates, taking precautions to inhibit cellular phosphatases during homogenization and sample preparation. Perhaps the best-characterized substrate for this type of assay is the so-called MARCKS protein, an acronym standing for myristoylated alanine rich C kinase substrate (Niedel and Blackshear, 1986; Woodgett et al., 1987). This protein has been studied intensively as a substrate for the PKC mixture that exists in rat brain, from which conventional PKC is usually purified. It is an extremely high-affinity substrate for this mixture of kinases, with K_m in the 20 to 100 nM range (Graff et al., 1991). In intact cells, it can be phosphorylated within seconds of hormone stimulation, and its dephosphorylation occurs rapidly upon addition of receptor antagonists in some systems (Rodriguez-Pena et al., 1986). So far as is known at the time of this writing, it is a specific substrate for PKC in intact cells; although it can be phosphorylated, for example, by the cyclic AMP-dependent protein kinase *in vitro;* this occurs with a very low affinity (Graff *et al.,* 1991), and the protein does not appear to be a substrate for this kinase in intact cells. The protein contains no tyrosines in its primary sequence, so it is not a substrate for cellular tyrosine kinases, and it has been shown to be a relatively poor or nonsubstrate for the calcium/calmodulin-dependent kinases, cyclic GMP-dependent protein kinase, and a variety of others (Graff *et al.,* 1991). Obviously, given the rapid discovery of many new serine/threonine kinases, it can still not be concluded safely that it is not a substrate for any other protein kinases; however, at the time of this writing, it appears to be specific for PKC in cells.

A more serious problem, which has not been resolved as of this writing, is the behavior of this protein as a substrate for each of the PKC isozymes. It can be phosphorylated by PKC-ϵ with nearly identical affinity to that seen with the mixture of brain isozymes commonly employed (Blackshear *et al.,* 1991); if PKC-ϵ is taken as a representative of the calcium-independent species of PKC, it suggests that MARCKS is a substrate for most, if not all, of the subspecies. We are currently attempting to determine precise kinetics of phosphorylation for each of the purified and/or expressed PKC isozymes, but these data are not yet available for all of the subspecies. For these reasons, as alluded to above, a positive result can be viewed with confidence as an indicator of PKC activation but a negative result is not conclusive as to lack of activation of PKC in a given circumstance. However, taken together with data on the absence of phosphatidylinisitol (PI) turnover and the absence of generation of IP_3 and/or DAG, a negative result in these phosphorylation assays generally can be assumed to represent lack of PKC activation in intact cells.

Another potential problem with using a specific protein such as MARCKS as an indicator of PKC activation is that certain cell types of interest do not express the protein at all. In our hands, mouse L-cells and rat hepatoma cells of the H4 or H35 series do not express detectable levels of the mRNA or protein. However, in most circumstances, this can be circumvented using either transient or stable transfection protocols, in which expression vectors are transfected into the cells and the expressed protein is then used as an indicator of PKC activation. An example of this transfection approach is described in Blackshear *et al.* (1991).

In general, two different approaches can be used to separate the phosphorylated MARCKS protein from other cellular proteins. The most straightforward involves two-dimension gel electrophoresis to clearly separate the protein from other cellular phosphoproteins; given the characteristic appearance of this protein, its characteristic anomalous migration on SDS gels, its extremely acidic isoelectric point, and its heat stability, the protein is easily separated from other phosphoproteins and its phosphorylation status can readily be assessed by this means (Blackshear et al., 1985, 1986). However, two-dimensional gels are somewhat cumbersome to run, particularly in experiments in which large numbers of samples are required; in addition, it is often difficult to compare the phosphorylation state of the stimulated protein with that of the control protein, unless the spots are cut out of the gel and counted, or sophisticated densitometry is available for the analysis of two-dimensional autoradiographs. For this reason, immunoprecipitation of the phosphorylated protein is preferred, but obviously this requires the availability of a specific antibody. A number of anti-MARCKS antibodies have been developed in our laboratory and others, including polyclonal antisera to the intact chicken protein (Graff et al., 1989); a monoclonal antibody to the human protein, which does not react with proteins from most other species; and a polyclonal antipeptide antiserum raised against conserved domains of the protein, which are identical in most animal species (Lobaugh and Blackshear, 1990). Such antibodies can be used to immunoprecipitate phosphorylated protein from virtually all cell types. To our knowledge, none of these antibodies is at present commercially available. For this reason, I describe both the two-dimensional electrophoretic approach as well as the immunoprecipitation protocol used in our laboratory for determining phosphorylation of this protein. It should be emphasized that this is only one of a number of potential substrates for protein kinase C suitable for study in this way; another example is a recently described MARCKS homologue, F52 (or MacMARCKS), which shares a number of primary sequence similarities with the MARCKS protein but which seems to be expressed at a somewhat lower level in some of the cell types that we have studied (Umekage and Kato, 1991; Blackshear et al., 1992; Li and Aderem, 1992). However, it has been estimated that PMA stimulates the phosphorylation of at least 120 proteins in NIH 3T3 cells, for example, many of which are probably direct or primary substrates for PKC; any one of these could serve as an assay for PKC activation in intact cells, provided that a number of studies are carried out to determine its specificity for PKC, sensitivity, ease of separation from other cellular phosphoproteins, presence in a wide variety of cells, and other characteristics which make the MARCKS protein particularly suitable for this type of assay.

Our protocol for serum deprivation is the same as that described above for PKC enzyme assays. After 16 to 18 hr in serum-free medium, monolayer cells are generally washed three times at 37°C with 4 ml of Krebs buffer (118 mM NaCl, 4.7 mM KCl, 1.4 mM CaCl$_2$, 1.2 mM MgSO$_4$, 25 mM NaHCO$_3$, 10 mM Hepes (pH 7.5), and 25 mM D-glucose) and then incubated for 2–4 hr in the same buffer containing 1% bovine albumin and between 0.1 and 1 mCi/ml of ^{32}P. The BSA used in this buffer should be the same growth factor-free lyophilized and crystallized product mentioned above. In addition, the use of ^{32}P-free medium is recommended in most circumstances to increase the specific activity of the phosphoproteins; some cells do not tolerate this, in which case small amounts of phosphate can be included in the Krebs incubation medium. Following the incubation with ^{32}P, to the cells is added the agonist of interest without changing the medium, since abrupt changes in pH, temperature, or tonicity can cause activation of PKC. Usually the additions are made in a small volume of Krebs incubation buffer, and concurrent controls are included in which a small volume of the buffer is added without agonists. After a variable period of time, usually 15 min for most agonists, the ^{32}P-containing medium is rapidly aspirated, and the cells are washed quickly two or three times with ice-cold phosphate-buffered saline or ice-cold Krebs buffer without BSA. At the end of the final wash, to the cells is added one of two homogenization buffers. If the cells are to be used for two-dimensional gel electrophoresis, the buffer includes 50 mM Tris–HCl (pH 7.0), 10 mM benzamidine, 1 mM EGTA, 100 mM NaF, 5 mM dithiothreitol, 1.5 mM MgCl$_2$, 0.25 M sucrose, 2% (v/v) Triton X-100. In this case, the cells are scraped into a small volume (usually less than 1 ml for a 10-cm dish of cells) and either homogenized or incubated on ice for 30–45 min with frequent mixing. The Triton X-100 in the incubation medium suffices both to lyse the cells and to release all membrane-associated MARCKS protein into the supernatant. The samples are then centrifuged at 68,000g for 45 min at 4°C. To an aliquot of the supernatant is then added (final concentration) 20–25% (w/v) TCA; this is incubated on ice for a further 30 min, followed by another centrifugation step (12,000g for 15 min at 4°C). The TCA pellets are then washed with acetone at −20°C two or three times to remove residual TCA; the precipitated proteins are then resuspended in a 1:1 mixture of 9 M urea:9 M urea, 2% Nonidet-P40 (NP-40), 5% 2-mercaptoethanol, 2% ampholytes in the pH range desired. Using a small Teflon pestle and conical centrifuge tube, the TCA-precipitated proteins are dissolved into this two-dimensional gel buffer; often mild heating (i.e., to 30–35°C) will help this solubilization. After the pellets are solubilized and the samples are centrifuged briefly to remove further insoluble contaminants, the supernatants are matched for TCA-precipitable radioactivity, since the urea and detergent will interfere with a simple protein determination. For 2D gel electrophoresis, careful matching of the TCA-precipitable radioactivity is one of the most important steps in determining whether a significant stimulation of protein phosphorylation has occurred. Following this matching, the samples are loaded onto one-dimensional isoelectric focusing gels according to standard techniques. Following the two-dimensional electrophoresis, usually conducted on a 7.2% SDS–polyacrylamide gel, the gels are stained and/or dried directly and used for autoradiography. The MARCKS protein can be identified by comigration with authentic standards; its characteristic appearance, anomalous migration on SDS gels, heat stability, acidic isoelectric point, and other properties usually suffice to make unambiguous identification possible without standards.

If samples are to be prepared for immunoprecipitation, then the cells are scraped into a RIPA buffer consisting of 50 mM Tris–HC1 (pH 8.3), 1% NP-40, 5 mM EDTA, 0.1 mM NaC1, 0.05 mM NaF. They are once again incubated on ice in a volume of 0.5 to 1 ml/10-cm plate for 30–45 min with frequent mixing followed by centrifugation at 68,000g for 30 min at 4°C. The supernatant of this centrifugation is then used for immunoprecipitation of the MARCKS proteins. At this point, matching of the samples is necessary before addition of antibody; this can best be done by matching TCA-precipitable radioactivity. We routinely add 3–5 μl of sample to two or three tubes containing 0.4 ml of 1 mg/ml bovine γ-globulin, then add 0.1 ml of 100% (w/v) TCA, mix, and incubate on ice for 30 min. The samples are then centrifuged (12,000g for 15 min at 4°C), and the pellets counted for Cerenkov radioactivity. All samples are then adjusted to equal numbers of TCA-precipitable cpm by adding additional RIPA buffer. Matching can also be performed using protein concentration; however, this is less accurate, given the effects of the BSA in the incubation buffer on the protein concentration, as well as the fact that the detergents used in the RIPA buffer often interfere with protein determination. Once the supernatant fractions are matched exactly for TCA-precipitable radioactivity, then a preclearing step is usually employed, in which preimmune or nonimmune serum or nonimmune ascites is used together with Pansorbin (Calbiochem) or a second antibody. The preclearing step is usually accomplished by incubation with the sample at 4°C for 0.5 to 1 hr followed by addition of Pansorbin or second antibody-coated Pansorbin and centrifugation. To the second supernatant is then added the bona fide MARCKS antiserum or monoclonal antibody, followed usually by an overnight incubation at 4°C. To the immune complexes is then added Pansorbin with or without second antibody; the complexes are precipitated by centrifugation, washed several times in an ice-cold buffer consisting of 50 mM Tris–HCl (pH 8.3), 0.10 M NaCl, 0.05 M NaF, 1 mM EDTA, and 0.5% NP-40; the precipitates are then solubilized into SDS sample buffer with frequent mixing, as well as boiling for 5 min. The last step is centrifugation at 12,000g for 5 min prior to loading the supernatants on one-dimensional SDS gels, usually 7.2% acrylamide. If the samples are carefully matched for TCA-precipitable radioactivity prior to the beginning of the first immunoprecipitation step, then any difference in MARCKS radioactivity in the immunoprecipitated samples indicates activation of PKC. Since rather long incubation periods are required with the antibody, even at 4°C, phosphatase inhibitors (50 mM sodium fluoride) must be included in the incubation buffers; for some cell types that are particularly prone to proteolysis, inclusion of a variety of protease inhibitors is also recommended, although this is not generally necessary for fibroblasts or similar cells if the temperature is kept at 4°C.

II. Down-Regulation of Protein Kinase C

One of the unusual features of many PKC isozymes which can be exploited in cell incubation studies is the phenomenon known as down-regulation of PKC in response to phorbol esters. This appears to be due primarily to enhanced proteolytic degradation of PKC species in response to high concentrations of phorbol esters. In many situations it is a convenient way to rid cells of most, if not all, of certain PKC isozymes, so that actions of specific hormones may be attributed to non-PKC signal transduction mechanisms. However, many of the same caveats noted above for various PKC isozymes apply to the down-regulation protocols used. Consequently, there are many disparate reports in the literature of differential susceptibility of different isozymes to down-regulating stimuli. In addition, many reports in the literature do not validate the down-regulation protocol in each individual cell type under the experimental conditions used, so that the extent of down-regulation of each isozyme is often not clear in a published report.

We have used essentially the same protocol in many cell types for the last 7 or 8 years and have found that this protocol consistently decreases PKC enzyme activity, as assayed by the histone NBS assay described above, to less than 5–10% of control in both membrane and cytosolic fractions. Similar decreases have been noted in specific phorbol ester binding, as well as PKC immunoreactivity with isotype-specific antibodies.

This protocol involves incubating cells with serum-free medium with 1% BSA for 18 hr as described above. During the last 16 hr of the incubation, to the cells is added a final concentration of 16 μM PMA in 0.1% dimethyl sulfoxide (DMSO). In practice this is easily accomplished by keeping frozen stock solutions of 16 mM PMA in DMSO; to the cell incubation medium is added a 1:1000 dilution of this concentration to make the final concentrations 16 μM and 0.1%, respectively. These conditions have resulted in nearly 100% down-regulation of PKC enzyme activity, assayed by the histone NBS method, PKC immunoreactivity, and specific phorbol ester binding. A major exception to this rule has been 3T3-L1 adipocytes, which contain extremely large amounts of lipid; we postulate that the hydrophobic phorbol esters bind to the lipid in these cells and thus are not available at equivalent concentrations to stimulate proteolysis of the PKC.

This protocol has been used so frequently in many types of cells (listed in Blackshear *et al.,* 1991) that it is possible, if it is followed identically, to assume that most PKC species have been down-regulated by this protocol. However, it is important for new cell types under study to validate the technique by one of several of methods; in addition, as novel PKC species are discovered, it is important to validate the downregulation protocol on each novel immunoreactive species. Enzyme activity can be checked using the histone NBS assay described above; in this case, the samples are homogenized in assay buffer containing 1% Triton X-100, which will solubilize essentially all PKC immunoreactivity of all species that we have checked. This supernatant fraction is then used in the high lipid assay to measure protein kinase C activity at least against this substrate. It is also possible to measure phorbol ester binding in the cells, which should pick up all known isozymes except PKC-ζ. However, when this technique is employed, it is important that phorbol dibutyrate (PDBu) (20 μM) be used as a down-regulating stimulus rather than PMA. If identical concentrations and conditions are employed PDBu can be washed out of cells following the down-regulation stimulus by six washes in incubation buffer without PDBu at 37°C over the span of approximately 1 hr, as originally described by Rozengurt and colleagues (Collins and Rozengurt, 1984; Rodriguez-Pena and Rozengurt, 1984). This will remove essentially all of the PDBu used as the down-regulating stimulus; however, similar extensive washing is not effective in removing PMA from cells. Following removal of the PDBu the cells can be used for determination of specific radiolabeled PDBu binding as described (Collins and Rozengurt, 1984; Rodriguez-Pena and Rozengurt, 1984). This is a rapid and accurate means of determining whether total cellular phorbol ester binding activity is markedly decreased following the PDBu preincubation. Since additional proteins besides PKC have been shown to bind phorbol esters in recent years, there are a number of theoretical difficulties with this procedure; nonetheless, following the down-regulation protocol described above, values of less than 3% of original phorbol ester binding are usually recorded in most cell types employed in our laboratory.

Probably the best way to evaluate down-regulation in the current era is to use isotype-specific antibodies to PKC. In this case, subtle changes in cellular translocation are not being investigated; for this reason, the cells can be homogenized directly in boiling SDS and a crude cellular homogenate used to determine the extent of down-regulation of PKC isozymes with the specific antibodies referred to above. In typical fibroblasts and several other cell types that we have studied, the down-regulation protocol described above causes a virtually complete disappearance of immunoreactive PK-α,-δ, and -ϵ; however, protein kinase C-ζ, which is widely expressed, does not seem to be affected at all by the down-regulation protocol described above. This is not particularly surprising, since this isotype does not appear to bind either phorbol esters or diacylglycerols (Nakanishi and Exton, 1992; Ways et al., 1992). Therefore, it would probably not be detected by the specific phorbol ester binding protocol described above, and perhaps not even by the PKC enzyme assay. In addition, since it cannot be activated by diacylglycerols, it is presumably not involved in PKC signaling by diacylglycerols generated in response to inositol phospholipid turnover. At the time of this writing, the activating ligand is not known, nor are preferred cellular substrates for this isotype known. It may be that this should not be included as a PKC subspecies, given that it is not a phorbol ester receptor; other kinases, such as the *raf*-1 protooncogene kinase, which contain cysteine-rich zinc finger structures but do not appear to bind phorbol esters, may be more suitable candidates for cellular relatives. In any case, it should be borne in mind that the downregulating protocol used here has not, in our hands, resulted in detectable changes in the cellular content of PKC-ζ.

References

Bell, R. M., and Burns, D. J. (1991). *J. Biol. Chem.* **266**, 4661–4664.
Blackshear, P. J., Witters, L. A., Girard, P. R., Kuo, J. F., and Quamo, S. N. (1985). *J. Biol. Chem.* **260**, 13304–13315.
Blackshear, P. J., Wen, L., Glynn, B. P., and Witters, L. A. (1986). *J. Biol. Chem.* **261**, 1459–1469.
Blackshear, P. J., Haupt, D. M., and Stumpo, D. J. (1991). *J. Biol. Chem.* **266**, 10946–10952.
Blackshear, P. J., Verghese, G. M., Johnson, J. D., Haupt, D. M., and Stumpo, D. J. (1992). *J. Biol. Chem.* **267**, 13540–13546.
Blank, J. L., Ross, A. H., and Exton, J. H. (1991). *J. Biol. Chem.* **266**, 18206–18216.
Borner, C., Nichols Guadagno, S., Fabbro, D., and Weinstein, I. B. (1992). *J. Biol. Chem.* **267**, 12892–12899.
Collins, M. K. L., and Rozengurt, E. (1984). *J. Cell. Physiol.* **118**, 133–142.
Glynn, B., Colliton, J., McDermott, J., and Witters, L. A. (1985). *Biochem. J.* **231**, 489–492.
Graff, J. M., Gordon, J. I., and Blackshear, P. J. (1989). *Science* **246**, 503–506.
Graff, J. M., Rajan, R. R., Randall, R. R., Nairn, A. C., and Blackshear, P. J. (1991). *J. Biol. Chem.* **266**, 14390–14398.
Halsey, D. L., Girard, P. R., Kuo, J. F., and Blackshear, P. J. (1987). *J. Biol. Chem.* **262**, 2234–2243.
Li, J., and Aderem, A. (1992). *Cell* **70**, 791–801.
Lobaugh, L. A., and Blackshear, P. J. (1990). *J. Biol. Chem.* **265**, 18393–18399.
Meisenhelder, J., Suh, P.-G., Rhee, S. G., and Hunter, T. (1989). *Cell* **57**, 1109–1122.
Nakanishi, H., and Exton, J. H. (1992). *J. Biol. Chem.* **267**, 16347–16354.
Niedel, J. E., and Blackshear, P. J. (1986). *In* "Receptors and Phosphoinositides" (J. Putney, ed.), pp. 47–88. Alan R. Liss, New York.
Nishizuka, Y. (1988). *Nature (London)* **334**, 661–665.
Rodriguez-Pena, A., and Rozengurt, E. (1984). *Biochem. Biophys. Res. Commun.* **120**, 1053–1059.
Rodriguez-Pena, A., Zachary, I., and Rozengurt, E. (1986). *Biochem. Biophys. Res. Commun.* **140**, 379–385.
Stabel, S., and Parker, P. J. (1991). *Pharmacol. Ther.* **51**, 71–95.
Taylor, S. J., Chae, H. Z., Rhee, S. G., and Exton, J. H. (1991). *Nature (London)* **350**, 516–518.
Umekage, T., and Kato, K. (1991). *FEBS Lett.* **286**, 147–151.

Wahl, M. I., Nishibe, S., Suh, P.-G., Rhee, S. G., and Carpenter, G. (1989). *Proc. Natl. Acad. Sci. U.S.A.* **86**, 1568–1572.

Ways, D. K., Cook, P. P., Webster, C., and Parker, P. J. (1992). *J. Biol. Chem.* **267**, 4799–4805.

Wetsel, W. C., Khan, W. A., Merchenthaler, I., Rivera, H., Halpern, A. E., Phung, H. M., Negro-Vilar, A., and Hannun, Y. A. (1992). *J. Cell Biol.* **117**, 121–133.

Woodgett, J. R., Hunter, T., and Gould, K. L. (1987). *In* "Cell Membranes: Methods and Reviews" (E. L. Elson, W. A. Frazier, and L. Glazer, eds.), Vol. 3, pp. 215–340. Plenum, New York.

18

Identification and Quantification of G-Proteins

Ravi Iyengar

I. Introduction
II. Receptor-Stimulated GTPase
III. Guanine Nucleotide Binding Assays
   A. Receptor-Stimulated [^{35}S]GTPγS Binding
   B. [^{35}S]GTPγS Binding in the Absence of Receptor Agonists
IV. Labeling of G-Proteins by Bacterial Toxins
   A. Cholera Toxin
   B. Pertussis Toxin
V. Immunoblotting with Sequence-Specific Antisera
   References

I. Introduction

Heterotrimeric G-proteins are ubiqutious cell surface signal-transducing proteins. The unique character of the individual G-proteins is conferred by the α subunit of the G-protein, even though there is considerable molecular heterogeneity of β and γ subunits. Close to 20 α subunits have been cloned (Simon et al., 1991). Many of the assays for measurement of G-proteins are based on the properties of the α subunits and hence are common for all G-proteins. Some assays, such as effector stimulation, are relatively specific and allow for the identification of various families of G-proteins. The most specific assays are those using sequence-specific antisera to identify the individual G-proteins. Several assays for G-proteins are described in this chapter. However, the reader is cautioned that not all of the assays described here may work as such for every membrane system and some adjustments may be necessary. Parameters that can be adjusted are indicated for the individual assays.

II. Receptor-Stimulated GTPase

The GTPase activity of G-proteins can be estimated by measurement of $^{32}P_i$ released from the hydrolysis of [γ-^{32}P]GTP. For this, 5–25 µg of membrane proteins is incubated in 100 µl with 25 mM Na Hepes (pH 7.5), 2 mM $MgCl_2$, 1 mM dithiothreitol (DTT), 0.1 mM EGTA, 1.0 mM adenylylimidodiphosphate, 0.1 mM ATP, a nucleoside triphosphate regenerating system of 10 mM creatine phosphate and 0.2 mg/ml creatine phosphokinase, 2 mg/ml bovine serum albumin (BSA), and the substrate, 100 nM [γ-^{32}P]GTP (~200,000 cpm). Typically the activity is measured for 15–20 min at 30°C. The reaction is stopped by the addition of 900 µl of ice-cold sodium phosphate (pH 2.0) containing 5% (w/v) activated charcoal. The mixture is then centrifuged at 10,000g for 30 min at 4°C. Then, 750 µl of the supernatant is counted in a liquid scintillation counter.

Since the plasma membranes contain many nucleoside triphosphate-hydrolyzing enzymes, it is necessary to distinguish between the various activities. This can be done by utilizing the high affinity and specificity that G-proteins show for GTP. Thus the adenylylimidodiphosophate serves to block a very substantial portion of the low-affinity nucleoside triphosphatase activity. The concentration of the adenylylimidodiphosphate can be varied in the range of 1–2 mM to obtain optimal supression of the low-affinity activity. However, such supression is never complete. Hence the low-affinity activity should always be measured in the presence of 50 µM GTP and subtracted from the total activity. In membrane systems typically a 50–200% stimulation by receptor agonists is seen. This procedure has been adapted from the protocols of Gierschik *et al.* (1989).

III. Guanine Nucleotide-Binding Assays

The nucleotide of choice for measurement of guanine nucleotide binding is [^{35}S]GTPγS. This assay is useful for measuring guanine nucleotide binding to G-proteins in general but does not yield any information about the identity of individual G-proteins. Further, in the absence of a receptor agonist, the binding of GTPγS is dependent on the off-rate of the bound GDP. If the G-protein has very tightly bound GDP, as is the case with G_q (Pang and Sternweis, 1990), then it cannot be identified by this method. Considerable caution should be exercised in analyzing the GTPγS-binding data to estimate G-proteins, since reliability in membrane systems is dependent upon the masking of the low-affinity binding sites, as well as the relative abundance of G-proteins in the membranes being analyzed. It is generally easy to obtain unequivocal binding data from neuronal membranes since the data are quite abundant in G-proteins. In contrast, GTPγS binding to G-proteins in liver membranes is quite difficult to measure because G-proteins are in low abundance and there is a substantial amount of nucleoside pyrophosphatase activity.

A. Receptor-Stimulated (^{35}S)GTPγS Binding

Typically 5–25 μg of membranes are incubated in 100 μl of a solution containing 25 mM Na Hepes (pH 7.5), 1 mM EDTA, 1 mM DTT, 1 μM GDP, 1 mM adenylylimidodiphosphate, varying concentrations (0.1–5.0 mM) of Mg ions added as $MgCl_2$, and 50 nM [^{35}S]GTPγS (~ 500,000 cpm). In addition the reaction mixture may contain the required concentration of receptor agonist. After a 20- to 60-min incubation at 30°C the reaction is stopped by the addition of 2 ml of ice-cold 25 mM Tris–HCl (pH 7.5), 25 mM $MgCl_2$, and 100 mM NaCl. The samples are then filtered through nitrocellulose filters. The filters are washed twice with 10 ml of ice-cold stopping solution, dried, and counted in a liquid scintillation counter. Nonspecific binding is measured in the presence of 10 μM GTPγS. This protocol has been adapted from Hilf *et al.* (1989).

B. (^{35}S)GTPγS Binding in the Absence of Receptor Agonists

In the absence of receptor agonists, the dissociation of bound GDP and the association of the GTPγS are completely regulated by Mg ion concentration. If high concentrations of Mg ions are not able to promote GDP dissociation prior to thermal denaturation of a particular G-protein then the binding assay will not be a useful measure of that G-protein. However, many G-proteins, including G_s, G_i, G_o, are able to release and bind nucleotides in a receptor-independent manner and hence the [^{35}S]GTPγS is often useful as a fast assay during purification. The assay mixture contains 25 mM Na Hepes (pH 7.5), 1 mM EDTA, 1 mM DTT, 100 mM NaCl, 0.1% (w/v) Lubrol-PX, 2-25 mM $MgCl_2$, and 50 nM [^{35}S]GTPγS (~ 200,000 cpm). Nonspecific binding is measured in the presence of 10 μM GTP or GTPγS. The reaction containing 5–25 μg protein is incubated at 30°C for the desired time (10–30 min) and then stopped and processed as described above. This procedure has been successfully used to estimate and purify brain G-proteins in our laboratory (Carty *et al.*, 1990; Padrell *et al.*, 1991).

IV. Labeling of G-Proteins by Bacterial Toxins

ADP-ribosylation of α subunits of G-proteins by bacterial toxins has proven to be a very useful tool for identifying G-proteins and their involvement in transducing receptor signals. The ADP-ribosylation reaction involves the transfer of the ADP-ribose moiety from NAD to an acceptor site on the α subunit. The α_1 subunit of the bacterial toxins are ADP-ribosyltransferases that catalyze this reaction. Since bacterial toxins are multimers, the free and enzymatically active α_1 subunit is generated in cell-free systems by treatment of the holotoxin with DTT.

A. Cholera Toxin

Cholera toxin transfers ADP-ribose to an arginine residue within the GTP hydrolysis domain of the α subunit. This results in blockade of GTP hydrolysis and consequently the protein is locked in a persistently active state. Though this arginine is conserved in all the α subunits only, G_s-α appears to be the natural substrate for cholera toxin (Gill and Woolkakis, 1991). For efficient ADP-ribosylation of G_s by cholera toxin an additional cellular cofactor called ARF is necessary (Kahn and Gilman, 1986). Membrane preparations from many cell types contain ARF and hence addition of exogenous ARF is not necessary. However, since ARF is a small GTP-binding protein, it needs to be occupied by a nonhydrolyzable analog of GTP for it to be functional in the ADP-ribosylation reaction. If ARF is not added then addition of excess nonhydrolyzable analogs of GTP is often useful. The analog that appears most beneficial is Gpp(NH)p.

Holocholera toxin is activated by incubation of 0.5 mg/ml toxin with 20 mM DTT for 20 min at 32°C. After the activation the toxin is kept on ice and used within 1–2 hr.

For the labeling of G_s the 50-μl reaction mixture contains 200 mM potassium phosphate (pH 7.5), 1 mM DTT, 10 mM thymidine, 10 mM ATP, 0.1 mM Gpp(NH)p, 100 mM NaCl, 2 mg/ml BSA, 10 μg/ml activated toxin, and 5 μM [^{32}P]NAD (10,000–50,000 cpm/pmol). Typically 5–20 μg of protein is used. The amount of membrane protein used and the concentration of the NAD has to be worked out for each membrane system since it will vary depending on the amount of NAD-glycohydrolase activity present in the membranes. The labeling reaction is incubated for 20 min at 30°C. The reaction is terminated by the addition of 0.5 ml of ice-cold trichloroacetic acid. After 10 min on ice the samples are centrifuged at 5000g for 20 min at 4°C. The pellets are washed once with 1.0 ml of 10% trichloroacetic acid and then washed again with 1.0 ml of cold (-20°C) ether to remove traces of the trichloroacetic acid. The pellets are dried at room temperature and then dissolved 25 ml in SDS-gel electrophoresis buffer. It is most useful to analyze the sample in 10% SDS-polyacrylamide slab gels.

If the membranes are to be treated for the measurement of adenylyl cyclase activity then GTP should be used instead of Gpp(NH)p so as not to persistently activate the enzyme. Also, since the catalytic activity of adenylyl cyclase is labile it is best protected by the addition of ATP and a nucleoside triphosphate-regenerating system and $1-2$ mM $MgCl_2$. At the end of the treatment, the membranes are diluted 20-fold with 25 mM Na Hepes (pH 7.5) and 1 mM EDTA and washed by centrifugation at 10,000g for 20 min before assaying for adenylyl cyclase activity.

B. Pertussis Toxin

This toxin catalyzes the transfer of ADP-ribose from NAD to a cysteine residue in the fourth position from the carboxy terminus of α subunits that are members of the G_i family (West et al., 1985). The carboxy-terminal region is involved in receptor contact, and this ADP-ribosylation results in uncoupling of the receptor from the rest of the system and the consequent loss of receptor stimulation of the pathway (Sullivan et al., 1987). Pertussis toxin treatment does not inactivate G-proteins or affect their GTPase activity. Pertussis toxin-modified G-proteins are capable of being stimulated by nonhydrolyzable analogs of GTP.

The heterotrimeric form of the α subunit is the substrate for pertussis toxin, and addition of EDTA and GTP (which when bound is hydrolyzed to GDP) promotes the reaction by stabilizing the inactive state of the protein. Also addition of 0.1% Lubrol-PX greatly increases the efficiency of labeling.

A typical pertussis toxin ADP-ribosylation reaction mixture contains 25 mM Na Hepes (pH 7.5), 1 mM EDTA, 10 mM thymidine, 0.1% Lubrol-PX, 1 mM ATP, 0.1 mM GTP, 0.02% BSA, 10 μM [^{32}P]NAD (10,000-50,000 cpm/pmol), and $1-5$ μg/ml activated pertussis toxin. The reaction is initiated by the addition of $1-5$ μg of membrane protein. The labeling mixture (final volume 50 μl) is incubated for 30 min at 30°C. The reaction is then stopped and processed as for the cholera toxin labeling. The conditions for activation of pertussis toxin are identical to that for cholera toxin except that 0.1 mg/ml toxin is used.

Since most of the known pertussis toxin substrates are in the 39- to 41-kDa range modification of the SDS-polyacrylamide gel electrophoresis system to yield better resolution is often useful. One modification first reported by Touant et al. (1987) is to alter the ratio of bisacrylamide to acrylamide to 0.4:30. This lowered ratio increases resolution. Inclusion of

4–8 M urea gradients or just 4 M urea in the resolving region of the gel also greatly enhances the separation between the various pertussis toxin substrates. We now often include 4 M urea in our SDS–polyacrylamide gels.

If the functional consequences of cholera or pertussis toxin treatment are to be studied, it is best to treat the intact cells or tissues with holotoxin. It is necessary to use the holotoxin because the B subunits are required for entry of the toxin into the cell. Generally injection of animals with toxins makes them quite sick, and such experiments should be undertaken only after due consideration. Treatments of cells in culture is often a very useful approach. Typically treatment of cells with 10–100 ng/ml toxin overnight is sufficient to obtain a full effect. If necessary the toxin concentration can be raised to 1 μg/ml and the length of treatment varied. Modification of substrates within the cell can be subsequently confirmed by loss of ^{32}P incorporation into membranes in a labeling reaction.

V. Immunoblotting with Sequence-Specific Antisera

The use of sequence-specific antipeptide antisera to identify the various α subunits is currently the most unequivocal method available. A battery of antipeptide antisera to the various unique regions of the α subunits has been raised by Gilman and Mumby as well as by Speigel and co-workers. The antisera raised by the Mumby and Gilman (1991) have been described in detail. The antisera raised by Speigel's group are now commercially available through New England Nuclear–DuPont. For immunoblotting the proteins are resolved by SDS–PAGE (Laemmli, 1970). Typically 50–150 μg of membrane proteins are resolved on 9–10% polyacrylamide gels. The resolved proteins are electrophoretically transferred to nitrocellulose paper by the method of Towbin *et al.* (1979).

After the transfer, the protocols of Kriegler (1990) are being used currently in our laboratory since our detection system is based on chemiluminesence and uses the Amersham ECL system. The nitrocellulose papers are treated with blocking buffer (1.0 M glycine, 5% (w/v) nonfat dry milk (Carnation), 5% fetal calf serum, 1% ovalbumin) for 1–3 hr at room temperature to block all additional protein sites. After this the filters are washed at room temperature with the wash solution (25 mM phosphate buffer, pH 6.8, 120 mM NaCl, 1% fetal calf serum, 0.1% Tween-20, 0.1% dry milk, 0.1% ovalbumin) three times for 20 min each with 10 ml of wash solution each time. Then the primary antibody is added at the appropriate dilution (1:100 to 1:1000) in the wash buffer and the filters are incubated overnight at room temperature. After incubation with the primary antibody, the filters are washed three times for 15 min with 20 ml wash solution. The second antibody used is horseradish peroxidase coupled with goat anti-rabbit IgG. The nitrocellulose filters are incubated with a 1:3000 dilution of the second antibody for 2 hr at room temperature. After this incubation the filters are washed three times (20 ml, each time) for 15 min with 25 mM phosphate buffer, 120 mM NaCl, and 0.1% Tween at room temperature. The filters are incubated with the ECL reagents according to the instructions supplied with the ECL reagents (Amersham) and the bands are visualized by use of a sensitized film (Hyperfilm,Amersham) that detects the chemiluminesence produced by horseradish peroxidase-catalyzed oxidation. We have found that this method is much faster and has a sensitivity similar to that obtained in our laboratory with ^{125}I-labeled second antibody (Carty *et al.*, 1991). Hence we are currently using this method of detection.

When using the sequence-specific antisera to detect the various G-protein α subunits the following concerns should be noted. If the antiserum is made against a sequence from one species then it may not interact with or poorly interact with the same protein from another species. This depends on the epitope within the peptide against which the antibody is reactive. If crude membranes (10–40,000g pellets) are used for immunoblotting then it is possible that the less-abundant G-proteins will not be detected. In such cases it may be useful to prepare purified plasma membranes by sucrose density gradient centrifugation.

References

Carty, D. J., Padrell, E., Codina, J., Birnbaumer, L., Hildebrandt, J. D., and Iyengar, R. (1990). *J. Biol. Chem.* **265**, 6228.

Carty, D. J., Premont, R. T., and Iyengar, R. (1991). *In* "Adenylyl Cyclase, G Proteins, and Guanylyl Cyclase" (R. A. Johnson and J. D. Corbin, eds.), Methods in Enzymology, Vol. 195, p. 302. Academic Press, San Diego.

Gierschik, P., Sidropolous, D., Steisslinger, M., and Jakobs, K. H. (1989). *Eur. J. Pharmacol.* **172**, 481.

Gill, M. D., and Woolkakis, M. J. (1991). *In* "Adenylyl Cyclase, G Proteins, and Guanylyl Cyclase" (R. A. Johnson and J. D. Corbin, eds.), Methods in Enzymology, Vol. 195, p. 267. Academic Press, San Diego.

Hilf, G., Gierschik, P., and Jakobs, K. H. (1989). *Eur. J. Biochem.* **186**, 725.

Kahn, R., and Gilman, A. G. (1986). *J. Biol. Chem.* **259**, 6228.

Kriegler, M. (1990). "Gene Transfer and Expression. A Laboratory Manual," pp. 218–219. Stockton Press, New York.

Laemmli, U. K. (1970). *Nature (London)* **227**, 680.

Mumby, S. M., and Gilman, A. G. (1991). *In* "Adenylyl Cyclase, G Proteins, and Guanylyl Cyclase" (R. A. Johnson and J. D. Corbin, eds.), Methods in Enzymology, Vol. 195, p. 215. Academic Press, San Diego.

Padrell, E., Carty, D. J., Moriarty, T. M., Hildebrandt, J. D., Landau, E. M., and Iyengar, R. (1991). *J. Biol. Chem.* **266**, 9771.

Pang, I.-H., and Sternweis, P. C. (1990). *J. Biol. Chem.* **265,** 18707.
Simon, M. I., Strathman, M. P., and Gautam, N. (1991). *Science* **252,** 802.
Sullivan, K. A., Miller, R. T., Masters, S. B., Beiderman, B., Heideman, W., and Bourne, H. R. (1987). *Nature (London)* **330,** 758.
Touant, M., Annis, D., Bockaert, J., Homberger, V., and Ronot, B. (1987). *FEBS Lett.* **215,** 339.
Towbin, H., Staehlin, T., and Gordon, J. (1979). *Proc. Natl. Acad. Sci. U.S.A.* **76,** 4350.
West, R. E., Moss, J., Vaughan, M., Liu, T., and Liu, T. Y. (1985). *J. Biol. Chem.* **260,** 14428.

19

Determination of Adenylyl Cyclase Catalytic Activity

Roger A. Johnson and Yoram Salomon

I. Considerations for Establishing Reaction Conditions
   A. Requirements for Metal-ATP and Divalent Cations
   B. Contaminating Enzyme Activities
   C. Enzyme Concentrations, Reaction Times, and Temperatures
   D. Guanine Nucleotides
II. Radioactive Substrates: [^{3}H]ATP vs [α-^{32}P]ATP
   A. [^{3}H]ATP Advantages
   B. [^{3}H]ATP Disadvantages
   C. [α-^{32}P]ATP Advantages
   D. [α-^{32}P]ATP Disadvantages
III. Stopping the Reaction
   A. Stopping with Zn(Ac)$_2$/Na$_2$CO$_3$/cAMP or/[^{3}H]cAMP
   B. Stopping with ATP/SDS/cAMP ± [^{31}I]cAMP
   C. Stopping with HCl, HCl/cAMP, or HCl/[^{3}H]cAMP
IV. Chromatographic Alternatives
   A. Sequential Chromatography on Dowex-50 and Alumina
   B. Single Alumina Column
   C. Disposal of Waste Isotope
V. Data Analysis
   References

Adenylyl cyclase (ATP-pyrophosphate lyase, cyclizing, EC 4.6.1.1) is a family of membrane-bound enzymes that exhibit inactive and active configurations resulting from the actions of a variety of agents, acting indirectly and directly on the enzyme. Enzyme activity may be increased or decreased by stimulatory or inhibitory hormones/neurotransmitters acting via specific hormone receptors coupled to the enzyme's catalytic unit by the respective guanine nucleotide-dependent regulatory proteins (G_s and G_{i2}, respectively). The G-proteins are also activated by aluminum fluoride and are targets for

ADP-ribosylation by specific bacterial toxins. The catalytic moiety of adenylyl cyclase from most tissues is stimulated by the diterpene forskolin and is inhibited by adenosine 3'-phosphates, the most potent of which is 2'-deoxyadenosine-3'-monophosphate, and the enzyme from some tissues is also stimulated directly by Ca^{2+}/calmodulin. The nature of some of these agents and the range of resulting activities can influence the assay conditions used for determining the enzyme's catalytic activity.

Catalytic activity of adenylyl cyclase is determined by methods that either rely on the measurement of cAMP formed from unlabeled substrate, with cAMP-binding proteins or radioimmunoassay procedures, or rely on radioactively labeled substrate followed by isolation and determination of the radioactively labeled product. The two different approaches have different purposes, different sensitivities, and different ease of use. The method of choice will depend in part on the facilities and orientation of a given laboratory. The procedures described here focus on the use of radioactively labeled substrate and isolation of the labeled product. Additional detailed considerations for the assay of adenylyl cyclase by these procedures can be found in the review by Salomon (1979).

I. Considerations for Establishing Reaction Conditions

A. Requirements for Metal-ATP and Divalent Cations

Both ATP and divalent cations (Mg^{2+} or Mn^{2+}) are required for adenylyl cyclase-catalyzed formation of cAMP (Rall and Sutherland, 1958; Sutherland *et al.*, 1962). The enzyme conforms to a bireactant sequential mechanism in which metal-ATP^{2-} is substrate and free divalent cation is a requisite cofactor (Garbers and Johnson, 1975). Thus, adenylyl cyclase requires divalent cation in excess of the ATP concentration. The concentrations of both substrates may be varied for determining kinetic constants. But because of the association constants for divalent cation and ATP^{2-} [ATP · Mg(65000/M); ATP · Mn(353000/M); Garbers *et al.*, 1975] the concentration of free Mg^{2+} or Mn^{2+} must be fixed at a concentration above the total ATP concentration (Cleland, 1970; Garbers *et al.*, 1975). This will maintain divalent cation concentrations essentially constant even though ATP concentrations are being varied. Kinetic constants can then be calculated by linear regression analysis of the slopes and intercepts of secondary plots as suggested by Cleland (1970). Buffers, such as Tris–Cl, that can significantly affect concentrations of free divalent cation (especially Mn^{2+}) in the reaction mixture should be avoided. Triethanolamine · HCl does not have this problem (Garbers and Johnson, 1975). Examples of K_m values for adenylyl cyclases have been reported as

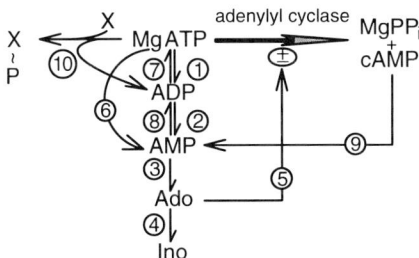

Figure 1

follows: detergent-dispersed enzyme from rat brain, $K_{m(MnATP)} \sim 7-9$ μM; $K_{mMn2+} \sim 2-3\ \mu M$; $K_{m(MgATP)} \sim 30-36\ \mu M$; $K_{mMg2+} \sim 800-900\ \mu M$ (Garbers and Johnson, 1975); human platelets, $K_{mMg2+} \sim 1100\ \mu M$ and $K_{m(MgATP)} \sim 50\ \mu M$ (Johnson et al., 1979) and liver, $K_{m(MnATP)}$ and $K_{m(MgATP)}$ were similar ($\sim 50\ \mu M$) (Londos and Preston, 1977).

B. Contaminating Enzyme Activities

Adenylyl cyclase is a membrane-bound protein of very low abundance and exists in an environment rich in contaminating enzyme activities (Fig. 1), including a number of nucleotide phosphohydrolases (1, 2, 3, and 6 in Fig. 1), cyclic nucleotide phosphodiesterases (9 in Fig. 1), and ATP-utilizing kinases (10 in Fig. 1). Thus, adenylyl cyclase in membrane preparations competes with other enzymes for ATP and the cAMP formed is readily hydrolyzed to 5′-AMP. Analogously, GTP, required for hormone-induced activation or inhibition of adenylyl cyclases, is also metabolized. Consequently, the use of regenerating systems to counteract this alternative metabolism of ATP (7 and 8 in Fig. 1) and/or GTP and the use of inhibitors of cAMP-phosphodiesterases are nearly unavoidable.

1. Cyclic Nucleotide Phosphodiesterases

cAMP is effectively inactivated through the hydrolysis of its 3′-phosphate bond, yielding 5′-AMP (9 in Fig. 1). Since cyclic nucleotide phosphodiesterase activity is substantial in most membrane preparations, these enzymes must be inhibited to measure accurately the rate of formation of cAMP by adenylyl cyclase. This is usually accomplished by the use of unlabeled cAMP in the reaction mixture or use of inhibitors of the enzyme, such as papaverine or alkyl xanthines (e.g., 3-isobutyl-1-methylxanthine; IBMX). IBMX also potently blocks adenosine A_1 and A_2 receptors and is thereby additionally useful. One must be judicious in the selection of a phosphodiesterase inhibitor, though, in that some agents do not block all cAMP phosphodiesterases. For example, the sole use of the imidizolidinone derivative Ro-20-1724 [4-(3-butoxy-4-methoxybenzyl)-2-imidazolidinone] is not recommended. This compound has been useful in the study of adenosine receptor-mediated effects on adenylyl cyclases because it is not an adenosine receptor antagonist as are the alkylxanthines. Although it may substantially suppress the hydrolysis of cAMP in preparations from some tissues (pig coronary arteries), it does so incompletely in others (e.g., platelets). In preparations of human platelet membranes, for example, the addition of a second phosphodiesterase inhibitor is required. In our hands the most effective combinations of agents for inhibition of hydrolysis of labeled cAMP produced by adenylyl cyclase have been 1 mM IBMX; cAMP (100 μM) ± papaverine (100 μm); 100 μM anagrelide [6,7-dichloro-1,5-dihydroimidazo-[2,1-6]quinazolin-one monohydrochloride (BL-4162A)]; or anagrelide (100 μM) + Ro-20-1724 (500 μM) (E. A. Martinson and R. A. Johnson, unpublished observations).

2. ATP-Regenerating Systems

The accurate determination of adenylyl cyclase activities is adversely affected by the hydrolysis of ATP between the β- and γ-phosphates (1 in Fig. 1), due to various membrane-bound ATPases, nonspecific phosphohydrolases, and flux through membrane-bound kinases (10 in Fig. 1) and phosphatases. Cleavage of the bond between the α- and β-phosphates occurs for ADP (2 in Fig. 1) by membrane phosphohydrolases and for ATP by nucleotide pyrophosphatase (6 in Fig. 1). Whether by step 2 or by step 6, the result is 5'-AMP, which is rapidly hydrolyzed by 5'-nucleotidase (3 in Fig. 1) to adenosine (Ado). Adenosine can stimulate or either inhibit (5 in Fig. 1) adenylyl cyclase, indirectly via inhibitory (A_1) or stimulatory (A_2) receptors, or inhibit the enzyme directly via the "P"-site, through which adenosine 3'-phosphates inhibit. These reactions can be counteracted and their influence minimized by use of an ATP-regenerating system. The most common systems have used creatine kinase and creatine phosphate or pyruvate kinase and phosphoenolpyruvate to counteract hydrolysis of ATP between the β- and γ-phosphates by catalyzing the conversion of ADP to ATP (7 in Fig. 1), but to counteract hydrolysis between the α- and β-phosphates myokinase is used to convert 5'-AMP to ADP (8 in Fig. 1). Since the action of 5'-nucleotidase cannot be reversed, the influence of the formed adenosine can be minimized by the use of adenosine deaminase (4 in Fig. 1), as the product inosine is without effect on adenylyl cyclase. In adenylyl cyclase reaction mixtures effective concentrations of these enzymes are myokinase (100 μg/ml; Boehringer-Mannheim, ammonium sulfate suspension from rabbit muscle) and adenosine deaminase (5 U/ml; Sigma, ammonium sulfate suspension, Type VIII from calf intestinal mucosa).

The influence of nucleotide pyrophosphatase and to a lesser extent 5′-nucleotidase can be further minimized by pretreatment of membranes with 5 mM EDTA and 3 mM dithiothreitol (Johnson and Weldon, 1977; Johnson, 1980). The necessity for the additions to the assay or the effectiveness of the membrane pretreatment with chelator and/or dithiothreitol depends on the source and purity of the adenylyl cyclase being studied. Adenosine deaminase and myokinase are effective in enhancing apparent adenylyl cyclase activity. Myokinase enhances adenylyl cyclase activity by recycling 5′-AMP and thereby helping to maintain ATP levels during the reaction. Adenosine deaminase presumably removes inhibitory levels of adenosine (Johnson, 1980).

Although both creatine kinase and creatine phosphate or pyruvate kinase and phosphoenolpyruvate have been utilized as the basis of ATP-regenerating systems, neither is without its pitfalls. Both of these enzymes bind and utilize adenosine phosphates and both creatine phosphate and phosphoenolpyruvate form weak complexes with divalent cations. Moreover, phosphoenolpyruvate has been shown to cause both stimulatory and inhibitory effects on the enzyme from liver and to inhibit the enzyme from heart (Johnson and Garbers, 1977) and contaminants in creatine phosphate have been found to cause both stimulatory and inhibitory effects on adenylyl cyclases (Johnson, 1980). For these reasons the preferable ATP-regenerating system is creatine kinase (100 μg/ml; Boehringer-Mannheim from rabbit muscle) and creatine phosphate. Creatine phosphate should be used at concentrations low enough (e.g., 2 mM) to minimize the influence of the contaminants, but high enough to allow ATP concentrations to be maintained for the duration of the incubation. If higher concentrations (e.g., 10 mM) are found to be necessary for the linear formation of cAMP with time, it may be necessary to purify creatine phosphate before use, e.g., by anion-exchange chromatography (Johnson, 1980).

C. Enzyme Concentrations, Reaction Times, and Temperatures

Three additional factors that obviously influence adenylyl cyclase-catalyzed formation of cAMP interdependently are enzyme concentration, time, and incubation temperature.

1. Enzyme Concentration

Adenylyl cyclase is more active in crude membrane preparations from some tissues than from others and the levels of the various contaminating enzymes that utilize adenine nucleotides (Fig. 1) also vary. Consequently, under any given set of reaction conditions it is imperative to establish (i) that sufficient enzyme is used to catalyze the formation of measurable amounts of [^{32}P]cAMP or [^{3}H]cAMP, (ii) that concentrations of both ATP and creatine phosphate are adequate for sustaining stable ATP concentrations, and (iii) that formation of cAMP is linear with respect to enzyme concentration. Formation of measurable amounts of [^{32}P]cAMP or [^{3}H]cAMP is improved with increasing specific radioactivity of labeled ATP and increasing enzyme concentration. However, increasing concentrations of crude membrane preparations of adenylyl cyclase typically also require increased amounts of creatine phosphate or proportionally decreased incubation times. Hence, enzyme concentrations, ATP-specific radioactivity, and incubation times must be adjusted to allow linear formation of cAMP with respect to both protein concentration and time. Useful ranges of specific radioactivity of [α-^{32}P]ATP are 10 to 200 cpm/pmol, depending principally on enzyme source. For studies of enzyme kinetics with respect to metal-ATP the range of [α-^{32}P]ATP-specific activity will be greater than this. Examples of, but not necessarily upper limits for, protein concentrations yielding linear product formation at 30°C with 0.1 mM ATP and 5 mM creatine phosphate, 100 μg creatine kinase/ml, 100 μg myokinase/ml, 5 U adenosine deaminase/ml, and 10 mM MnCl$_2$ or MgCl$_2$, would be (in mg/ml) as follows: heart, 0.6; liver, 1.2; kidney, 0.7; skeletal muscle, 0.3; adiocytes, 0.2; spleen, 0.5; human platelets, 0.2; bovine sperm particles, 1.0; washed particles from brain, 0.2; detergent-solubilized brain, 0.2.

2. Time

Incubation time for adenylyl cyclase reactions is dictated by a balance between rates of formation of cAMP from ATP, hydrolysis of cAMP to 5'-AMP by contaminating cyclic nucleotide phosphodiesterases, hydrolysis of ATP by a number of membrane-bound phosphohydrolytic enzymes (Fig. 1), inactivation of adenylyl cyclase by regulatory components, and denaturation of the enzymes. It is essential that linearity of product formation with respect to time be established. The reaction is typically linear with respect to time for crude membrane preparations, with the conditions given above (Section I,C,1, Enzyme Concentration), for 2 to 15 min and for purified enzyme for 60 min.

3. Incubation Temperature

Temperature can be used to advantage to exhibit certain characteristic behaviors of adenylyl cyclases as well as to modify rates of alternative substrate utilization and enzyme denaturation. Formation of cAMP is linear with respect to time for a longer period at 30°C than at a more physiological 37°C.

However, in the absence of protective agents the catalytic moiety is readily inactivated by exposure to heat for a short time. For example, exposure of adenylyl cyclases from platelets and from S49 lymphoma wild-type and cyc⁻ cells for 8 min at 35°C causes 70 to 75% inactivation (Awad *et al.*, 1983; Florio and Ross, 1983). Comparable inactivation of adenylyl cyclases from bovine sperm and detergent-dispersed porcine brain occurred by exposure at 45°C for 8 and 4 min, respectively (Awad *et al.*, 1983). Partial protection against thermal inactivation is afforded by forskolin (200 μM), metal-ATP (<mM), the P-site agonist 2'5'-dideoxy-adenosine (250 μM), guanine nucleotides (μM), and, for the Ca^{2+}/calmodulin-sensitive form of adenylyl cyclase, Ca^{2+}/calmodulin (50 μM/5 μM) (Harwood *et al.*, 1973; Brostrom *et al.*, 1978; Salter *et al.*, 1981; Awad *et al.*, 1983; Florio and Ross, 1983). In addition, adenylyl cyclase reactions conducted at different temperatures can be used to enhance selective regulatory properties of the enzyme. For example, inhibition of adenylyl cyclase mediated by guanine nucleotide-dependent regulatory protein (G_i), whether by hormone or stable guanine nucleotide, is more readily shown experimentally at lower temperatures (e.g., 24°C), whereas activation, mediated by the stimulatory G-protein (G_s), is evident at higher temperatures (e.g., 30°C) (Cooper and Londos, 1979).

D. Guanine Nucleotides

GTP is required for G_s- and G_{i2}-mediated activation and inhibition of adenylyl cyclases (Rodbell *et al.*, 1971; Jakobs *et al.*, 1978; Londos *et al.*, 1978; Cooper *et al.*, 1979; Perez-Reyes and Cooper, 1986). The more stable GTP analogs, guanosine-5' (β,γ-imino) triphosphate [GPP(NH)P] and guanosine 5'-0-(3-thiotriphosphate) (GTPγS), can substitute for GTP. Effects of these analogs are evident after a distinct lag phase and preincubation of the enzyme with them will result in persistently activated or inhibited enzyme, depending on incubation conditions. In addition, the effectiveness of G_s and G_{i2} to regulate adenylyl cyclase is further influenced by divalent cation (type and concentration) and by membrane perturbants (e.g., Mn^{2+} and detergents obliterate G_{i2}-mediated inhibition). Half-maximal stimulation of adenylyl cyclase in the presence of hormones is usually observed with 10 to 50 nM GTP, or with 50 to 100 nM GPP(NH)P or GTPγS. Maximal stimulation occurs with >1 to 10 μM GTP, GPP(NH)P, or GTPγS. Half-maximal inhibition by hormones occurs with 100 to 500 nM GTP, 10 to 100 nM GPP(NH)P, or 1 to 10 nM GTPγS. Maximum inhibition occurs with GTPγS > 100 nM, GPP(NH)P > 10 nM, and GTP > 1 μM. Consequently, even in relatively pure membrane preparations enzyme activity may be increased somewhat by stimulatory hormones due to endogenously present GTP (e.g., as contaminant of

ATP; GTP-free ATP can be purchased from Sigma). The addition of GTP enhances stimulation further. By comparison, GTP must be added to demonstrate hormonal inhibition of adenylyl cyclase and GTP-dependent inhibition is often best elicited with enzyme that has been stimulated by forskolin or a stimulatory hormone. In addition, the concentrations of guanine nucleotides necessary for regulation of adenylyl cyclase activity are dependent on enzyme source and incubation temperature and are influenced by the relative activities and abundances of G_s and G_{i2}.

II. Radioactive Substrates: (^{3}H)ATP vs (α-^{32}P)ATP

[^{3}H]ATP and [α-^{32}P]ATP are commonly used as labeled substrate for measuring adenylyl cyclase catalytic activity. The use of each has both advantages and disadvantages, some of which are described below.

A. (^{3}H)ATP Advantages

The main advantage to the use of [^{3}H]ATP as labeled substrate is its long half-life (~12.3 years). This allows the nearly complete usage of purchased isotope without regard to loss through decay, thereby being cost effective per assay tube. Thus, low usage rates may adequately compensate for its being initially substantially more expensive than [α-^{32}P]ATP. A second advantage is due to the low energy of the tritium β emission, obviating use of cumbersome thick Lucite shielding usually used with ^{32}P-labeled compounds.

B. (^{3}H)ATP Disadvantages

There are several significant disadvantages to the use of [^{3}H]ATP as substrate in adenylyl cyclase reactions.

1. Tritium-labeled adenine nucleotides and nucleotides are chemically unstable in that the C(8)-tritium exchanges with water, especially under alkaline conditions. This results in a continuous loss of tritium to ^{3}H$_2$O that can occur at the rate of several percent per month. Consequently, for accurate estimations of substrate-specific activity the ^{3}H$_2$O must be removed periodically, either chromatographically or by lyophilization. Both procedures necessitate undue handling of and exposure to moderate quantities of isotope and have the potential of major isotope spills in a laboratory environment.

2. The low energy of tritium's β decay necessitates the use of scintillation cocktails to detect [^{3}H]cAMP and tritium's long half-life means that large volumes of liquid radioactive waste, which necessarily also contains large quantities of organic solvents, must be disposed rather than be allowed to dissipate through radioactive decay as would be the case with ^{32}P. Disposal of radioactive waste, especially mixed with scintillation cocktail, is an expensive and undesirable consequence of the use of tritium-labeled substrate.

3. The low energy of tritium decay makes it more difficult to detect if there are inadvertent spills or contamination in a laboratory and could thereby lead to undue exposure of laboratory personnel to low-energy radiation.

4. Breakdown products of [^{3}H]ATP or [^{3}H]cAMP include various nucleotides and nucleosides, as well as xanthine, hypoxanthine, and others, that are also labeled. Chromatographic techniques for the separation of [^{3}H]ATP and [^{3}H]cAMP must take this into consideration. The several breakdown products and the continuous formation of ^{3}H$_2$O from tritium-labeled adenine nucleotides contribute to blank values (counts per minute for samples in the absence of enzyme) with the Dowex-50/Al$_2$O$_3$ column system described below showing substantially higher values than those obtained with [α-^{32}P]ATP as substrate. With adenylyl cyclases of low specific activity in crude membrane preparations such high blank values may constitute a substantial percentage of the [^{3}H]cAMP formed enzymatically.

C. (α-^{32}P)ATP Advantages

There are several important advantages to the use of [α-^{32}P]ATP as substrate for adenylyl cyclase reactions.

1. The specificity of labeling of [α-^{32}P]ATP, which is dictated by the enzymic means typically used for its synthesis (Johnson and Walseth, 1979), means that only α-phosphates are labeled and, since the α-phosphate of ATP is not readily transferred to other compounds, that only purine nucleotides immediately derivable from ATP will be labeled. Additional products that could result from contaminating activities in crude membrane preparations, e.g., [α-^{32}P]ADP, [α-^{32}P]AMP, [α-^{32}P]IMP, and ^{32}P$_i$ are all readily separated from [^{32}P]cAMP due to differences in their ionic properties.

2. ^{32}P is a high-energy β emitter that allows detection by Cherenkov radiation and obviates use of scintillation cocktails, i.e., it can be detected in aqueous solutions with efficiencies approaching that of tritium in scintillation cocktails, but with little influence by agents that typically quench detection of tritium. The high energy of the β emission also allows easy detection of

inadvertent spills with a Geiger/Muller detector and thereby actually enhances laboratory safety due to increased awareness.

3. The short half-life of ^{32}P($\sim$14.3 days) allows waste to be decayed off before disposal, effectively eliminating expensive or awkward disposal of radioactive waste, whether solid or liquid.

D. (α-^{32}P)ATP Disadvantages

1. The short half-life of ^{32}P implies that the usefulness of the isotope is often lost to decay before the [α-^{32}P]ATP is fully utilized. Consequently, if usage rates are low the decay of the isotope may result in the cost of [α-^{32}P]ATP approaching that of [^{3}H]ATP. However, the cost of [α-^{32}P]ATP can be substantially reduced if it is enzymatically synthesized in the laboratory (Johnson and Walseth, 1979).

2. Blank values can depend on the quality of substrate, even with the double-column procedures described below. The quality can vary substantially among different suppliers and in different batches from a given supplier. Blank values for the adenylyl cyclase assay may be supplied on the product data sheet. The quality of [α-^{32}P]ATP can be assured by its purification before use or through its enzymatic synthesis in the laboratory from carrier-free ^{32}P$_i$ (Johnson and Walseth, 1979), a process which also includes its purification.

III. Stopping the Reaction

There are several good methods for stopping adenylyl cyclase reactions. The choice depends on whether [^{3}H]ATP or [α-^{32}P]ATP is used as substrate, on the method used for estimating loss of labeled cAMP during its purification, and on the chromatographic system used for separating labeled product from labeled substrate. Some of these considerations are dealt with below.

A. Stopping with Zn(Ac)$_2$/Na$_2$CO$_3$/cAMP, or /(^{3}H)cAMP

The use of coprecipitation or absorption of nucleotides with inorganic salts dates from an early assay for adenylyl cyclase developed by Krishna *et al.* (1968) who used a combination of column chromatography on Dowex-50 and precipitation with ZnSO$_4$ and Ba(OH)$_2$, yielding the insoluble salts of BaSO$_4$ and Zn(OH)$_2$, which absorb phosphomonoesters and polyphosphates but not cyclic nucleotides. A disadvantage in the use of ZnSO$_4$/Ba(OH)$_2$ is that cAMP may be formed nonenzymatically from ATP at alkaline pH, especially at elevated temperatures, leading to variable and high blank values. This problem is circumvented by the use of other salt combinations or by the use of acidic inactivation of adenylyl cyclase. The effectiveness of a variety of combinations of inorganic salts, e.g., $ZnSO_4/Na_2CO_3$, $CdCl_2/Na_2CO_3$, $ZnSO_4/BaCl_2$, $BaCl_2/Na_2CO_3$ to bind labeled ATP, ADP, AMP, cAMP, and adenosine has been cataloged previously (Chan and Lin, 1974). Since comparable separation of ATP and cAMP can be achieved with columns packed with $ZnCO_3$ (Chan and Lin, 1974) or Al_2O_3 (White and Zenser, 1971), it is likely that absorption to the insoluble inorganic salts, rather than coprecipitation with them, is the basis of the separation of cAMP from the multivalent nucleotides and hence the basis of their usefulness in assays of adenylyl or guanylyl cyclases. It is important to emphasize that none of these salt combinations alone will separate cAMP from adenosine or inosine. The co-elution of [^{3}H]cAMP, formed via adenylyl cyclase, and [^{3}H]nucleoside contaminants (e.g., [^{3}H]adenosine, [^{3}H]inosine, [^{3}H]xanthine, [^{3}H]hypoxanthine) is effectively circumvented (i) by use of a two-step chromatographic procedure, (ii) by the acidification of samples before chromatography on alumina, and/or (iii) by the use of [α-^{32}P]ATP instead of [^{3}H]ATP as substrate.

The characteristics of the insoluble inorganic salts are taken advantage of in the following procedure, adapted from Jakobs *et al.* (1976).

Reagents

Zinc acetate/cAMP. A total of 120 mM [$Zn(C_2H_3O_2)_2 \cdot 2H_2O$; FW 219.49] is prepared in deionized, double-distilled, or Millipore-grade water that has been boiled and then cooled to remove dissolved carbon dioxide. cAMP is then added (165 mg/liter → 0.5 mM). This is kept refrigerated and tightly capped between uses to minimize precipitation of atmospheric CO_2.

Zinc acetate/[^{3}H]cAMP. Prepared as above except the tritiated cAMP, from which ^3H_2O has been removed, is added to an amount of $Zn(C_2H_3O_2)_2$ needed for a given assay to yield approximately 10,000 to 20,000 cpm/ml, when counted in the same volume of eluate used for samples (see Section V.)

Sodium carbonate. A total of 144 mM (Na_2CO_3, anhydrous; FW 106.0). Both $Zn(C_2H_3O_2)_2$ and Na_2CO_3 solutions are stored in and dispensed from glass repipettors.

PROCEDURE

1. With [α-^{32}P]ATP as substrate adenylyl cyclase reactions, typically 50 to 200 µl in 1.5-ml plastic Eppendorf tubes, is terminated by the addition of 0.6 ml of 120 mM $Zn(C_2H_3O_2)_2$/cAMP or $Zn(C_2H_3O_2)_2$/[^{3}H]cAMP. If [^{3}H]ATP is used as substrate, tritiated

cAMP cannot be used for determination of recoveries. Unlabeled cAMP, [^{32}P]cAMP, or [^{14}C]cAMP would have to be used. Aliquots of these stopping solutions are taken for determining absorbance (A_{259nm}) or radioactivity, as appropriate; these values will be used for quantitating sample recovery (see Section V.)

2. A total of 0.5 ml of 144 mM Na_2CO_3 is added to precipitate $ZnCO_3$ and multivalent adenine nucleotides.
3. Samples are placed on ice or can be kept refrigerated or frozen overnight. The $ZnCO_3$ precipitate is sedimented by centrifugation in a bench-top centrifuge. Pellets of frozen samples are smaller and heavier than those of unfrozen samples.
4. The supernatant fractions are decanted onto columns for purification of sample cAMP.
5. Assay blanks are prepared by substituting enzyme buffer for enzyme.
6. A potential disadvantage of this method is that if [^{3}H]cAMP is used, it becomes necessary to use, and hence eventually dispose of, scintillation cocktails for quantitating [^{32}P]cAMP and its recovery.
7. An advantage of this procedure that has led to its use in many laboratories is that >98% of all multivalent nucleotides, i.e., substrate [α-^{32}P]ATP, [α-^{32}P]ADP, [^{32}P]AMP, as well as any $^{32}P_i$, are retained in the capped Eppendorf assay tubes in the $ZnCO_3$ precipitate. The waste radioactivity is thus highly confined, occupies little volume, and can be allowed to decay off and then be dealt with as normal solid waste.

B. Stopping with ATP/SDS/cAMP ± (^{3}H)cAMP

This method (Salomon, 1979; Salomon et al., 1974) relies on sodium dodecyl sulfate to inactivate adenylyl cyclase and on unlabeled ATP and unlabeled cAMP to overwhelm adenylyl cyclase and cAMP phosphodiesterase with substrate and thereby effectively prevent the further formation or degradation of [^{32}P]cAMP.

Reagents

"Stopping solution". To monitor recovery of [^{32}P]cAMP, 2% (w/v) sodium dodecyl sulfate, 40 mM ATP, 1.4 mM cAMP, pH 7.5, and approximately 100,000 cpm [^{3}H]cAMP/ml are used. Alternatively, [^{3}H]cAMP could be omitted from the stopping solution and added separately.

PROCEDURE

1. Adenylyl cyclase reactions, typically 50 to 200 μl in 13 × 65-mm glass or plastic tubes or in 1.5-ml plastic Eppendorf tubes, are terminated by the addition of 100 μl of the stopping solution.
2. To achieve full membrane solubilization in cases of high membrane content, it is advisable to boil the test tubes for 3 min at this state. This also facilitates the rate of chromatography. Hence, use heat-stable tubes.
3. The mixtures in the reaction tubes are then diluted and decanted onto chromatography columns for purification of sample cAMP.
4. Assay blanks are prepared by omitting enzyme or by adding enzyme after the stopping solution.
5. A disadvantage of this procedure is that all radioactive compounds, including unused substrate [α-^{32}P]ATP, [α-^{32}P]ADP, [^{32}P]AMP, ^{32}P$_i$, as well as degradation products of [^{3}H]cAMP, are passed with the labeled cAMP onto the chromatography column and are typically eluted in a fall-through fraction that must be collected and then dealt with as a large volume of liquid radioactive waste. To minimize this waste see Section IV,C, Disposal of Waste Isotope, below.
6. A disadvantage of either stopping procedure when [^{3}H]cAMP is used to monitor recovery of sample [^{32}P]cAMP is that scintillation cocktails must be used and consequently disposed.

C. Stopping with HCl, HCl/cAMP, or HCl/(^{3}H)cAMP

Since cAMP can be formed nonenzymatically from ATP at alkaline pH, especially in the presence of Mn^{2+}, lower blank values can sometimes be obtained by stopping the adenylyl cyclase reaction with acid. The essence of this procedure was first reported by Nakai and Brooker (1975) and has since been verified and modified somewhat by Counis and Mongongu (1978) and by Alvarez and Daniels (1990, 1992).

Reagents

HCl (2.2 N). To monitor recovery of [^{32}P]cAMP, add either unlabeled cAMP (165 mg/liter → 0.5 m*M*) or [^{3}H]cAMP (10,000 to 20,000 cpm per sample) to the hydrochloric acid just before use. When kept cold, cAMP and [^{3}H]cAMP are stable in acid against chemical degradation and measurable isotopic exchange.

PROCEDURE

1. Adenylyl cyclase reactions, typically 50 µl in 13 × 65-mm glass or plastic tubes or in 1.5-ml plastic Eppendorf tubes, are terminated by the addition of sufficient volume of the HCl solution to give 0.2 to 0.5 N HCl (e.g., 10 µl of 2.2 N HCl to a 100-µl reaction volume). Concentrations less than 0.1 N do not result in markedly lower blank values and concentrations greater than 1 N cause degradation of [^{32}P]cAMP when samples are heated (Nakai and Brooker, 1975; Counis and Mongongu, 1978). If [^{3}H]ATP is used as a substrate, adenine (0.1 mM) may be included in the reaction mixture to reduce the specific activity of nonphosphorylated metabolites of [^{3}H]ATP generated during the reaction (Alvarez and Daniels, 1992).

2. The test tubes are then placed in a water bath at 90–95°C for 4 to 8 min. Examples are 90°C for 8–10 min in 0.95 N HCl (Nakai and Brooker, 1975), 4 min at 95°C in 0.165 N HCl (Counis and Mongongu, 1978), or 95°C for 10 min in 0.2 N HCl (Alvarez and Daniels, 1990). The heat step hydrolyzes ATP and unknown substances that contribute to assay blanks if [α-^{32}P]ATP is used as substrate. If [^{3}H]ATP is used as substrate, the reaction should not be terminated by heating as this increases the assay blank (Alvarez and Daniels, 1992).

3. The mixtures in the reaction tubes are then either diluted and decanted onto chromatography columns for purification of sample cAMP or neutralized and precipitated by the subsequent addition of $Zn(C_2H_3O_2)_2/Na_2CO_3$ (see Section III,A, procedure step 1). Chromatography is either on sequential Dowex-50 and alumina columns or on single alumina columns (see Section IV,B, Single Alumina Column).

4. Assay blanks are prepared by omitting enzyme, by adding enzyme after the HCl, or by prior heat denaturation of the enzyme.

5. A disadvantage of each stopping procedure when [^{3}H]cAMP is used to monitor recovery of sample [^{32}P]cAMP is that scintillation cocktails must be used and consequently disposed.

IV. Chromatographic Alternatives

The characteristic property of neutral alumina and other insoluble inorganic salts to bind multivalent nucleotides but not cAMP is the central feature of a number of variations of assays for adenylyl and guanylyl cyclases. White and Zenser (1971) passed reaction mixtures over columns of neutral alumina that were equilibrated and then developed with neutral buffer. Assay blanks with this procedure were variable and depended highly on the radiochemical purity of the α-^{32}P-labeled substrate and on the quality of the alumina. Salomon et al. (1974) and later Wincek and Sweat (1975) showed that sequential chromatography on Dowex-50 and alumina produced an assay for adenylyl cyclase that was more consistent than alumina columns alone, a combination that has also been utilized for the assay of guanylyl cyclase (Nesbitt et al., 1976). Additional variations on this procedure have been reported by a number of investigators. For example, nearly quantitative separation of cAMP from ATP was achieved by a combination of precipitation with inorganic salts [Zn(acetate)$_2$CO$_3$] followed by chromatography on alumina (Jakobs et al., 1976). To minimize the influence of variations in the quality of [α-^{32}P]ATP, variations in the behavior of various sources of alumina, and the co-elution of potential contaminants, the method of choice has become sequential chromatography on Dowex-50 and then alumina columns (Salomon, 1979; Salomon et al., 1974). This preference notwithstanding, an effective alternative single alumina column procedure (Alvarez and Daniels, 1990, 1992) with its inherent ease of use is also presented.

A. Sequential Chromatography on Dowex-50 and Alumina

Reagents

Dowex-50. H$^+$-form (e.g., Bio-Rad AG50X8, 100–200 mesh). Before use the Dowex-50 is washed sequentially with approximately 6 vol each of 0.1 N NaOH, water, 1 N HCl and water. The Dowex-50, in an approximately 1:1 slurry, is then poured into columns (~0.6 × 4 cm). After each use, Dowex-50 columns are regenerated by washing with 5 ml 1 N HCl and stored until reused. Before use the columns are then washed with one 10-ml wash with water. Between uses columns are covered with a dust cover. Columns can be reused many dozens of times, though additional resin may need to be added occasionally. If flow rates decrease they should be regenerated with NaOH, water, and HCl as above.

Alumina. Neutral (e.g., Bio-Rad AG7, 100–200 mesh; Sigma WN-3; ICN Alumina N, Super I). The source of Al$_2$O$_3$ is less critical with the two column procedure than if it is used alone, or if used alone and washed with acid before elution of cAMP with buffer (see below). The alumina (ca. 1 g) may be poured dry into columns (e.g., with a plastic scoop or a large disposable plastic syringe from which the alumina is allowed to drain), or an RCBS Uniflow adjustable gun powder measure (Omark Industries, Oroville, CA) as suggested by Alvarez and Daniels (1992).

Elution buffers. 100 mM imidazole [pH 7.5, as per the original method of Salomon *et al.* (1974)]. An equally effective and cheaper alternative is 100 mM Tris–Cl, pH 7.5, and a more efficient and consistent elution has been reported with 0.1 M ammonium acetate (Alvarez and Daniels, 1990). The purpose of the buffer is to elute cyclic nucleotides. Since eluate from the Dowex-50 columns is acidic, which enhances absorption of cyclic nucleotides to alumina, elution of cyclic nucleotides is achieved principally through the increase in pH of the buffer as well as through increased ionic strength.

Column Care Before initial use alumina columns *must* be washed once with elution buffer, either 10 ml 100 mM Tris–Cl or 10 ml 1 M imidazole, pH 7.5, otherwise the procedure does not work right away. After each use the columns are washed with 10 ml 100 mM Tris–Cl or 10 ml 100 mM imidazole, pH 7.5. Alumina columns may be reused virtually indefinitely, though additional alumina may need to be added occasionally.

Apparatus

Rapid flow rates for the alumina columns, and consequently short chromatography times, are achieved with glass columns with a large cross-sectional area and a coarse scintered glass plug to retain the alumina (Fig. 2). Satisfactory dimensions are alumina to ∼1 cm in a column 11 mm i.d. by 4 to 9 cm attached to a 2- to 4-cm glass funnel (24 mm i.d.) [smaller columns (e.g., ∼0.6 × 2 cm, alumina) while allowing satisfactory chromatographic performance are slow]. It is important that the volume above the alumina be sufficient to hold all the buffer necessary for elution of the cAMP.

The alumina may clog the scintered glass plug in time. This can be minimized by placing a glass-fiber disc on the scintered glass plug before adding alumina. The filters (Whatman GF/D) are cut to size with the plastic lip of a Sarstedt polypropylene tube (No. 72-693; 10.8 mm diameter). If it becomes necessary a clogged scintered glass plug can be freed and restored to initial flow rates by sonication in 6 N nitric acid for 30 min, followed by reverse flushing with water.

Both Dowex-50 and alumina columns are most conveniently used if they are mounted in racks (e.g., Lucite) with spacing the same as that of the racks of scintillation vials to be used. The design of the racks supporting the Dowex-50 columns should be such that the columns can be conveniently mounted above the alumina columns so that the eluate of the Dowex-50 columns can drip directly onto the alumina columns.

Similarly, the design of the racks supporting the alumina columns should allow the eluate from them to drip directly into scintillation vials.

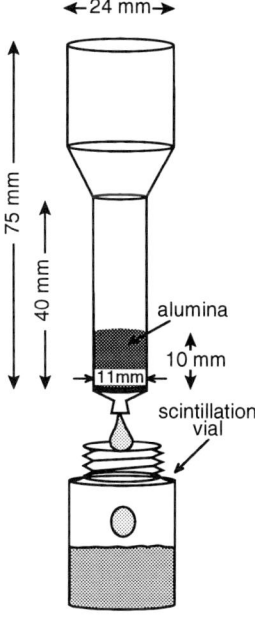

Figure 2

PROCEDURE

Two alternative procedures are described, the use of which depends on the quality of ^{32}P-labeled substrate. The first procedure should be adequate with all but the poorest quality substrate and the second procedure should lower blank values further if necessary.

1. Water elution of Dowex-50
   a. Whether reactions are stopped by the $Zn(acetate)_2/Na_2CO_3$, ATP/SDS/cAMP, or HCl method, the samples are decanted directly onto the Dowex-50 columns.
   b. The Dowex-50 columns are then washed with ca. 3 ml water. [The actual volume necessary for this step may vary slightly from batch to batch or with the age of Dowex-50 and should be determined (Salomon, 1979).] The eluate from this wash contains $^{32}P_i$ and $[\alpha\text{-}^{32}P]$ATP and -ADP and should be disposed of by the procedure described below (Section IV,C, Disposal of Waste Isotope).

c. The Dowex-50 columns are then mounted above a comparable number of alumina columns so that the eluate drips directly onto the alumina. Dowex-50 columns are washed with 8 ml water. The eluate from the Dowex-50 columns is slightly acidic and causes cAMP to be retarded on the alumina column.

d. After the eluate from the Dowex-50 columns has dripped onto and through the alumina columns, the alumina columns are placed over scintillation vials.

e. cAMP is eluted from the alumina columns directly into scintillation vials. The volume of elution buffer used depends on whether unlabeled cAMP or [^{3}H]cAMP is used to quantitate sample recovery and on the types of vials used in the scintillation counter. It is important to use sufficient buffer to elute all the cAMP as well as to optimize counter efficiency, which is dictated by the geometry of the counter's phototubes. Two examples follow: (i) If recovery is monitored with unlabeled cAMP and [^{32}P]cAMP is determined by Cherenkov radiation in 20-ml counting vials, [^{32}P]cAMP is eluted from alumina columns with 8 ml 100 mM Tris–Cl. Smaller volumes do not give optimal counting efficiency. Following counting, absorbance at 259 nm is determined on an aliquot of the sample to quantitate recovery of unlabeled cAMP. (ii) If recovery is monitored with [^{3}H]cAMP, both [^{3}H]cAMP and [^{32}P]cAMP are eluted with 4 ml 100 mM imidazole into 10-ml vials to which 5 ml scintillation cocktail is then added. The smaller vials spare expensive cocktail. Sample recovery is determined from dual-channel counting.

f. Scintillation counting of [^{32}P] by Cherenkov radiation is achieved with a single channel with wide-open windows.

g. Counting of samples containing both [^{3}H] and [^{32}P] can be achieved in a two-channel scintillation counter with windows adjusted such that there is zero crossover of [^{3}H] into the [^{32}P] window and small but measurable crossover of [^{32}P] into the [^{3}H] window.

2. Acid elution of Dowex-50, adapted from White and Karr (1978).

a. Before use the Dowex-50 columns are washed with 10 ml 0.01 N HClO$_4$.

b. Whether reactions are stopped by the Zn(acetate)$_2$/Na$_2$CO$_3$, ATP/SDS/cAMP, or HCl method, the samples are decanted directly onto the Dowex-50 columns.

c. Dowex-50 columns are then washed with 6 ml 0.01 N HClO$_4$. (The actual volume necessary for this step may vary from batch to batch or with the age of Dowex-50 and should be determined.) The eluate from this wash contains $^{32}P_i$, [α-^{32}P]ATP and -ADP, and should be disposed of by the procedures described below (Section IV,C, Disposal of Waste Isotope).

d. The Dowex-50 columns are then mounted directly above a comparable number of alumina columns and are washed with 8 ml 0.01 N HClO$_4$, which is allowed to drain through both columns.

e. The alumina columns are then washed with 10 ml water. This eluate is discarded.

f. The alumina columns are mounted above a rack of scintillation vials and the cAMP is eluted as in IV,A, procedure 1, step e, above.

NOTE Dowex-50 is slowly decomposed by HClO$_4$ and therefore cannot be used too many times. This problem is not apparent with HCl and it may be substituted for perchloric acid.

B. Single Alumina Column

Use of a single alumina column for the separation of labeled cAMP from labeled substrate and its various degradation products has been reported by a number of investigators. Typically, the problems with the procedure have been (a) that there was substantial variation in the behavior of alumina from various sources; (b) that assay blank was considerably higher than that for the double-column procedure above; (c) that assay blank was dependent on the source of [^{32}P]ATP; and (d) that when [^{3}H]ATP was used as substrate or [^{3}H]cAMP was used to monitor recovery enzymically derived breakdown products, e.g., [^{3}H]xanthine and [^{3}H]hypoxanthine, would coelute with [^{3}H]cAMP, resulting in erroneous values for [^{3}H]cAMP. These concerns have been considerably reduced by two modifications to the original methods: the use of acid to stop the adenylyl cyclase reaction and improved elution characteristics when columns are first washed with dilute acid and when ammonium acetate is used instead of Tris–Cl (Nakai and Brooker, 1975; Counis and Mongongu, 1978; Alvarez and Daniels, 1990, 1992). Thus, the best variations on this method are to stop the reaction in acid, followed by chromatography (a) on the sequential Dowex-50/alumina system above or (b) on a single alumina column first washed in acid before elution of cAMP with ammonium acetate (Counis and Mongongu, 1978; Alvarez and Daniels, 1990, 1992).

Reagents

Alumina. The source of Al_2O_3 is not critical. Alumina (ca. 1 g) may be poured dry into columns (see Section IV,A, reagents).

Elution buffers.

0.005 N HCl

100 mM ammonium acetate, pH 7 (F.W. 77.08; 7.7 g/liter)

Apparatus

Alumina columns are the same as those described above (Section IV,A, Sequential Chromatography on Dowex-50 and Alumina and Fig. 2).

PROCEDURE

1. Samples, from adenylyl cyclase reactions stopped by the addition of HCl are centrifuged (10 min/3000g, bench-top) to sediment debris. Supernatant fractions are decanted directly onto prewashed alumina columns. If the recovery of labeled cAMP is not determined, a known aliquot of the sample (e.g., 100 µl) is applied to the column. The ratio of applied volume to total volume is used in the calculation of reaction velocities (see Section V, Calculation 1).
2. Alumina columns are washed with 8 ml 0.005 N HCl followed by 1 ml 100 mM ammonium acetate. The radioactive effluent is collected and is discarded safely (Section IV,C, Disposal of Waste Isotope).
3. The rack(s) of alumina columns are placed over rack(s) of scintillation vials and the labeled cAMP is eluted.
4. For [α-^{32}P]ATP as substrate
    a. If Cherenkov radiation will be measured, add 8 ml 100 mM ammonium acetate to elute [^{32}P]cAMP directly into the vials and determine radioactivity in a liquid scintillation counter.
    b. If unlabeled cAMP is used to monitor sample recovery, after counting, determine the absorbance at 259 nm of the stopping solution and of each sample. Use these values to determine sample recovery.
    c. If [^{3}H]cAMP is used to monitor sample recovery, add 3.5 ml 100 mM ammonium acetate to elute both [^{32}P]cAMP and [^{3}H]cAMP into vials, add scintillation cocktail, and determine [^{32}P]cAMP and [^{3}H]cAMP by dual-channel counting in a liquid scintillation counter.

5. For [³H]ATP as substrate
   a. If [³²P]cAMP or [¹⁴C]cAMP is used to monitor sample recovery, add 3.5 ml 100 mM ammonium acetate to elute labeled cAMP into vials, add scintillation cocktail, and determine [³H]cAMP and recovery tracer by dual-channel counting in a liquid scintillation counter.
   b. If unlabeled cAMP is used to monitor sample recovery, add 4 ml per 100 mM ammonium acetate to elute [³H]cAMP into vials, remove an aliquot of known volume (e.g., 1.0 ml) from each sample. Determine the absorbance of 259 nm of the stopping solution and on these aliquots from each sample. Use these values to determine sample recovery. Add scintillation cocktail to the vials and determine [³H]cAMP in a liquid scintillation counter.
6. For each elution method, Section V, Calculation 1 shows how these values are used in the analysis of data.

Column Care Since a precipitation step (e.g., with $Zn(C_2H_3O_2)_2$/Na_2CO_3) is not used before samples are applied to the alumina columns, the columns absorb virtually all the radioactivity in the sample (excepting [³H]H$_2$O and ³H-labeled products from [³H]cAMP). Consequently, radioactivity accumulates very rapidly with frequent use, probably to unacceptable levels as far as lab safety is concerned. It will likely also lead to unacceptable increases in blank values. If usage is infrequent, columns may be reused if washed after each use with 8 ml 100 mM ammonium acetate followed by 8 ml 0.005 N HCl. Otherwise, the accumulation of radioactivity and its untoward effects can be circumvented either by repouring the columns before each use (Alvarez and Daniels, 1990, 1992) or by periodically washing columns with 8 ml 1 N NaOH to elute >95% bound ³²P-labeled products, followed by 8 ml water and then 8 ml 0.005 N HCl. This radioactive alkaline wash is then neutralized or acidified before disposal on the alumina–charcoal filters (Section IV,C, Disposal of Waste Isotope). Aside from the problem of accumulated radioactivity alumina columns may be reused virtually indefinitely, though additional alumina may need to be added occasionally.

C. Disposal of Waste Isotope

To avoid contamination of waste water, all radioactive waste from the above procedures is collected and pooled. It is poured onto a large Buchner funnel, containing perhaps 250 g alumina (e.g., Fisher, Al_2O_3, anhydrous), attached in series to four parallel 100-g carbon filters and a flask attached to a water aspirator. (The purpose of the flask is to allow an aliquot of the filtered waste to be monitored for radioactivity before the waste is discarded down the drain.) By use of alumina and carbon filters, virtually no ^{32}P-radioactivity is discarded in waste water, though some ^{3}H$_2$O will be lost if tritiated nucleotides are used ([^{3}H]ATP or [^{3}H]cAMP). An important additional advantage of this is that absorbed radioactive waste can then be treated as compact solid waste. This is especially useful for tritiated nucleosides and nucleotides. ^{32}P-Labeled solid waste can be allowed to decay off. The alumina can be used almost indefinitely, whereas the carbon filters tend to clog with prolonged use and need to be replaced periodically, e.g., annually.

The Zn(acetate)$_2$/CO$_3$ precipitation step offers an advantage with regard to waste disposal. Whether used to stop the reaction (Section III,B, Stopping with ATP/SDS/cAMP ± [^{3}H]cAMP), precipitation with Zn(acetate)$_2$/Na$_2$CO$_3$ traps most of the radioactivity in the ZnCO$_3$ pellet, including unused and unhydrolyzed substrate [α-^{32}P]ATP, [α-^{32}P]ADP, [^{32}P]AMP, ^{32}P$_i$. This prevents most of the radioactivity from being applied to the chromatography columns and allows it to be treated immediately as solid waste. Otherwise, with the sequential Dowex-50/alumina procedure radioactive substances are eluted in a large wash fraction, and with the single alumina column most of the radioactivity is absorbed, causing excessive accumulation of radioactivity on these columns. The investigator should weigh these alternatives for the individual laboratory and application.

Apparatus

Carbon filters for the removal of absorbable radioactive materials from column eluates are from Gelman (No. 12011) "Carbon Capsule," each containing 100 g activated charcoal.

V. Data Analysis

Calculation of adenylyl cyclase activities determined with radioactive substrates is straightforward, but must take into consideration the loss of sample cAMP that occurs during its chromatographic purification. For this reason volumes applied to columns must be known or sample recovery must be determined, either with unlabeled cAMP or with cAMP that is labeled with a second isotope. In addition, since many adenylyl cyclases exhibit low activities, especially under basal assay conditions, the radioactivity measured in the sample in the absence of enzyme (no enzyme blank) can represent a sizeable percentage of that measured with enzyme. Consequently, it becomes important to consider how this value is to be treated in the calculation of activity. If there is measurable nonenzymatic formation of cAMP from ATP, as may be the case under alkaline assay conditions, especially in the presence of manganese, the labeled cAMP must be corrected for sample loss during purification. However, if it can be established that the sample radioactivity in the absence of enzyme is due to ^{32}P-labeled contaminants in the sample, that is ^{32}P-labeled compounds not absorbed by alumina (White and Karr, 1978) or as determined through alternative chromatographic techniques, the blank value should not be corrected for sample recovery. Such a correction would give rise to an erroneously high blank value and the apparent enzyme activity would be lower than it should be. Both sample recovery and assay blank adjustments to the determination are readily made with programmable calculators, though calculations are more conveniently done with a computer since programs can easily be written to accommodate variable amounts of protein, substrate concentrations, assay times, and volumes; can be extended to the computation of enzyme kinetic constants; and can be interfaced with graphic plotters. In addition, scintillation counters may be attached directly to such computers to enhance data acquisition and processing. Examples of these calculations are given below:

1. The calculation of adenylyl cyclase activities without determination of sample recovery is simplest and is the same whether [^{3}H]ATP or [α-^{32}P]ATP is used as substrate. The example below assumes [α-^{32}P] is substrate.

Velocity = (sample [^{32}P]cpm − no enzyme [^{32}P]cpm ∗ ATP concentration ∗ reaction volume/([α-^{32}P]ATP cpm − no enzyme [^{32}P]cpm)/ fraction of sample applied to column/fraction of sample counted/ time/protein.

Fraction of sample applied to the column is determined from the volume of the reaction divided by the volume of the reaction plus the stopping solution (e.g., HCl in the single-column procedure).

2. For the calculation of adenylyl cyclase activities with [^{3}H]cAMP used for sample recovery, the assumption is made that the windows for the ^{32}P- and ^{3}H-channels of the scintillation spectrometer have been set so there is zero crossover of ^{3}H-cpm into the P^{32}-channel. The calculation compensates for crossover of ^{32}P-cpm into the ^{3}H-channel.

Velocity = (sample[^{32}P]cpm - no enzyme [^{32}P]cpm) ∗ ATP concentration ∗ reaction volume/fraction of sample counted/([α-^{32}P]ATP cpm - no enzyme [^{32}P]cpm) ∗ [^{3}H]cAMP std cpm/(sample [^{3}H]cpm − ((sample [^{32}P]cpm − no enzyme [^{32}P]cpm) ∗ [^{32}P]cpm in [^{3}H]channel/[α-^{32}P]ATP - cpm)/time/protein).

[^{3}H]cAMP - std - cpm is the value that would represent 100% recovery of the added [^{3}H]cAMP, e.g., the total [^{3}H]cpm in the 0.6 ml of Zn (acetate)$_2$

containing [^{3}H]cAMP used to stop the reaction, counted under quench conditions comparable to those used to count the sample.

3. An analogous though simpler calculation is used for activities when unlabeled cAMP is used for sample recovery and is the same whether [α-^{32}P]ATP or [^{3}H]ATP is used as substrate tracer.

Velocity = (sample cpm − no enzyme cpm) * ATP concentration * reaction volume/fraction of sample counted/(substrate cpm − no enzyme cpm) * cAMP standard-A_{259}/sample-A_{259}/time/protein.

cAMP standard-A_{259} is the optical density at 259 nm that would represent 100% recovery of the added unlabeled cAMP. This value usually also includes a factor to compensate for the volume of the final sample. In the example given here for samples chromatographed first on Dowex-50 than on Al_2O_3 columns, samples are 8 ml. For example, the optical density (A_{259}) of the 0.6 ml of Zn (acetate)$_2$ containing unlabeled cAMP is typically determined on an aliquot diluted 40-fold in 100 mm Tris–Cl, pH 7.5, and gives a value of approximately 0.2. In this example, 0.2 * 40 * 0.6 ml/8 ml → 0.6 for the cAMP standard-A_{259}.

Velocities are in nmol cAMP formed (min · mg protein)$^{-1}$ when substrate concentration is entered as micromolar, time is minutes, protein is microgram per tube, and reaction volume is microliters. The value for "fraction of sample counted" is usually 1 and would be less than 1 only if an aliquot of the sample were used for some other purpose. If protein is not known or it is not desirable to normalize to protein, a value of 1 is used and velocities are pmol cAMP formed (min · tube)$^{-1}$. For Calculations 2 and 3 the determinations of velocity assume that [^{32}P]cpm observed in the absence of enzyme is *not* cAMP and no correction is made for loss during purification of those samples. This is an important assumption only in instances when enzyme activity is low and the counts per minute observed in the absence of enzyme represent a sizeable percentage of sample counts per minute.

Acknowledgments

Yoram Salomon is the Charles and Tillie Lubin Professor of Hormone Research. Research in R.A.J.'s laboratory was supported by NIH Grant DK38828.

References

Alvarez, R., and Daniels, D. V. (1990). *Anal. Biochem.* **187**, 98–103.
Alvarez, R., and Daniels, D. V. (1992). *Anal. Biochem.* **203**, 76–82.

Awad, J. A., Johnson, R. A., Jakobs, K. H., and Schultz, G. (1983). *J. Biol. Chem.* **258**, 2960-2965.
Brostrom, M. A., Brostrom, C. O., and Wolff, D. J. (1978). *Arch. Biochem. Biophys.* **191**, 341-350.
Chan, P. S., and Lin, M. C. (1974). *In* "Hormone Action, Part C: Cyclic Nucleotides" (J. Hardman and B. O'Malley, eds.), Methods Enzymology, Vol. 38, pp. 38-41. Academic Press, New York.
Cleland, W. W. (1970). *In* "The Enzymes" (P. E. Boyer, ed.), 3rd Ed., Vol. 2, pp. 1-65. Academic Press, New York.
Cooper, D. M. F., and Londos, C. (1979). *J. Cyclic Nucleotide Res.* **5**, 289-302.
Cooper, D. M. F., Schlegel, W., Lin, M. C., and Rodbell, M. (1979). *J. Biol. Chem.* **254**, 8927-8931.
Counis, R., and Mongongu, S. (1978). *Anal. Biochem.* **84**, 179-185.
Florio, V. A., and Ross, E. M. (1983). *Mol. Pharmacol.* **24**, 195-202.
Garbers, D. L., and Johnson, R. A. (1975). *J. Biol. Chem.* **250**, 8449-8456.
Garbers, D. L., Dyer, E. L., and Hardman, J. G. (1975). *J. Biol. Chem.* **250**, 382-387.
Harwood, J. P., Low, H., and Rodbell, M. (1973). *J. Biol. Chem.* **248**, 6239-6245.
Jakobs, K. H., Saur, W., and Schultz, G. (1976). *J. Cyclic Nucleotide Res.* **2**, 381-392.
Jakobs, K. H., Saur, W., and Schultz, G. (1978). *FEBS Lett.* **85**, 167-170.
Johnson, R. A. (1980). *J. Biol. Chem.* **255**, 8252-8258.
Johnson, R. A., and Garbers, E. L. (1977). *In* "Receptors and Hormone Action" (B. W. O'Malley and K. Birnbaumer, eds.), Vol. 1, pp. 549-572. Academic Press, New York.
Johnson, R. A., and Walseth, T. F. (1979). *Adv. Cyclic Nucleotide Res.* **10**, 135-167.
Johnson, R. A., and Welden, J. (1977). *Arch. Biochem. Biophys.* **183**, 2176-2227.
Johnson, R. A., Saur, W., and Jakobs, K. H. (1979). *J. Biol. Chem.* **254**, 1094-1101.
Krishna, G., Weiss, B., and Brodie, B. (1968). *J. Pharmacol. Exp. Ther.* **168**, 379-385.
Londos, C., and Preston, M. S. (1977). *J. Biol. Chem.* **252**, 5957-5961.
Londos, C., Cooper, D. M. F., Schlegel, W., and Rodbell, M. (1978). *Proc. Natl. Acad. Sci. U.S.A.* **75**, 5362-5366.
Nakai, C., and Brooker, G. (1975). *Biochim. Biophys. Acta* **391**, 222-239.
Nesbitt, J. A., III, Anderson, W. B., Miller, Z., Pastan, I., Russell, T. R., and Gospodarowicz, D. (1976). *J. Biol. Chem.* **251**, 2344-2352.
Perez-Reyes, E., and Cooper, D. M. F. (1986). *J. Neurochem.* **46**, 1508-1516.
Rall, T. W., and Sutherland, E. W. (1958). *J. Biol. Chem.* **232**, 1065-1076.
Rodbell, M., Birnbaumer, L., Pohl, S. L., and Krans, H. M. J. (1971). *J. Biol. Chem.* **246**, 1877-1882.
Salomon, Y. (1979). *Adv. Cyclic Nucleotide Res.* **10**, 35-55.
Salomon, Y., Londos, C., and Rodbell, M. (1974). *Anal. Biochem.* **58**, 541-548.
Salter, R. S., Krinks, M. H., Klee, C. B., and Neer, E. J. (1981). *J. Biol. Chem.* **256**, 9830-9833.
Sutherland, E. W., Rall, T. W., and Menon, T. (1962). *J. Pharmacol. Chem.* **237**, 1220-1227.
White, A. A., and Karr, D. B. (1978). *Anal. Biochem.* **85**, 451-460.
White, A. A., and Zenser, T. V. (1971). *Anal. Biochem.* **41**, 372-396.
Wincek, T. J., and Sweat, F. W. (1975). *Anal. Biochem.* **64**, 631-635.

20

Intracellular Mediators of Peptide Hormone Action: Glycosyl Phosphatidylinositol/Inositol Phosphoglycan System

Isabel Varela-Nieto, Luis Alvarez, and José M. Mato

I. Introduction
II. Methods for the Purification and Characterization of the Glycosyl Phosphatidylinositol/Inositol Phosphoglycan System
   A. Purification of Rat Liver Glycosyl Phosphatidylinositol
   B. Metabolic Labeling of Cells and Glycosyl Phosphatidylinositol Purification
   C. Chemical Labeling of Cells with Isethionyl Acetimidate
   D. Labeling of Purified Glycosol Phosphatidylinositol with Galactose Oxidase/$NaB[^3H]_4$ and Generation of [Galactose-^3H] Inositol Phosphoglycan
   E. Glycosyl Phosphatidylinosital Characterization
   F. Purification and Characterization of IPG
III. Biological Activity of Inositol Phosphoglycan
   A. Insulin-Like Effects of Inositol Phosphoglycan
   B. Regulation of Gene Expression by Inositol Phosphoglycan
   References

I. Introduction

Glycosyl phosphatidylinositols are found in a diversity of organisms from bacteria and yeast to mammalian cells. In 1986, Saltiel and Cuatrecasas first reported that the addition of insulin to BC3H1 myocytes produced a fast and transient hydrolysis of a glycosyl phosphatidylinositol (GPI) with generation of diacylglycerol and the polar headgroup of the lipid, an inositol phosphoglycan (IPG) (for reviews see Mato and Varela, 1990; Larner *et al.*, 1990; Saltiel, 1991). Since then, several laboratories have reported a similar effect of insulin on GPI hydrolysis using hepatoma cells, hepatocytes, adipocytes, lymphocytes, and Chinese hamster ovary (CHO) cells. The detailed structure of IPG is not known but it has been reported to contain *myo*- and *chiro*inositol, glucosamine, galactosamine, galactose, and mannose (Mato and Varela, 1990; Saltiel, 1991; Larner et al., 1990). A variety of results indicate a strong structural similarity between IPG and the glycan moiety of glycosyl phosphatidylinositol molecules that serve as protein anchors. Thus, an antibody raised against the glycan moiety of *Trypanosome brucei* variant surface glycoprotein (VSG) (Romero et al., 1990) has been found to cross-react with IPG obtained from both rat liver and chick embryo GPI. This antibody is specific and did not react with inositol, inositol phosphate, glucosamine, galactose, or mannose (Represa et al., 1991).

In addition to insulin, other growth factors and hormones have been found to stimulate GPI hydrolysis in target cells. The receptors for these ligands fall into two categories: the receptor protein-tyrosine kinase family, formed by the insulin, insulin-like growth factor-I, epidermal growth factor, nerve growth factor, and interleukin-2 receptors, and the group of receptors with seven transmembrane domains, formed by the adrenocorticotropin and thyroglobin receptors. These results suggest the existence of two mechanisms leading to GPI hydrolysis, one dependent and the other independent of protein-tyrosine kinase activity.

Insulin binds to the α subunit of the insulin receptor which stimulates the protein-tyrosine kinase activity of the β subunit. The mechanism by which the insulin receptor protein-tyrosine kinase and GPI hydrolysis are linked is not known. Hydrolysis of GPI in response to insulin is reduced in CHO cells bearing protein-tyrosine kinase-deficient insulin receptors (Villalba et al., 1990). Cells carrying normal human receptors hydrolyzed up to 70% of their GPI within 2 min of the addition of 0.1 nM insulin, whereas parental cells and cells expressing the mutant receptor hydrolyzed only 20–30% in response to 100 nM insulin. These results indicate that the receptor protein-tyrosine kinase is necessary to transduce efficiently the effect of insulin on GPI hydrolysis. Insulin-stimulated GPI hydrolysis can be decreased in myocytes by treatment with cholera toxin (Luttrell et al., 1990). Thus, it is possible that G-proteins are an intermediate in the signal cascade from receptor to phospholipase activation.

The role of IPG as insulin mediator has been postulated on the basis of, first, insulin-dependent regulation of GPI turnover (Saltiel et al., 1986; Mato et al., 1987a; Romero et al., 1988; Varela et al., 1990b); second, anti-IPG antibodies that selectively block some of the actions of insulin (Romero et al., 1990); and third, the ability of IPG to mimic the short-term effects of the hormone (see, e.g., Kelly et al., 1987; Alvarez et al., 1987; Bruni et al., 1990, 1991; Machicao et al., 1990). However, the role of the GPI/IPG system as a signal transduction pathway for hormones and growth factors is still subjected to debate. The controversy arises partially from the experimental difficulties in the purification and full characterization of these molecules in order to study their biological activity.

The present chapter deals with the procedures currently used to address these points. An assay to test the biological effects of IPG on gene expression is described in detail.

II. Methods for the Purification and Characterization of the Glycosyl Phosphatidylinositol/Inositol Phosphoglycan System

A. Purification of Rat Liver Glycosyl Phosphatidylinositol

This procedure is schematized in Fig. 1. Thirty rat livers (about 300 g of tissue) are homogenized in 0.9% NaCl (40 ml/g liver). A membrane fraction is prepared by centrifugation at 100,000g for 1 hr at 4°C of the supernatant at 12,000g (10 min). Membranes are extracted with 1.5 liter of chloroform/methanol (1:2) containing 0.05 N HCl. After standing for 30 min at room temperature, 500 ml of chloroform plus 500 ml of 0.1 M KCl are added to form two phases and then shaken for 30 min. The sample is centrifuged at 2800g for 5 min and the aqueous phase is removed. The organic extract is adsorbed to silica G60 (50 g) (preactivated at 110°C for 60 min) and contaminants are removed first with 500 ml of chloroform and then the pellet from a 2800g (5 min) spin is washed twice with 1 liter of chloroform:methanol:HCl (300:50:3). Polar lipids are extracted with 500 ml of methanol and evaporated to dryness in a rotary evaporator and the GPI is purified by sequential thin-layer chromatography (TLC). The dried samples are dissolved in chloroform:methanol (2:1) and spotted on silica gel G60 TLC plates. The plates are developed twice in chloroform:acetone:methanol:glacial acetic acid:water (50:20:10:10:5) (solvent 1). A 1-cm region around the origin is eluted with methanol at 37°C and this fraction rechromatographed in chloroform:methanol:NH_4OH:water (45:45:4:10) (solvent 2). The purified GPI has an R_F value of 0.5; finally this fraction is scraped, eluted three times with 20 ml of methanol, and kept at $-20°C$ until it is used. This procedure can also be used to purify GPI from fresh bovine liver. The glycophospholipid isolated by this method is free of contamination by all major phospholipids, including the polyphosphoinositides. The presence of specific sugars can be confirmed by gas chromatography/mass spectrometry analysis (Mato et al., 1987b).

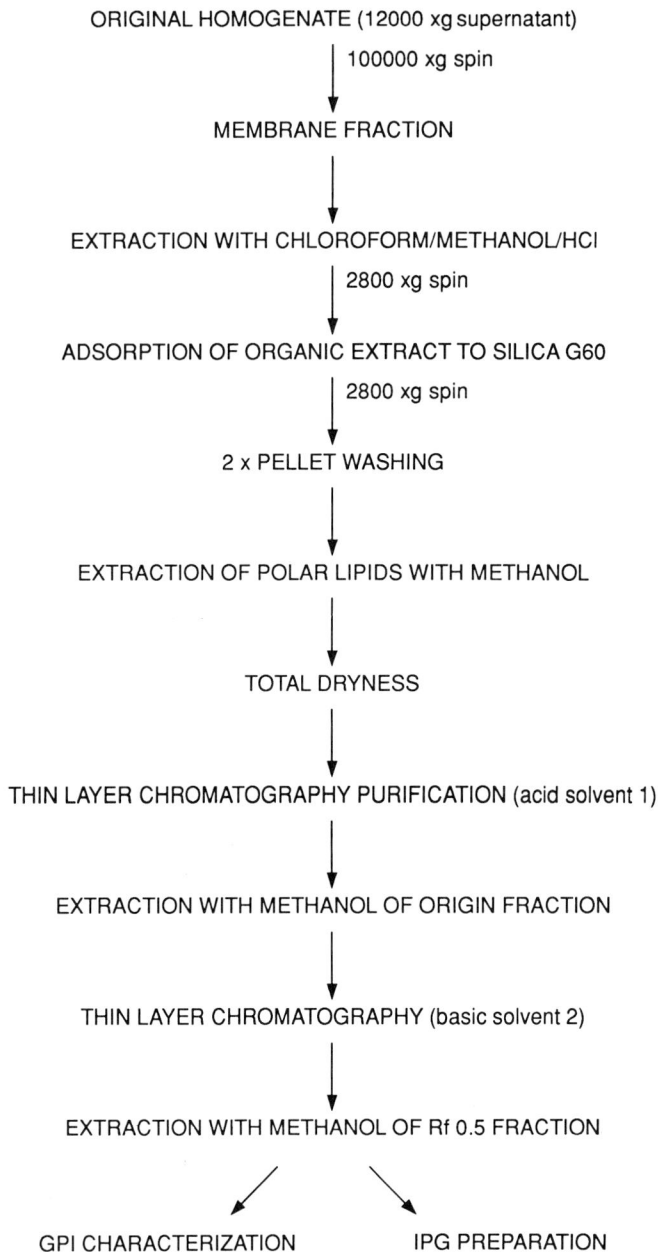

Figure 1 Scheme of the purification procedure of glycosyl phosphatidylinositol.

Table I
Regulation of GPI Hydrolysis

Signal	Effect	Cell type	References
Insulin	Stimulation	BC$_3$H-1 myocytes	Saltiel et al. (1986); Lutrell et al. (1988); Larner et al. (1990)
		H35 hepatoma	Mato et al. (1987b)
		T-lymphocytes	Gaulton et al. (1988); Avila et al. (1992)
		Rat hepatocytes	Alvarez et al. (1988)
		Rat adipocytes	Macaulay and Larkins (1990)
		CHO cells	Villalba et al. (1990)
IGF-I[a]	Stimulation	BC$_3$H-1 myocytes	Farese et al. (1988); Suzuki et al. (1991)
EGF	Stimulation	BC$_3$H-1 myocytes	Farese et al. (1988)
NGF	Stimulation	PC-12 cells	Chan et al. (1989)
		CVG	Represa et al. (1991)
ACTH	Stimulation	Adrenal glomerulosa cells	Cozza et al. (1988)
TSH	Stimulation	Pig thyroid cells	Martiny et al. (1990)
IL-2	Stimulation	T- and B-lymphocytes	Eardley and Koshland (1991); Merida et al. (1990)
IL-1	Stimulation	Fibroblasts	Dobson and Brown (1990)
IL-4	Blockade[b]	B-lymphocytes	Eardley and Koshland (1991)

[a] IGF-I, insulin-like growth factor-I, EGF, epidermal growth factor; NGF, Nerve growth factor, CVG, chicken embryo cochleovestibular ganglia; ACTH, adrenocorticotropic hormone; TSH, thyroid-stimulating hormone; IL, interleukin.
[b] Blockade of IL-2-dependent GPI hydrolysis.

B. Metabolic Labeling of Cells and Glycosyl Phosphatidylinositol Purification

This method has been extensively used to examine the presence and chemical composition of GPI in different cell lines, as well as in studies of the sensitivity of GPI turnover to hormones and growth factors (Table I).

Cells in exponential growth phase can be labeled for periods ranging from 24 to 72 hr in the presence of 5 μCi/ml of the various labels known to be precursors of GPI such as [^{3}H]glucosamine, [^{3}H]galactose, [^{3}H]myristic acid, or [^{3}H]palmitic acid. Metabolic labeling with myo-[^{3}H]inositol (10 μCi/ml) has rendered different results depending on the cell system studied. While no incorporation of myo-[^{3}H]inositol is observed in H35 hepatoma cells (Mato et al., 1987a), a positive result has been reported with BC$_3$H-1 myocytes (Saltiel and Cuatrecasas, 1986; Farese et al., 1988).

Cellular lipids are extracted and the GPI is purified as follows. At the end of the incubation period, cells are collected and resuspended in 2 ml of phosphate-buffered saline and 2 ml of ice-cold 10% trichloroacetic acid added to each sample. After standing for 15 min at 4°C, cells are scraped and transferred to glass tubes; wells are washed once with 1 ml of 5% trichloroacetic acid and the washes pooled. After centrifugation for 10 min at 2000 rpm at 4°C the supernatant is discarded and the pellet extracted with 1.5 ml of chloroform:methanol (1:2) containing 0.05 N HCl. After standing for 15 min at room temperature, the sample is centrifuged and the supernatant transferred to a clean tube. The pellet is reextracted with 0.75 ml of chloroform:methanol (1:2), 0.05 N HCl and centrifuged and the supernatant pooled. A total of 0.75 ml of chloroform plus 0.75 ml of 0.1 M KCl is added to the pooled supernatants and the organic and aqueous phases are separated by centrifugation at 4°C. The aqueous phases are treated with 0.75 ml of chloroform and centrifuged; then the organic phases are pooled. Finally, 0.75 ml of 0.1 M KCl in 50% methanol is added to the organic phases and, after standing at $-20°C$ for 30 min, it is collected and evaporated to dryness in an Speed-Vac concentrator. The dried samples are dissolved into 50 μl of chloroform:methanol (2:1) and spotted on a silica gel G60 TLC. The plate is developed twice with solvent 1, and the material which remained at the origin of the plate (0.5 cm below to 1 cm above the origin) is scraped and extracted three times with 1 ml methanol at 37°C. The acidic plate extract is evaporated as above and loaded onto a second silica gel G60 plate, which is developed once in solvent 2. One-centimeter regions are scraped and assayed for radioactivity in a β-scintillation counter.

In some instances, labeled GPI peaks from either the acidic or the basic plates can be further analyzed by two-dimensional thin-layer chromatography. Plates are then developed first in chloroform:methanol:water (10:10:3) and second in solvent 2. Radioactivity associated to GPI is evaluated by fluorography after the plates are sprayed with En^{3}Hance (New England Nuclear).

C. Chemical Labeling of Cells with Isethionyl Acetimidate

Hepatocytes, adipocytes, erythrocytes, fibroblasts, and lymphocytes are among the cell types that can be labeled by this method (Alvarez *et al.*, 1988; Varela *et al.*, 1990b). The free amino group of the residue of glucosamine of the GPI can be amidinated with the imidoesters [1-^{14}C]isethionyl acetimidate, which is unable to penetrate intact cells, or [1-^{14}C]ethyl acetimidate, which is able to do it. [1-^{14}C]Isethionyl acetimidate is available from Amersham only upon request.

D. Labeling of Purified Glycosyl Phosphatidylinositol with Galactose Oxidase/NaB(^{3}H)$_4$ and Generation of (Galactose-^{3}H) Inositol Phosphoglycan

GPI is purified from rat liver membranes by sequential TLC as described in Section II, A, Purification of Rat Liver Glycosyl Phosphatidylinositol. Purified GPI from 20 rat livers is resuspended, using sonication, into 0.3 ml of 50 mM phosphate buffer, pH 8.0, and reacted with 5 U galactose oxidase from *Doctylium dendroides* and 2 mCi of NaB[^{3}H]$_4$ (15 Ci/mmol) for 1 hr at 37°C. The mixture is then dried under vacuum, dissolved in chloroform:methanol (2:1, v:v), and purified by TLC on silica gel G60. The plate is developed twice in solvent 1. A 1-cm region around the origin is eluted twice with 2 ml methanol at 37°C, and this fraction is rechromatographed in solvent 2. One-centimeter regions are then scraped, lipids are eluted twice with 2 ml methanol at 37°C; and the radioactivity associated with each fraction is determined by counting an aliquot. Labeled GPI is then dissolved into 0.75 ml chloroform:methanol (2:1, v:v) containing 0.03 N HCl and 0.2 ml water is added to form two phases. After vigorous shaking, the organic phase is retained, the water phase washed once with 0.3 ml chloroform:methanol (2:1, v:v), and the organic phases combined (Alvarez *et al.*, 1991a). This method offers the advantage of generating unmodified labeled GPI molecules first by oxidation with galactose oxidase treatment, quickly followed by borohydride reduction. This is a standard protocol in sugar chemistry studies (Steck and Dawson, 1974).

The efficacy of the purification procedure can be assessed by HPLC of the [galactose-^{3}H]GPI using a silica column (Ultrasil-Si, 10 μm, 4.6 × 250 mm, Beckman). A 20-min linear gradient from chloroform:methanol:glacial acetic acid (70:10:5, v:v) to chloroform:methanol:glacial acetic acid:water (40:45:10:2, v:v) is used. The flow rate recommended is 1 ml/min and the elution of the labeled GPI is detected by measuring the amount of radioactivity in each fraction.

[Galactose-^{3}H]IPG can be prepared from purified labeled GPI by treatment with phosphatidylinositol-phospholipase C (PI-PLC)-specific phospholipase C from *Bacillus cereus* as described under Section II, E, 2, Phospholipase C activity. [Galactose-^{3}H]IPG is potentially an important tool for research in this field, and so far it has facilitated the study of an IPG transport system in rat hepatocytes (Alvarez *et al.*, 1991a).

E. Glycosyl Phosphatidylinositol Characterization

The presence of GPI is typically confirmed by demonstrating that the glucosamine molecule is covalently linked to PI (Saltiel *et al.*, 1986; Mato *et al.*, 1987a).

1. Nitrous Acid Deamination

Purified GPI can be cleaved by nitrous acid deamination with generation of PI, indicating the presence of inositol monophosphate linked to non-*N*-acetylated glucosamine. To determine the nitrous acid sensitivity of glycolipids, samples of purified [^{3}H]GPI, obtained from the different sources, are dried under a stream of nitrogen and resuspended in 0.1 ml of 50 mM sodium acetate, pH 3.5. After the addition of 0.1 ml of 0.33 M NaNO$_2$ samples are incubated for 5 hr at room temperature. Lipids are then extracted and the amount of remaining GPI in the organic phase is determined as described above. Alternatively, the cleavage by nitrous acid can be estimated by measuring radioactivity recovered in the aqueous phase. A blank sample containing GPI but not NaNO$_2$ should be included in each test to evaluate the rate of basal hydrolysis.

2. Phospholipase C Assay

Sensitivity to PI-PLC is assessed using either a commercial PI-PLC purified from *B. cereus* (Boehringer–Mannhein) or a PI-PLC purified from *Bacillus thuringiensis* (strain 11607) by ammonium sulfate precipitation and cation-exchange chromatography on a Whatman CM-52 (Udenfriend *et al.*, 1991). PI-PLC can be also purified from *B. cereus T* following the procedure of Kominami (Kominami *et al.*, 1985). Samples of purified labeled glycolipid are resuspended by ultrasonication (1 min, three times) in 0.2 ml of a 20 mM sodium borate, pH 7.4, buffer containing 0.16% (w/v) sodium deoxycholate, 50 μg of phosphatidylcholine (PC), and 50 μg phosphatidylethanolamine (PE). This sample is treated with 1 unit of either PI-PLC for 2 hr at 37°C. One unit of PI-PLC activity is defined as that amount of enzyme that will hydrolyze 0.8 nmol of phosphatidylinositol in 1 min at 37°C. Reactions are terminated by the addition of chloroform:methanol, and the amount of remaining [^{3}H]GPI is determined.

The sensitivity to both treatments varies significantly depending on the source of the GPI tested, the purification procedure, and the assay performed (see, e.g., Gaulton *et al.*, 1988; Eardley and Koshland, 1991; Avila *et al.*, 1992).

F. Purification and Characterization of IPG

1. *In Vitro* Purification of IPG

IPG is prepared by treating purified liver GPI with bacterial PI-PLC. Typically GPI purified from 30 rat livers is resuspended as above in 0.2 ml of 20 mM borate buffer, pH 7.4, and treated with 5 U/ml of PI-PLC at 37°C for 10 hr. The treated lipid is then extracted with 0.75 ml of chloroform:methanol:1 N HCl (2:1:0.03, v:v), the organic phase is reextracted with 0.5 ml of 0.005 M NaCl in 50% methanol, and the upper aqueous phases are pooled, evaporated to remove methanol, dissolved in distilled water, and lyophilized. When required, IPG can be further purified by HPLC using a SAX-column and the following gradient: from 0 to 12 min (100% Buffer A: KH_2PO_4, 7 mM; KCl, 7 mM, pH 4.0), 12 to 17 min (50% Buffer A), 17 to 32 min (100% Buffer B: KH_2PO_4, 250 mM, KCl, 500 mM, pH 5.0). A total of 3 ml/min flow and fractions of 1.5 ml are recommended. IPG is eluted from the HPLC column with a retention time of about 8 min. It should be mentioned that the purification of two distinct peaks using this protocol have been reported (Saltiel and Sorbara-Cazan, 1987).

The concentration of IPG can be calculated by following any of three different procedures. Bear in mind that, at present, the detailed structure of IPG is not known, so the molecular weight estimated is inexact. The first method involves the measurement of free amino groups considering that an IPG molecule contains one glucosamine molecule (Merida *et al.*, 1988). Basically, non-*N*-acetylated glucosamine is determined in the sample by reacting its non-*N*-acetylated amino group with fluorescamine. IPG sample and known concentrations of glucosamine are dissolved in 0.1 ml distilled water. After addition of 1.4 ml of 0.2 M sodium borate, pH 9.0, 0.5 ml fluorescamine (0.3 mg/ml) is added with constant vigorous shaking. Fluorescence detection is then measured with a spectrofluorometer (excitation, 390 nm; emission, 4.75 nm). This procedure offers the advantages of being both fast and economical in terms of IPG spent. Second, the concentration of IPG can be quantified by measuring the amount of organic phosphate by the classical method of Barlett (Barlett, 1959). In this procedure it is assumed that each molecule of IPG contains three phosphate groups. This method is not convenient for routine purposes since it consumes an elevated amount of compound. Finally, IPG concentration can be evaluated by its effect on protein kinase cAMP-dependent activity (Villalba *et al.*, 1988). A standard curve is made by plotting the percentage of remaining protein kinase A activity vs IPG doses calculated by any of the methods described above. This method is reliable and recommended when performing *in vivo* studies because it gives the best reference for both IPG concentration and activity. However, bear in mind that the activity of the protein kinase A can be inhibited by other factors such as salt concentration, pH changes, or uncontrolled impurities which would render a wrong estimation of the amount of IPG present in the assay. To avoid this pitfall a parallel assay of IPG activity can be performed; for example, the stimulation of cell proliferation in fibroblasts as it is described later in the chapter can be tested. Alternatively, a nitrous acid deamination of IPG, it can be performed, as has been described for the GPI deamination, followed by an assay of relative loss of the inhibitory effect on protein kinase A activity.

2. *In Vivo* Purification of Inositol Phosphoglycan

Insulin-generated IPG can be purified from [^{3}H]glucosamine-labeled hepatoma cells treated for 10 min with 100 nM insulin. After insulin treatment cells are pelleted and extracted with chloroform:methanol:1 N HCl (2:1:0.03, v:v). The organic phase is reextracted as already described and the pool of aqueous phases is freeze-dried. Insulin-generated [^{3}H]IPG is then purified by SAX HPLC (Saltiel and Sorbara-Cazan, 1987). A similar protocol can be followed to purify IPG from different unlabeled cells upon insulin (Alvarez *et al.*, 1991b) or TSH (Martiny *et al.*, 1990) stimulation.

We have described here the procedures to purify and characterize both GPI and IPG molecules. Section III, Biological Activity of Inositol Phosphoglycan contains an example of evaluating the biological activity of IPG purified as above.

III. Biological Activity of Inositol Phosphoglycan

A. Insulin-Like Effects of Inositol Phosphoglycan

The number of insulin-like effects of IPG on both intact cells and cellular extracts has grown rapidly since it was first described in 1986 (Table II).

Usually, the biological activity of a new batch of IPG is assessed *in vitro* by testing its capacity to inhibit the phosphorylation of histone IIA by the catalytic subunit of the cAMP-dependent protein kinase (Villalba *et al.*, 1988; Larner *et al.*, 1988). Insulin-like effects of IPG in intact cells are easily evaluated by checking its ability to stimulate [U-^{14}C]glucose incorporation into glycogen (Cabello *et al.*, 1990), amino acid transport measured as [^{14}C]AIB uptake in isolated rat hepatocytes (Varela *et al.*, 1990a), or [^{3}H]thymidine incorporation into DNA in NIH 3T3 fibroblasts (Alvarez *et al.*, 1991b). Typically when IPG is added to intact cells it is neutralized with concentrated NaOH, diluted in a convenient buffer, and passed through a Dynagard 0.2-μm syringe filter.

Recently, it has been reported that IPG modulates gene expression in an

Table II
Insulin-Like Effects of IPG in Intact Cells and in Cellular Extracts

Intact Cells

Biological Activity	Effect	Reference
Lipolysis	Inhibition	Kelly et al. (1987)
Lipogenesis	Stimulation	Saltiel and Sorbara-Cazan (1987); Machicao et al. (1990)
Phospholipid methyltransferase	Inhibition	Kelly et al. (1987)
Glucose transport	No effect[a]	Kelly et al. (1987); Saltiel and Sorbara-Cazan (1987)
Acetil-CoA carboxilase activity	Stimulation	Witters and Watt (1989)
Glycogen fosforilase activity	Inhibition	Alvarez et al. (1987)
Pyruvate kinase activity	Stimulation	Alvarez et al. (1987)
Glucose oxidation	Stimulation	Saltiel and Sorbara-Cazan (1987)
Lactate accumulation	Stimulation	Bruni et al. (1990)
Glucogen synthesis	Stimulation	Alvarez et al. (1991a)
Tyrosine aminotransferase activity	No effect	Witters and Watts (1989)
Protein phosphorylation	Stimulation/inhibition	Alemany et al. (1987)
cAMP levels	Inhibition	Alvarez et al. (1987); Larner et al. (1990)
Fructose-2,6-2P levels	Stimulation	Bruni et al. (1990)
Amino acid transport	Stimulation	Varela et al. (1990a)
Protein synthesis	Stimulation	Varela et al. (1990a)
Specific mRNA levels	Stimulation	Alvarez et al. (1991b); Sato et al. (1988)
	Inhibition	Alvarez et al. (1991b)
DNA synthesis	Stimulation	Witters and Watts (1989); Varela-Nieto et al. (1991)
Cellular proliferation	Stimulation	Varela-Nieto et al. (1991)
Insulin secretion	Inhibition	Albor et al. (1989)

Cellular Extracts

Enzymatic activity/ phosphorylation	Effect	Reference
cAMP phosphodiesterase	Stimulation	Saltiel et al. (1986)
Pyruvate dehydrogenase	Stimulation	Saltiel et al. (1986); Larner et al. (1988)
Adenylate cyclase	Inhibition	Saltiel et al. (1986)
cAMP-kinase	Inhibition	Villalba et al. (1988); Larner et al. (1988)
Casein kinase II	Stimulation/inhibition	Alemany et al. (1987)

[a] Actovegin is a plasma-derived compound that contains IPG-like molecules and stimulates glucose transport (Machicao et al., 1990).

insulin-like manner (Alvarez *et al.*, 1991b). This biological action of IPG appears to be highly significant since any putative signaling mechanism has to compensate for insulin's ability to stimulate the expression of some genes while inhibiting others and for the time required for observing a maximal effect of the hormone on transcription, which vary from minutes to hours.

B. Regulation of Gene Expression by Inositol Phosphoglycan

The ability of IPG to mimic insulin effects on the regulation of the expression of specific mRNAs can be studied in several cellular systems. Research has been focused in the model of isolated hepatocytes from either normal or diabetic rats, as well as in the H4IIE hepatoma cell line.

Isolation of rat hepatocytes from either normal or diabetic animals, culture of H4IIE rat hepatoma cells, RNA extraction and Northern analysis, and nuclear *in vitro* transcription assay were performed as described previously with some modifications as detailed in Alvarez *et al.* (1991b).

1. Determination of mRNA Levels

IPG action on the expression of phosphoenolpyruvate carboxykinase (PEPCK) and α_2-microglobulin has been examined. These are two typical target genes of insulin action. The enzyme phosphoenolpyruvate carboxykinase (GTP; PEPCK, E.C. 4.1.1.32) plays a key regulatory role in the gluconeogenic pathway and its synthesis rate is acutely regulated by a number of hormones (Loose *et al.*, 1985). Insulin decreases the level of PEPCK mRNA in rat liver *in vivo* and *in vitro*, antagonizing the stimulatory effects of cAMP. This effect of insulin has been studied in detail and it occurs at the transcriptional level. α_2-Microglobulin gene expression is another well-defined example of multihormonal regulation (Mira and Castaño, 1989). Insulin increases the transcriptional rate of α_2-microglobulin gene in livers from diabetic animals.

The effect of IPG on α_2-microglobulin mRNA levels was studied in parallel with its effect on PEPCK mRNA levels in diabetic rats. Figure 2 shows that IPG is able concomitantly to increase the expression of α_2-microglobulin mRNA and decrease the levels of PEPCK mRNA, similarly to insulin. IPG produces within 15 min a dose-dependent 4-fold increase in the levels of α_2-microglobulin mRNA and a 2.5-fold decrease in PEPCK mRNA levels. Nitrous acid-inactivated IPG and GPI are used as controls to determine the specificity of the effect of IPG.

Parallel experiments with hepatocytes isolated from normal rats and with the H4IIE hepatoma cell line indicate that, as described for insulin, within 90 min IPG is able to antagonize cAMP prestimulation of PEPCK mRNA.

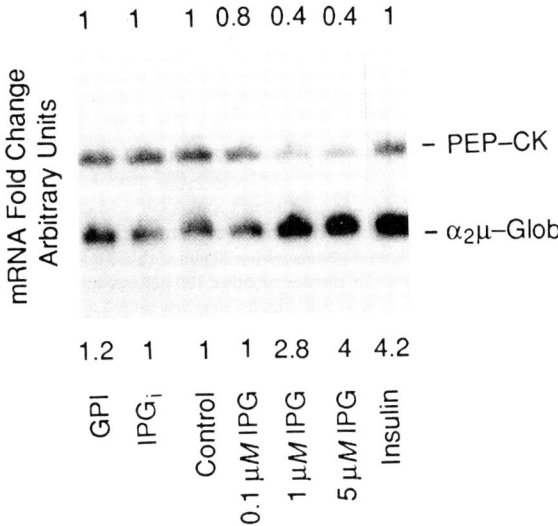

Figure 2 Effects of IPG on α_2-microglobulin and PEPCK mRNA expression on hepatocytes from diabetic rats. Isolated hepatocytes from diabetic animals were incubated for 15 min with saline (control), 5 μM nitrous acid-inactivated inositol phosphate-glycan (IPG$_i$), 5 μM glycosyl phosphatidylinositol (GPI), 100 nM insulin, or the indicated concentrations of IPG. Total RNA was hybridized first to the α_2-microglobulin ($\alpha_2\mu$-glob) probe, the filter was washed but this probe was not removed before adding the second probe (PEPCK). The resulting autoradiogram was quantitated and data are expressed as mRNA fold change over the control that was set at a value 1. [From Alvarez et al. (1991b), courtesy of Molecular Endocrinology. Reprinted from Mol. Endocrinol. 5, 1062–1068 (1991).]

2. *In Vitro* Transcription Assay

To study whether the inhibition of PEPCK mRNA levels by IPG is controlled at a transcriptional level, an *in vitro* transcription assay in isolated H4IIE hepatoma cell nuclei has been performed. This system was chosen since it has been extensively used to study the effect of insulin on PEPCK gene expression.

Run-on experiments have been carried out in nuclei isolated from serum-deprived H4IIE hepatoma cells pretreated for 30 min with 8-Br-cAMP (0.1 mM) and then incubated for another 30 min in the presence or absence of either insulin (100 nM) or IPG (1 μM). The relative rate of PEPCK mRNA synthesis has been determined by hybridization of the ^{32}P-labeled nuclear RNA transcripts to immobilized DNA probes. The transcription rate of the PEPCK gene increases 4.3-fold in the presence of 8-Br-cAMP relative to serum-deprived control cells (Table III). When insulin is added to the medium

Represa, J., Avila, M. A., Miner, C., Giraldez, F., Romero, G., Clemente, R., Mato, J. M., and Varela-Nieto, I. (1991). *Proc. Natl. Acad. Sci. U.S.A.* **88**, 8016–8019.

Romero, G., Lutrell, L., Rogol, A., Zeller, K., Hewlett, E., and Larner, J. (1988). *Science* **240**, 509–511.

Romero, G., Gámez, G., Huang, L. C., Lilley, K., and Luttrell, L. (1990). *Proc. Natl. Acad. Sci. U.S.A.* **87**, 1476–1480.

Saltiel, A. R. (1991). *J Bioenerg. Biomembr.* **23**, 29–41.

Saltiel, A. R., and Cuatrecasas, P. (1986). *Proc. Natl. Acad. Sci. U.S.A.* **83**, 5793–5797.

Saltiel, A. R., and Sorbara-Cazan, L. R. (1987). *Biochem. Biophys. Res. Commun.* **149**, 1084–1092.

Saltiel, A. R., Fox, J. A., Sherline, P., and Cuatrecasas, P. (1986). *Science* **233**, 967–972.

Sato, T., Villar Palasi, C., Huang, L., Tang, G., Larner, A. C., and Larner, J. (1988). *Endocrinology (Baltimore)* **123**, 1559–1564.

Steck, T. L., and Dawson, G. (1974). *J. Biol. Chem.* **249**, 2135–2142.

Suzuki, S., Satoh, Y., and Toyota, T. (1991). *J. Biol. Chem.* **266**, 8115–8119.

Udenfriend, S., Micanovic, R., and Kodukula, K. (1991). *Cell Biol. Int. Rep.* **15**, 739–759.

Varela, I., Avila, M., Mato, J. M., and Hue, L. (1990a). *Biochem. J.* **267**, 541–544.

Varela, I., Alvarez, J. F., Ruiz-Albusac, J. M., Clemente, R., and Mato, J. M. (1990b). *Eur. J. Biochem.* **188**, 213–218.

Varela-Nieto, I., Represa, J., Avila, M. A., Miner, C., Mato, J. M., and Giraldez, F. (1991). *Dev. Biol.* **143**, 432–435.

Villalba, M., Kelly, K., and Mato, J. M. (1988). *Biochim. Biophys. Acta* **968**, 69–76.

Villalba, M., Alvarez, J. F., Russell, D., Mato, J. M., and Rosen, O. (1990). *Growth Factors* **2**, 91–97.

Witters, L. A., and Watts, T. D. (1989). *J. Biol. Chem.* **263**, 8027–8036.

21

Free Calcium Measurements: Down to the Single-Cell Level

Antonio Sanchez-Bueno and Peter H. Cobbold

I. Introduction
II. Single Cell versus Population to Measure Ca^{2+}
III. Properties of Aequorin
IV. The Aequorin Technique
   A. Aequorin Preparation
   B. Preparation of Cells for Microinjection
   C. Pipette Filling
   D. Microinjection of the Cells
   E. Signal Detection
   F. Signal Normalization and Calibration
V. Examples of Cytosolic Ca^{2+} Measurements in Single Cells Using Aequorin
References

I. Introduction

The well-known observation made by Ringer (1883) of the essential role of calcium ions in frog heart contraction focused attention on Ca^{2+} as an important regulator of physiological processes. Cytosolic free Ca^{2+} functions as an important intracellular signaling mechanism whereby agonists regulate many different cellular processes such as muscle contraction (Ridgeway and Ashley, 1967), nervous transmission (Reuter, 1983), hormone secretion (Dean and Matthew, 1970), oocytes fertilization (Cuthbertson and Cobbold, 1985), and hepatic glycogenolysis (Keppens *et al.*, 1977). Thus, many efforts have been made to measure Ca^{2+} inside the cell.

The first attempt to measure free Ca^{2+} in living cells was made by Pollack (1928), using alizarin sulfonate injected into an amoeba. But it was not until the late 1960s that the first real measurements of free Ca^{2+} were made, with aequorin in giant muscle cells (Ridgway and Ashley, 1967). In the 1970s two further methods were developed, arsenazo-absorbance dyes (mainly arsenazo

II) and calcium-selective microelectrodes (Ashley and Campbell, 1979). In 1980 R. Y. Tsien introduced a new generation of fluorescent dyes (Tsien, 1980); first quin-2, then fura-2, and later indo-1, and made cytosolic free Ca^{2+} measurements easily feasible (Cobbold and Rink, 1987; Thomas and Delaville, 1991).

During the past decade our lab has developed a technique to measure free Ca^{2+} in single mammalian cells by injecting the cell with aequorin and then simply measuring the light emitted from the cell with a photomultiplier. The technique has been used in different cells such as amoeba (Cobbold, 1980), mouse oocytes (Cuthbertson and Cobbold, 1985), heart cells (Cobbold and Bourne, 1984), hepatocytes (Woods et al., 1986), adrenal gland (Cobbold et al., 1987), aorta vascular smooth cells (Cobbold et al., 1989), and osteoblasts (Schöfl et al., 1991). This technique has been recently explained in great detail by Cobbold and Lee (1991). In this chapter we present the rationale for measuring free Ca^{2+} in single cells, provide a brief description of aequorin technique, and supply some examples of cytosolic Ca^{2+} measurements using aequorin.

II. Single Cell vs Population to Measure Ca^{2+}

Single-cell free Ca^{2+} measurements have a number of advantages. Heterogeneous cell populations will suffer from large aequorin signals arising from a disproportionately small number of dying cells, particularly, we suspect, when aequorin is loaded into cell population [e.g., by various reversible permeabilization techniques (Cobbold and Rink, 1987)]. Population measurements will also suffer from cell-to-cell asynchrony; this is particularly important because in the past few years a large number of cell types have been shown to generate a transient free Ca^{2+} response, when measurements were made at single cell level (Berridge et al., 1988). This is relevant because the kinetics of free Ca^{2+} changes may also be distorted, since the fast fluctuations in free Ca^{2+} at the single-cell level can be masked in a population study. Single-cell Ca^{2+} measurements can yield information about the spatial and temporal Ca^{2+} changes within the cell (although photoproteins are not suitable for imaging studies, except in really giant cells). Clearly, cell population measurements are easier to carry out and no specialized instrumentation is required. However, it seems that microinjection is the only way to introduce sufficient aequorin to allow signals to be detected from a single cell (Cobbold and Rink, 1987).

III. Properties of Aequorin

The bioluminescent jellyfish *Aequorea Victoria* possesses in the outer margin of its umbrella cells containing a Ca^{2+}-binding protein, aequorin, which emits light upon reacting with Ca^{2+} (Shimomura et al., 1962). A similar photoprotein, obelin, is found in the hydroid *Obelina geniculata* (Morin and Hastings, 1971; Campbell, 1974) and in the jellyfish *Clytia* (Inouye and Tsuji, 1993).

1. *Molecular mass.* The molecular mass of aequorin is 21,400 Da. The photoprotein must be microinjected (Section IV,D) or introduced by temporary permeabilization. Once in the cytosol, aequorin appears to remain in the cell for many hours after microinjection.

2. *Luminescence mechanism.* The photoprotein consists of three components: an apoprotein (apoaequorin), molecular oxygen and a chromophore. The chromophore is made up of coelenterazine. One of the oxygens of molecular oxygen is attached to the C-2 carbon of coelenterazine in the form of a peroxide (Musicki *et al.*, 1986). When Ca^{2+} binds to aequorin, an intramolecular reaction takes place in which coelenterazine is oxidized to coelenteramide by the bound oxygen, yielding, as products, light (λ_{max}, 470 nm), CO_2, and a blue fluorescent protein (Johnson and Shimomura, 1978). The blue fluorescent protein consists of coelenteramide attached to apoaequorin; the excited-state coelenteramide is the emitter in the reaction (Shimomura and Johnson, 1973). Coelenterazine can be chemically synthesized (Inouye *et al.*, 1975) and is now available from Molecular Probes (Eugene, Oregon). However, coelenterazine usually need not be supplied to the cell since native aequorin binds and retains coelenterazine over years of storage.

3. *Calcium sensitivity.* There is a steep relationship between luminescence and calcium concentration due to the fact that aequorin contains three Ca^{2+}-binding sites, which are homologous to the 2^+ binding sites of calmodulin (Inouye *et al.*, 1985).

4. *Luminescence kinetics.* The kinetics of the luminous reaction are reasonably fast, with a pseudo first-order rate constant of about 100 sec^{-1}.

5. Mg^{2+} *concentrations.* Since millimolar concentrations of Mg^{2+} depress the sensitivity to Ca^{2+}, the intracellular Mg^{2+} concentration needs to be known to improve the precision of calibration.

6. *Inactivators.* Care must be taken with Sr^{2+} and La^{3+} (and probably other lanthanides), because they will also discharge aequorin. Also, traces of silver and mercury can inactivate aequorin; solutions that contact aequorin should not have reference electrodes that might leak those ions immersed in them.

7. *pH.* pH (6.6–7.4) has no important effect on calcium sensitivity.

8. *Ionic strength.* Ionic strength influences Ca^{2+} sensitivity; so calibration needs to be carried out at the correct ionic strength. Ca^{2+} chelators such as EDTA and EGTA, at low ionic strength, depress the aequorin sensitivity to Ca^{2+}. However, at mammalian physiological ionic strength they have no direct effect on aequorin.

IV. The Aequorin Technique

A. Aequorin Preparation

Aequorin is available from Dr. J. R. Blinks (Friday Harbour, P.O. Box 3050, WA 98250) or Prof. O. Shimomura (Marine Biological Laboratory, Woods Hole, MA 02543). It can be supplied as vials of 1-mg quantities freeze-dried from 100 μl of 10 mg ml^{-1} solution of aequorin in isotonic KCl buffer. The powder is dissolved in ca. 7 μl of Ca^{2+}-free buffer (EDTA, 10 mM; Pipes, 10 mM; pH 7.0) to give a stock solution of ca. 150 mg ml^{-1} aequorin, which is aliquoted into ca. ten 60-mm plastic tissue culture petri dishes under a 3-mm depth of liquid paraffin (Merck–BDH; spectroscopic grade) and stored at $-70°$C. The day before an experiment a small aliquot (ca. 200 nl) of this stock aequorin solution is removed and dialyzed on a microscale to reduce the concentrations of KCl and EDTA (Cobbold *et al.*, 1983). Dialysis is carried out across a 180-μm bore microdialysis tubule, ca. 5-mm-long "microtubing" (Medicell International, London) with a cutoff of 5 kDa. The tubule is held between two Teflon-coated wires. Prior to introducing the aequorin stock solution, the dialysis tubule should be washed with injection buffer (KCl, 150 mM; Pipes, 1 mM; EDTA, 100 μM; EGTA, 25 μM; pH 7.2). This buffer is kept in a plastic flask and pH is raised by measurements of aliquots of ca. 10 ml, which are then discarded to avoid silver contamination of the buffer (see Section III, Properties of Aequorin). The dialysis is carried out against ca. 20 ml injection buffer with the inclusion of 25 mg ml^{-1} of polyvinylpyrrolidine (PVP, 40 kDa) and dithiothreitol 1 mM. PVP is an osmotic buffer which allows dialysis to be conducted without attention, overnight, at 4°C. Without PVP, the aequorin solution swells initially and needs to be removed from the injection buffer periodically and allowed to evaporate back to the original size. At the end of dialysis the aequorin sample is removed from the tubule and a droplet is placed under ca. 4 mm of liquid paraffin in a 60-mm tissue culture petri dish. For all these procedures, wash out the dialysis tubule and introducing and remove the aequorin sample from the tubule using a blunt borosilicate glass pipette (Camblab, Cambridge, UK) with a bore ca. 50 μm connected to a rubber bulb is used; all these procedures are done under a stereomicroscope ($\times$20).

B. Preparation of Cells for Microinjection

A diluted suspension of freshly isolated cells is kept in an appropriate culture medium containing 1% (w/v) of Type IX agarose (Sigma) held in an incubator at 37°C. Preparation of agarose medium can be done either by boiling the culture medium until the agarose powder is dissolved or (preferably) by boiling the agarose at double concentration in distilled water and then adding double-strength culture medium.

Optically flat glass capillaries, "microslides" (Camlab; Cambridge, UK), with path length (floor to ceiling) of 0.1 mm are cut in ca. 5-mm lengths, washed with nitric acid, and stored in absolute ethanol. For cell preparation, a flame-dried microslide is transferred onto a rectangular piece of coverslip (22 × 10 mm), which is affixed to the edge of a glass microscope slide with silicone grease. The free end of the coverslip piece is cantilevered out and tilted slightly downward. The microslide is viewed through a stereomicroscope (×32). A small amount of 1% Type VII agarose in culture medium containing 0.05% bovine serum albumin (the concentration of bovine serum albumin present is critical for successful microinjection and subsequent survival of cells) is applied to the edge of the microslide using a snapped-off tip soda glass micropipette (Clark Electromedical Instruments, Reading, U.K.), with a tip bore of ca. 50 μm. The agarose flows by capillarity between the microslide and coverslip and serves to anchor the microslide in place. This Type VII agarose stock is kept at 37°C in a block heater, with an overlay of liquid paraffin (heavy grade) to reduce loss of CO_2. The microslide is filled with this agarose solution. Once filled, the microslide is immediately covered with liquid paraffin (heavy grade) to prevent evaporation. The microscope should be held at 30–37°C to prevent premature gelling of the agarose. The petri dish of cells is held on the stage of an adjacent stereomicroscope (×28–×180), and a cell of healthy appearance (no membrane blebs and distinct nuclei) is taken up by capillarity into a blunted soda glass pipette containing Type VII agarose. Immediately after the cell enters, the pipette is withdrawn from the petri dish and the cell is expelled into the microslide. The meniscus left when filling the microslide will reduce the tendency of the microslide to overflow with medium. The cell should be positioned within 50 μm of the meniscus. Once the cell is in the microslide, it is refrigerated at 4°C for ca. 2 min to gel the agarose, after which it is returned to the incubator at 37°C until required for microinjection.

C. Pipette Filling

Borosilicate glass pipettes (length 100 mm, bore 1 mm, wall 0.5 mm) are pulled on a simple vertical puller to yield pipettes with long tapering tips (ca. 11 mm from tip to full diameter) and a tip bore of ca. 0.3 μm. Pipettes are filled by simply dipping the tip through the liquid paraffin and then into the dialyzed aequorin droplet for a few seconds, so that a short column of each is entered. This procedure is conducted under a stereomicroscope (×80).

D. Microinjection of the Cells

1. Instrumentation

An inverted microscope (Nikon Diaphot) with Nomarski optics is used, giving adequate image size for a hepatocyte using a 40X objective and 15X eyepieces. Mechanical fittings are attached to the microscope stage, controlling the position of a thin brass platform mounted on the free end of a piezoelectric bender element. A voltage-reversing switch controls the piezoelectric element via a variable supply, providing up to 15 V in 1.5 V increments, and is set to give ca. 2–5 μm movement of the brass platform. A second switch provides 45 V in the reverse direction, allowing the platform to be rapidly shifted away from the pipette tip. The microscope stage is held at 25–30°C. A mechanical micromanipulator (Leitz) is mounted to the right of the microscope and controlled with a joystick by the right hand. The filled pipette is held in a brass holder which is clamped horizontally onto the micromanipulator. The pipette holder is connected to a supply of nitrogen at 4 atm. The pressure reaching the pipette is directed through one of two regulators by operating a solenoid valve through a footswitch. Before injection, the valve is closed and a pressure of ca. 0.3–0.5 atm is transmitted to the pipette to prevent entry of medium into the tip, which would discharge the aequorin. Once the cell had been impaled (see Section IV,D), the footswitch is depressed and pressure diverted through the second regulator. The pressure provided is regulated by a handle operated by the left hand and increased until sufficient pressure is applied to expel the aequorin from the pipette.

2. Microinjection Procedure

The coverslip holding the microslide and cell is taken from the incubator attached by a film of grease to the brass platform on the microscope stage. The filled micropipette (see Section IV,C) mounted on the micromanipulator is positioned alongside the microslide edge and adjusted until the two are parfocal, using a 20X objective. The tip is then aligned with the microslide lumen. The pipette is repositioned to point at the cell, and the gas pressure of 0.5 atm is applied before the tip enters the agarose. The vertical positioning of the tip is adjusted until the cell equator is parafocal with the pipette tip; the pipette is then advanced until the cell surface is slightly dimpled. Cell penetration is achieved using the piezoelectric bender, which suddenly throws the cell ca. 2–4 μm onto the pipette tip. The solenoid valve is opened, and the gas pressure is slowly increased to expel the aequorin to ca. 0.5–1% of the cell volume. A swirl of cytoplasm will be detected as the aequorin enters the cell. The cell is then thrown off the pipette by applying the reverse voltage to the piezoelectric element, and the pipette is removed from the agarose. After the microinjection procedure it is convenient to watch the cell for a minute or so to see if the cell is damaged. Of course, success of microinjection depends on the size of the cell; for an experienced person the success rate with heart or liver cells is easily 60–70%.

E. Signal Detection

After injection, the cell in its microslide is transferred to the cavity of a stainless-steel perfusion chamber $8 \times 4 \times 2$ mm deep, which is sealed with a circular glass coverslip over a film of liquid paraffin. The microslide chamber is positioned immediately under a photomultiplier (E.M.I. 9789), with a 10-mm-diameter, low-noise, bialkali photocathode. The photomultiplier is cooled to ca. $0-4°C$ by a constant temperature incubator or refrigerator. Tubes with ca. a 1-mm bore allow perfusion of culture medium through the chamber at required flow, controlled by a peristaltic pump. The chamber is maintained at $37°C$ by a flowing water jacket. Swift changes of media are done by syringe suction from a flask to a "T"-piece just beneath the chamber. The photomultiplier is supplied with high voltage, and the current generated by aequorin luminescence passes through an amplifier adjacent to the photomultiplier housing and then to an amplifier discriminator. The brief (ca. 5 nsec) pulses of current generated, each of which corresponded to a single photon of aequorin luminescence, are captured by a photon counter (SR400). These are deposited into 50-ms bins which are then transferred to an IBM-compatible computer which has sufficient memory capacity to store 20 samples sec^{-1} for ca. 3 hr. A slow data file is also collected at 1 sample sec^{-1}. The SR400 also provides a chart recorder signal for on-line visualization of the response of the cell.

F. Signal Normalization and Calibration

The total number of photon counts recorded from a single cell is about 3×10^5. At resting Ca^{2+} (ca. 200 nM), about 10^{-6} of the aequorin is consumed per second, generating a "resting signal" of ca. 2 counts sec^{-1} (cps) detectable over the background count of ca. 1 cps (Woods et al., 1987). When Ca^{2+} rises to 10^{-6} M the signal from a hepatocyte will be 10^3-fold greater, ca. 300 cps, detectable above background after only a fraction of a second. At the end of an experiment, the total aequorin content of each cell is determined by discharging the aequorin by lysing the cell with distilled water. The signal is normalized retrospectively by calculating cps divided by the total counts remaining; this process is carried out by computer.

In vitro data of the aequorin signals in various batches of free Ca^{2+} are made by determining the rate of consumption of aequorin in Ca^{2+}-EGTA buffers mimicking the intracellular milieu, at 37°C with Ca^{2+} ranging from 10^{-8} to 10^{-5} M (Cobbold *et al.*, 1983; Cobbold and Rink, 1987). The data are incorporated into the data analysis (off-line) program, enabling the computed fractional rate of aequorin consumption in the cell to be plotted as Ca^{2+} concentration. The calibration values assumes 1 mM intracellular free magnesium as determined from studies on hepatocyte populations (Murphy *et al.*, 1980).

Data analysis is carried out using software developed by Dr. K. S. R. Cuthbertson. The programme allows a number of parameters associated with individual transients to be calculated. These include the duration, peak Ca^{2+}, rate of rise, falling time constant, and period between transients. Data are plotted out using exponential smoothing and because of the large dynamic range of the original signals, the program allows the time constant of the plot

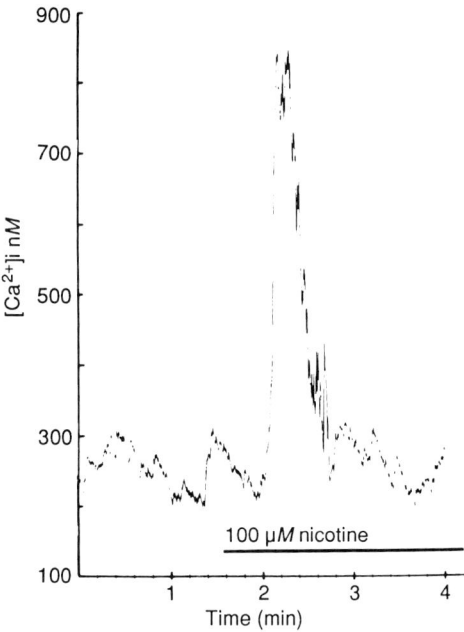

Figure 1 Aequorin signals calibrated as Ca^{2+} free (nM) from a single chromaffin cell exposed to 100 μM nicotine. Time constants: 7 sec for resting levels, 1 sec for transient. (Reproduced with permission from Cobbold *et al.*, 1987.)

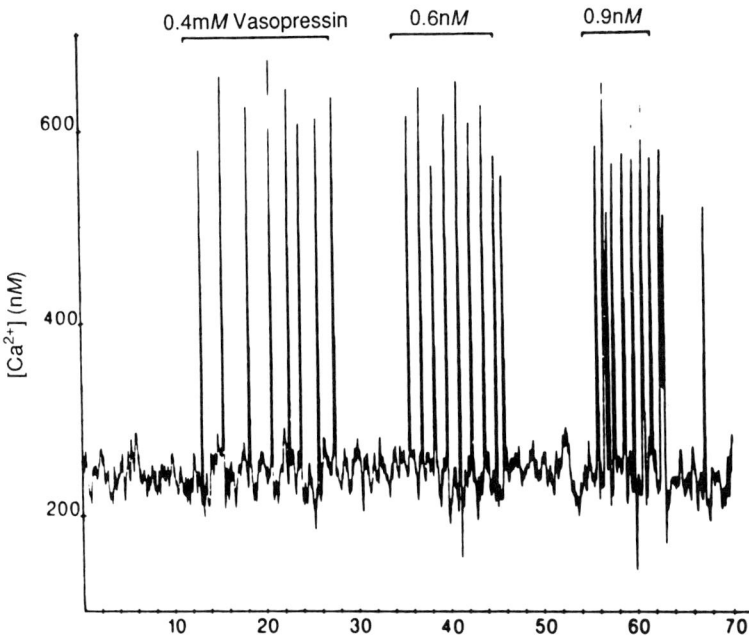

Figure 2 Aequorin signals plotted as free Ca^{2+} (nM) from a single hepatocyte exposed to various concentrations of [Arg^8]vasopressin (0.4, 0.6, 0.9 nM) which induced free Ca^{2+} transients at increasing frequencies (periods were 120, 80, and 60 sec, respectively). Before the 0.4 nM recording, the cell had failed to generate transients when exposed to 0.2 nM vasopressin for 8.5 min. Time constants: 15 sec for resting levels, 1 sec for transients. (Reproduced with permission from Woods et al., 1986.)

to be varied. For example, resting signals are plotted with a long time constant (typically 10-20 sec), which switch to a much shorter value (typically 0.4-1 sec) when the signal exceeds a predetermined rate during the transient rises in Ca^{2+}. Once the signal falls below a second threshold value the plot is resumed with the longer time constant. This allows transients to be faithfully presented without excessive noise between transients when Ca^{2+} is at resting level.

V. Examples of Cytosolic Ca^{2+} Measurements in Single Cells Using Aequorin

In this section we show some examples of Ca^{2+} measurements using aequorin carried out in our laboratory in different cell types. Figure 1 shows the increase in free Ca^{2+} induced by nicotine (100 μM) in a single chromaffin

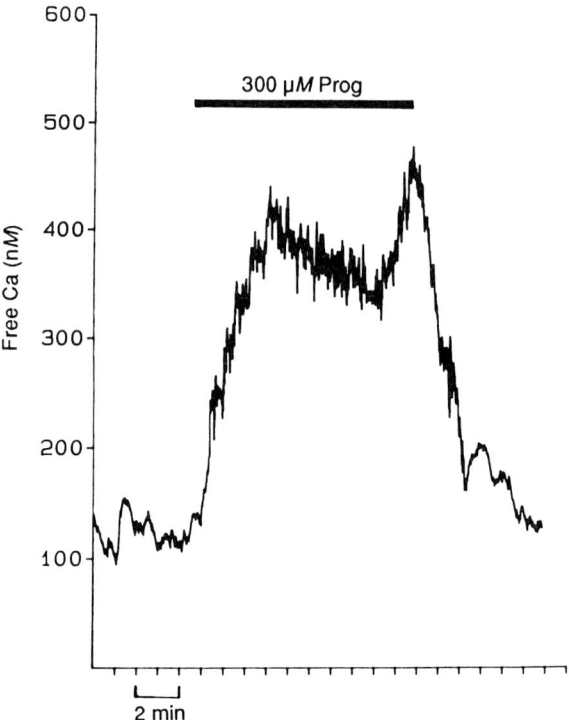

Figure 3 A single rat hepatocyte was microinjected with aequorin and perfused with 300 μM progesterone (Prog). Time constant: 20 sec for resting concentration of free Ca^{2+}, and 2 sec for increases. (Reproduced with permission from Sanchez-Bueno et al., 1991.)

cell from bovine adrenal medulla. The rise in free Ca^{2+} in a single chromaffin cell in response to nicotine is transient but it is not repetitive, even in the continued presence of the agonist. Figure 2 shows repetitive free Ca^{2+} transients (spikes) induced by different concentrations of phenylephrine in a nonexcitable cell: a single hepatocyte. These spikes show very interesting patterns. During an individual transient, free Ca^{2+} rise suddenly from a resting level of ca. 200 nM, to a peak of ca. 600–800 nM, before falling back to the resting level. The frequency of these transients is also dependent upon agonist dose, an increase in agonist concentration yielding transients at a greater frequency without affecting the time course of individual transients. High concentrations of some agonists (vasopressin, angiotensin II) lead to a sustained elevation of free Ca^{2+} rather than to spiking (Woods et al., 1986). The time course of individual spikes does not change with agonist concentration for a given agonist. However, the time course depends on the agonist species.

While the rise time and amplitude do not change with the agonist, the falling time, and hence the spike duration, is markedly dependent upon the type of agonist, in the same individual hepatocyte. Thus, phenylephrine consistently induces transients with an overall duration of ca. 7 sec, while the transients induced by vasopressin and angiotensin II are of longer duration: ca. 11 and 15 sec, respectively (Woods et al., 1987; Cobbold et al., 1991). Similar agonist-specific transient shapes have been reported in fura-2-loaded hepatocytes (Kawanishi et al., 1989; Rooney et al., 1989). However, not all single hepatocytes calcium responses are oscillatory; Fig. 3 shows a sustained rise in free Ca^{2+} in response to progesterone (300 μM) reminiscent of that induced by ethanol (Sanchez-Bueno et al., 1990). It seems that free Ca^{2+} responses have only an oscillatory behavior when the hepatocyte is stimulated with agonists which act through the phosphatidylinositol cycle (Berridge et al., 1988).

Acknowledgments

We thank Dr. F. I. Tsuji for his comments on the manuscript. We are grateful for funding from the Basque Government (A.S-B.) and The Wellcome Trust.

References

Ashley, C. C., and Campbell, A. K., eds. (1979). "Detection and Measurements of Free Ca^{2+} in Cells." Elsevier/North-Holland, Amsterdam.
Berridge, M. J., Cobbold, P. H., and Cuthbertson, K. S. R. (1988). *Philos. Trans. R. Soc. London Ser. B* **320**, 325–343.
Campbell, A. K. (1974). *Biochem. J.* **143**, 411–418.
Cobbold, P. H. (1980). *Nature (London)* **285**, 441–446.
Cobbold, P. H., and Bourne, P. K. (1984). *Nature (London)* **312**, 444–446.
Cobbold, P. H., and Lee, J. A. C. (1991). In "Cellular Calcium: A Practical Approach" (J. G. MacCormack and P. H. Cobbold, eds.), pp. 55–81. IRL Press, Oxford.
Cobbold, P. H., and Rink, T. J. (1987). *Biochem. J.* **248**, 313–328.
Cobbold, P. H., Cuthbertson, K. S. R., Goyns, M. H., and Rice, V. (1983). *J. Cell Sci.* **61**, 123–136.
Cobbold, P. H., Cheek, T. R., Cuthbertson, K. S. R., and Burgoyne, R. D. (1987). *FEBS Lett.* **211**, 44–48.
Cobbold, P. H., Daly, M., Dixon, J., and Woods, N. M. (1989). *Biochem. Soc. Trans.* **17**, 9–10.
Cobbold, P. H., Sanchez-Bueno, A., and Dixon, J. (1991). *Cell Calcium* **12**, 87–95.
Cuthbertson, K. S. R., and Cobbold, P. H. (1985). *Nature (London)* **316**, 541–542.
Dean, P. M., and Matthew, E. K. (1970). *J. Physiol. (London)* **210**, 255–264.
Inouye, S., and Tsuji, F. I. (1993). *FEBS Lett.* **315**, 343–346.
Inouye, S., Sugiura, S., Kakoi, H., Hasizume, K., Goto, T., and Lio, H. (1975). *Chem. Lett.*, 141–144.
Inouye, S., Noguchi, M., Sakaki, Y., Miyata, T., Iwanaga, S., Miyata, T., and Tsuji, F. I. (1985). *Proc. Natl. Acad. Sci. U.S.A.* **82**, 3154–3158.
Johnson, F. H., and Shimomura, O. (1978). *In* "Bioluminescence and Chemiluminescence"

(M. DeLuca, ed.), Methods in Enzymology, Vol. 57, pp. 271-291. Academic Press, New York.
Kawanishi, T., Blank, L. M., Harotunian, A. T., Smith, M. T., and Tsien, R. Y. (1989). *J. Biol. Chem.* **264**, 12859-12866.
Keppens, S., Vanderheede, J. R., and De Wulf, H. (1977). *Biochim. Biophys. Acta* **496**, 448-457.
Morin, J., and Hastings, J. W. (1971). *J. Cell. Comp. Physiol.* **77**, 305-311.
Murphy, E., Coll, K., Rich, T. L., and Williamson, J. R. (1980). *J. Biol. Chem.* **255**, 6600-6608.
Musicki, B., Kishi, Y., and Shimomura, O. (1986). *J. Chem. Soc. Chem. Commun.*, 1566-1568.
Pollack, H. (1928). *J. Gen. Physiol.* **11**, 539-545.
Reuter, H. (1983). *Nature (London)* **301**, 569-574.
Ridgway, E. B., and Ashley, J. C. (1967). *Biochem. Biophys. Res. Commun.* **29**, 229-234.
Ringer, S. (1883). *J. Physiol. (London)* **4**, 29-42.
Rooney, T. A., Sass, E. J., and Thomas, A. P. (1989). *J. Biol. Chem.* **264**, 17131-17141.
Sanchez-Bueno, A., Dixon, J., Woods, N. M., Cuthbertson, K. S. R., and Cobbold, P. H. (1990). *Adv. Second Messenger Phosphoprotein Res.* **24**, 115-121.
Sanchez-Bueno, A., Sancho, M. J., and Cobbold, P. H. (1991). *Biochem. J.* **280**, 273-276.
Schöfl, C., Cuthbertson, K. S. R., Gallagher, J. A., Pennington, S. R., Cobbold, P. H., Brabant, G., Hesch, R. D., and Muhlen, A. (1991). *Biochem. J.* **274**, 15-20.
Shimomura, O., and Johnson, F. H. (1973). *Biochem. Biophys. Res. Commun.* **53**, 490-494.
Shimomura, O., Johnson, F. H., and Saiga, Y. (1962). *J. Cell. Physiol.* **59**, 223-240.
Thomas, A. P., and Delaville, F. (1991). *In* "Cellular Calcium: A Practical Approach" (J. G. McCormack and P. H. Cobbold, eds.), pp. 1-54. IRL Press, Oxford.
Tsien, R. Y. (1980). *Biochemistry* **19**, 2396-2404.
Woods, N. M., Cuthbertson, K. S. R., and Cobbold, P. H. (1986). *Nature (London)* **319**, 541-542.
Woods, N. M., Cuthbertson, K. S. R., and Cobbold, P. H. (1987). *Cell Calcium* **8**, 79-100.

PART IV

Molecular Techniques and Specific Model Systems

22

Approaches for the Purification, Quantitation, and Analysis of Hormone and Receptor mRNAs

Martin L. Adamo, Bethel Stannard, Derek LeRoith, and Charles T. Roberts, Jr.

I. General Introduction
II. RNA Extraction
    A. Principles and Applications
    B. LiCl Method
    C. CsCl Method
    D. Poly(A)$^+$ Selection
    E. Assessment of RNA Quantity and Integrity
III. Northern Blot Hybridization Analysis
    A. Transfer Techniques
    B. Crosslinking
    C. Probe Labeling
    D. Hybridization and Wash Conditions
IV. Solution Hybridization/RNase Protection Assay
V. Primer Extension
    References

I. General Introduction

Any complete understanding of the regulation of the expression of genes encoding peptide hormones and growth factors, as well as hormone receptors and binding proteins, includes the analysis of the mRNAs which encode them. RNA preparations are the starting materials required for cDNA cloning and obviously provide reagents for the analysis of factors regulating the abundance of specific mRNAs in normal endocrine physiology and in endocrinopathies. cDNA probes can be used to obtain genomic clones and, together or separately, these reagents can be used to determine the precise structure of mature mRNAs and the parameters which show how this structure is achieved, i.e., transcriptional start sites, patterns of exon splicing, and 3' end formation. In this chapter, we discuss the principles and applications of RNA isolation and analysis, and we give detailed protocols of methods we and others have used for these purposes.

II. RNA Extraction

A. Principles and Applications

The principal goal of total RNA extraction from tissues or from cultured cells is preservation of its integrity. This, in turn, is dependent primarily on the inactivation of endogenous or introduced RNases, the former depending on the particular source of the RNA. A secondary consideration is isolating the RNA as free as possible from DNA contamination. Finally, it is important to obtain a reasonable yield of total RNA, so that multiple experiments and further manipulations (e.g., poly(A)$^+$ selection) may be performed. In order to meet the two primary goals, disruption of tissues and cells is performed in the presence of a chaotropic agent and a strong reductant (as RNase inhibitors) with a mechanical homogenizer designed to shear genomic DNA. Subsequently, the RNA either is precipitated by addition of a salt such as LiCl, which allows DNA and protein to remain in solution (Cathala *et al.*, 1983), or is pelleted through a CsCl gradient (Chirgwin *et al.*, 1979). The chaotropic agent of choice is guanidinium thiocyanate (GTN) while β-mercaptoethanol (β-ME) has typically been used as the reductant. Preparations of total RNA can be used directly for the analysis of a specific mRNA in the sample or they can be enriched for poly(A)$^+$ RNA, which in turn can be used for similar analysis or for the preparation of cDNA libraries.

B. LiCl Method

This method, although time consuming and labor intensive, is preferred for the preparation of RNA from whole tissues. The method utilizes multiple extraction steps in the presence of RNase inhibitors (i.e., GTN and β-ME, and, later, vanadyl ribonucleoside complex, or VRC) and thus results in more stable RNA preparations, which, if handled properly, can be used for months or even years. In addition, it does not require an ultracentrifuge, only a preparative centrifuge. As a result, it is a more flexible method, allowing the simultaneous processing of more samples using more readily available equipment. The method given below is based on a modification (Lowe et al., 1987a) of the original method of Cathala et al. (1983).

PROTOCOL FOR RNA ISOLATION FROM TISSUES USING LiCl

Day 1

1. Prepare
    a. Guanidinium stock solution (5 M GTN, 10 mM EDTA, 25 mM Tris, pH 7.5–8.0)
        177 g GTN (Fluka)
        15 ml 0.5 M Tris, pH 7.5–8.0
        6 ml 0.5 M EDTA, pH 8.0
        Autoclaved H_2O to bring solution up to 270 ml
        Heat to 60°C to dissolve and filter through a 0.45-μM filter. Do not autoclave final solution. This solution may be stored indefinitely.
    b. Homogenization solution (GTB; guanidinium stock solution with 8% β-ME)
        92 ml guanidinium stock solution
        8 ml β-ME (add fresh just before use)
        Volumes other than 100 ml may be prepared provided that the ratio of GTN to β-ME is maintained at 92:8.
2. Wash Polytron before use by pulsing (i.e., turning on at full power for a few sec) with
        20 ml 1 N NaOH in sterile 50-ml centrifuge tube
        20 ml sterile H_2O (×6) (discard the first three H_2O washes)
        5 ml GTB
3. Weigh 0.6–0.8 g of tissue in 50-ml sterile centrifuge tube (keep tube on dry ice until weighed).
4. Add 7 × weight of homogenization solution (GTB); e.g., 7 × 0.6 = 4.2 ml.
5. Polytron on high setting until tissue is totally homogenized.
6. Transfer 4.0 ml to a 30-ml glass Corex tube.
7. Add equal volume isoamyl alcohol:$CHCl_3$ (1:24).
8. Polytron (setting 2) for 5 sec.

9. After each sample, wipe off probe with a clean utility wipe and wash polytron by pulsing briefly in 20 ml H_2O ($\times$3), followed by a pulse in 5 ml GTB.
10. Spin (6000 rpm) for 20 min at 10–15°C.
11. Transfer top (aqueous) layer to sterile 30-ml Corex tube.
12. Add 20 ml 4 M LiCl (autoclaved) (i.e., ~5 vol), mix thoroughly. Precipitate overnight at −4 to −20°C.

Day 2

1. Prepare
   a. RNA resuspension buffer (10 mM Tris, pH 7.5–8.0, 0.3% Triton X-100)
      10 ml 0.5 M Tris, pH 7.5–8.0
      1.5 ml 100% Triton X-100
      Sterile H_2O to bring buffer up to 500 ml
      Autoclave and filter. This solution is stable.
   b. RNA suspension buffer with 10 mM VRC (prepare fresh)
      84 ml RNA resuspension buffer
      4.42 ml 200 mM VRC stock
      Smaller volumes may be prepared as required, provided that the final VRC is 10 mM.
2. Spin at 8500 rpm in, e.g., the SS34 rotor in Sorvall RCB-6 for 90 min at 4°C. Decant or aspirate supernatant (DNA, etc.) and discard. Invert tubes to drain, taking care that pellet does not slide out.
3. Wash polytron before use with
   a. 20 ml 1 N NaOH in sterile 50 ml centrifuge tube (first wash only)
   b. 20 ml sterile H_2O ($\times$6) (discard the first three H_2O washes)
   c. 5 ml 3 M LiCl
4. Add 8 ml 3 M LiCl (i.e., 0.33 $\times$ the total volume of the LiCl precipitation step) and loosen pellet from sides with pipette.
5. Polytron sample until precipitate is thoroughly homogenized.
6. Wash polytron between samples by wiping probe with a clean, new utility wipe and pulsing in 20 ml sterile H_2O ($\times$3) followed by 3 M LiCl.

22. Hormone and Receptor mRNAs

7. Spin at 8500 rpm (e.g., with SS34 rotor in the Sorvall RCB-6 centrifuge) for 60 min at 4°C.
8. Decant supernatant and discard. Invert tubes to drain.
9. Wash Polytron with
   a. 20 ml sterile H_2O ($\times 6$)
   b. 5 ml RNA resuspension buffer *without* VRC
10. Add 4.8 ml RNA resuspension buffer with VRC (i.e., 0.2 × the volume of the initial LiCl precipitate) and loosen pellet from sides.
11. Polytron sample.
12. Wash polytron between samples by wiping probe with a new utility wipe and pulsing with 20 ml H_2O ($\times 3$) followed by RNA resuspension buffer.
13. Add 4.8 ml IAA (isoamylalcohol):$CHCl_3$ (1:24), and vortex.
14. Spin at 6000 rpm (e.g., SS34 rotor in the Sorvall RCB-6 centrifuge) for 20 min at 4°C.
15. Transfer top layer to sterile Corex tube.
16. Repeat steps 13–15.
17. Add 1.2 ml 0.5 M EDTA, pH 8.0, and keep 10–15 min at room temperature.
18. Add 0.6 ml 6 M ammonium acetate (autoclaved).
19. Add 13.2 ml 100% ethanol, vortex.
20. Precipitate RNA overnight at $-20°C$.

Day 3

1. Spin at 6000 rpm (e.g., with SS34 rotor in the Sorvall RCB-6 centrifuge) for 20–30 min at 4°C and decant supernatant.
2. Add 500 µl 70% ethanol to the pellet and resuspend by pipetting up and down.
3. Transfer to 1.5-ml screw-cap tube.
4. Wash Corex tube with additional 500 µl 70% ethanol and add to screw-cap tube.
5. Store at $-20°C$ for 30 min.
6. Spin at 12,000 rpm (in microcentrifuge) at 4°C for 30 min.
7. Decant and wipe with cotton swab. Repeat twice. After third wash, dry pellet.

8. Add 400 μl sterile H$_2$O to pellets and dissolve.
9. Add 40 μl 6 M ammonium acetate.
10. Add 880 μl ethanol, vortex.
11. Precipitate overnight at −20°C.

Day 4

1. Microfuge for 30 min at 4°C. Decant, rinse pellet with 70% sterile ethanol, swab, and dry completely.
2. Redissolve in 40–400 μl sterile H$_2$O. Store aqueous RNA solutions at −80°C. (Final volume depends on tissues, e.g., use 400 μl for liver and 40 μl for heart and skeletal muscle.)

C. CsCl Method

This is the preferred method for preparing total RNA from cells in culture. Like the LiCl method, it involves vigorous homogenization in the presence of GTN and β-ME to inactivate RNases. It also takes advantage of the fact that RNA will pellet through a CsCl gradient, while sheared DNA and protein will not (Chirgwin et al., 1979). However, the technique uses ultracentrifugation to pellet the RNA and to leave the DNA and protein banded either in the CsCl cushion or at the CsCl–GTN interface. As a result, separation of RNA from DNA and protein is effectively done in a single step, rather than in multiple extractions as with the LiCl method. Such a purification scheme apparently works best when the total protein and DNA content of a sample is low, as is generally the case with cell cultures. In fact, when extracting RNA from tissues, considerable amounts of protein (and possibly DNA) often precipitate with the RNA in the initial step(s). The LiCl method takes advantage of the protein as a carrier to maximize the yield of RNA, with the protein being removed in subsequent extractions. This does not occur with homogenates of cells and, consequently, the LiCl method with its large dilutions results in poor yields when preparing RNA from cultured cell homogenates. Conversely, preparation of tissue RNA with the CsCl method may result in considerable protein contamination, which must be removed by a subsequent extraction step. Furthermore, RNA prepared from tissues by CsCl gradient centrifugation is often found to be unstable upon prolonged storage, unless an RNase inhibitor is included. The method described below is based on methods described by Chirgwin et al. (1979) and Davis et al. (1986).

PROTOCOL FOR RNA ISOLATION FROM CELLS USING CsCl

Day 1

1. Prepare
   a. 3 M sodium acetate, pH 6.0; add 40.82 g sodium acetate · H_2O (FW 136 · 08) to sterile H_2O; pH with glacial acetic acid, make up to 100 ml and autoclave.
   b. Guanidinium stock solution (4 M guanidinium thiocyanate, 25 mM sodium acetate)
      189 g guanidinium thiocyanate (Fluka)
      3.34 ml 3 M sodium acetate, pH 6.0
      Sterile H_2O to bring solution up to 400 ml
      It may be necessary to heat to 60°C to dissolve the GTN. Sterilize with 0.2 μM filter and store at room temperature (do not autoclave). This solution is stable.
   c. Homogenization solution (GIT; guanidinium stock solution with 0.835% β-ME)
      50 ml guanidinium stock solution
      0.42 ml β-ME (added just before using)
      Smaller volumes may be prepared.
   d. CsCl solution (5.7 M CsCl, 25 mM sodium acetate, pH 6.0)
      95.97 g CsCl
      0.83 ml 3 M sodium acetate, pH 6.0
      Sterile H_2O to bring solution up to 100 ml; filter sterilize
2. Wash Polytron with
   20 ml 1 N NaOH in sterile 50-ml centrifuge tube
   20 ml sterile H_2O ($\times 6$) (discard the first three washes)
   5 ml GIT
3. Lyse cells by adding GIT to monolayers on tissue culture plates (after removing tissue culture medium). Harvest lysates with a cell lifter or scraper. Use ≈ 4 ml/100-mm plate. This can be serially transferred to replicate plates. After scraping the last plate, dispense lysate into a 50-ml conical tube. Vortex and freeze at $-80°C$ or proceed with homogenization.
4. Add additional GIT if needed.
5. Polytron (setting 4) for 15 sec (setting 6 for 10 sec with small probe).
6. After each sample wash Polytron by wiping probe with new utility wipe and pulsing in 20 ml H_2O ($\times 3$) followed by a brief pulse in GIT.

7. Transfer 7.5 ml of homogenate to a 12-ml polyallomer tube containing 4 ml 5.7 M CsCl. For smaller preps, layer 3.5 ml homogenate onto 1.5 ml CsCl in 5-ml polyallomer tubes.
8. Balance tubes in swinging buckets using additional sample or GIT.
9. Spin at 32,000 rpm for >16 hr (i.e., 16–21 hr) at 24°C in Sorval OTD75B using TH-641 rotor for 12-ml tubes or AH-650 rotor for 5-ml tubes (or equivalent centrifuge and rotors).

Day 2

1. Carefully pipette or decant supernatant and drain tube. Cut off tube with alcohol-washed scissors ~1 cm above bottom.
2. Resuspend RNA pellet in 200 µl sterile H_2O and transfer to sterile screw-cap microfuge tube.
3. Wash tube with additional 200 µl sterile H_2O and pool with first 200 µl.
4. Add 40 µl 6 M ammonium acetate.
5. Add 880 µl cold ethanol.
6. Precipitate overnight or longer at −20°C.

Day 3

1. Microfuge for 20 min, 4°C.
2. Decant, swab the walls of the tube, and dry.
3. Redissolve in 40–200 µl H_2O and store aqueous RNA solutions at −70 to −80°C.

D. Poly(A)$^+$ Selection

Many, probably most, mRNAs encoding peptide hormones and receptors are polyadenylated. Over 90% of a total RNA preparation, however, consists of ribosomal RNA. Thus, detection of rare transcripts in a total RNA sample, particularly by Northern blot hybridization (see below), is often impractical. Affinity chromatography using oligo deoxythymidylate (dT) immobilized on cellulose has been used to enrich samples in polyadenylated RNA and, therefore, in low-abundance mRNAs (Aviv and Leder, 1972). Note that we use the term "enrichment," since this affinity step, or even two sequential steps, does not completely separate poly(A)$^+$ RNA from ribosomal RNA, but rather increases its relative percentage in a preparation.

The chromatography is simple in principle, in that poly(A)$^+$ RNA binds to oligo(dT) in the presence of a high salt concentration and detergent and, after extensive washing to remove nonpolyadenylated material, it is eluted from the oligo(dT) cellulose by lowering the salt concentration. In practice, the problem usually faced is obtaining an adequate yield of material sufficiently enriched in the mRNA of interest. In addition to allowing the detection of low-abundance mRNAs, poly(A)$^+$-enriched preparations of RNA are preferable as starting materials for primer extension analyses and various cloning strategies, such as cDNA library preparation, reverse transcriptase-PCR and RACE techniques (see accompanying chapters).

PROTOCOL FOR POLY(A)$^+$ PREPARATION USING OLIGO(dT) CELLULOSE

1. Prepare
   a. Binding buffer (0.5 M NaCl, 10 mM Tris–HCl, pH 7.5, 1 mM EDTA, pH 8.0, 0.5% SDS). Warm solution to 37°C to keep SDS in solution.
   b. Elution buffer (10 mM Tris–HCl, pH 7.5, 1 mM EDTA, pH 8.0, 0.2% SDS).
2. Suspend 0.3 g oligo(dT) cellulose in 5–10 ml ethanol at room temperature in 15-ml conical tube.
3. Mix, let settle ~3 min, and remove fines.
4. Repeat steps 2–3 four more times.
5. Pipette into disposable, autoclaved column.
6. Pack column while dripping. Prepare a bed volume of ~1 ml.
7. Wash with 40 ml of elution buffer.
8. Equilibrate with 40 ml of binding buffer.
9. Dilute RNA up to 400 μl with sterile H$_2$O (for a 1-ml column, use, e.g., 1 mg total RNA).
10. Heat to 70°C for 1 min, immediately chill on ice for 30 sec.
11. Add equal volume (400 μl) of 2× binding buffer (kept at 40°C to keep SDS in solution).
12. Apply total RNA to column, collect flow-through and recycle five or six times.
13. Wash with 40 ml binding buffer. Check A_{260} of the final 1 ml of the wash to be certain that protein and poly(A)$^-$ RNA have been maximally removed.
14. Elute with 4 ml elution buffer.
15. Collect ten 400-μl fractions in sterile 1.5-ml microfuge tubes.

16. Measure A_{260} of 30-fold dilution.
17. Precipitate RNA by adding 100 μl 5 M NaCl and 1 ml 100% ethanol and keeping at −20°C overnight.
18. Spin (12,000 g, 30 min, 4°C in a microfuge), swab the walls, and dry.
19. Combine pellets from all 10 tubes into 1 tube by quantitatively transferring in a total volume of 300 μl of autoclaved H_2O.
20. Precipitate by adding 75 μl 5 M NaCl and 750 μl ethanol.
21. Store the tubes overnight at −20°C.
22. Centrifuge at 12,000 g for 30 min, at 4°C, decant supernatant, and dry pellet completely and dissolve in small volume of autoclaved H_2O (e.g., 15–50 μl). Store at −70 to −80°C.

E. Assessment of RNA Quantity and Integrity

1. Absorbance Measurements

The concentration of RNA, in either a total RNA preparation or one enriched in poly(A)⁺, is initially determined by use of absorbance measurements taken at 260 nM. An extinction coefficient of 40 mg/ml per one A_{260} is used for calculation (Sambrook et al., 1989). Absorbance measurements can also be helpful in assessing the quality of an RNA preparation. Specifically, absorbance measurements are taken on the same sample at both 260 and 280 nM. Theoretically, the ratio of A_{260} to A_{280} for a pure RNA should be 2.0. Using the methods described above for preparation of total RNA, we have found ratios to range from ∼1.6 to ∼2.0, with no apparent correlation with RNA integrity or quality. This is further illustrated by the fact that favorable ratios can be obtained with DNA preparations or even nucleoside triphosphates (NTPs). Therefore, while absorbance measurements are almost always the initial means of quantitation, visualization of RNA on agarose gels is ultimately required for confirmation of (relative) quantitation and integrity.

2. Agarose/Formaldehyde Gel Electrophoresis

Electrophoresis of total RNA through a denaturing agarose gel followed by ethidium bromide (EtBr) staining is used to separate and visualize the 28 and 18 S ribosomal RNAs (Lowe et al., 1987a; Graham et al., 1984; Sambrook et al., 1989). For most cellular mRNAs, 1.25 or 1.5% agarose is sufficient for adequate separation. RNA samples (generally 10–30 μg of total RNA) are denatured by heating for 10 min at 65°C before being loaded on the gel. In preparing the gel, precautions used are similar to those for RNA preparation, i.e., avoidance of RNase contamination. Although commercial RNA size markers are available, many investigators use the mobility and known size of the 28 and 18 S rRNAs (e.g., 4.6 and 1.9 kilobase (kb) in the rat) to calculate the size of specific mRNA bands on subsequent Northern hybridizations. Several methods may be used for EtBr staining, including addition of the EtBr directly to each sample, addition to the electrophoresis buffer, or staining the gel with EtBr after electrophoresis. We prefer the latter two techniques, especially when the gel is to be used for quantitation, as they obviously provide for exposure of RNAs throughout the gel to equivalent concentrations of EtBr. Postelectrophoresis staining has the added advantage of generating a smaller volume of toxic chemical waste. Quantitation of the RNA is achieved in a relative sense. Thus, based on A_{260} measurements, equal microgram quantities are electrophoresed and stained with EtBr. Visual inspection is used to determine the equivalence of staining, and if unequal staining is observed, corrections to the volume of RNA solution analyzed can be made and a subsequent gel run to confirm any such adjustments. EtBr staining also allows for the assessment of the integrity of the RNA. Clearly distinct 28 and 18 S bands should be observed and should be present in a ratio of approximately 2:1. In addition, smears or streaking of EtBr-stained material or multiple bands ("ladder" effect), especially if migrating faster than the 18 S band, may be indicative of degradation. It should be noted that these relationships do not necessarily hold for EtBr-stained denaturing gels of poly(A)$^+$ RNA. In such samples, one has hopefully removed a large percentage of the ribosomal RNA, and thus the 28 and 18 S bands generally appear weak, and not necessarily in a 2:1 ratio. Furthermore, one really cannot use equivalence of residual 28 and 18 S rRNA staining to perform relative quantitation of multiple poly(A)$^+$ RNA samples. However, some 28 and 18 S material is usually present, and this is taken as evidence for intactness. Although not presented in this chapter, glyoxyl agarose gels have also been used to size-separate RNA (Sambrook *et al.*, 1989) and have the advantage that DNA size markers can be employed.

PROTOCOL FOR AGAROSE GEL ELECTROPHORESIS OF RNA

1. Prepare
   a. 10× Mops
      81.8 g Mops
      13.6 g sodium acetate
      20 ml 0.5 M EDTA, pH 8.0
      Autoclaved H$_2$O to bring buffer up to 1 liter.

Sterilize with 0.45 μM filter, do not autoclave. This solution may be stored at 4°C.
2. Soak the gel apparatus, including the comb and the casting tray, in 0.5 NaOH for ~15 min.
3. Wash the gel apparatus thoroughly with double-distilled H_2O. Residual NaOH will destroy RNA.
4. Prepare the first part of a 75-ml, 1.25% agarose gel in a 100-ml glass bottle by adding 0.937 g agarose to 37.5 ml H_2O.
5. Autoclave for 15 min and keep at 60°C for not longer than 1 hr. This is solution A.
6. In a 50-ml sterile tube, add 7.5 ml 10× Mops, 17.25 ml sterile double-distilled H_2O, and 12.75 ml formaldehyde.
7. Mix and incubate 60°C for 10 min. This is solution B.
8. Tape ends of casting tray and place in level position in hood.
9. Mix solutions A and B. Wait for bubbles to clear, then pour into the casting tray. Remove bubbles with sterile pipette. Wait 30 min or more before using.
10. Add running buffer (1× Mops containing 25 μl/liter of a 10 mg/ml stock solution of EtBr in H_2O) to apparatus before loading gel.
11. Prepare RNA samples by mixing 3.5 μl RNA in H_2O (generally one can visualize as little as 2 μg; do not run more than 20 μg), 5 μl formamide, 1.5 μl 10× Mops, and 2 μl formaldehyde.
12. Vortex, spin, and heat samples to 60°C for 10 min. Chill on ice. Add 2 μl sample dye (50% glycerol, 1 mM EDTA, pH 8.0, 0.25% bromophenol blue) to each sample, vortex, and spin.
13. Remove comb from gel using both hands and add samples just above well.
14. Run samples from negative to positive at ~60 V for ~3 hr at room temperature or 20 V overnight.
15. Continue electrophoresis until the bromophenol blue dye is ~$\frac{3}{4}$ of the way into the gel. Visualize EtBr-stained RNA using an ultraviolet (UV) transilluminator.

NOTE The conditions given above are for a 75-ml gel volume. For smaller or larger gels, these can be adjusted accordingly.

III. Northern Blot Hybridization Analysis

Once RNA samples have been size separated on denaturing agarose gels and their quantity and integrity confirmed, detection of specific mRNAs is done. RNA must be quantitatively transferred from the gel to a solid support (generally a nitrocellulose or nylon membrane) and fixed by baking *(in vacuo)* or by UV crosslinking, and then the membrane is incubated with a radioactively labeled nucleic acid probe, which is complementary to some stretch of sequence in the RNA of interest. After this binding (or hybridization reaction), the blot is washed in appropriate buffers to remove nonspecifically bound radioactivity, and the blot is then autoradiographed to visualize specific hybridizing bands (e.g., Fig. 1). The size, or sizes, of these bands can be calculated by comparing their position on the blot to the position of RNA size markers or 28 and 18 S ribosomal bands and is generally reported as bases (b) or kb. As discussed below, the conditions for hybridization and washing (i.e., salt concentration and temperature) are critical variables in reducing or removing nonspecific hybridization and simultaneously ensuring quantitative specific hybridization. These variables in turn depend on the type and sequence of probe used. We prefer double-stranded homologous DNA probes or, if necessary, complementary single-stranded oligonucleotides. If RNA–RNA hybridization is desired, then solution hybridization/RNase protection is the assay of choice. We now discuss in detail each of the steps required for successful Northern blot hybridization.

A. Transfer Techniques

The primary consideration when transferring RNA is obvious, i.e., that it be done in a quantitative manner and preserves the integrity of the RNA. There are four methods commonly used for RNA transfer: passive or capillary transfer, electroblotting, vacuum blotting, and pressure blotting. Similarly, a number of solid supports have been used as transfer membranes including nylon and nitrocellulose membranes. Based on our experience, we prefer pressure blotting onto a nylon membrane. Passive transfer has the advantage of simplicity, i.e., it requires no manufactured piece of equipment. However, we have found that this method does not reliably result in transfer of larger molecular-weight RNAs, e.g., those of 7 kb and greater. We have found both electroblotting and pressure blotting to be effective in transferring larger RNAs; our experience with vacuum blotting is too limited to make any general statements. Similarly, we have found both nylon and nitrocellulose membranes to be good transfer membranes, and our use of nylon is really a matter of convenience, i.e., this same type of membrane can be used for DNA and protein transfer. At the conclusion of the transfer period, it is necessary to view the gel and the membrane by UV transillumination to ensure complete RNA transfer (as ribosomal RNA) and also to ensure that no "transfer artifacts" have occurred. These include such things as bubbles, which will have

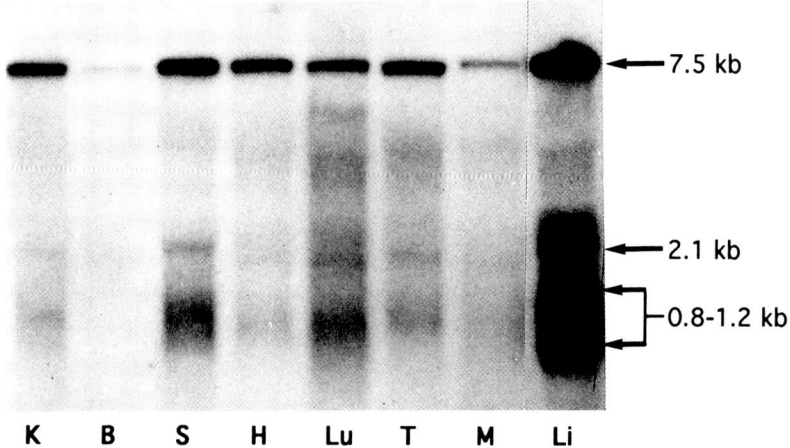

Figure 1 Northern blot hybridization of IGF-I mRNA from various rat tissues. Aliquots (20 μg) of total RNA isolated (by the LiCl method) from rat kidney (K), brain (B), stomach (S), heart (H), lung (Lu), testes (T), skeletal muscle (M), and liver (Li) were electrophoresed through a 1.25% agarose/2.2 M formaldehyde gel. After visualization of EtBr-stained 28 and 18 S rRNA bands, the RNA was transferred to a nylon membrane and crosslinked. After prehybridization, the membrane was hybridized with a ^{32}P-labeled (by random priming) rat IGF-I cDNA fragment, washed, and autoradiographed, all as described in the text. The arrows to the right represent the specifically hybridized IGF-I mRNA species.

prevented RNA transfer in the area where they occurred and thus obviously interfere with subsequent hybridization. Below we give a protocol used for pressure blot transfer of RNA, using the Stratagene apparatus and the protocol basically as recommended by the manufacturer.

PROTOCOL FOR PRESSURE BLOTTING OF NORTHERN BLOTS

1. Prepare
    a. 12.5 mM phosphate buffer (in 2 liters final volume)
        3.44 g $NaH_2PO \cdot 4H_2O$ (monobasic)
        6.7 g $Na_2HPO_4 \cdot 7H_2O$ (dibasic)
    b. Buffer A (150 mM NaCl, 50 mM NaOH)
    c. Buffer B (150 mM NaCl, 100 mM Tris–HCl, pH 7.5)
    d. 2× SSPE (1× SSPE is 180 mM NaCl, 10 mM sodium phosphate, pH 7.7, and 1 mM EDTA).

2. Stain gel in EtBr in 1× TBE for 10 min.
3. Destain gel in H$_2$O for a few minutes.
4. Photograph.
5. Denature gel by shaking gently in buffer A for 30 min.
6. Neutralize gel by shaking gently in buffer B for 30 min.
7. Cut mask 2–3 mm smaller than gel, cut nylon membrane slightly larger than gel, and cut 3MM paper larger than membrane.
8. Soak membrane and 3MM paper in H$_2$O for ~10 min.
9. Wet sponge (first in H$_2$O for new sponge) in 12.5 mM phosphate buffer.
10. Assemble as recommended by the manufacturer. Roll out bubbles. Always keep sponge in horizontal position.
11. Attach hose from pressure control station (PCS) to blotter inlet port on top. Run at 75 mm Hg for 3 hr.
12. Turn off PCS and disconnect hose.
13. Photograph ribosomal RNA bands on RNA side.
14. Briefly rinse membrane in 2× SSPE.

B. Crosslinking

After confirming the efficacy of the transfer, it is necessary to irreversibly bind the RNA to the membrane. Two techniques are available for this. The first is simple heating, whereby the blot is incubated at 80°C in a vacuum oven for 2 hr (Lowe et al., 1987a; Graham et al., 1984). The second is UV crosslinking. The advantage of the first method is obvious, i.e., it does not require a dedicated piece of equipment. The advantage of the second method is speed as well as the observation by some investigators that hybridization of UV crosslinked blots results in a reduced background.

C. Probe Labeling

Two types of radioactively (i.e., ^{32}P) labeled probes are generally used for Northern hybridizations: double-stranded DNA or single-stranded complementary oligonucleotides. Generally, hybridization and wash conditions are more straightforward (i.e., predictable and reproducible) using double-stranded DNA probes, which should range in length from approximately 200 base pairs (bp) to 1000 bp. Longer probes do not work as well. The great advantage of oligonucleotide probes is that workers do not need to either themselves clone or obtain from others the DNA of interest. Rather, an oligonucleotide can be designed by simply examining a published sequence. Of course, researchers must either have access to an oligonucleotide synthesizer or be in a position to purchase oligonucleotides. As described above, the use of labeled RNA probes as hybridization probes on Northern blots is not recommended.

Several methods are available for labeling DNA probes, including nick translation (Rigby *et al.*, 1977), random priming (Feinberg and Vogelstein, 1983), and end labeling (Harrison and Zimmerman, 1986). For oligonucleotides, end labeling by phosphorylating the 5' base is the most common method. Another advantage of using oligonucleotide probes becomes apparent during the labeling technique, i.e., end labeling with [γ-^{32}P]-ATP and polynucleotide kinase is a much more rapid technique (30 min) than random priming of a double-stranded DNA (several hours). After the labeling reaction, it is necessary to separate the labeled DNA or oligonucleotide from unincorporated radioactivity. Several methods are available for this, including gel filtration (on conventional or spin columns) and use of binding resins. We prefer the latter, using the "Elu-tip-D" columns from Schleicher and Schuell because the technique is rapid. After the separation step, if a resin-type separation is used, DNA probes are generally ethanol precipitated. Oligonucleotides can be ethanol precipitated, especially if buffer exchange is required (as for primer extension probes, see below), but because this precipitation can be inefficient, it is often not done.

PROTOCOL FOR RANDOM PRIMING OF DNA FRAGMENTS

1. Prepare
   a. 0.1 M glycine, pH 9.2
      5 ml 0.2 M glycine
      1.2 ml 0.2 N NaOH
      3.8 ml H_2O
   b. 2× random priming buffer (88.5 mM glycine, pH 9.2, with 10 mM $MgCl_2$, 20 mM β-ME, 40 μM each dATP, dTTP, and dGTP)
      886.5 μl 0.1 M glycine
      100 μl 0.1 M $MgCl_2$
      1.5 μl β-ME

4 μl 10 mM dATP
4 μl 10 mM dTTP
4 μl 10 mM dGTP
store in aliquots at −20°C

c. 20× random primer mix (0.1 A_{260} U/μl pdN6; P. L. Biochemical Cat. No. 27-2166-01)
  50 U pdN6
  500 μl H_2O
  Store in aliquots at −20°C

d. Medium-salt buffer (0.4 M NaCl, 20 mM Tris, 1 mM EDTA)
  20 ml 5 M NaCl
  5 ml 1 M Tris, pH 7.4
  0.5 ml 0.5 M EDTA
  H_2O to bring buffer up to 250 ml

e. High-salt buffer (1 M NaCl, 20 mM Tris, 1 mM EDTA)
  50 ml 5 M NaCl
  5 ml 1 M Tris, pH 7.4
  0.5 ml 0.5 M EDTA
  H_2O to bring buffer up to 250 ml.

2. Mix in 1.5-ml microfuge tube
   1 μl DNA (50–200 ng; linear, <1 kb)
   1 μl 20× random primer mix
   2 μl H_2O
3. Boil for 2 min.
4. Spin.
5. Add
   10 μl 2× random priming buffer
   5 μl [^{32}P]dCTP (3000 Ci/mmol)
   1 U Klenow enzyme
6. Incubate 2–3 hr at room temperature.
7. Wash Elu-tip with 3.5 ml high-salt buffer (using syringe).
8. Wash Elu-tip with 3.5 ml medium-salt buffer.
9. Add 480 μl medium salt buffer to 20 μl labeled probe. Add to Elu-tip syringe and press slightly to start; discard flow-through.
10. Wash Elu-tip with 3.5 ml medium-salt buffer.

11. Elute probe with 500 μl high-salt buffer and collect in microcentrifuge tube.
12. Add
    a. 1 μl of a 5 μg/μl solution of tRNA
    b. 1 ml cold 100% ethanol (2 vol)
13. Store for 30 min at −20 to −30°C.
14. Spin for 30 min in microfuge, decant, swab, and vacuum dry.
15. Count radioactivity.
16. Dissolve pellet in 100 μl TE (10 mM Tris, 1 mM EDTA, pH 7.5–8.0).

PROTOCOL FOR END LABELING OF OLIGONUCLEOTIDES

1. Prepare
   a. Low-salt buffer (0.1 M NaCl, 20 mM Tris, 1 mM EDTA)
      5 ml 5 M NaCl
      5 ml 1 M Tris, pH 7.4
      0.5 ml 0.5 M EDTA
2. H_2O to bring buffer to 250 ml.
3. Mix in a 1.5-ml microcentrifuge tube
   a. 5 μl 0.2 M glycine, pH 9.2
   b. 2 μl 0.1 M $MgCl_2$
   c. 2 μl 0.1 M dithiothreitol (DTT)
   d. 5 μl oligonucleotide (50–300 ng)
   e. 2.5 μl [γ-^{32}P]ATP (10 Ci/ml; 6000 Ci/mmol)
   f. H_2O to bring solution up to 19 μl
   g. 1 μl polynucleotide kinase
4. Incubate 30 min at 37°C.
5. Add 80 μl low-salt buffer.
6. Apply slowly to Elu-tip attached to a 3-ml syringe (previously equilibrated with 3.5 ml high-salt buffer followed by 3.5 ml of low-salt buffer).
7. Wash Elu-tip with 3.5 ml low-salt buffer.
8. Elute labeled oligonucleotide with 150 μl high-salt buffer into a clean 1.5-ml Eppendorf tube.

NOTE If the oligonucleotide is to be ethanol precipitated, it may be eluted in a larger volume.

9. Count a 1-μl aliquot to determine radioactivity. If desired, labeled oligonucleotides can be precipitated with ethanol. In this case, elute with 200 μl high-salt buffer and add 1 μl of tRNA (5 mg/ml) as a carrier and 500 μl 100% ethanol.

D. Hybridization and Wash Conditions

Technically, Northern hybridizations are quite simple: incubate the filter with a ^{32}P-labeled probe under proper conditions of salt and temperature in a closed container for a number of hours to allow hydrogen bonding, then wash the blot under conditions of appropriate salt and temperature to remove probe nonspecifically bound to the filter or nonspecifically hybridized to RNA, while retaining specifically bound probe. In practice, several modifications are utilized in order to effect maximal specific hybridization (Lowe et al., 1987a; Graham et al., 1984). These include prehybridization of the filter in the absence of probe and of the probe to a "blank" filter and use of components in prehybridizations and hybridizations such as a heterologous DNA (e.g., salmon sperm DNA), dextran sulfate dissolved in formamide, and Denhardt's solution (Denhardt, 1966). Still, the most important considerations for both hybridization and washing conditions are the salt concentration and temperature. Collectively, these variables contribute to what is termed "stringency," i.e., maintaining specific hybridization while reducing or eliminating nonspecific binding and hybridization. Two types of salt formulations have been popularized: SSC, which is a combination of NaCl and sodium citrate, and SSPE, which is a combination of NaCl, sodium phosphate, and EDTA (Sambrook et al., 1989). Lowering the concentrations of these salts and increasing the temperature results in increased stringency and vice versa. Practically, the stringency of hybridization and wash must be determined empirically. However, some general guidelines can be given here. First, use of double-stranded DNA probes as opposed to complementary oligonucleotides requires a higher stringency. Second, formulae are available which compute the melting temperature, T_m, for single-stranded oligonucleotide hybridized to target nucleic acids which take into account the length and specific composition (i.e., GC content) of oligo probes (Sambrook et al., 1989). Third, stringency must usually be lowered if a heterologous DNA probe (i.e., that from a different animal species than that of the target RNA being probed) is utilized.

Oftentimes this is necessitated by the unavailability of homologous DNA probes.

After hybridization and washing, filters are exposed for autoradiography. Exposure at $-70°C$ and use of two intensifying screens are recommended. The size of hybridizing RNAs is usually calculated by comparison to the migration of the 28 and 18 S ribosomal RNA bands. If desired, blots can by "stripped," i.e., incubated under conditions which will remove specifically hybridized probe but leave the RNA on the filter, and subsequently hybridized to a different probe. The methods given below are based on procedures described in Graham et al. (1984) and Lowe et al. (1987a).

PROTOCOL FOR NORTHERN HYBRIDIZATION

1. Prepare Northern hybridization buffer (i.e., for 30 ml)
    a. 15 ml formamide containing 20% dextran sulfate (prepared by grinding 10 g dextran sulfate with a mortar and pestle and adding to 50 ml, final vol, formamide. Heat at 60°C overnight to solubilize. Mix to homogenize solution. Force this solution through a 0.45-µm nitrocellulose filter and store at 4°C.)
    b. 7.5 ml 20× SSPE
    c. 1.5 ml 2% (also referred to as 100×) Denhardt's (2% Ficoll, 2% bovine serum albumin, 2% polyvinylpyrrolidone)
    d. 1.5 ml 20% SDS
    e. 4.2 ml H_2O
    f. 300 µl salmon sperm DNA (10 mg/ml in H_2O, boiled for 5 min before adding)

NOTE 10 mg/ml salmon sperm DNA stock solution is itself prepared by boiling until DNA is in solution. This also shears DNA to appropriate size for use as a heterologous nucleic acid blocking agent.

2. Place hybridization buffer in 50°C water bath until ready to use.
3. Preincubate RNA filter by placing in dry plastic bag; seal all but one side of bag close to filter.
4. Add an appropriate volume of hybridization buffer to the bag (e.g., 10 ml for a small filter or 25 ml for a large filter).
5. Squeeze out bubbles and seal top of bag, leaving sufficient space for subsequent opening, addition of probe, and resealing (i.e., leave space between filter and seal).

22. Hormone and Receptor mRNAs

6. Incubate filter for 2 hr in 50°C water bath.
7. Denature probe by boiling labeled probe for 5 min.
8. Add the probe in hybridization buffer to blank filter at a final concentration of 1–2 million cpm/ml, remove bubbles, and seal bag.
9. Incubate blank filter with probe in 50°C water bath for 2 hr.
10. Carefully recover probe solution from bag containing blank filter and heat in tube to 80–90°C for 10 min. Discard blank filter and bag.
11. Add probe to RNA filter, remove bubbles, and seal bag two times close to filter. *Mix well.* Assume that 50–75% of the labeled probe is recovered from the probe prehybridization. Therefore, adjust final vol in bag containing RNA filter so that the final probe concentration remains at $1-2 \times 10^6$ cpm/ml.
12. Incubate in 50°C water bath overnight.
13. Wash hybridized filter twice in 250 ml of 2× SSPE, 0.2% SDS (heated to 50°C) in Pyrex container on shaker for 15 min.
14. Wash filter once in 250 ml 0.1× SSPE heated to 60°C in container on shaker for 15 min.
15. Wash filter once more in 250 ml 0.1× SSPE at 60°C in container on shaker for 5 min.
16. Seal wet filter in plastic bag and expose to film overnight or longer at −70°C. Do not let filter dry out or subsequent removal of probe (if desired) will be difficult.
17. When appropriate, strip probe from filter by incubating 10 min in H_2O at 60°C on shaking platform.
18. Subsequently, incubate blot 30 min in 95% formamide, 5 mM Tris, pH 8, 5 mM EDTA at 60°C on shaking platform.
19. Autoradiograph for 24–28 hr at −70°C to verify probe removal.

IV. Solution Hybridization/RNase Protection Assay

This technique takes advantage of the high degree of specificity of RNA:RNA hybrids and the advantages of solution chemistry to provide a means of determining both the amount and the structure of an RNA. A given DNA sequence is cloned into one of a number of available plasmid vectors containing bacteriophage RNA polymerase promoters, the vector is linearized, and the cloned insert is transcribed into complementary RNA (i.e., antisense RNA) in the presence of a radioactive NTP. The resulting "run-off" transcript is hybridized to RNA in solution, and nonhybridized RNA is subsequently removed by RNase digestion. Hybridized probe fragment(s) are collected by ethanol precipitation. If the sequence complementary to the antisense RNA (either all or part) is present in the preparation of RNA being analyzed, then those sequences will hybridize and thus be protected from RNase digestion. Radiolabeled protected probe fragments resulting from hybridization are detected by electrophoresis on denaturing polyacrylamide gels and autoradiography. Because the antisense radioactive RNA probe is always present in excess during the hybridization, the intensity of the protected band on autoradiographs (as determined by one of a number of techniques, including scanning densitometry, laser densitometry, or phosphorimaging) should be proportional to the concentration of the particular RNA of interest. An example of a solution hybridization/RNase protection assay is shown in Fig. 2.

In addition to providing an excellent means for quantitating RNA, RNA structure and heterogeneity in RNA structure can also be assessed with this method. If alternative splicing occurs in an mRNA, the pattern of splicing and the relative levels of splicing variants can be assessed by using an antisense RNA probe complementary to one of the variants. For example, in the rat insulin-like growth factor-I (IGF-I) gene, alternate splicing results in two forms of mRNA, one of which contains a 52-b insert, and one lacking this insert (Roberts et al., 1987). By using an antisense RNA probe containing the insert (Lowe et al., 1988), specific RNA hybridization and RNase digestion results in three bands. The longest results from protection of the full length of the probe insert, and thus represents mRNAs containing the insert, while the two shorter bands result from protection of probe sequence upstream of the 5' end of the 52-b insert and downstream of its 3' end. Thus, the absence of the 52-b insert leaves a corresponding gap in the probe:mRNA hybrid, resulting in RNase digestion of that gapped (i.e., single-stranded) portion of the probe. The actual sizes of the protected bands depends on how long the probe insert is; it becomes obvious that if a probe is being used to detect and quantitate heterogeneity in RNAs such as described above, then the region of probe used must not extend to other possible sources and regions of mRNA heterogeneity. These would include alternate splicing involving other exons, or different 5' or 3' ends in mRNAs. Such heterogeneity must be assessed separately by utilizing antisense RNA probes complementary to these regions of the mRNA.

As alluded to above, another powerful application of solution hybridization/RNase protection assay is determining the end structure of RNAs. For example, RNAs containing different 5' ends can be detected and quantitated

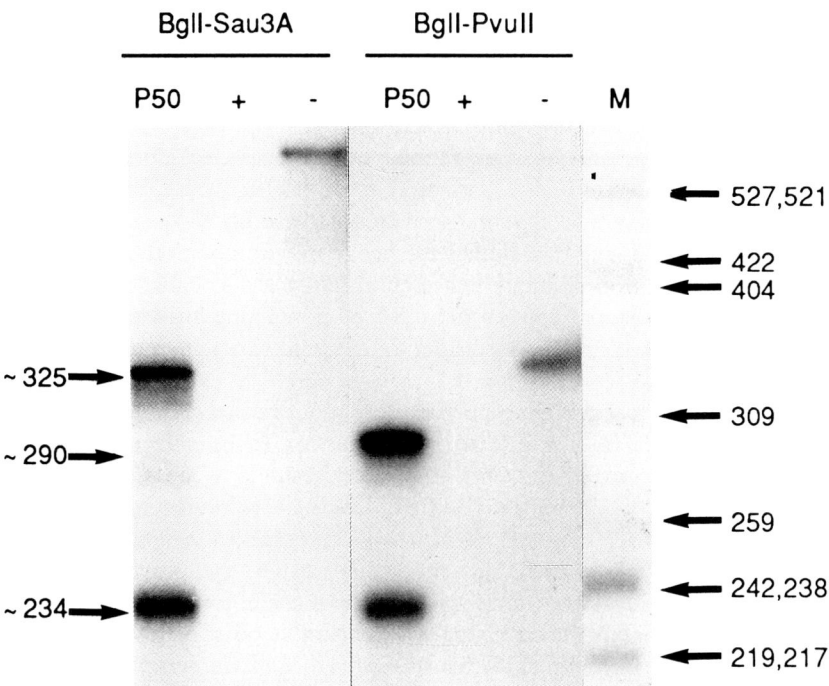

Figure 2 Solution hybridization/RNase protection assay of rat liver IGF-I mRNA. Two ^{32}P-labeled, antisense RNA probes (complementary to part of the first exon of the rat IGF-I gene) were prepared as described in the text and were hybridized to 20 μg of total RNA from the liver of a 50-day-old rat (P50), after which RNase digestion, electrophoresis, and autoradiography of protected probe bands were conducted as described in the text. Lanes (+) and (−) represent probe carried through the assay and incubated with or without RNases, respectively. Lane M contains ^{32}P-labeled pBR325 DNA markers, the sizes of which are shown by the arrows to the right. The arrows to the left indicate protected probe bands resulting from hybridization to specific rat liver IGF-I mRNAs. The two probes were identical at their 3' ends, but differed at their 5' ends, i.e., the BglII-Sau3A probe extended ~205 b further 5' than did the BglII-PvuII probe. Therefore, the 325- and 234-b protected bands derived from the BglII-Sau3A probe represent IGF-I mRNAs with 5' ends separated by 100 b. The shorter RNA also protected the same 234 b in the BglII-PvuII probe, while the longer RNA protected the entire length of this probe, resulting in a protected band of 290 b. The difference in length between this band and the labeled probe band [i.e., lane (−)] reflects transcribed vector sequence.

using protection assays. The rat IGF-I system again provides an excellent example of this application (Adamo et al., 1991a,b). In this case, two types of probes were utilized. One type is represented by the RNase protection assay shown in Fig. 2, in which use of a leader exon-specific probe yielded multiple protected bands, all of which were shorter than the labeled probe. The difference in length between the various protected bands and the probe presumably reflects the position at which transcription was initiated, i.e., the different lengths of protected bands reflect 5' end heterogeneity. As discussed below, assignment of transcription start sites based upon generation of these bands in protection assays is confirmed by primer extension analysis. To confirm that protection of less than the full length of the probe indeed reflected 5' end heterogeneity, and to increase the utility of the probe, it was lengthened at the 3' end to include sequences contiguous with the leader exon sequence. In addition to bands representing different 5' ends of the leader exon, additional bands representing mRNAs containing an alternate leader exon or splicing within the first leader exon could simultaneously be visualized.

A few practical guidelines for solution hybridization/RNase protection assays include the following: first, it is necessary to be able to accurately resolve protected probe bands on 6 to 8% acrylamide gels, meaning that these bands should be > 100b and < 1000b in length. This in turn means that if a probe insert is chosen such that mRNAs will protect its entire length, it should not exceed ~1 kb, and preferably should be smaller, e.g., between 200 and 600 b. Second, when constructing plasmids, it is useful to clone inserts into a site or sites of the polylinker as far downstream as possible from the phage RNA polymerase promoter. In this way, the run-off transcript will include vector sequence and will be significantly longer than the cloned insert. If the insert is chosen such that mRNAs will protect its entire length, then the difference in size between the run-off transcript and the full-length protected band resulting from hybridization of the mRNA of interest is easy to detect and is an assurance of specificity. If probe inserts are chosen such that they are much longer than the longest protected band (as with the example of multiple, widely spaced transcription start sites in one of the leader exons of the rat IGF-I gene) then this consideration is less important. Third, if multiple specifically protected bands appear on autoradiograms as a result of alternative splicing or different end structure, and if relative quantitations of the variant mRNAs resulting in different protected bands is desired, then it is necessary to correct for differences in the specific radioactivity of the various protected bands. The specific radioactivity of a band depends on the number of radioactive bases present in that band. The band is assumed to be uniformly labeled. For example, if two protected bands are obtained, [^{32}P]UTP was the radioactive NTP used, and the longer band contains 80 U residues, the shorter band contains 50 residues, and the insert contains 100 U residues, it is necessary to multiply the intensity of the longer band by 1.25 and the shorter band by 2 in order to normalize the specific radioactivities. Fourth, the specificity of the assay is usually controlled by including a sample of RNA in which the particular RNA of interest is not represented; in such a case, the autoradiogram of this sample should indicate no protected probe bands. Finally, it is necessary to demonstrate empirically that the probe is present in excess during the hybridization. This can be done in two ways: first, the reaction should be essentially first order with respect to test RNA as reflected by linearly increasing intensity of protected bands as increasing quantities of RNA are hybridized; second, the reaction should be zero order with respect to probe such that addition of increasing amounts of probe to a fixed amount of RNA does not alter the intensity of protected bands. The methods below are taken from protocols described by Melton et al. (1984) Starksen et al. (1986), and Lowe et al. (1987b) and reagents were supplied by Promega (Madison, WI).

PROTOCOL FOR SOLUTION HYBRIDIZATION/RNase PROTECTION ASSAY

Day 1

1. Precipitate overnight at $-20°C$ total RNA by mixing
   20 μg total RNA
   H_2O up to 18 μl
   2 μl 5 M NaCl
   40 μl 100% ethanol

Day 2

1. Prepare 4× hybridization buffer (80 mM Tris, pH 7.5, 4 mM EDTA, 1.6 M NaCl, 0.4% SDS)
   472 μl sterile H_2O
   160 μl 0.5 M Tris, pH 7.5
   8 μl 0.5 M EDTA, pH 8.0
   320 μl 5.0 M NaCl
   40 μl 10% SDS

   Store at room temperature, heat to 40–50°C (to dissolve SDS) before using.

2. Label probe by mixing in 1.5-ml sterile microcentrifuge tube
   4 μl 5× transcription buffer (from Promega kit)
   2 μl 100 mM DTT

      0.8 μl 25–40 U/μl RNasin
      4 μl of a 2.5 mM mixture of ATP, GTP, and CTP
      2.4 μl 200 μM UTP (12 μM minimum final concentration)
      1 μl (>200 ng) DNA template
      5 μl 10 mCi/ml [α-^{32}P]UTP (800 Ci/mmole)
      1 μl of appropriate RNA polymerase
   Vortex and incubate 1 hr at 37–40°C.

3. Add
      1 μl VRC (200 mM)
      1 μl DNase I (1 μg/μl)
      1.5 μl tRNA (5 μg/μl)
   Vortex and incubate 15 min at 37°C.

4. Add
      12 μl phenol (0.5 vol)
      24 μl CHCl$_3$: IAA (24 : 1)
   Vortex and microfuge 2 min at room temperature

5. Transfer 22 μl of top layer to new tube. Add 22 μl CHCl$_3$: IAA (24 : 1). Vortex and microfuge 2 min at room temperature.

6. Transfer 20 μl of top layer to new tube. Add
      1.5 μl 3 M sodium acetate, pH 6.0
      50 μl ethanol
   Vortex and precipitate 30 min at −20°C.

7. Microfuge 30 min at 4°C, pipette off supernatant, and vacuum dry.

8. Resuspend in
      30 μl sterile H$_2$O
      1.5 μl 3 M sodium acetate, pH 6.0
      1 μl tRNA
      65 μl cold 100% ethanol (2 vol)
   Vortex and precipitate 30 min at −20°C.

9. Microfuge 30 min at 4°C, pipette off supernatant, vacuum dry, and count radioactivity in tube.

10. Resuspend in sterile H$_2$O to a final concentration of 400,000 cpm/μl.

11. Start hybridization by microfuging RNA precipitated on Day 1 for 30 min at 4°C, pipette off supernatant, and vacuum dry.

12. Resuspend RNA [plus two additional tubes without exogenous RNA (C1 and C2)] in 29.5 μl hybridization buffer (one part 4× hybridization buffer to three parts deionized formamide) and add 0.5 μl probe (≈ 200,000 cpm).
13. Heat 5 min at 85°C.
14. Incubate overnight at 45°C.

Day 3

1. Prepare RNase digestion buffer (10 mM Tris, pH 7.6, 5 mM EDTA, 300 mM NaCl, 40 μg/ml RNase A, 2 μg/ml RNase T1). For 3 ml (i.e., for 10 samples)
    60 μl 0.5 M Tris, pH 7.6
    30 μl 0.5 M EDTA, pH 8.0
    180 μl 5 M NaCl
    2.73 ml H$_2$O
    12 μl 10 mg/ml RNase A
    1.5 μl 4 mg/ml RNase T1
2. Add 270 μl RNase digestion buffer to hybridized samples and C1. Add 270 μl buffer *without* RNases to C2. Incubate for 1 hr at 30°C.
3. Add
    20 μl 10% SDS
    2.5 μl 20 mg/ml proteinase K
    Incubate for 15 min at 37°C.
4. Add
    150 μl phenol (0.5 vol)
    300 μl CHCl$_3$: AA (1 vol)
    Microfuge 2 min at room temperature.
5. Transfer 320 μl of top layer to new sterile screw cap tube. Add
    4 μl 5 μg/μl tRNA carrier
    1.5 μl 3 M sodium acetate
    700 μl cold 100% ethanol
    Precipitate 30 min at −20 to −30°C and microfuge 30 min at 4°C. Decant, swab, and dry.

6. Prepare
   a. Dyes for sample buffer
      10 mg xylene cyanol
      10 mg bromophenol blue
      3 ml H_2O
   b. Sample buffer
      950 μl (deionized) formamid
      20 μl 0.5 M EDTA
      30 μl dyes in H_2O
   c. 40% acrylamide stock
      380 g acrylamide
      20 g bisacrylamide
      Double-distilled H_2O to bring buffer up to 1000 ml.
   d. Instagel
      100 ml 40% acrylamide
      250 g urea
      25 ml 10× TBE
      Double-distilled H_2O to bring gel up to 500 ml.
   e. Running buffer (0.5× TBE)
7. Resuspend samples and C1 in 12.5 μl of a 1:4 mixture of H_2O and sample buffer. Transfer 6.5 μl of each sample and C1 to sterile 500-μl tubes (optional).
8. Resuspend C2 in 150–200 μl H_2O. Transfer 1 μl to a 500-μl sterile tube. Add 5.5 μl sample buffer.
9. Add 0.5–1 μl native probe to 99 μl H_2O. Transfer 1 μl to a sterile 500-μl tube. Add 5.5 μl sample buffer.
10. Mix 1 μl of a suitable DNA marker (^{32}P-labeled, ~5000 cpm) with 5.5 μl sample buffer.
11. Wash electrophoresis apparatus (instructions here refer to the Sequi-Gen system from Bio-Rad) plates and comb with glass cleaner, H_2O, 70% ethanol, 100% ethanol. Put plates together with two spacers. Clamps go on with male plug on top and widest part on same side as top reservoir of plates. Plug up upper reservoir with paper towel. Press firmly so bottom of clamps and glass plates are even. Put rubber piece and blotter paper (3M) in bottom of holder.
12. Weigh out 0.026 g ammonium persulfate (AP) in 15-ml tube (for plug) and 0.036 g AP in 15-ml tube (for gel). For plug, add ~0.5 ml

H₂O to 0.026 g AP, and add AP solution to 15 ml 8% instagel in 50-ml tube. Add 75 μl TEMED, mix, and quickly pour onto blotter, then press gel plates into blotter. Hold for 45–60 sec until plug begins to gel. Screw bottom stand into place. Wait 5 min.

13. To pour gel, add ~0.5 ml H₂O to 0.036 g AP, and add AP to 40 ml 8% instagel in 50-ml tube. Add 14 μl TEMED, mix, and pipette into plates.
14. Place plates flat on bench and insert comb between plates with top of comb ~2 mm below top of longest plate. Wait 30 min–1 hr.
15. Preelectrophorese after adding 950 ml 0.5× TBE to top and bottom reservoirs and washing wells. Turn on switch on back of power supply and turn voltage and current knobs to maximum. Push wattage switch and set to 40 W. Push set button so light goes off. Plug in leads and turn on toggle switch and run until gel temperature is ~50°C.
16. Add more 0.5× TBE to top reservoir and wash wells before adding samples.
17. Heat samples to 95°C for 3 min, chill on ice, and load 6.5-μl aliquots.
18. Run gel at 40 W ~1–2 hr, i.e., when bromophenol blue has run off and xylene cyanol has migrated $\frac{3}{4}$ way to bottom of gel.
19. Dry gel and autoradiograph with two intensifying screens at −70°C for required time (usually overnight to several days, depending on abundance of mRNA of interest).

V. Primer Extension

Primer extension analysis is used in conjunction with solution hybridization/RNase protection assays for determining 5' ends of RNA (Fig. 3). As detailed in the previous section, use of RNA probes complementary to sequences encompassing the putative cap site(s) of an RNA can yield protected probe bands whose size indicates the location of the cap site(s). However, this analysis alone is not sufficient for assigning transcription start sites for the following reasons: first, the 5' end of a protected band merely indicates where the RNA sequence diverged from the probe sequence. This divergence may in fact, reflect the 5' end of the RNA or it may reflect a point of splicing of divergent upstream sequence. Second, it is generally impractical to calculate the exact size of a protected band from an autoradiogram of a protection assay gel and thus assign an exact cap site. In primer extension, the size of a primer extension product may be correlated with the size of a protected band from RNase protection assay, and thus the two independent methods can corroborate one another for purposes of assigning transcription start sites. Third, primer extension allows for a precise assignment of cap site(s), since sequencing reactions can be done using the unlabeled primer and any DNA plasmid containing the sequence encompassing the putative cap site(s). The reactions are run on gels in lanes adjacent to those on which are run the primer extension assays, which utilized the end-labeled primer. Upon autoradiography, the band representing labeled primer extension product runs at the same location as the base representing its 5′ terminus on the sequencing ladder, and therefore that base is assigned as the cap site. Obviously, one of the major assumptions of such an analysis is that the primer has been extended all the way to the 5′ terminus of a given RNA. If, however, secondary structure or some other complication prevents extension of the primer to the 5′ end of the RNA, then an incorrect "cap site" downstream of the actual cap site will be assigned. Thus, primer extension and solution hybridization/RNase protection assays should be used together (compare Figs. 2 and 3) for the determination of transcription start sites because each overcomes the inherent drawback of the other: protection assays (because of stringent denaturing and hybridization conditions) are relatively independent of RNA secondary structure, while primer extension will obviously occur regardless of whether the contiguous sequence of the RNA upstream of that complementary to the primer (or probe) is known. As a practical matter, a rational approach to mapping 5′ ends is to first approximate a putative 5′ end using solution hybridization/RNase protection assays. Subsequently, an antisense primer should be placed approximately 30 b downstream of the putative cap site. If the primer is, e.g., a 22-mer, then a ∼52-bp primer extension product, whose exact size can be determined by comparison to the sequencing ladder as described above, should be generated. As an additional control, it is useful to utilize at least two downstream primers, to generate differently sized primer extension products whose 5′ ends should, however, be identical.

In designing and performing primer extension experiments, some important points should be kept in mind: first, as stated above, primers of about 22 b, complementary to RNA sequences 30–60 b downstream of the predicted cap site, should be used. Second, the GC content of the primers should be ∼50%. Third, as stated above, several primers should be used in primer extension analysis to verify a cap site. Fourth, it is essential that suitable negative controls be included in the hybridization and primer extension steps. Thus, at the very least, tubes containing no RNA (to control for "primer dimer" formation) and tRNA (to control for nonspecific hybridization of the primer to unrelated RNA sequence) should be included in the assay along with tubes containing the test RNA. However, it has been our experience that a

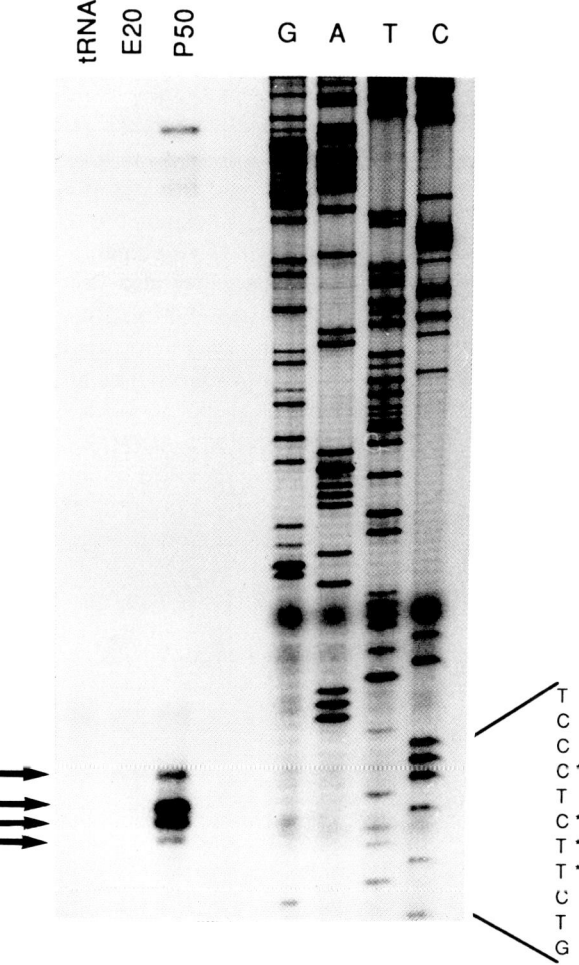

Figure 3 Primer extension analysis of rat liver IGF-I mRNA. An antisense oligonucleotide complementary to a 22-b sequence approximately 30 b downstream of the second transcription start site identified in Fig. 2 was end labeled and hybridized to 3 μg of tRNA or poly(A)⁺ RNA from the livers of 20-day-old fetal rats (E20) or a 50-day-old postnatal rat (P50). Primer extension was carried out as described in the text, and primer extension products were electrophoresed alongside sequencing reactions of a rat IGF-I exon 1 plasmid using the same (unlabeled) primer. The arrows indicate the primer extension products whose 5' termini correspond to the nucleotides denoted by asterisks. The two sets of primer extension products correspond to the two protected bands observed in Fig. 2 and thus corroborate the assignment of two major transcription start sites.

given primer may hybridize to an unrelated sequence present in the test RNA sample. The only way to control for this is to include a tube containing a sample of test RNA which is absolutely or relatively depleted of the RNA of interest. Again using the rat IGF-I mRNA as an example (Adamo et al., 1991a), we used liver RNA from 50-day postnatal rats as the test RNA, and, as a negative control, we used liver RNA from Day 20 fetal rats, because the expression of IGF-I mRNAs is 100 times higher in P50 rat liver than in E20 rat liver (Fig. 3) (Adamo et al., 1991b). Fifth, our experience has been that better results are obtained using poly(A)$^+$-enriched RNA preparations, although total RNA can be used. Finally, our experience has also been that the reverse transcription step, using AMV RT, can be performed successfully at 50°C. This may have the advantage of reducing secondary structure inhibitory to synthesis of full-length primer extension products. The method given below was used to map transcription start sites in the rat IGF-I receptor gene (Werner et al., 1990) and the rat IGF-I gene (Adamo et al., 1991a).

PROTOCOL FOR PRIMER EXTENSION

Day 1

1. Prepare

    5× primer hybridization buffer (50 mM Pipes, pH 6.4, 2 M NaCl, 5 mM EDTA)

    100 μl 0.5 M Pipes, pH 6.4

    400 μl 5 M NaCl

    10 μl 500 mM EDTA

    490 μl sterile H$_2$O

2. Label probe (single-stranded oligonucleotides which have been gel purified on a 20% polyacrylamide/8 M urea gel) by mixing in a 1.5-ml microcentrifuge tube

    1 μl oligo (50–200 ng)

    4 μl 0.1 M MgCl$_2$

    4 μl 0.1 M DTT

    10 μl 0.2 M glycine

    4 μl [γ-^{32}P]ATP (10 mCi/ml; 6000 Ci/mmol)

    1.5 μl PNK (polynucleotide kinase)

    Sterile H$_2$O to bring total up to 40 μl.

3. Incubate for 30 min at 37°C.
4. Wash Elu-tip with 3.5 ml high-salt buffer (see previous sections), using syringe.
5. Wash Elu-tip with 3.5 ml low-salt buffer.

6. Add 460 μl low-salt buffer to 40 μl labeled oligo reaction. Add to Elu-tip syringe and press slightly to start; discard flow-through.
7. Wash Elu-tip with 3.5 ml low-salt buffer.
8. Elute oligo with 200 μl high-salt buffer and collect in 2-ml microcentrifuge tube.
9. Add

    20 μl 3 M sodium acetate

    1 μl tRNA (5 μg/μl)

    550 μl 100% ethanol (2.5 vol)

    Precipitate on dry ice for 30 min, incubate at room temperature for 2 min, spin 30 min, 4°C, decant, swab, and vacuum dry.
10. Count and redissolve in H_2O to 1,000,000 dpm/4.3 μl.
11. In 1.5-ml screw-top tubes, mix

    3–5 μg of poly(A)$^+$ RNA or tRNA (no RNA for reagent blank; use more RNA if this is low-abundance message).

    8 μl 5× hybridization buffer

    4.3 μl ^{32}P-labeled primer (1 × 10^6 cpm)

    sterile H_2O to bring solution up 40 μl total
12. Put samples in bath at 75°C, immediately adjust dial to 42°C. Incubate 16–18 hrs.

Day 2

1. Prepare 10× primer extension buffer (500 mM Tris, pH 8.0, 1 M KCl, 100 mM MgCl$_2$)

    0.5 ml 1 M Tris, pH 8.0

    0.5 ml 2 M KCl

    0.1 ml 1 M MgCl$_2$

2. Precipitate samples by adding 1 μl tRNA and 100 μl 100% ethanol (2.5 vol). Keep on dry ice for 30 min and then at room temperature for 2 min. Spin at 4°C, pipette off supernatant, wash with cold 80% ETOH, dry, and count.
3. Resuspend pellets in 39 μl of the following mix

    27 μl sterile H_2O

    4 μl 10× primer extension buffer

    2 μl 10 mM dATP

    2 μl 10 mM dCTP

    2 μl 10 mM dGTP

    2 μl 10 mM dTTP

4. Add 1 µl avian myeloblastosis virus (AMV) reverse transcriptase (~20 units).
5. Incubate 42°C for 1 hr (temperature can be increased to 50°C).
6. Add
   15 µl 0.1 M EDTA
   60 µl phenol
   60 µl $CHCl_3$-IAA
   Vortex, spin, and save top layer (e.g., 45 µl).
7. Add
   1 µl tRNA (5 µg/µl)
   5 µl 3 M sodium acetate (final 0.3 M)
   128 µl ethanol
   Keep on dry ice for 30 min, spin, wash with 70 or 80% ethanol, and dry.
8. Resuspend pellet in 10 µl RNA loading dye. Heat 3 min at 95°C, then chill on ice. Load 5-µl samples on gel.
9. Run with appropriate sequencing ladder and labeled DNA markers on 8% gel for 45–75 min. See solution hybridization and gel electrophoresis protocols above.

References

Adamo, M. L., Ben-Hur, H., LeRoith, D., and Roberts, C. T., Jr. (1991a). *Biochem. Biophys. Res. Commun.* **176**, 887–893.

Adamo, M. L., Ben-Hur, H., Roberts, C. T., Jr., and LeRoith, D. (1991b). *Mol. Endocrinol.* **5**, 1677–1686.

Aviv, H., and Leder, P. (1972). *Proc. Natl. Acad. Sci. U.S.A.* **69**, 1408–1412.

Cathala, G., Savouret, J.-F., Mendez, B., West, B. L., Karin, M., Marial, J. A., and Baxter, J. D. (1983). *DNA* **2**, 329–336.

Chirgwin, J. M., Przybyla, A. E., MacDonald, R. J., and Rutter, W. J. (1979). *Biochemistry* **18**, 5294–5300.

Davis, L. G., Dibner, M. D., and Battey, J. F. (1986). "Basic Methods in Molecular Biology," pp. 130–135. Elsevier, New York.

Denhardt, D. T. (1966). *Biochem. Biophys. Res. Commun.* **13**, 641–646.

Feinberg, A. P., and Vogelstein, B. (1983). *Anal. Biochem.* **132**, 6–13.

Graham, D. E., Medina, D., and Smith, G. (1984). *J. Virol.* **49**, 819–827.

Harrison, B., and Zimmerman, S. B. (1986). *Anal. Biochem.* **158**, 307–313.

Lowe, W. L., Jr., Schaffner, A. E., Roberts, C. T., Jr., and LeRoith, D. (1987a). *Mol. Endocrinol.* **1**, 181–187.

Lowe, W. L., Jr., Roberts, C. T., Jr., Lasky, S. R., and LeRoith, D. (1987b). *Proc. Natl. Acad. Sci. U.S.A.* **84**, 8946–8950.

Lowe, W. L., Jr., Lasky, S. R., LeRoith, D., and Roberts, C. T., Jr. (1988). *Mol. Endocrinol.* **2,** 528-535.

Melton, D. A., Krieg, P. A., Rebagliati, M. R., Maniatis, T., Zinn, K., and Green, M. R. (1984). *Nucleic Acids Res.* **12,** 7035-7056.

Rigby, P. W. J., Dieckmann, M., Rhodes, C., and Berg, P. (1977). *J. Mol. Biol.* **113,** 237-250.

Roberts, C. T., Jr., Lasky, S. R., Lowe, W. L., Jr., Seaman, W. T., and LeRoith, D. (1987). *Mol. Endocrinol.* **1,** 243-248.

Sambrook, J., Fritsch, E. F., and Maniatis, T. (1989). "Molecular Cloning: A Laboratory Manual." Cold Spring Harbor Lab. Press, Cold Spring Harbor, New York.

Starksen, N. F., Simpson, P. C., Bishopric, N., Coughlin, S. R., Lee, W. F., Escobedo, J. A., and Williams, L. T. (1986). *Proc. Natl. Acad. Sci. U.S.A.* **83,** 8348-8350.

Werner, H., Stannard, B., Bach, M. A., LeRoith, D., and Roberts, C. T., Jr. (1990). *Biochem. Biophys. Res. Commun.* **169,** 1021-1027.

23

The Polymerase Chain Reaction: Applications to Endocrine Research

Alan R. Shuldiner and Riccardo Perfetti

    I. Introduction
   II. Polymerase Chain Reaction: Theory of the Method
  III. General Protocol: The Polymerase Chain Reaction
  IV. Comments: General Polymerase Chain Reaction Protocol
   V. Reverse Transcription-Polymerase Chain Reaction (RT-PCR or RNA-PCR)
  VI. Protocol for Reverse Transcription
 VII. Comments: Reverse Transcription
VIII. Quantitative Reverse Transcription-Polymerase Chain Reaction
  IX. Contamination and Polymerase Chain Reaction False Positives
   X. RNA Template-Specific Polymerase Chain Reaction (RS-PCR)
  XI. Subcloning Polymerase Chain Reaction Products
 XII. Direct Sequencing of Polymerase Chain Reaction Products
XIII. Some Commonly Encountered Polymerase Chain Reaction Artifacts
     A. Unexpected PCR Products
     B. Taq Polymerase Misincorporation Errors
     C. Hybrid DNA Artifact
XIV. Conclusions
     References

I. Introduction

The polymerase chain reaction (PCR) is a simple and versatile method to amplify *in vitro* a specific segment of DNA even in the presence of a very large number of unrelated sequences (Saiki *et al.*, 1985, 1988; Mullis and Faloona, 1987; Mullis *et al.*, 1986). The great utility of the method lies in its exquisite sensitivity and specificity. Typically, as few as 1 to 100 copies of DNA can be

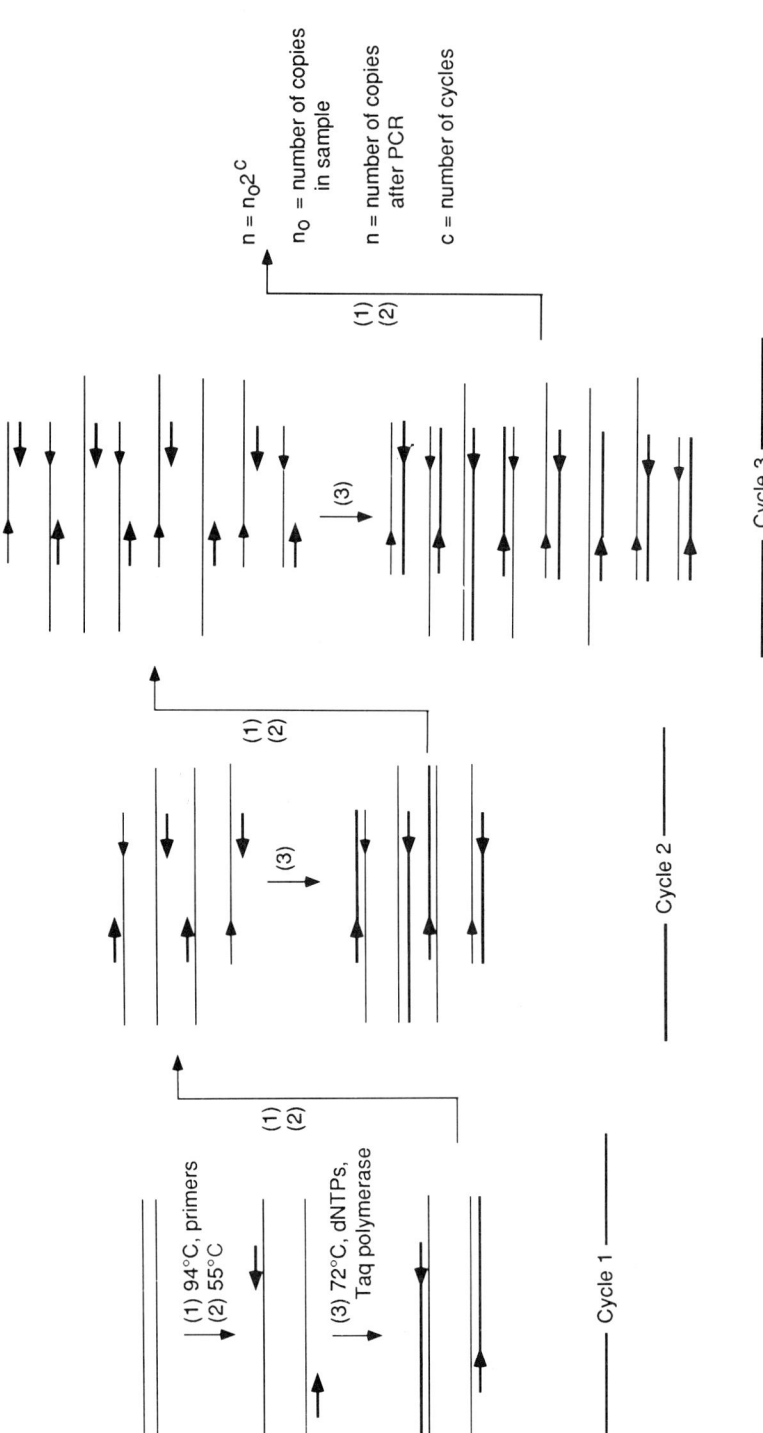

Figure 1 Schematic of the polymerase chain reaction. Primers are denoted by horizontal arrows, DNA templates are denoted by narrow solid lines, and extending DNA strands are denoted by bold solid lines. Note that double-stranded short products accumulate logarithmically from the third cycle onward.

amplified by a factor of 10^5 to 10^6 resulting in nanogram to microgram quantities of the DNA product. Since its invention in 1985, the PCR method has gained widespread use in virtually all fields of biology, especially molecular biology and biomedical research (Eisenstein, 1990; Erlich et al., 1991; Tompkins, 1992; Arnheim and Erlich, 1992; Reiss and Cooper, 1990). Since it is not possible to review in detail the multitude of applications of PCR to endocrine research, this chapter provides a practical guide in which we (i) review the theory of the PCR method, (ii) provide basic protocols that may be adapted by the investigator for specific applications, (iii) discuss common pitfalls and limitations, and (iv) provide useful troubleshooting suggestions.

II. Polymerase Chain Reaction: Theory of the Method

The first step in PCR is denaturation of the DNA template (i.e., genomic DNA, plasmid DNA, single-stranded or double-stranded cDNAs) into single strands by heating to 94–95°C (Fig. 1). Synthetic oligonucleotides (primers) whose sequences hybridize to opposite strands of the DNA template a predetermined distance from one another (optimally 100 to 1000 bp) are added to the denatured DNA template, and the mixture is cooled to 45–60°C. At this temperature, the primers anneal to their respective complementary regions on the DNA template and effectively ignore noncomplementary sequences. The reaction mixture is heated to 72°C, and in the presence of the four deoxyribonucleotide triphosphates (dATP, dTTP, dCTP, and dGTP) and Taq DNA polymerase (Gelfand and White, 1990; Chien et al., 1976; Lawyer et al., 1988; Bloch, 1991), a thermostable DNA polymerase, the oligonucleotides are extended resulting in two double-stranded DNA molecules. These steps, denaturation (94°–95°C), annealing (45°–60°C), and extension constitute a "cycle," and are repeated many times. During the first and second cycles, only long PCR products are generated, and during subsequent cycles, both long and short PCR products are generated (Figure 1). Since the long PCR products accumulate linearly, while the short PCR products accumulate logarithmically, after 25 cycles, one can expect as great as 10^5 amplification of almost exclusively short products, i.e., DNA of a specific size and sequence defined by the primer pair (Fig. 2).

PCR can be adapted for a great number of applications in endocrine research (Table I). For example, since Taq DNA polymerase will tolerate one or more mismatches between the primer sequences and their templates, if redundant (or consensus) primers are used that would have a high probability of hybridizing to unknown related target sequences, homologous sequences of the same gene between different species or related genes within a species may be amplified by PCR and characterized (Lee et al., 1988; Shuldiner et al.,

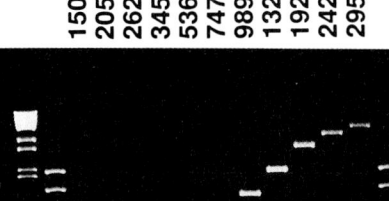

Figure 2 Examples of PCR products. PCR amplification of the human β-globin gene was accomplished from genomic DNA with primer pairs that gave expected products ranging in size from 150 to 2951 bp. The PCR products were electrophoresed on a 1.6% agarose gel and visualized by ethidium bromide staining and UV transillumination. The markers are BstEII-digested phage and HaeIII-digested φX174-RF. (After Saiki, 1990, with permission.)

1990b; Scavo *et al.*, 1991; Leibrock *et al.*, 1989; Batzer *et al.*, 1991; Zhang and Goldstein, 1991; Mack and Sninsky, 1988; Gould *et al.*, 1989). In addition, PCR may be used as a powerful tool in a variety recombinant DNA strategies including cDNA and genomic DNA library construction and screening (Ludecke *et al.*, 1989; Belyavsky *et al.*, 1989; Gussow and Clackson, 1989; Ochman *et al.*, 1988; Frohman *et al.*, 1988; Loh *et al.*, 1989; Jain *et al.*, 1992; Aslanidis and DeJong, 1991; Guzzetta *et al.*, 1991; Nelson *et al.*, 1991; Mares *et al.*, 1991; Triglia *et al.*, 1988; Roux and Dhanarajan, 1990; Wesley *et al.*, 1990; Brunet *et al.*, 1991; Timblin *et al.*, 1990; Zaraisky *et al.*, 1992; Ko, 1990; Patanjali *et al.*, 1991; Wieland *et al.*, 1990), generation of cDNA probes (Scully *et al.*, 1990; Jansen and Ledley, 1989; Lion and Haas, 1990; Emanuel, 1991; Jackson, 1991; Hayashi *et al.*, 1989), subcloning (Scharf, 1990; Scharf *et al.*, 1987; Shuldiner *et al.*, 1991b; Shuldiner and Tanner, 1993; Jones and Howard, 1990; Liu and Schwartz, 1992; Kaufman and Evans, 1990; Bhat *et al.*, 1991; Starr and Quaranta, 1992; Mitchell *et al.*, 1992), DNA sequencing (Beven *et al.*, 1992; Gyllensten and Erlich, 1988; Innis *et al.*, 1988; Stoflet *et al.*, 1988; Mazars *et al.*, 1991; Lee, 1991; Ruano and Kidd, 1991; Engelke *et al.*, 1988; Hultman *et al.*, 1989; Paabo *et al.*, 1988), and site-directed mutagenesis (Higuchi *et al.*, 1988; Horton *et al.*, 1989; Kadowaki *et al.*, 1989; Zhou *et al.*, 1991; Cadwell and Joyce, 1992). Other applications include the rapid screening of candidate genes for mutations (Lerman and Silverstein, 1985; Sheffield *et al.*, 1989; Weinstein *et al.*, 1991; Hayashi, 1991; Orita *et al.*, 1989;

Table I
Applications of PCR to Endocrine Research

	Reference
PCR-assisted cloning	
Construction of cDNA and genomic libraries	Ludecke et al. (1989); Belyavsky et al. (1989); Wesley et al. (1990); Brunet et al. (1991)
Construction of subtraction libraries	Timblin et al. (1990); Zaraisky et al. (1992)
Generation of probes for screening	Scully et al. (1990); Hayashi et al. (1989); Lion and Haas (1990)
Screening of DNA libraries	Gussow and Clackson (1989)
Cloning across phylogenic boundaries or related genes within a species	Shuldiner et al. (1990b); Scavo et al. (1991); Zhang and Goldstein (1991); Mack and Sninsky (1988); Gould et al. (1989)
Amplification of DNA sequences adjacent to a known sequence	
Anchored PCR or rapid amplification of DNA ends	Frohman et al. (1988); Loh et al. (1989); Jain et al. (1992)
Inverse PCR	Ochman et al. (1988); Triglia et al. (1988)
Alu PCR	Mares et al. (1991); Nelson et al. (1991); Guzzetta et al. (1991)
Site-directed mutagenesis	Higuchi et al. (1988); Horton et al. (1989); Kadowaki et al. (1989)
Random mutagenesis	Zhou et al. (1991); Cadwell and Joyce (1992)
Subcloning DNA fragments	Scharf (1990); Kaufman and Evans (1990); Bhat et al. (1991); Shuldiner et al. (1991b)
In vivo DNase footprint analysis	Krummel (1990); Mueller and Wold (1989); Pfeifer and Riggs (1991)
Screening for new mutations	
Denaturing gradient gel electrophoresis (DGGE)	Sheffield et al. (1989); Weinstein et al. (1991); Barbetti et al. (1992)
Heteroduplex analysis	Nagamine et al. (1989); Keen et al. (1991)
Single-stranded conformational polymorphism (SSCP) analysis	Orita et al. (1989); O'Rahilly et al. (1991); Kishimoto et al. (1992)
Diagnostic screening for known mutations	
Allele specific oligonucleotide hybridization (ASO)	Kogan et al. (1987); Saiki et al. (1986); Ma et al. (1991)
PCR amplification of specific alleles (PASA)	Sommer et al. (1991); Wu et al. (1989)
Ligase chain reaction	Barany (1991a,b); Wu and Wallace (1989)
DNA sequencing	Bevan et al. (1992); Gyllensten and Erlich (1988); Lee (1991); Innis et al. (1988)
Measuring gene expression	
Reverse transcription-PCR (RT-PCR)	Kawasaki et al. (1988); Kawasaki (1990); Rappolee et al. (1988); Gilliland et al. (1990a,b); Feere (1992)
RNA template-specific–PCR (RS-PCR)	Shuldiner et al. (1991a,c, 1993)
Alternative splicing of mRNA	Moller et al. (1989); Nakabeppu and Nathans (1991); Foulkes et al. (1991)
In situ PCR	Nuovo et al. (1991a,b), Nuovo (1992)

Kadowaki et al., 1992; O'Rahilly et al., 1991; Dean and Berrard, 1991; Suzuki et al., 1990; Nagamine et al., 1989; Kwok et al., 1990; Keen et al., 1991; Barbetti et al., 1992; Cotton, 1989; Kogan et al., 1987; Saiki et al., 1986; Sommer et al., 1991; Ma et al., 1991; Wu et al., 1989; Barany, 1991a,b), in vivo DNase footprinting (Krummel, 1990; Mueller and Wold, 1989; Pfeifer and Riggs, 1991) and characterization of RNA- and DNA-binding proteins (Blackwell and Weintraub, 1990; Triesen and Bach, 1990; Mavrothalassitis et al., Tuerk and Gold, 1990). Finally, RNA-derived sequences may be amplified by PCR if the RNA is first reverse transcribed into cDNA followed by amplification of the cDNA (Kawasaki, 1990; Kawasaki et al., 1988; Rappolee et al., 1988; Feere, 1992; Gilliland et al., 1990a,b; Buck et al., 1991; Ballagi-Pordany et al., 1991; Robinson and Simon, 1991; Wang et al., 1989; Kellogg et al., 1990; Robinson and Simon, 1991; Gaudette and Crain, 1991; Shuldiner et al., 1991a,c, 1993; Nakabeppu and Nathans, 1991; Foulkes et al., 1991; Moller et al., 1989). This approach, designated reverse transcription-PCR (RT-PCR or RNA-PCR) may be used to detect, quantitate, and localize very low-abundance mRNAs from single cells or small numbers of cells, as well as to study mRNA splicing. A very recent and exciting methodologic advance is the adaptation of RT-PCR for histological specimens (in situ PCR; Nuovo et al., 1991a,b; Nuovo, 1992).

III. General Protocol: The Polymerase Chain Reaction

We perform PCR in a final volume of 50 μl. Since multiple samples are often amplified in the same experiment, we recommend that a PCR master mix be prepared and that the PCR reaction mixture contain 45 μl of PCR master mix and 5 μl of the DNA template in water. The recipe for PCR master mix (for 10 tubes) is as follows:

Ingredient	Amount
10× PCR buffer [KCl (500 mM), Tris–HCl (100 mM, pH 8.3, at 25°C), MgCl$_2$ (15 mM), gelatin (0.1 mg/ml)]	50 μl
dATP (10 mM)	10 μl
dCTP (10 mM)	10 μl
dGTP (10 mM)	10 μl
dTTP (10 mM)	10 μl
Upstream primer (5 μM)	50 μl
Downstream primer (5 μM)	50 μl
Sterile water	257 μl
Taq DNA polymerase (5 units/μl)	3 μl
Total	450 μl

To each reaction, 45 μl of PCR master mix is added to 5 μl of the DNA template in water or TE (10 mM Tris-HCl, 1 mM EDTA, pH 7.4), and the reaction mixture is covered with approximately 50 μl of paraffin oil. Note that final concentrations after the DNA template is added are PCR buffer [1× = KCl (50 mM), Tris (10 mM, pH 8.3, at 25°C), $MgCl_2$ (1.5 mM), gelatin (0.01 mg/ml)], dNTPs (200 μM each), primers (0.5 μM each), and Taq DNA polymerase (1.5 units/tube).

PCR may be performed manually with three water baths preequilibrated at the appropriate temperatures, or with an automated thermocycler. A typical temperature profile is as follows:

	Temperature (°C)	Time	Cycles
Initial denaturation	94	5 min	1
Annealing	45–55	1 min	
Extension	72	1 min	25–50
Denaturation	94	1 min	
Final annealing	45–55	1 min	1
Final extension	72	10 min	1

Note that the initial denaturation time is usually prolonged to ensure efficient denaturation of the DNA template and that the final extension time is prolonged to ensure complete extension by Taq DNA polymerase.

After thermocycling is completed, an aliquot of the PCR mixture (usually 10 to 20 μl) is run on a composite gel consisting of 1% agarose and 1 to 3% NuSieve GTG (FMC Bioproducts, Rockland, ME) in 1× TBE, or on a 5–8% polyacrylamide gel. The PCR product may be visualized by ethidium bromide staining and UV transillumination or by Southern blot analysis using established procedures (Southern, 1975; Sambrook *et al.*, 1989). If Southern blot analysis is performed, the probe should correspond to a sequence internal to the PCR primers since, by definition, all PCR products irrespective of whether they are the correct PCR product will contain the primer sequences and will hybridize to probes that encompass these sequences. The remainder of the PCR mixture may be stored indefinitely at −20°C for subsequent use.

IV. Comments: General Polymerase Chain Reaction Protocol

1. *Optimization of PCR reagents.* PCR conditions must be optimized for each application. Generally, the concentrations of the primers can range from 0.1 to 1 μM, the dNTP concentrations from 50 to

400 µM, and Taq DNA polymerase from 0.5 to 2.5 units/50 µl reaction (Innis and Gelfand, 1990; Saiki, 1990). Too much Taq polymerase may result in an increase in unwanted low-molecular-weight species that include primer dimers (Watson, 1989), primer artifacts, and nonspecific PCR products, as well as the polymerization of the desired PCR product into higher molecular-weight species. Too little Taq polymerase can result in a low yield of the desired PCR product. Taq polymerase activity is markedly affected by the magnesium concentration (Gelfand and White, 1990; Saiki, 1990; Innis and Gelfand, 1990; Chien et al., 1976; Lawyer et al., 1988; Bloch, 1991). While 1.5 mM Mg is usually optimal, it is often advantageous to optimize the Mg concentration by performing a titration curve from 1 to 9 mM $MgCl_2$ Similarly, pH 8.3 is usually optimal, however sometimes optimization of pH may be useful.

2. *Optimization of PCR reaction times, temperatures, and cycle number.*
   a. *Reaction times.* Generally, denaturation and annealing times may vary from 15 sec to 1.5 min, and extension times can vary from 15 sec to 3 min.
   b. *Reaction temperatures.* Alterations in reaction temperatures may have profound effects on yield and specificity. For denaturation, 94–95°C is commonly used since, at temperatures greater than 95°C, Taq polymerase rapidly loses its activity (Gelfand and White, 1990; Innis and Gelfand, 1990; Chien et al., 1976; Lawyer et al., 1988; Bloch, 1991). For extension, 72°C is commonly used since this is close to the optimal temperature for Taq polymerase. At this temperature, 35 to 100 nucleotides/sec may be added to the growing chain (Innis and Gelfand, 1990). The annealing temperature can vary greatly (Innis and Gelfand, 1990; Saiki, 1990; Rychlik et al., 1990; Wu et al., 1991; Bloch, 1991). If the annealing temperature is too low, nonspecific primer annealing may occur resulting in extraneously amplified products. If the annealing temperature is too high, the yield of the desired product may decrease dramatically. If the primers are known to be an exact match with the DNA template, an annealing temperature 5°C below the theoretical melting temperature is recommended. Generally, for an oligonucleotide 20 nucleotides in length that is approximately equally rich in the four nucleotides, an annealing temperature of 50–55°C may be used. If the oligonucleotides are longer, i.e., 25 nucleotides or greater, or if the GC content is high, the annealing temperature may be increased to 55–65°C. For some applications, larger oligonucleotides (i.e., 30 b or greater)

may be used, and the annealing and extension cycles may be combined and performed at 68–72°C. When using redundant oligonucleotides with possible mismatches, or oligonucleotides with known mismatches, the annealing temperature may be decreased to 37–45°C.

c. *PCR cycles.* Generally, the number of cycles needs to be determined empirically and can vary from 25 to 50 cycles depending on the amount of starting template and the desired sensitivity (Innis and Gelfand, 1990; Saiki, 1990; Bell and DeMarini, 1991). Too few cycles will result in a low yield of the desired product. Too many cycles will result in accumulation of unwanted side-products without a proportional increase in the desired PCR product including primer dimers and other primer artifacts. This phenomenon, the so-called plateau effect, is caused by multiple factors and occurs primarily during later PCR cycles (Innis and Gelfand, 1990).

3. *Oligonucleotides.* Selection of oligonucleotides for PCR primers is somewhat empirical and they may vary in length from 20 to 60 b (Innis and Gelfand, 1990; Saiki, 1990; Lowe *et al.,* 1991). Generally, at least 20 b at the 3′ ends of the primers must be complementary to the sequence of interest. If desired, sequences that are not complementary to the DNA template may be added at the 5′ end of the primer without adversely affecting amplification (Higuchi *et al.,* 1988; Shuldiner *et al.,* 1990b, 1991b,c; Scharf, 1990; Jones and Howard, 1990; Stoflet *et al.,* 1988). This will result in a PCR product with a specific sequence at its 5′ ends. For example, a restriction endonuclease recognition site may be added at the 5′ ends of the primers for subsequent subcloning (Scharf, 1990; Kaufman and Evans, 1990; Shuldiner *et al.,* 1990b), or an RNA polymerase promoter sequence may be added at the 5′ end for subsequent *in vitro* transcription (Stoflet *et al.,* 1988; Higuchi *et al.,* 1988). The primer sequences should be designed to amplify a segment of DNA, optimally between 100 to 1000 bp in length, although amplification of PCR products of up to 10 kb have been reported (Erlich *et al.,* 1991). Shorter PCR products (<100 bp) are often difficult to discern from the primers or primer artifacts, and longer PCR products (>1000 bp) often amplify inefficiently. The primer sequences should be approximately equally rich in the four nucleotides, should contain no discernable secondary structure, and the 3′ ends of the two primers should not be complementary to each other (Innis and Gelfand, 1990; Saiki, 1990; Lowe *et al.,* 1991). Some have touted having at least one C or G at the

3′ end of the primer to promote annealing and extension by Taq polymerase.

4. *DNA templates.* The DNA template may be genomic DNA (0.1 to 1 μg; 1 μg of human genomic DNA contains approximately 3×10^5 copies of a single-copy gene), plasmid DNA (0.1–1 ng), or phage DNA (0.5 to 5 ng). Note that there is no need to linearize supercoiled plasmid DNA since the resulting short PCR products will be linear. In general, protein contaminants in crude DNA preparations do not interfere with the amplification. If the genomic DNA is of very high molecular weight (i.e., >50 kb), denaturation may not occur efficiently, and the PCR yield may be low. In these circumstances, it may be advantageous to first cut the genomic DNA with a restriction enzyme that does not cut within the target sequence, or to physically shear the DNA by sonication or by heating to 100°C for 2 to 5 min.

5. *Preparative PCR.* The reaction volume can be increased to as much as 100 μl/tube for preparative PCR. If the volume is increased above 100 μl, the required temperature transitions do not occur effectively, and the yield may decrease dramatically.

6. *Difficult amplifications.* For amplification of GC-rich sequences, addition of 5–10% dimethyl sulfoxide (DMSO) (Smith *et al.*, 1990; Masoud *et al.*, 1992), addition of 10% glycerol (Smith *et al.*, 1990), or mixing of 7-deaza-2-deoxyguanosine triphosphate (c^7dGTP) with dGTP in a 3:1 ratio may be useful (Innis, 1990; McConlogue *et al.*, 1988).

7. *PCR product detection.* In addition to ethidium bromide staining and Southern blot analysis with an internal radiolabeled probe, the PCR product may be detected by incorporation of $[\alpha\text{-}^{32}\text{P}]$dCTP (use 0.1 to 0.5 μl/tube; sp act 3000 Ci/mmol). Alternatively, ^{32}P-radiolabeled primer(s) may be used to label the PCR product (Hayashi *et al.*, 1989). Following gel electrophoresis, the PCR product can be visualized by autoradiography. Alternatively the PCR product may be labeled with nonradioisotopic groups if fluorescent, digoxigenin, or biotinylated primers or dNTPs are used (Jackson, 1991; Syrajanen *et al.*, 1991; Lion and Haas, 1990; Emanuel, 1991).

V. Reverse Transcription–Polymerase Chain Reaction

Standard molecular biological techniques to detect and quantitate mRNA include Northern blot analysis, which has a limit of detection of approximately 10^6 to 10^7 mRNA copies, and solution assays such as S_1 nuclease or RNAse protection which have a limit of detection of approximately 10^5 to 10^6 copies. The PCR method may be used to measure low-level gene expression by the amplification (and quantitation) of as few as 1 to 10 mRNA copies from single cells or small numbers of cells (Kawasaki, 1990; Kawasaki *et al.*, 1988; Rappolee *et al.*, 1988; Feere, 1992; Gilliland *et al.*, 1990a,b; Buck *et al.*, 1991; Ballagi-Pordany *et al.*, 1991; Robinson and Simon, 1991; Wang *et al.*, 1989; Kellogg *et al.*, 1990; Gaudette and Crain, 1991; Shuldiner *et al.*, 1991a,c, 1993; Nakabeppu and Nathans, 1991; Foulkes *et al.*, 1991; Moller *et al.*, 1989). Since RNA is not a very effective template for Taq DNA polymerase, the mRNA must first be reverse transcribed into single-stranded cDNA that may then be used as a template in the PCR. During the first cycle of PCR, the second strand is synthesized, and during subsequent cycles, the RNA-derived DNA is amplified logarithmically.

VI. Protocol for Reverse Transcription

We perform RT in a final volume of 20 µl. Since multiple samples are often amplified in the same experiment, we recommend that a 2× RT master mix be prepared. Ten microliters of the 2× RT master mix is then added to 10 µl of the RNA sample in water. The master mix is prepared as follows (for 10 tubes):

Ingredient	Amount
10× "PCR" buffer [KCl (500 mM), Tris (100 mM, pH 8.3, at 25°C), MgCl$_2$ (15 mM), gelatin (0.1 mg/ml)]	20 µl
dATP (10 mM)	4 µl
dCTP (10 mM)	4 µl
dGTP (10 mM)	4 µl
dTTP (10 mM)	4 µl
Downstream primer (5 µM)	20 µl
Sterile water	24 µl
RNAsin (40 units/µl)	10 µl
Avian myeloblastosis virus-reverse transcriptase (AMV-RT) (~7 units/µl)	10 µl
Total	100 µl

Note that final concentrations after the RNA template is added are PCR buffer [1× = KCl (50 mM), Tris-HCl (10 mM, pH 8.3, at 25°C), MgCl$_2$ (1.5 mM), gelatin (0.01 mg/ml)], dNTPs (200 μM each), primer (0.5 μM), RNasin (2 units/μl), and AMV-RT (approximately 0.35 units/μl).

Add 10 μl of the reverse transcription master mix to 10 μl of the RNA sample (1 to 5 μg). Incubate at 37°C for 1 hr. The reaction mixture may then be used directly for PCR (we usually use 5 μl of reverse transcription reaction mixture in a final PCR volume of 50 μl. See general PCR protocol above).

VII. Comments: Reverse Transcription

1. *RNA Preparation.* Most methods for the preparation of RNA work well. Since the size of the PCR product is usually relatively small (i.e., <1000 bp), efficient amplification may still result from RNA preparations that are somewhat degraded. Since the RT-PCR method is extraordinarily sensitive, it is not necessary to enrich for poly(A)$^+$ mRNA.

2. *Reverse Transcription Primer Design.* The reverse transcription primer may be a specific downstream (antisense) primer of 20 nucleotides in length or greater that will prime only the mRNA of interest. If a specific primer is used, it may be the same downstream primer that will be used in the PCR, or a different primer that is further 3' than the downstream PCR primer. The later experimental design, often referred to as the "nested primer approach" has the theoretical advantage that the amplification should be more specific since sequences corresponding to all three primers (rather than two) must be present in the RNA template for efficient amplification to occur. Alternatively, dT$_{20}$ or random hexamers may be used as the reverse transcription primer; however, if this approach is used, we recommend that the RT reaction mixture be precipitated or subjected to ultrafiltration to remove excess primers prior to PCR.

3. *Reverse Transcription.* Mouse moloney leukemia virus reverse transcriptase (MMLV-RT) (100–200 units/tube) may be substituted for AMV reverse transcriptase. If the RNA template has appreciable secondary structure, the efficiency of reverse transcription (usually approximately 10 to 20%) can be enhanced by first heating the RNA sample to 70°C for 10 min followed by cooling to 37°C over 5 to 10 min. After reverse transcription, it is generally not necessary to digest the RNA template before amplification, although some have observed more efficient PCR amplification by treating the re-

verse transcription reaction mixture with RNase H prior to PCR amplification.

VIII. Quantitative Reverse Transcription–Polymerase Chain Reaction

Since small differences in the efficiency of amplification at each cycle can result in large differences in the yield of the final PCR product, it is very difficult to obtain quantitative data from conventional PCR or RT-PCR (Feere, 1992). Since the quantitation of mRNA levels of rare transcripts from single cells or small numbers of cells is often desirable, several quantitative RT-PCR approaches have been developed (Gilliland *et al.*, 1990a,b; Buck *et al.*, 1991; Ballagi-Pordany *et al.*, 1991; Robinson and Simon, 1991; Wang *et al.*, 1989; Kellogg *et al.*, 1990; Gaudette and Crain, 1991; Shuldiner *et al.*, 1991b, 1993). The most rigorous quantitative RT-PCR assays make use of an internal standard that is introduced into the reaction mixture in known quantities at the beginning of the RT (if the internal standard is RNA), or PCR (if the internal standard is DNA). Since the internal standard is amplified with the same primers as the endogenous mRNA, their amplification efficiencies are similar, and direct comparison of the two PCR products allows for the accurate quantitation of the endogenous mRNA. Usually, a curve is generated in which increasing amounts of the internal standard are introduced into replicates of the RNA sample to be quantitated (Fig. 3). The PCR product derived from the internal standard may be differentiated from the PCR product derived from the endogenous mRNA by (i) gel electrophoresis if the PCR product derived from the internal standard differs in size, (ii) restriction enzyme digestion if the internal standard sequence is altered such that a restriction enzyme recognition site is introduced or eradicated, or (iii) slot blot analysis with internal oligonucleotide probes that hybridize specifically to each product.

IX. Contamination and Polymerase Chain Reaction False Positives

Due to the extraordinary sensitivity of the PCR method, many laboratories have found that false positives caused by inadvertent contamination with minute quantities of DNA (e.g., cDNAs, plasmid DNAs, genomic DNA or PCR carryover) is a major shortcoming of the PCR method even when meticulous laboratory technique is employed (Kwok and Higuchi, 1989; Lo *et al.*, 1988; Sarkar and Sommer, 1990, 1991; Shuldiner *et al.*, 1990a, 1991a,c, 1993). For example, assuming a sensitivity of 100 molecules, accidental con

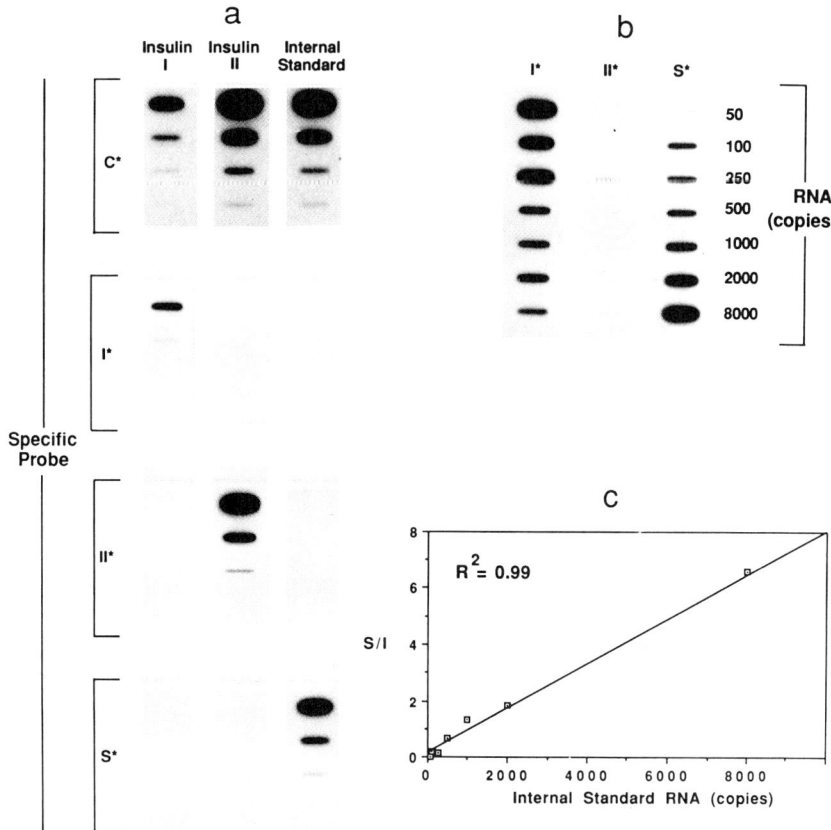

Figure 3 Example of quantitative (competitive) RT-PCR. (a) Serial dilutions (1:4) of *Xenopus* insulin I DNA, *Xenopus* insulin II DNA, or mutated *Xenopus* insulin internal standard DNA were subjected to slot blot hybridization in quadruplicate with oligonucleotide probes that specifically recognized each sequence (probes I*, II*, or S*) as well as to a common probe (probe C*). (b) *Xenopus* oocyte RNA (one oocyte equivalent) was spiked with increasing quantities (50–8000 copies) of the mutated *Xenopus* insulin internal standard RNA, and RS-PCR was performed as described. The RS-PCR products were subjected to slot blot hybridization in triplicate with the three sequence-specific probes described in a (probes I*, II*, or S*). (c) Plot of the ratio of the signal intensities of the internal standard/*Xenopus* insulin I (S/I)(ordinate) as a function of the number of copies of mutated internal standard RNA (abscissa). Data are corrected for background and for the specific activities of the probes (see a). When the signals from the internal standard and the endogenous mRNA are equal (i.e., the ratio on the ordinate is 1), extrapolation to the abscissa reflects the number of endogenous *Xenopus* insulin I mRNA molecules. Thus, there are 1000 copies of insulin I mRNA and less than 50 copies of insulin II mRNA in a single *Xenopus* oocyte. (After Shuldiner *et al.*, 1991a.)

tamination with only $1 \times 10^{-17} \mu l$ of a 1 mg/ml plasmid solution containing the target sequence would result in a positive signal. False positives are particularly troublesome when low-abundance mRNAs are being amplified and the extreme sensitivity of the PCR method is being exploited (Shuldiner *et al.*, 1990b, 1991b,c, 1993). The following recommendations may aid in reducing the frequency of false positives:

1. Designate a "clean area" to set up PCR reactions. Dedicate reagents, pipettes, autoclaved tips, tubes, ice baths, and other necessary supplies for PCR setup only. If possible, this area should be enclosed (e.g., laminar flow hood with the fan turned off, or enclosed work station)

2. To avoid contamination from the inside barrel of the pipette device use positive displacement pipettes (Rainin, Inc., Woburn, MA) or conventional pipettes with plugged tips.

3. Use positive controls judiciously. Use only enough DNA template to give a weak positive signal.

4. Use gloves and change them often.

5. Do not produce aerosols (i.e., open Eppendorf tubes carefully, do not use vortexers or centrifuges, carefully eject used pipette tips from the pipette device).

6. Prealiquot all reagents and store them safely in a clean "PCR-only" dedicated area of the freezer.

7. Add the DNA sample last (pipette through the oil layer), and immediately cap the tube.

8. Use multiple negative controls. Negative controls may include a "no template" control. To confirm that amplification is originating from an RNA template in RT-PCR, use an RNase control in which an aliquot of the RNA sample is digested with RNase A (0.1 μg) for 30 min at 37°C prior to reverse transcription. Since Taq polymerase has weak reverse transcriptase activity, a "no RT" negative control may result in a faint positive signal despite the absence of DNA contamination.

9. Since contamination with PCR products from a previous experiment may result in false positives (so-called carryover contamination), for the characterization of PCR products, a separate area from the PCR setup area with dedicated supplies and reagents should be used.

10. RNA preparations may contain small amounts of genomic DNA that may act as a template for amplification with the appropriate primers and give a false positive signal. Therefore, when performing RT-PCR, it is advantageous to design primers that span one or more exon–exon splice junctions so that mRNA-derived sequences (which lack the introns) will result in a PCR

product of the desired length, while amplification of genomic DNA-derived sequences will result in an easily discernible larger PCR product, or no product if the intron is very large and thus not amplified effectively. If amplifications across an exon–exon splice junction is not possible, the RNA sample may be pretreated with RNase-free DNase to remove genomic DNA (Grillo and Margolis, 1990). DNase may be subsequently inactivated by heating to 70–80°C for 15 min. Another option to prevent false positives from genomic DNA, or even plasmid DNAs or cDNAs, is to use a novel modification of the RT-PCR method developed in our laboratory, designated RNA template-specific PCR (RS-PCR) (see below; see also Shuldiner et al., 1990a, 1991a,c, 1993). Still others have touted UV irradiation (Sarkar and Sommer, 1990, 1991) or use of uracil-N-glycosylase (Longo et al., 1990) to reduce the frequency of false positives.

X. RNA Template-Specific Polymerase Chain Reaction

RS-PCR is a modification of the conventional RT-PCR method that dramatically reduces the frequency of false positives without sacrificing sensitivity (Shuldiner et al., 1990a, 1991a,c, 1993). With RS-PCR, total RNA is reverse transcribed using an oligonucleotide primer 47 nucleotides in length (designated primer $d_{17}t_{30}$) whose sequence contains 17 b at its 3′ end (segment d_{17}) that are complementary to a region of the target mRNA, and 30 b at its 5′ end (segment t_{30}) that are unique in sequence (Fig. 4). Thus reverse transcription yields single-stranded DNA that contains a unique 30 base "tag" (segment t_{30}) at its 5′ end. The second strand of DNA is synthesized during the first cycle of PCR with primers u_{30} and t_{30} (Fig. 4). Upstream (sense) primer u_{30} is a 30-mer whose sequence corresponds to the target RNA a predetermined distance (optimally 200 to 500 b) upstream from reverse transcription primer $d_{17}t_{30}$, while downstream (antisense) primer t_{30} is a 30-mer whose sequence is identical to segment t_{30} of reverse transcription primer $d_{17}t_{30}$. With these primers, sequences derived from RNA that had been tagged with unique sequence t_{30} during reverse transcription are amplified preferentially, while contaminating DNA, lacking the unique tag, is not amplified.

Note that the lengths of the segments of reverse transcription primer $d_{17}t_{30}$ were derived empirically. The 17-b d_{17} segment hybridizes efficiently with its RNA during reverse transcription at 37°C, but does not hybridize to contaminating DNA that lacks the unique tag (t_{30}) during the PCR in which an elevated annealing temperature (65–72°C) is used; PCR primers of about 30 b in length (primers u_{30} and t_{30}) retain efficient hybridization at the higher PCR annealing temperature.

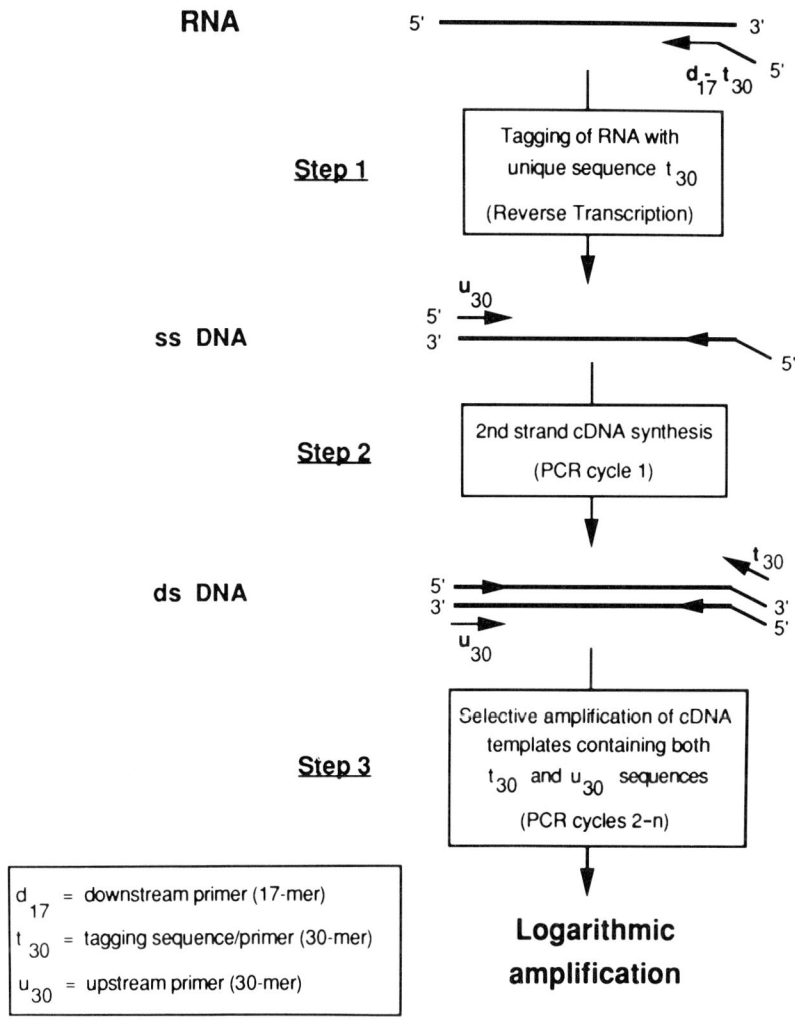

Figure 4 Schematic of RNA template-specific PCR (RS-PCR). With RS-PCR, sequences derived from RNA that are tagged with the unique sequence (t_{30}) during reverse transcription (step 1) are amplified preferentially during PCR with primers t_{30} and u_{30} (steps 2 and 3). (After Shuldiner et al., 1991c.)

XI. Subcloning Polymerase Chain Reaction Products

For some applications, it is advantageous to subclone the PCR product into a plasmid vector for subsequent replication in bacteria. Subcloning the PCR product into a plasmid vector has several advantages: (i) the amplified fragment usually can be sequenced with greater reliability, (ii) only one allele is sequenced per clone, and (iii) the vector containing the PCR product may be used for other molecular biological experiments, e.g., *in vitro* transcription, radiolabeling, and further amplification in bacteria. Conventional strategies such as blunt-end or sticky-end subcloning may be employed (Sambrook *et al.*, 1989). If blunt-end ligation is used, (i) ragged ends of the PCR product must first be "filled in" with Klenow, (ii) the 5' ends of the PCR product must be phosphorylated with polynucleotide kinase, and (iii) the PCR product must be ligated into a vector, that has been linearized previously with a restriction enzyme that leaves blunt ends, and dephosphorylated with alkaline phosphatase (Sambrook *et al.*, 1989). Blunt-end subcloning is nondirectional, which is a disadvantage of this strategy.

Sticky-end ligation requires that restriction enzyme recognition sequences be incorporated onto the 5' ends of the primers. Since some restriction enzymes will not efficiently cut at the extreme end of a DNA fragment, two to four additional bases should be incorporated (Scharf, 1990; Shuldiner *et al.*, 1990b; Kaufman and Evans, 1990). (For example, 5'-gcacGAATTC . . . -3' or 5'-gcacGGATCC . . . -3' added to the 5' end of the PCR primer would introduce an *Eco*R1 or *Hin*dIII restriction endonuclease recognition site, respectively.) After PCR amplification with the 5' extended primers, the PCR product may be digested with the appropriate restriction enzyme and ligated into a vector, that has been previously linearized with the corresponding restriction enzyme, and dephosphorylated with alkaline phosphatase (Sambrook *et al.*, 1989). If different restriction enzyme recognition sites are incorporated onto each primer, subcloning is directional. Naturally, if the same restriction enzyme recognition sequence(s) are present within the PCR product, this strategy cannot be used to subclone the full-length product.

Often, PCR products are difficult to subclone using the above conventional methods. We have devised a versatile new strategy to directionally subclone PCR products in a single day without the use of DNA ligase (Fig. 5) (Shuldiner *et al.*, 1991b; Shuldiner and Tanner, 1993). PCR products up to 1.7 kb have been subcloned with this method (E. Friedman, personal communication). With ligase-free subcloning, sequences are incorporated onto the 5' ends of the primers that are identical to the 3' ends of the recipient linearized plasmid. The PCR product and the desired recipient linearized plasmid now contain 3' ends that are complementary to each other, and may be "ligated"

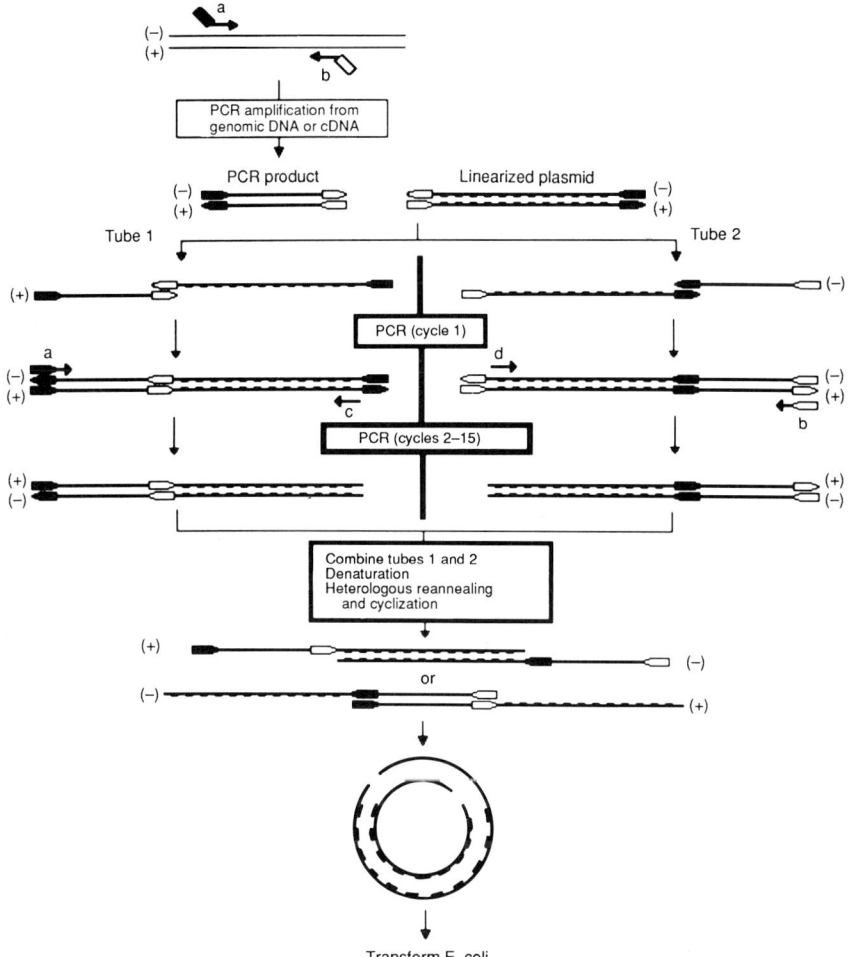

Figure 5 Schematic of ligase-free subcloning. The 5' addition sequences of primers a and b are designated by clear and filled boxes, respectively. DNA sequences corresponding to the PCR-amplified product are shown as straight lines while DNA sequences corresponding to the plasmid vector are shown as hatched lines.

together in a second PCR amplification in which Taq polymerase extends the overlapping 3' ends (i.e., ligation by overlap extension). The linear PCR products that have been ligated at one end (tube 1) or the other end (tube 2) may then be amplified with the appropriate primers during subsequent PCR cycles (tube 1, primers a and c; tube 2, primers b and d; see Fig. 5). The linear

PCR products from tubes 1 and 2 are combined, denatured, and reannealed resulting in cyclized products that contain two nicks (noncovalently linked ends). The cyclized products may be transformed into competent *Escherichia coli* where the nicks are covalently joined, and the recombinant plasmid containing the PCR insert is replicated (see Shuldiner and Tanner, 1993, for a detailed protocol of ligase-free subcloning).

Recently, a rapid one-step PCR product subcloning strategy has been developed that takes advantage of the fact that Taq polymerase often leaves 3' adenosine overhangs (nontemplate additions) on the PCR product (Clark, 1988). With this approach, the PCR product may be ligated directly into a linearized vector that has a 5' thymidine overhang at each end (TA Cloning, Invitrogen, San Diego, CA; Stratagene, La Jolla, CA). In our hands, this is a very reliable method for subcloning most PCR products, but has the limitation that only specific vectors may be used.

PROTOCOL FOR BLUNT-END SUBCLONING

The following protocol is an example of blunt-end subcloning into the *Sma*I site of pGem4Z (Promega Biotech, Madison, WI).

1. Optimize PCR conditions to obtain a PCR product that is as free from undesired products as possible. After PCR amplification is complete, add 1 μl of Klenow (approximately 1 to 5 units) to the PCR reaction mixture and incubate at 37°C for 1 hr to fill in ragged ends.
2. Phenol chloroform extract twice, and ethanol precipitate the PCR product in the presence of carrier tRNA (10 μg). Dissolve the pellet in 10.5 μl of water and phosphorylate as follows:

Ingredient	Amount
PCR product	10.5 μl
10× Kinase buffer [Tris-HCl (500 mM, pH 7.6 at 25°C), MgCl$_2$ (100 mM), dithiothreitol (50 mM), spermidine (1 mM), EDTA (1 mM)]	1.5 μl
ATP (10 mM)	2 μl
Polynucleotide kinase (10 units/μl)	1 μl
Total	15 μl

   Incubate at 37°C for 1 hr.
3. Add 35 μl of water to the reaction mixture, phenol chloroform extract twice, and ethanol precipitate. Dissolve the pellet in 10 μl of

water. Estimate the amount of blunt-ended-phosphorylated PCR product by running 2 μl on a composite gel consisting of 1% agarose and 2% NuSeive GTG with a known quantity of a HaeIII digest of PhiX174 size standard followed by ethidium bromide staining and UV transillumination (Sambrook et al., 1989).
4. Ligate approximately equal quantities of the PCR product and pGem4Z that has been previously linearized with SmaI and dephosphorylated with alkaline phosphatase as follows:

Ingredient	Amount
PCR product (10 to 100 ng)	—
SmaI linearized, dephosphorylated pGem4Z (10 to 100 ng)	—
10× Ligation buffer [Tris–HCl (500 mM, pH 7.4 at 25°C), MgCl$_2$ (100 mM), dithiothreitol (100 mM), spermidine (10 mM), ATP (10 mM), BSA (1 mg/ml)]	1.5 μl
Water	qs to 14 μl
T4 DNA ligase (approximately 4 units/μl)	1 μl
Total	15 μl

Incubate at 4°C overnight.

5. Transform competent DH5-α *E. coli* (Bethesda Research Laboratories; Gaithersburg, MD) according to the manufacturer's directions.

PROTOCOL FOR STICKY-END LIGATION

The following protocol provides an example of directional subcloning of a PCR product into the HindIII and EcoR1 sites of pGem4Z (Promega Biotech, Madison, WI).

1. Amplify the desired PCR product with primers that contain the appropriate restriction endonuclease recognition site(s) at their 5' ends (see above; see also Kaufman and Evans, 1990). Optimize PCR conditions to obtain a PCR product that is as free from undesired products as possible. Phenol chloroform extract twice, and ethanol precipitate the PCR product in the presence of carrier tRNA (10 μg). Dissolve the pellet in 17 μl of water, and generate sticky ends as follows:

Ingredient	Amount
PCR product	17 μl
10× digestion buffer [NaCl (1000 mM), Tris-HCl (100 mM, pH 8.0 at 37°C, MgCl$_2$ (50 mM), 2-mercaptoethanol (10 mM)]	2 μl
EcoR1 (approx. 10 units/μl)	0.5 μl
HirdIII (approx. 10 units/μl)	0.5 μl
Total	20 μl

Incubate at 37°C for 4 to 16 hours.

2. Ligate approximately equal quantities of the PCR product and pGem4Z that has been previously linearized with EcoR1 and HindIII and dephosphorylated with alkaline phosphatase as described in steps 3 and 4 of the blunt-end subcloning protocol (Section XII).
3. Transform competent DH5-α E. coli (Bethesda Research Laboratories; Gaithersburg, MD) according to the manufacturer's directions.

XII. Direct Sequencing of Polymerase Chain Reaction Products

For some applications, it may be desirable to obtain the nucleotide sequence of the PCR product. A faster alternative to subcloning the PCR product into a plasmid vector for sequencing using conventional methods is to sequence the PCR product directly (Bevan et al., 1992; Gyllensten and Erlich, 1988; Innis et al., 1988; Stoflet et al., 1988; Mazars et al., 1991; Lee, 1991; Engelke et al., 1988; Hultman et al., 1989; Paabo et al., 1988). Linear double-stranded DNA is normally difficult to sequence with the conventional dideoxy method due to the tendency of the denatured single strands to rapidly reanneal to each other excluding efficient access of the sequencing primer to its complementary sequence. Thus one approach to directly sequence PCR products requires a second PCR amplification in which one of the PCR primers is used in great excess (asymmetric PCR) to generate large amounts of a single-stranded product that may be sequenced more reliably. The conditions for asymmetric PCR, especially the relative amounts of the two primers, the amount of starting double-stranded template, and the number of cycles must be optimized for each application. For direct sequencing, the PCR product should be less than 500 bp in length.

Recently, a reliable method to sequence directly double-stranded PCR products without antecedent asymmetric PCR has been devised (Lee, 1991;

Ruano and Kidd, 1991). The method uses Taq polymerase and multiple cycles of PCR in the presence of ^{32}P-labeled sequencing primer, dNTPs, and the appropriate dideoxyribonucleotide triphosphate. Several manufacturer's supply kits and protocols for cycle sequencing of double-stranded PCR products, and thus this method will not be discussed further (dsDNA Cycle Sequencing System, Bethesda Research Laboratories, Gaithersburg, MD; fmol DNA Sequencing System, Promega, Madison, WI; AmpliTaq Cycle Sequencing, Perkin Elmer Cetus, Norwalk, CT).

GENERAL PROTOCOL FOR SEQUENCING BY ASYMMETRIC PCR

1. *Symmetric PCR.* The initial amplification is performed with equal concentrations of the two primers as described above (see Section III, General Protocol: The Polymerase Chain Reaction). It is crucial to optimize conditions to yield a single PCR product while minimizing primer artifacts. If side products are unavoidable, the PCR product to be sequenced may be cut out from a gel, the gel slice melted, and 1–5 μl of the melted agarose used directly for asymmetric PCR.

2. *Asymmetric PCR.* One microliter of the PCR product from step 1 is amplified in a second PCR reaction in which a 50- to 100-fold excess of one of the two primers is used. For a typical 100-μl PCR reaction, use 0.5 to 1 pmol of the limiting primer, and 50 pmol of the nonlimiting primer. Thirty to forty cycles of PCR are performed. (In some protocols, a second asymmetric PCR reaction is performed to maximize the amount of single-stranded PCR product.) An aliquot of the product of the asymmetric PCR may be run on an agarose gel to assess whether single-stranded DNA was generated. (Note that single-stranded DNA migrates more slowly than double-stranded DNA in a nondenaturing agarose gel and takes up ethidium bromide less efficiently.) Excess primers and dNTPs are then removed by ultrafiltration (Centricon 100; Amicon, Danvers, MA; or Millipore MC; Bedford, MA) according to the manufacturer's recommendations.

3. *Dideoxy sequence analysis.* The asymmetric PCR product may be sequenced using the conventional dideoxy chain termination method, e.g., Sequenase (U.S. Biochemicals, Cleveland OH), GemSeq K/RT (Promega), TaqTrack (Promega), according to the manufacturer's recommendations. Usually, the sequencing primer (the limiting PCR primer) is end labeled with γ-[^{32}P]ATP and polynucleotide kinase. Alternatively, [α-^{35}S]ATP may be used for direct incorporation during the dideoxy sequence reactions.

XII. Some Commonly Encountered Polymerase Chain Reaction Artifacts

A. Unexpected PCR Products

Often, PCR products other than the desired product may be generated. These unexpected products are usually the result of "mispriming" of sequences that are partly complementary to one or both primers. When the unexpected PCR product is different in size than the desired product, it is easy to recognize. Occasionally, the unexpected product may be a size similar to the desired product, and other methods to distinguish it from the desired product may be required such as Southern blot analysis with an internal probe or direct sequence analysis. Usually, when the PCR conditions are optimized (see above), unwanted side products can be minimized. Some have touted the use of "hot start" techniques to minimize annealing and mispriming at the start of the PCR when the reaction temperature is low (D'Aguila et al., 1991; Nuovo, 1992; Chou et al., 1992; Hart et al., 1990). Use of nested primers is another approach to minimize unexpected PCR products (Mullis and Faloona, 1987; Kai et al., 1991; Brytting et al., 1991).

B. Taq Polymerase Misincorporation Errors

The point mutation rate of Taq polymerase has been reported to be approximately 2×10^{-4} to 2×10^{-5} mutations per nucleotide per cycle (Tindall and Kunkel, 1988; Keohavong and Thilly, 1989; Lundberg et al., 1991; Eckert and Kunkel, 1990, 1991; Saiki et al., 1988). The frequency of frame shift mutations is estimated to be approximately 3×10^{-5} mutations per nucleotide per cycle. These frequencies may be overestimated since these studies were accomplished with magnesium and dNTP concentrations that were greater than those used in conventional PCR. With a dNTP concentration of 200 μM and Mg concentration of 1.5 mM, the error frequency of Taq polymerase was estimated to be approximately 5×10^{-6} mutations per nucleotide per cycle (Fucharoen et al., 1989). Nucleotide misincorporations by Taq polymerase are usually AT to GC substitutions (Keohavong and Thilly, 1989).

Since the locations of nucleotide misincorporations are usually random and infrequent, for a given nucleotide position most of the PCR product will be of the correct sequence, and direct sequencing will reveal the correct sequence. However, when a PCR product is being subcloned, and the subclone is then sequenced, misincorporation errors may be problematic. Thus, we recommend that the sequence of a subcloned PCR product be confirmed by at least two clones isolated from independent PCR reactions.

C. Hybrid DNA Artifact

When amplifying DNA sequences from a sample that also contains closely related sequences (e.g., allelic genes from a heterozygote, homologous nonallelic genes, or closely related members of a gene family), hybrid PCR products may be obtained (Shuldiner *et al.*, 1990c; Paabo and Wilson, 1988; Paabo *et al.*, 1990). Hybrid DNA artifact, sometimes referred to as "jumping PCR" may occur if there is incomplete extension during one cycle, and then during a subsequent cycle the partially extended PCR product anneals to a closely related sequence and extension is completed. The resulting hybrid molecule which contains the sequence of one gene at its 5' end, and the sequence of the other gene at its 3' end, is then amplified logarithmically during subsequent cycles. PCR hybrid artifact may be particularly common if the extension time is too short, the DNA template is partially degraded (Paabo and Wilson, 1988), or appreciable secondary structure exists in the DNA template that causes Taq polymerase to stutter or stop (A. R. Shuldiner, unpublished observations).

Hybrid DNA artifact may easily be identified if the sequences of the two genes are known; however, this is not always the case. Since hybrid DNA formation is usually infrequent, most of the PCR product is of the correct sequence, and direct sequencing will reveal the correct sequence. However, when a PCR product is being subcloned, and the subclone is being sequenced, hybrid DNA artifact may be problematic. Thus, to exclude the possibility of hybrid DNA artifact, we recommend that the sequence be confirmed by at least two clones isolated from independent PCR reactions.

XIV. Conclusions

Since its conception in 1985, the PCR method has revolutionized the practice of molecular biology and biomedical research. Its simplicity both conceptually and in practice has promoted its widespread use and its modification for literally hundreds of applications. In addition to its use in biomedical research, PCR has important applications to medical diagnostics which will soon bring the PCR method to the bedside. Clearly, an intimate understanding of this powerful method as well as its limitations will soon be essential not only for the basic scientist, but also for the clinician of the near future.

References

Arnheim, N., and Erlich, H. (1992). *Annu. Rev. Biochem.* **61,** 131.
Aslanidis, C., and DeJong, P. J. (1991). *Proc. Natl. Acad. Sci. U.S.A.* **88,** 6765.
Ballagi-Pordany, A., Ballagi-Pordany, A., and Funa, K. (1991). *Anal. Biochem.* **196,** 89.
Barany, F. (1991a). *Proc. Natl. Acad. Sci. U.S.A.* **88,** 189.
Barany, F. (1991b). *PCR Methods Appl.* **1,** 5.
Barbetti, F., Gejman, P. V., Taylor, S. I., Raben, N., Cama, A., Bonora, E., Pizzo, P., Moghetti, P., Muggeo, M., and Roth, J. (1992). *Diabetes* **41,** 408.
Batzer, M. A., Carlton, J. E., and Peininger, P. L. (1991). *Nucleic Acids Res.* **19,** 5081.
Bell, D. A., and DeMarini, D. M. (1991). *Nucleic Acids Res.* **19,** 5079.
Belyavsky, A., Vinogradova, T., and Rajewsky, K. (1989). *Nucleic Acids Res.* **17,** 2919.
Bevan, I. S., Rapley, R., and Walker, M. R. (1992). *PCR Methods Appl.* **1,** 222.
Bhat, G. J., Lodes, M. J., Myler, P. J., and Stuart, K. D. (1991). *Nucleic Acids Res.* **19,** 338.
Blackwell, T. K., and Weintraub, H. (1990). *Science* **250,** 1104.
Bloch, W. (1991). *Biochemistry* **30,** 2735.
Brunet, J.-F., Shapiro, E., Foster, S. A., Kandel, E. R., and Iino, Y. (1991). *Science* **252,** 56.
Brytting, M., Sundqvist, V. A., Stalhandske, P., Linde, A., and Wahren, B. (1991). *J. Virol. Methods* **32,** 127.
Buck, K. J., Harris, R. A., and Sikela, J. M. (1991). *BioTechniques* **11,** 636.
Cadwell, R. C., and Joyce, G. F. (1992). *PCR Methods Appl.* **2,** 28.
Chien, A., Edgar, D. B., and Trela, J. M. (1976). *J. Bacteriol.* **127,** 1550.
Chou, Q., Russel, M., Birch, D., Raymond, J., and Bloch, W. (1992). *Nucleic Acids Res.* **20,** 1717.
Clark, J. M. (1988). *Nucleic Acids Res.* **16,** 9677.
Cotton, R. G. H. (1989). *Biochem. J.* **263,** 1.
D'Aguila, R. T., Bechtel, L. J., Videler, J. A., Eron, J. J., Gorcqyca, P., and Kaplan, J. C. (1991). *Nucleic Acids Res.* **19,** 3749.
Dean, M., and Gerrard, B. (1991). *BioTechniques* **10,** 332.
Eckert, K. A., and Kunkel, T. A. (1990). *Nucleic Acids Res.* **18,** 3739.
Eckert, K. A., and Kunkel, T. A. (1991). *PCR Methods Appl.* **1,** 17.
Eisenstein, B. I. (1990). *N. Engl. J. Med.* **322,** 178.
Emanuel, J. R. (1991). *Nucleic Acids Res.* **19,** 2790.
Engelke, D. R., Hoener, P. A., and Collins, F. S. (1988). *Proc. Natl. Acad. Sci. U.S.A.* **85,** 544.
Erlich, H. A., Gelfand, D., and Sninsky, J. J. (1991). *Science* **252,** 1643.
Feere, F. (1992). *PCR Methods Appl.* **2,** 1.
Foulkes, N. S., Borrelli, E., and Sassone-Corsi, P. (1991). *Cell* **64,** 739.
Frohman, M. A., Dush, M. K., and Martin, G. R. (1988). *Proc. Natl. Acad. Sci. U.S.A.* **85,** 8998.
Fucharoen, S., Fucharoen, G., Fucharoen, P., and Fukumaki, Y. (1989). *J. Biol. Chem.* **264,** 7780.
Gaudette, M. F., and Crain, W. R. (1991). *Nucleic Acids Res.* **19,** 1879.
Gelfand, D. H., and White, T. J. (1990). *In* "PCR Protocols: A Guide to Methods and Applications" (M. A. Innis, D. H. Gelfand, J. J. Sninsky, and T. J. White, eds.), pp. 129–141. Academic Press, San Diego.
Gilliland, G., Perrin, S., and Bunn, H. F. (1990a). *In* "PCR Protocols: A Guide to Methods and

Applications" (M. A. Innis, D. H. Gelfand, J. J. Sninsky, and T. J. White, eds.), pp. 60–69. Academic Press, San Diego.
Gilliland, G., Perrin, S., Blanchard, K., and Bunn, H. F. (1990b). *Proc. Natl. Acad. Sci. U.S.A.* **87**, 2725.
Gould, S. J., Subramani, S., and Scheffler, I. E. (1989). *Proc. Natl. Acad. Sci. U.S.A.* **86**, 1934.
Grillo, M., and Margolis, F. L. (1990). *BioTechniques* **9**, 262.
Gussow, D., and Clackson, T. (1989). *Nucleic Acids Res.* **17**, 4000.
Guzzetta, V., Montes de Oca-Luna, R., Lupski, J. R., and Patel, P. I. (1991). *Genomics* **9**, 31.
Gyllensten, U. B., and Erlich, H. A. (1988). *Proc. Natl. Acad. Sci. U.S.A.* **85**, 7652.
Hart, C., Chang, S.-Y., Kwok, S., Sninsky, J., Ou, C.-Y., and Schochetman, G. (1990). *Nucleic Acids Res.* **18**, 4029.
Hayashi, K. (1991). *PCR Methods Appl.* **1**, 34.
Hayashi, K., Orita, M., Suzuki, Y., and Sekiya, T. (1989). *Nucleic Acids Res.* **17**, 3605.
Higuchi, R., Krummel, B., and Saiki, R. K. (1988). *Nucleic Acids Res.* **16**, 7351.
Horton, R. M., Hunt, H. D., Ho, S. N., Pullen, K., and Pease, L. R. (1989). *Gene* **77**, 61.
Hultman, T., Stahl, S., Hornes, E., and Uhlen, M. (1989). *Nucleic Acids Res.* **17**, 4937.
Innis, M. A. (1990). *In* "PCR Protocols: A Guide to Methods and Applications" (M. A. Innis, D. H. Gelfand, J. J. Sninsky, and T. J. White, eds.), pp. 54–59. Academic Press, San Diego.
Innis, M. A., and Gelfand, D. H. (1990). *In* "PCR Protocols: A Guide to Methods and Applications" (M. A. Innis, D. H. Gelfand, J. J. Sninsky, and T. J. White, eds.), pp. 3–12. Academic Press, San Diego.
Innis, M. A., Myambo, K. B., Gelfand, D. H., and Brow, M. A. D. (1988). *Proc. Natl. Acad. Sci. U.S.A.* **85**, 9436.
Jackson, M. P. (1991). *J. Clin. Microbiol.* **29**, 1910.
Jain, R., Gomer, R. H., and Murtagh, J. J., Jr. (1992). *BioTechniques* **12**, 58.
Jansen, R., and Ledley, F. D. (1989). *Gene Anal. Tech.* **6**, 79.
Jones, D. H., and Howard, B. H. (1990). *BioTechniques* **8**, 178.
Kadowaki, H. T., Kadowaki, T., Wondisford, F. E., and Taylor, S. I. (1989). *Gene* **76**, 161.
Kai, M., Kamiya, S., Sawamura, S., Yamamoto, T., and Ozawa, A. (1991). *Nucleic Acids Res.* **19**, 4562.
Kaufman, D. L., and Evans, G. A. (1990). *BioTechniques* **9**, 304.
Kawasaki, E. S. (1990). *In* "PCR Protocols: A Guide to Methods and Applications" (M. A. Innis, D. H. Gelfand, J. J. Sninsky, and T. J. White, eds.), pp. 21–27. Academic Press, San Diego.
Kawasaki, E. S., Clark, S. S., Coyne, M. Y., Smith, S. D., Champlin, R., Witte, O. N., and McCormick, F. P. (1988). *Proc. Natl. Acad. Sci. U.S.A.* **85**, 5698.
Keen, J., Lester, D., Inglehearn, C., Curtis, A., and Ghattacharya, S. (1991). *Trends Genet.* **7**, 5.
Kellogg, D. E., Sninsky, J. J., and Kwok, S. (1990). *Anal. Biochem.* **189**, 202.
Keohavong, P., and Thilly, W. G. (1989). *Proc. Natl. Acad. Sci. U.S.A.* **86**, 9253.
Kishimoto M., Sakura H., Hayashi, K., Akanuma Y., Yazaki, Y., Kasuga, M., Kadowaki, T. (1992). *Diabetologia* **74**, 1027.
Ko, M. S. H. (1990). *Nucleic Acids Res.* **18**, 5705.
Kogan, S. C., Doherty, A. B. M., and Gitschier, J. (1987). *N. Engl. J. Med.* **317**, 985.
Krummel, B. (1990). *In* "PCR Protocols: A Guide to Methods and Applications" (M. A. Innis, D. H. Gelfand, J. J. Sninsky, and T. J. White, eds.), pp. 184–188. Academic Press, San Diego.
Kwok, S., and Higuchi, R. (1989). *Nature (London)* **339**, 237.
Kwok, S., Kellogg, D. E., McKinney, N., Spasic, D., Goda, L., Lexenson, C., and Sninsky, J. J. (1990). *Nucleic Acids Res.* **18**, 999.
Lawyer, F. C., Stoffel, S., Saiki, R. K., Nyambo, K., Drummond, R., and Gelfand, D. H. (1988). *J. Biol. Chem.* **264**, 6427.

Lee, C. C., Wu, X., Gibbs, R. A., Cook, R. G., Muzny, D. M., and Caskey, C. T. (1988). *Science* **239**, 1288.
Lee, J. S. (1991). *DNA Cell Biol.* **10**, 67.
Leibrock, J., Lottspeich, F., Hohn, A., Hofer, M., Hengerer, B., Masiakowsik, P., Thoenen, H., and Barde, Y.-A. (1989). *Nature (London)* **341**, 149.
Lerman, L. S., and Silverstein, K. (1985). *In* "Recombinant DNA," Part F (R. Wu, ed.), Methods in Enzymology, Vol. 155. Academic Press, San Diego.
Lion, T., and Haas, O. A. (1990). *Anal. Biochem.* **188**, 335.
Liu, Z., and Schwartz, L. M. (1992). *BioTechniques* **12**, 28.
Lo, Y.-M., Mehal, W. Z., and Fleming, K. A. (1988). *Lancet* **ii**, 679.
Loh, E. Y., Elliot, J. F., Cwirla, S., Lanier, L. L., and Davis, M. M. (1989). *Science* **243**, 217.
Longo, M. C., Berninger, M. S., and Hartley, J. L. (1990). *Gene* **93**, 125.
Lowe, T., Sharefkin, J., Yang, S. Q., and Dieffenbach, C. W. (1991). *Nucleic Acids Res.* **18**, 1757.
Ludecke, H. J., Senger, G., Claussen, U., and Horsthemke, B. (1989). *Nature (London)* **338**, 348.
Lundberg, K. S., Shoemaker, D. D., Adams, M. W. W., Short, J. M., Sorge, J. A., and Mathur, E. J. (1991). *Gene* **108**, 1–6.
Ma, Y., Henderson, H. E., Murphy, V., Roederer, G., Monsalve, M. V., Clarke, L. A., Davignon, J., Lupien, P. J., Brunzell, J., and Hayden, M. R. (1991). *N. Engl. J. Med.* **324**, 1761.
Mack, D. H., and Sninsky, J. J. (1988). *Proc. Natl. Acad. Sci. U.S.A.* **85**, 6977.
Mares, A., Jr., Ledbetter, S. A., Ledbetter, D. M., Roberts, R., and Hejtmancik, J. F. (1991). *Genomics* **11**, 215.
Masoud, S. A., Johnson, L. B., and White, F. F. (1992). *PCR Methods Appl.* **2**, 89.
Mavrothalassitis, G., Beal, G., and Papas, T. S. (1990). *DNA Cell Biol.* **9**, 783.
Mazars, G. R., Moyret, C., Jeanteur, P., and Theillet, C. G. (1991). *Nucleic Acids Res.* **19**, 4783.
McConlogue, L., Brown, M. A. D., and Innis, M. A. (1988). *Nucleic Acids Res.* **16**, 9869.
Mitchell, D. B., Ruggli, N., and Tratschin, J.-D. (1992). *PCR Methods Appl.* **2**, 81.
Moller, D. E., Yokota, A., Caro, J. F., and Flier, F. S. (1989). *Mol. Endocrinol.* **3**, 1263.
Mueller, P. R., and Wold, B. (1989). *Science* **246**, 780.
Mullis, K. B., and Faloona, F. A. (1987). *In* "Recombinant DNA," Part F (R. Wu, ed.), Methods in Enzymology, Vol. 155, p. 335. Academic Press, San Diego.
Mullis, K., Faloona, F., Scharf, S., Saiki, R., Horn, G., and Erlich, H. (1986). *Cold Spring Harbor Symp. Quant. Biol.* **51**, 263.
Nagamine, C. M., Chan, K., and Lau, Y.-F. C. (1989). *Am. J. Hum. Genet.* **45**, 337.
Nakabeppu, Y., and Nathans, D. (1991). *Cell* **64**, 751.
Nelson, D. L., Ballabio, A., Victorka, M. F., Peiretti, M., Bies, R. D., Gibbs, R. A., Maley, J. A., Chinault, A. C., Webster, T. D., and Caskey, C. T. (1991). *Proc. Natl. Acad. Sci. U.S.A.* **88**, 6157.
Nuovo, G. J. (1992). "PCR *in situ* Hybridization. Protocols and Applications." Raven, New York.
Nuovo, G. J., Gallery, F., MacConnell, P., Becker, J., and Bloch, W. (1991a). *Am. J. Pathol.* **139**, 1239.
Nuovo, G. J., MacConnell, P., Forde, A., and Delvenne, P. (1991b). *Am. J. Pathol.* **139**, 847.
Ochman, H., Gerber, A. S., and Hartl, D. L. (1988). *Genetics* **120**, 621.
O'Rahilly, S., Choi, W. H., Patel, P., Turner, R. C., Flier, J. S., and Moller, D. E. (1991). *Diabetes* **40**, 777.
Orita, M., Suzuki, Y., Sekiya, T., and Hayashi, K. (1989). *Genomics* **5**, 874.
Paabo, S., and Wilson, A. C. (1988). *Nature (London)* **334**, 387.
Paabo, S., Gifford, J. A., and Wilson, A. C. (1988). *Nucleic Acids Res.* **16**, 9775.
Paabo, S., Irwin, D. M., and Wilson, A. C. (1990). *J. Biol. Chem.* **265**, 4718.
Patanjali, S. R., Parimoo, S., and Weissman, S. M. (1991). *Proc. Natl. Acad. Sci. U.S.A.* **88**, 1943.
Pfeifer, G. P., and Riggs, A. D. (1991). *Genes Dev.* **5**, 1102.

Rappolee, D. A., Brenner, C. A., Schultz, R., Mark, D., and Werb, Z. (1988). *Science* **241**, 1823.
Reiss, J., and Cooper, D. N. (1990). *Hum. Genet.* **85**, 85, 1.
Robinson, M. O., and Simon, M. I. (1991). *Nucleic Acids Res.* **19**, 1557.
Roux, K. H., and Dhanarajan, P. (1990). *BioTechniques* **8**, 48.
Ruano, G., and Kidd, K. K. (1991). *Proc. Natl. Acad. Sci. U.S.A.* **88**, 2815.
Rychlik, W., Spencer, W. J., and Rhoads, R. E. (1990). *Nucleic Acids Res.* **18**, 6409.
Saiki, R. K. (1990). *In* "PCR Protocols: A Guide to Methods and Applications" (M. A. Innis, D. H. Gelfand, J. J. Sninsky, and T. J. White, eds.), pp. 13–20. Academic Press, San Diego.
Saiki, R. K., Scharf, S., Faloona, F., Mullis, K. B., Horn, G. T., Erlich, H. A., and Arnheim, N. (1985). *Science* **230**, 1350.
Saiki, R. K., Bugawan, T. O., Horjn, G. T., Mullis, K. B., and Erlich, H. A. (1986). *Nature (London)* **324**, 163.
Saiki, R. K., Gelfand, D. H., Stoffel, S., Scharf, S. J., Higuchi, R., Horn, G. T., Mullis, K. B., and Erlich, H. A. (1988). *Science* **239**, 487.
Sambrook, J., Fritsch, E. F., and Maniatis, T. (1989). "Molecular Cloning: A Laboratory Manual," 2nd Ed. Cold Spring Harbor Lab. Press, Cold Spring Harbor, New York.
Sarkar, G., and Sommer, S. S. (1990). *Nature (London)* **343**, 27.
Sarkar, G., and Sommer, S. S. (1991). *BioTechniques* **10**, 591.
Scavo, L., Shuldiner, A. R., Serrano, J., De Pablo, F., and Roth, J. (1991). *Proc. Natl. Acad. Sci. U.S.A.* **88**, 6214.
Scharf, S. J. (1990). *In* "PCR Protocols: A Guide to Methods and Applications" (M. A. Innis, D. H. Gelfand, J. J. Sninsky, and T. J. White, eds.), Chap. 11. Academic Press, San Diego.
Scharf, S. J., Horn, G. T., and Erlich, H. A. (1987). *Science* **239**, 487.
Scully, S. P., Joyce, M. E., Abidi, N., and Bolander, M. E. (1990). *Mol. Cell. Probes* **4**, 485.
Sheffield, V. C., Cox, D. R., Lerman, L. S., and Myers, R. M. (1989). *Proc. Natl. Acad. Sci. U.S.A.* **86**, 232.
Shuldiner, A. R., and Tanner, K. (1993). *In* "Methods in Molecular Biology" (J. Walker, ed.). pp. 229–239. Humana Press, Clifton, New Jersey.
Shuldiner, A. R., Nirula, A., and Roth, J. (1990a). *Gene* **91**, 139.
Shuldiner, A. R., Nirula, A., Scott, L. A., and Roth, J. (1990b). *Biochem. Biophys. Res. Commun.* **166**, 223.
Shuldiner, A. R., Scott, L. A., and Roth, J. (1990c). *Nucleic Acids Res.* **18**, 1920.
Shuldiner, A. R., DePablo, F., Moore, C. A., and Roth, J. (1991a). *Proc. Natl. Acad. Sci. U.S.A.* **88**, 7679.
Shuldiner, A. R., Scott, L. A., Tanner, K., and Roth, J. (1991b). *Anal. Biochem.* **194**, 9.
Shuldiner, A. R., Tanner, K., Moore, C. A., and Roth, J. (1991c). *BioTechniques* **11**, 760.
Shuldiner, A. R., Perfetti, R., and Roth, J. (1993). *In* "Methods in Molecular Biology" (J. Walker, ed.). pp. 169–176. Humana Press, Clifton, New Jersey.
Smith, K. T., Long, C. M., Bowman, B., and Manos, M. M. (1990). *Amplifications* **5**, 16.
Sommer, S. S., Groszbach, A. R., and Bottema, C. D. K. (1991). *BioTechniques* **12**, 82.
Southern, E. (1975). *J. Mol. Biol.* **98**, 503.
Starr, L., and Quaranta, V. (1992). *BioTechniques* **13**, 612.
Stoflet, E. S., Koeberl, D. D., Sarker, G., and Sommer, S. S. (1988). *Science* **239**, 491.
Suzuki, Y., Orita, M., Shiraishi, H., Hayashi, K., and Sekiya, T. (1990). *Oncogene* **5**, 1037.
Thiesen, H.-J., and Bach, C. (1990). *Nucleic Acids Res.* **18**, 3203.
Timblin, C., Battey, J., and Kuehl, W. M. (1990). *Nucleic Acids Res.* **18**, 1587.
Tindall, K. R., and Kunkel, T. A. (1988). *Biochemistry* **27**, 6008.
Tompkins, L. S. (1992). *N. Engl. J. Med.* **327**, 1290.
Triglia, T., Peterson, M. G., and Kemp, D. J. (1988). *Nucleic Acids Res.* **16**, 8186.
Tuerk, C., and Gold, L. (1990). *Science* **249**, 505.

Wang, A. M., Doyle, M. V., and Mark, D. F. (1989). *Proc. Natl. Acad. Sci. U.S.A.* **86**, 9717.
Watson, R. (1989). *Amplifications* **2**, 5.
Weinstein, L. S., Shenker, A., Gejman, P. V., Mewrina, M. J., Friedman, E., and Spiegel, A. M. (1991). *N. Engl. J. Med.* **325**, 1688.
Wesley, C. S., Ben, M., Kreitman, M., Hagag, N., and Eanes, W. F. (1990). *Nucleic Acids Res.* **18**, 599.
Wieland, I., Bolger, G., Asouline, G., and Wigler, M. (1990). *Proc. Natl. Acad. Sci. U.S.A.* **87**, 2720.
Wu, D. Y., and Wallace, R. B. (1989). *Genomics* **4**, 560.
Wu, D. Y., Ugozzoli, L., Pal, B. K., and Wallace, R. B. (1989). *Proc. Natl. Acad. Sci. U.S.A.* **86**, 2757.
Wu, D. Y., Ugozzoli, L., Pal, B. K., Qian, J., and Wallace, R. B. (1991). *DNA Cell Biol.* **10**, 233.
Zaraisky, A. G., Lukyanov, S. A., Vasiliev, O. L., Smirnov, Y. V., Belyavsky, A. V., and Kazanskaya, O. V. (1992). *Dev. Biol.* **152**, 373.
Zhang, W. R., and Goldstein, B. J. (1991). *Biochem. Biophys. Res. Commun.* **187**, 1291.
Zhou, Y., Zhang, X., and Ebright, R. H. (1991). *Nucleic Acids Res.* **19**, 6052.

24

Detection of Mutations in Hormone Receptor Genes

Domenico Accili, Fabrizio Barbetti, Takashi Kadowaki, and Simeon I. Taylor

I. Introduction
II. Southern Blotting
III. Sequencing of Polymerase Chain Reaction-Amplified Genomic DNA and cDNA
   A. Enzymatic Amplification of Genomic DNA (First Polymerase Chain Reaction)
   B. Synthesis of Single-Stranded DNA (Second PCR)
IV. Allele-Specific Oligonucleotide Hybridization of Polymerase Chain Reaction Products
V. Denaturing Gradient Gel Electrophoresis
VI. Single-Stranded Conformation Polymorphism (SSCP)
   References

I. Introduction

A wide array of recombinant DNA techniques can be applied to detect mutations in genes encoding hormone receptors. Because no single technique to detect mutations has been reported to be unequivocally effective and generally applicable, a combination of different approaches is required in order to analyze candidate genes (Caskey, 1987; Rossiter and Caskey, 1990). In this chapter, we describe techniques that have been employed successfully in our laboratory for the detection and characterization of mutations in hormone receptor genes in humans (Taylor *et al.*, 1991). Because of the large number of techniques developed in recent years, this chapter is limited to those of general applicability.

II. Southern Blotting

In 1975, Edwin Southern described a technique for the detection of sequences of genomic DNA by size fractionation of total DNA using digestion with a restriction enzyme followed by electrophoresis on agarose gels and transfer ("blotting") to nitrocellulose filters (Southern, 1975). This technique has since been referred to as "Southern" blotting. The principle of Southern blotting is very simple: total genomic DNA is cleaved at specific sites using restriction endonucleases. This procedure will generate DNA fragments of varying size, the average length of which is determined by how frequently the restriction endonuclease cleaves DNA. For example, a restriction endonuclease with a 4-base pair (bp) recognition sequence, like Msp I or Taq I, will cleave DNA more frequently than a restriction endonuclease with an 8-bp recognition sequence, like Not I. Therefore, Msp I or Taq I will generate more numerous and smaller fragments than Not I. Following cleavage by the restriction enzyme, DNA fragments are separated according to their size by electrophoresis through an agarose gel and subsequently transferred to a solid support (nitrocellulose or nylon membrane) for hybridization to a labeled probe. The probe will base pair with homologous target sequences within the genome (hybrid formation or "hybridization"). Loosely bound sequences are removed by way of high-temperature/low-salt washes. A radioactive signal corresponding to the size band generated by the restriction endonuclease is thus obtained.

Southern blotting has been extensively applied to the diagnosis of human disease, as a way of detecting disease-associated genes (linkage analysis) and characterizing their structure. Two applications of Southern blotting are relevant to our discussion.

1. Because the human genome is diploid with respect to autosomes, it is possible to detect differences between two copies of the same gene if these differences occur at a restriction site. These variations are defined as allelic markers, because they facilitate the distinction between two alleles in humans. For example, they can be used to determine whether a patient is homozygote (i.e., possesses two identical copies of the same allele) or heterozygote (i.e., possesses two copies of the allele and differs at one or more epitopes or loci) for a given gene (Fig. 1) (Accili et al., 1989). Allelic differences are determined by a normal variation in the base composition of the human genome. It has been calculated that there is approximately 1 variable nucleotide every 1000 bp in humans (Gusella, 1986).

2. Mutations can affect the pattern on Southern blotting in one of the following ways:
    a. Point mutations may abolish an existing site or generate a new one (Fig. 2), thus yielding a characteristic pattern.

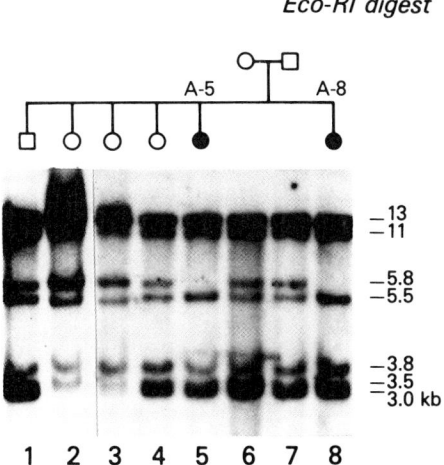

Figure 1 RFLP mapping of a mutant allele of the insulin receptor gene in a kindred with a genetic syndrome of insulin resistance. Two members of this consanguineous kindred are affected by type A extreme insulin resistance, a genetic disorder caused by mutations of the insulin receptor gene. The inheritance of alleles at the insulin receptor locus in this kindred is demonstrated using Southern blot analysis. Genomic DNA from each family member was analyzed by Southern blotting after restriction digestion with *Eco*RI. A partial insulin receptor cDNA clone was used as a probe to identify a polymorphism characterized by bands at 5.5 and 5.8 kbp, respectively. Both mother and father are heterozygotes for this polymorphism, as are the four unaffected siblings. In contrast, the two patients have inherited the same allele (characterized by the 5.5 kbp fragment) from both parents. This analysis is consistent with the notion that the two patients are homozygotes by descent at the insulin receptor locus. This conclusion has been validated by cloning of the mutant allele (Accili *et al.*, 1989).

b. Major rearrangements (i.e., deletions or insertions of several hundreds to thousands of nucleotides) may change the size of bands generated by restriction digestion. A special case should be made for large chromosomal deletions which may remove an entire gene on one allele. These types of deletions represent a problem, because they do not alter the Southern blot pattern, but rather reduce the intensity of the bands. They can be detected only by a careful quantitative analysis comparing the intensity of the same bands on a normal sample and a sample with a single-allele deletion. For example, it is possible to run scalar dilutions of two DNA samples side by side on the same gel. Alternatively, physical mapping methods are required in order to determine exactly the boundaries of a large chromosomal rearrangement. For example, very large DNA fragments (i.e., larger than 100 kbp) can be analyzed by pulse-field gel electrophoresis (PFGE).

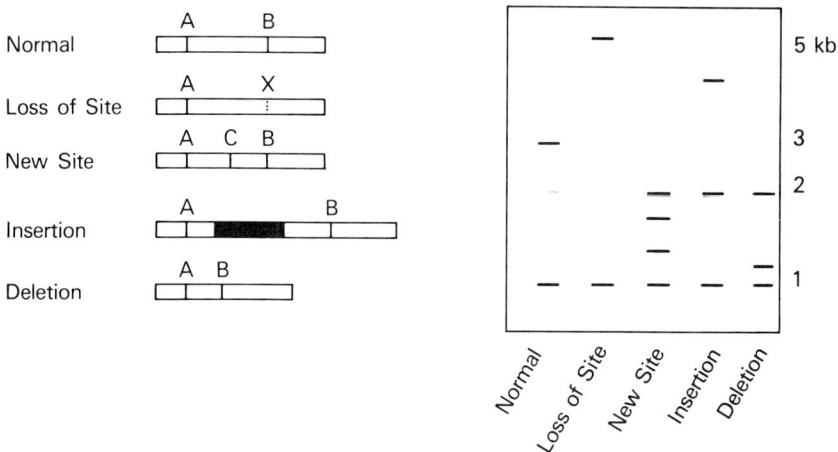

Figure 2 Causes of polymorphic patterns in human DNA. If a single base-pair change occurs at a recognition site for a restriction endonuclease, it may be detected by digestion of genomic DNA with that enzyme, followed by Southern blotting. Removal of a restriction site will generate a band, the size of which equals the length of the two original bands. On the other hand, when a restriction site is generated by a polymorphic base change, the original fragment will be divided into two fragments, the sizes of which will add up to that of the original fragment. Similar alterations in size will be caused by deletions and insertions.

PROTOCOL

1. *Restriction analysis.* In a final volume of 40 μl mix 5–10 μg of genomic DNA, 10–50 U of restriction endonuclease, and 4 μl of the appropriate buffer. Incubate in a water bath at 37°C for 4–6 hr. (It is sometimes helpful, especially when the DNA preparation is not very pure or protein-free, to add additional enzyme after 3-4 hr; certain enzymes are stabilized by the presence of 1 mM spermidine.)

2. *Agarose gel electrophoresis.* Add to the samples 4 μl of running buffer (0.02% bromophenol blue, 0.02% xylene cyanol, 1 mM EDTA, 50% glycerol); heat to 65°C for 10 min; cool on ice. Apply to a 0.8–1% agarose gel in 1× TBE (89 mM Tris; 89 mM boric acid; 2 mM EDTA, pH 8.3). Run for 20–24 hr at 30 V.

3. *Transfer.* Soak the gel in 0.2 N HCl for 10–15 min to destabilize purine residues of DNA ("depurination"); rinse briefly with distilled water, then transfer to a solution of 0.5 N NaOH 1.5 M NaCl for 45–50 min; then transfer to a 0.5 M Tris (pH 7.4), 1.5 M NaCl solution for 45–50 min. Equilibrate the gel in transfer buffer (20×

SSC for nitrocellulose filters, 10× SSC for nylon filters) for 15–20 min.

Set up a transfer wick by cutting a strip of Whatman 3MM paper of the same width as the gel; wet the paper in a tray with transfer buffer. Place a glass plate across the transfer tray. Place the paper on the glass plate with the edges dipping into the transfer solution. Place the gel upside down on the filter paper. Place a nitrocellulose or nylon membrane on the gel, taking care to squeeze out all the air bubbles with a glass rod or a pipette. Place two sheets of filter paper (Whatman 3MM) on the membrane. Place a 5-cm stack of adsorbent paper on the top. (Some people put a weight on top of the paper; in our experience of several years we have found this to be unnecessary.)

Allow transfer to proceed overnight. Dry the filter and bake in a vacuum oven for 2 hrs at 80°C. Alternatively, we have found UV crosslinking (using a Stratagene Stratalinker or similar apparatus) to be very helpful.

4. *Hybridization.* Prepare a solution of 50% formamide, 5× SSPE (1× SSPE is 180 mM NaCl, 10 mM sodium phosphate, pH 7.7, and 1 mM EDTA), 8× Denhardt's solution (1× is 0.02% BSA, 0.02% polyvinylpirrolidone, 0.02% Ficoll 400), and 0.1% SDS (0.5% when using nylon membranes). Sterilize by filtration, then add 0.1 mg/ml denatured salmon sperm DNA. Place the membrane in a plastic bag. Add enough solution to wet the membrane (6–8 ml for an 11 × 14-cm. filter). Incubate at 42°C for 4–6 hr.

DNA and RNA probes can be used for Southern hybridization. Radioactive labeling can be carried out by nick translation, random priming, or *in vitro* transcription. The use of oligonucleotide probes for genomic Southern blotting requires hybridization in the presence of tetramethylammonium chloride or high specific activity labeling by primer extension. Description of these techniques goes beyond the scope of this chapter (Nukiwa *et al.*, 1986; DiLella and Woo, 1987).

1. Heat-denature double-stranded probes. Add $5-10 \times 10^6$ cpm labeled probe/ml. of hybridization solution.
2. Incubate overnight at 42°C.
3. Remove the filter from the bag and place it in a tray containing 300 ml of 1× SSC, 0.1% SDS (0.5% for nylon). Rinse at room temperature for 5 min with gentle shaking. Repeat twice.
4. Wash at 60°C for 1 hr in 1× SSC, 0.1% SDS (0.5% for nylon). Wash at 60°C for 1 hr in 0.1× SSC, 0.1% SDS (0.5% for nylon). Do not let

the filter dry at any step, otherwise the probe becomes irreversibly bound to the DNA on the filter. To strip the probe and reuse the filter, boil 1 liter of distilled water containing 0.01% SDS. Pour 300 ml in a glass tray. Place the filter in the tray, and shake until temperature goes down to about 50°C. Repeat three times. Rinse with distilled water. Wrap in Saran wrap for storage.

III. Sequencing of Polymerase Chain Reaction-Amplified Genomic DNA and cDNA

Amplification of DNA and RNA sequence by thermostable DNA polymerases has rapidly become the method of choice for detection of sequence abnormalities in human DNA (Saiki, 1989). This technique can be combined with other methods to determine the exact nucleotide sequence of a DNA fragment. Alternatively, as described in other sections of this chapter, the amplified DNA product can be employed in a sequence-specific hybridization assay for the identification of known variations in a nucleotide sequence. Finally, the presence of base-pair changes of the DNA sequence can be detected by virtue of the fact that such changes will alter the electrophoretic mobility of a DNA fragment under denaturing conditions (see Section V). Direct sequence analysis of PCR products is to be preferred to sequencing of PCR products that have been cloned in a plasmid vector for a number of different reasons. First, it is more rapid and less costly. Second, it is less likely to detect artifacts related to the misincorporation of bases during enzymatic amplification at high temperature. [This phenomenon is due to the lack of "proofreading" activity by the Taq DNA polymerase, and is less pronounced with other thermostable DNA polymerases (Keohavong and Thilly, 1989).] Third, when more than one sequence is present in a DNA sample (i.e., when a patient is heterozygous at a given locus) direct sequencing allows for detection of sequences derived from both alleles in the same reaction (Kadowaki *et al.*, 1990; Hurley *et al.*, 1991). When this experiment is carried out with cDNA, it also allows estimation of the relative levels of mRNA derived from each allele (Kadowaki *et al.*, 1990).

Several techniques for direct sequencing of PCR products have been described (McMahon *et al.*, 1987; Innis *et al.*, 1988; Engelke *et al.*, 1988; Olsen and Eckstein, 1989; Casanova *et al.*, 1990; Bachmann *et al.*, 1990). We describe here a method based on the generation of an intermediate single-stranded sequencing template from the original PCR reaction (Kadowaki *et al.*, 1990).

A. Enzymatic Amplification of Genomic DNA (First Polymerase Chain Reaction)

Genomic DNA is amplified by the polymerase chain reaction (PCR) catalyzed by Taq DNA polymerase (Saiki, 1989). The reaction is carried out in a total volume of 0.1 ml containing the following additions:

1. Genomic DNA template (1 µg)
2. Upstream and downstream oligonucleotide primers (100 pmol of each)
3. 10 µl of 10× buffer [500 mM KCl; 100 mM Tris–HCl, pH 8.3, 15 mM MgCl$_2$, 0.1% (w:v) gelatin]
4. 16 µl of a solution containing dATP, dCTP, dGTP, and dTTP (each at concentrations of 1.25 mM)
5. Taq DNA polymerase (2.5 U) (Cetus-Perkin Elmer; Emerysville, CA)
6. Water to make total volume 0.1 ml

Amplification is typically carried out for 30 cycles, each cycle consisting of incubations for 60 sec at 94°C for denaturation, 90 sec at 55°C for annealing, and 90 sec at 72°C for primer extension. At the beginning of the first cycle, DNA is denatured for 5 min rather than 60 sec; at the end of the last cycle, the 72°C incubation proceeds for 5 min instead of 90 sec.

10 µl of the PCR product is analyzed by electrophoresis through a 1% agarose gel in 1× TBE for 2 hr at 6.5 V/cm.

AMPLIFICATION OF COMPLEMENTARY DNA (cDNA)

In this reaction, the template is represented by mRNA. Therefore, the first step requires reverse transcription of the mRNA molecule into cDNA.

1. In a 0.5-ml microcentrifuge tube mix together

    1 µg of total RNA

    2 µl of 10× PCR buffer (see above)

    1 µl of RNase inhibitor (RNasin from Promega Biotec, Madison, WI)

    10 ng of oligo (dT) primer or downstream primer of the PCR reaction to follow

    2 µl of a dNTPs mixture (1 mM each)

    Water to a final volume of 20 µl (Heat the reaction mixture at 70°C for 2 min; cool on ice)

    27 U (1 µl) of reverse transcriptase (Boehringer–Mannheim, Indianapolis, IN)

2. Incubate for 1 hr at 42°C.
3. Heat denature reverse transcriptase at 70°C for 15 min; 1 μl of this reaction can now be applied to a PCR reaction as described above. The remainder can be stored at −70°C.

B. Synthesis of Single-Stranded DNA (Second PCR)

The second PCR is carried out exactly as described above for the first PCR except for three changes. First, by including only one of the two oligonucleotide primers (either the upstream or the downstream primer), it is possible to amplify only one strand of DNA selectively. The sense strand is synthesized using the upstream primer, the antisense strand using the downstream primer. Second, in place of the genomic DNA, amplified DNA (1−5 ng) synthesized in the first PCR is used as template. Ordinarily, this represents approximately <1% (0.2−1 μl) of the DNA synthesized in the first PCR. As judged by the readability of the sequencing ladders, we have obtained the best results when smaller quantities ($\approx$1 ng) of amplified DNA are used in the second PCR. Finally, the second PCR reaction was carried out for 20−25 rather than 30 cycles.

The product of the second PCR is extracted once with chloroform (0.1 ml) and diluted to 2 ml with water. The single-stranded DNA is separated from the oligonucleotide primers and deoxynucleotide triphosphates by filtration through a Centricon 100 membrane according to the manufacturer's instructions (Amicon; Danvers, MA). A Millipore ultrafiltration membrane can also be used (Millipore, Bedford, MA). The retentate (45 μl) is dried by speed vacuum and dissolved in 7 μl of water immediately before use in the DNA sequencing reaction.

SEQUENCING OF AMPLIFIED SINGLE-STRANDED DNA

1. Mix amplified single-stranded DNA (7 μl) in a 0.5-ml microcentrifuge tube with 1 μl of primer ($\approx$25 pmol) and 2 μl of 5× Sequenase buffer (Sequenase, USB, Cleveland, OH).
2. After annealing for 10 min at 65°C, allow the mixture to cool down to 37°C.
3. Add 2 μl of the "labeling" mix (containing all four deoxyribonucleotide and dideoxyribonucleotide triphosphate), 1 μl of 0.1 M dithiothreitol (DTT), 0.5 μl of [α-^{35}S]dATP (sp act 1000Ci/mmol; Amersham, Arlington Heights, IL) plus 1 μl of diluted (1:7) Sequenase (1.2 units).

4. Incubate at room temperature for 2–5 min, then add 3.5-µl aliquots of the reaction to tubes containing 2.5 µl of "termination" mixtures preheated to 37°C.
5. Terminate the reaction by adding 3 µl of "stop" solution (95% formamide, 20 mM EDTA, 0.05% bromphenol blue, 0.05% xylene cyanol FF).
6. Analyze the samples by electrophoresis through a 6% polyacrylamide/8 M urea gel. Electrophoresis is carried out for 2.5–5 hr at 60 W.

IV. Allele-Specific Oligonucleotide Hybridization of Polymerase Chain Reaction Products

This technique can be combined with PCR to detect known point mutations in a gene sequence (Saiki *et al.*, 1986). It can be used to assess the prevalence of a given polymorphism or mutation in a control population or in other subjects at risk for the disease. It can also be used to resolve ambiguities in a DNA sequence and rule out PCR or cloning artifacts (Fig. 3). When the sequence of a particular point mutation is known, it is possible to design oligonucleotide probes that will detect this mutation with high specificity. Two probes are generally employed: one bearing the wild-type sequence and one with the mutant sequence. The oligonucleotide probes are then hybridized to the PCR-amplified DNA bound to a filter membrane. Following hybridization, the filters are washed under extremely stringent conditions. If the hybrid formed is a perfect match between the probe sequence and the target sequence, a melting temperature can be calculated which is proportional, among other things, to the length of the oligonucleotide probe as well as its base composition (T_m = temperature at which half of the probe is dissociated from the target sequence). However, if a single mismatch exists in a critical position (i.e., in the middle of the sequence of the probe), the melting temperature will be considerably lower. For example, if an 18-mer is used and the mismatch occurs at position 9, the melting temperature of the mismatched hybrid will be about 10°C lower. This technique is very simple, fast, and highly reproducible. We recommend choosing an oligonucleotide 18-20 nucleotides-long, and to place the mismatch exactly in the middle (Accili *et al.*, 1989).

PROTOCOL

Following amplification, PCR products can be applied to a nitrocellulose membrane using a blotting apparatus (e.g., Manifold, Schleicher and Schuell,

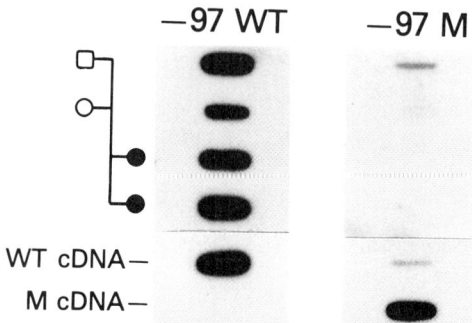

Figure 3 Allele-specific oligonucleotide hybridization. In the course of cloning the mutant allele of the insulin receptor of the two patients shown in Fig. 1, a substitution of T for C was identified at position −97 in the 5′ untranslated sequence of the patients' cDNA (Accili et al., 1989). We suspected that this base change may represent a cloning artifact, because it was not observed in other cDNA clones derived from the same patient. However, because the assumption that the patients were homozygotes was fundamental to our analysis, we investigated this base change by PCR amplification of genomic DNA from the two patients (third and fourth lane) and their parents (top two lanes). cDNA clones with the wild-type and mutant sequence were also included as controls. Hybridization to the wild-type probe (C −97) revealed a positive signal in the two patients, whereas the mutant sequence (T −97) was not detected. This experiment suggests that the observed variation has most probably occurred as a result of a cDNA cloning artifact.

Keene, NH) or can be analyzed on an agarose gel and then transferred according to the protocol in Section II.

1. For transfer using a blotting apparatus, prepare 5-μl aliquots of PCR products (equivalent to ∼ 50 ng) and denature by heating to 95°C for 5 min. Cool on ice. Add 200 μl of 20× SSC. Apply to the membrane. Bake the membrane or UV irradiate as described in Section II protocol, step 3, Transfer.

2. Prehybridize the membrane in 5× SSPE, 5× Denhardt's solution, and 0.1% SDS at 37°C for 30 min. If nylon membranes are used, it is advised that the concentration of SDS be raised to 0.5%. In fact, some of these membranes are positively charged in order to increase binding of negatively charged nucleic acids. Hence, it is possible to decrease background by increasing the concentration of negative charges in solution with SDS.

3. Add to the hybridization solution $1-2 \times 10^6$ cpm labeled probe/ml solution. Remove the prehybridization solution and add the hybrization solution to the bag. Incubate at 37°C for 4–6 hr. (Alternatively, it is possible to incubate overnight.)

4. Wash the filter in 300 ml of a solution containing 2× SSPE and 0.1% SDS for 5 min at room temperature twice. Wash at the higher temperature (see below) in 5× SSPE and 0.1% SDS for exactly 10 min. To determine the best temperature, we recommend an initial wash at 55°C followed by exposure (it should take few hours to detect the signal). If there is residual hybridization to the mismatched allele, the investigator can proceed to a new wash at a temperature 5°C higher and continue to increase until it disappears.

V. Denaturing Gradient Gel Electrophoresis

The differential melting properties of mismatched DNA/DNA hybrids (herein referred to as heteroduplexes) versus perfectly matched hybrids (homoduplexes) can be employed to detect variations in a DNA sequence (Myers *et al.*, 1985). The principle of the technique follows: the mobility of DNA electrophoresed in the presence of a gradient of denaturants (e.g., urea, formamide) is inversely related to its molecular weight until the DNA molecule reaches a region of the gel in which the concentrations of denaturants will cause the DNA to denature into its single-stranded form. The mobility of single-stranded DNA will then sharply decrease. It follows that a mid-melting-point temperature (MMPT: the temperature at which there is a 50% chance that the DNA will exist in its double-stranded conformation) can be calculated for a DNA fragment. The presence of a mismatch in the DNA sequence will change the MMPT, so that mutant and wild-type DNA can be distinguished by electrophoresis of PCR-amplified products (Myers *et al.*, 1989).

There are, however, two other major factors complicating this analysis: first, melting of a DNA sequence is not an all-or-none phenomenon; it rather occurs by way of melting of single, discrete domains in the DNA structure. Therefore, the presence of multiple melting domains in a DNA sequence may reduce the sensitivity with which mutations can be detected in a given fragment. This problem can be obviated by choosing relatively small fragments for denaturing gradient gel electrophoresis (DGGE) analysis (i.e., 200–300 bp), or by choosing a DNA fragment that is predicted to contain only one melting domain (Lerman and Silverstein, 1987). Furthermore, it is possible to analyze a DNA fragment in a way that will detect mutations even when there are multiple melting domains in the sequence. This procedure is referred to as "perpendicular" DGGE in order to distinguish it from "parallel" DGGE. In this variation of the technique, the gradient of denaturants is perpendicular to the electric field rather than parallel to it. DNA is electrophoresed on a gel in which the concentration of denaturants does not increase from top to bottom, but rather from the left to the right of the gel, so that DNA will migrate under a

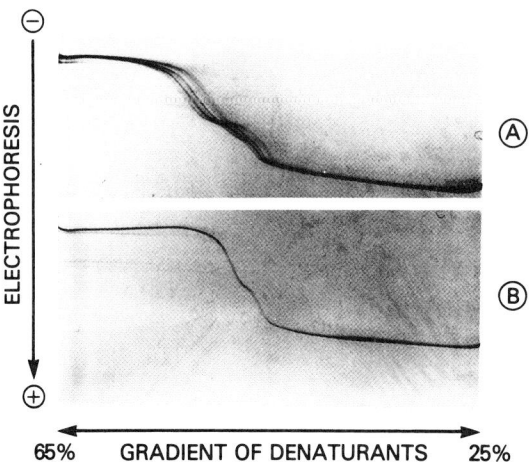

Figure 4 Perpendicular DGGE. Mutations of the insulin receptor gene have been detected with high sensitivity using the technique of denaturing gradient gel electrophoresis (Barbetti et al., 1992). Following amplification by PCR, DNA is analyzed through a denaturing urea/formamide/polyacrylamide gel. The concentration of denaturants increases from right to left, causing the DNA to migrate more slowly on the left of the gel, where the concentration of denaturants is highest. In fact, denatured DNA is single stranded and migrates more slowly than double-stranded DNA. Two typical transition curves can be visualized on each of the two gels shown (A and B). Each transition curve corresponds to a different melting domain. At each curve, four bands can be observed. One band corresponds to homoduplexes of wild-type DNA, two bands to heteroduplexes of wild-type and mutant DNA, and one band to homoduplexes of mutant DNA. (A) Amplified DNA from the father of an insulin-resistant patient with the genetic syndrome of leprechaunism (leprechaun Ver-1; Barbetti et al., 1992). Four bands can be distinguished at each transition curve, consistent with the presence of a sequence variation in one of the two alleles of the insulin receptor gene. Subsequent sequence analysis confirmed the presence of a single base-pair substitution in amplified DNA derived from exon 4 of the insulin receptor gene. In contrast, the patient's mother (B) is homozygous at this locus. Therefore, only one transition pattern can be detected.

constant concentration of denaturants. DNA samples to be analyzed are applied to a slot spanning the entire width of the gel. After electrophoresis, a characteristic pattern or "profile" can be detected (Fig. 4). On the left, where the concentration of denaturants is highest, DNA will migrate as denatured, single-stranded molecule and will be slowest. On the right, at the lowest concentrations of denaturants, DNA will maintain a double-helical conformation and migrate fastest. The presence of a melting domain is associated with the appearance of a steep melting curve. Each new melting domain will have a new melting curve, so that mutations in each of the domains can be detected by the appearance of multiphasic melting curves (Fig. 4).

A problem with DGGE analysis is represented by homozygous mutations in the DNA sequence. In fact, if a homozygous mutation is present, it will effectively migrate on the gel as a homoduplex which may not necessarily possess a different melting profile from a wild-type homoduplex. To increase the sensitivity of this detection system, it may then be necessary to mix together the test sample and a known wild-type sample. This will generate heteroduplexes in addition to wild-type and mutant homoduplexes. This pattern is reflected by the appearance of multiple melting curves on a perpendicular gel (see Fig. 4). Obviously, when the test sample contains a heterozygous mutation it will serve as its own control, and heteroduplexes will be readily detected in addition to homoduplexes. The technique of perpendicular DGGE approaches a 100% sensitivity in the detection of point mutations (Barbetti *et al.*, 1992).

A second problem that may be encountered in this type of analysis is related to the melting properties of the DNA fragment to be analyzed. In fact, the choice of the best concentration of denaturants required to detect a melting point difference may represent a time-consuming procedure. In our laboratory, we have applied a computer program originally described by Lerman and Silverstein (1987), which allows the prediction of the best window in the concentration of denaturants to resolve melting differences (Barbetti *et al.*, 1992).

Finally, it has been shown that attachment of a sequence rich in G and C at the 5' end of one of the PCR primers increases the sensitivity of this technique ("GC-clamp"; Sheffield *et al.*, 1989). To understand the principle of this modification, it should be kept in mind that if a DNA fragment possesses, for example, two melting domains, only mutations in the first domain will be detected. This limitation is due to the fact that, as DNA approaches total dissociation of the strands, its mobility on a denaturing gel becomes independent of the base composition. However, a GC-rich domain will not melt under electrophoresis conditions. The presence of a domain that virtually never melts will unveil other melting domains in the sequence. Therefore, all mutations contained in the test sequence will now become amenable to DGGE analysis.

PROTOCOL

1. PCR amplification is performed according to the protocol in Section III, A. However, when the patient is potentially homozygous for a mutant sequence, it is desirable to perform mixing experiments in which the patient's DNA is mixed together with a control

sample of known sequence. To maximize the probability of generating heteroduplexes between control DNA and test DNA, the two samples can be mixed together prior to PCR amplification.

2. Gel electrophoresis is performed as follows: Amplified samples are analyzed on a 6.5% polyacrylamide gel (acrylamide : bisacrylamide = 30 : 1) containing a continuous gradient of denaturants (urea or urea and formamide). The concentrations of denaturants can be chosen based on theoretical predictions (Lerman and Silverstein, 1987) or determined empirically by running preliminary experiments with a perpendicular gel. On a perpendicular gel, different melting domains can be visualized and the corresponding concentrations of denaturants can be chosen for analysis on a parallel gel. Denaturing conditions are maintained during the run by keeping the temperature of the gel between 55 and 65°C. For most purposes, a temperature of 60°C will be adequate. The gel should contain a concentration of denaturants about 10°C around the T_m (1°C corresponds approximately to 3% of the maximally effective concentrations of denaturants; 100% denaturing conditions correspond to 7 M urea and 40% formamide). The inclusion of 40% (v : v) formamide in the gel mix is recommended to resolve highly GC-rich sequences. Thus, if the calculated T_m for a DNA fragment is 75°C, the concentration of denaturants at 60°C will range between 30% (60°C + 10°C) and 60% (60°C + 20°C). The gradient is poured vertically using a gradient maker (Hoefer Scientific, San Francisco, CA). The orientation of the gradient with respect to the electric field can be parallel or perpendicular. Gels are electrophoresed in a Hoefer HE 600 vertical apparatus. A peristaltic pump is used to recirculate the buffer between the lower and the upper chamber. In perpendicular gels, the DNA in 70 μl is applied to a slot corresponding to the entire width of the gel. Electrophoresis is performed for 16 hr at 80 V for parallel gels and for 2/2.5 hr at 150 V for perpendicular gels.

Following the run, the gel is stained with ethidium bromide or silver stain.

VI. Single-Stranded Conformation Polymorphism (SSCP)

In addition to the methods described above, other PCR-based methods have been developed to detect single base-pair changes in a DNA sequence. The method of single-stranded conformation polymorphism relies on the observation that the mobility of single-stranded DNA electrophoresed under nondenaturing conditions is influenced by its base composition (Orita *et al.*,

1989). Therefore, if two DNA fragments differ by one or more bases, they will migrate as two distinct bands on a nondenaturing polyacrylamide/glycerol gel. While the theoretical basis of this phenomenon is not entirely clear, this method has been applied successfully to detect mutations associated with human genetic diseases (Kim et al., 1992). In this experimental system, genomic or complementary DNA is first amplified by the polymerase chain reaction. Subsequently, amplified DNA is heat denatured in the presence of formamide so as to obtain single-stranded molecules. Single-stranded DNA molecules are then separated according to their different migration patterns through a neutral polyacrylamide gel. Different conformations of single-stranded molecules are thought to cause the difference in the migration patterns. It appears, however, that these single-stranded conformations are not very stable; therefore, conditions required to detect a given polymorphism or mutation are less predictable than with the method of DGGE (see above). In most circumstances, it is advised to conduct preliminary experiments in which the same DNA fragment is electrophoresed at two different temperatures in the presence or absence of 10% glycerol in the polyacrylamide gel.

This method has been employed to detect mutations of the insulin receptor gene in patients with genetic forms of extreme insulin resistance and with type II diabetes (Kim et al., 1992). The accuracy of detection of known mutations and polymorphisms at this locus using this method has been reported. It is more difficult to estimate the success rate of this technique at detecting new mutations, because no controlled studies have been carried out to date. Such a study would indeed represent a useful contribution.

The method of SSCP is seemingly easier to apply than the method of DGGE. However, each sample requires analysis under multiple electrophoresis conditions, a factor which may complicate the analysis of large numbers of samples. Therefore, it is difficult to suggest *a priori* a method that will prove to be the best for every experimental condition. If the main aim of the study is detection of known sequence variations, then SSCP may have the advantage over DGGE of not requiring long oligonucleotide primers to be used in the amplification reaction and of not requiring an expensive electrophoresis setup. However, because the DNA to be amplified is radiolabeled in order to be autoradiographed, DGGE might be more easily employed in clinical laboratories where handling of radioactive waste is at times problematic.

PROTOCOL

1. *Primer labeling*
   a. In a 0.5-μl microcentrifuge tube mix
      0.4 μl of 10× kinase buffer (1× is 50 mM Tris–HCl, pH 8.3, 10 mM $MgCl_2$,

5 mM DTT)
   1 μl of upstream PCR primer (10 pmol/ml)
   1 μl of downstream PCR primer (10 pmol/ml)
   1.2 μl [γ-^{32}P]ATP (3000 Ci/mmol; NEN–DuPont)
   0.4 μl T4 polynucleotide kinase.
   b. Incubate at 37°C for 30 min.
   c. Add 120 μl of water, 20 μl of 10× PCR buffer (see above), and 20 μl of 1.25 mM of the four dNTPs.
   d. Mix by vortex (this mixture can be used for up to 40 samples). Prior to PCR, add 2.0 μl of Taq polymerase.
2. PCR. To a 0.5 μl microcentrifuge tube add in order
   1 μl of genomic DNA (50 ng/ml in water)
   4 μl of reaction mixture from point 1
   a. Overlay the reaction mixture with 10 μl of light mineral oil.
   b. PCR is preformed as described in Section III, A.
   c. PCR reaction is stopped by the addition of 45 μl of a solution containing 0.05% bromophenol blue, 0.05% xylene cyanol, 20 mM EDTA, and 95% formamide.
3. *Electrophoresis.* Each sample is analyzed under different conditions, in order to maximize the sensitivity of this technique to detect single base-pair changes. Therefore, duplicate sets of gels with and without 10% glycerol are prepared: one set is electrophoresed at room temperature and one at 4°C.
   a. Cast 5% polyacrylamide gel (acrylamide/bis acrylamide, 19:1) with and without 10% glycerol. Electrophoresis is carried out in 0.5× TBE (see Section III, A) at a constant power of 40 W for 2–4 hr in a sequencing gel apparatus, in which a thermal aluminum plate is in contact with the rear glass plate of the gel.
   b. Heat samples at 80°C for 5 min.
   c. Load 5 μl of the sample.
   d. Start electrophoresis at 40 W.
   e. Cool the electrophoresis apparatus using two laminar flow fans, one facing the front glass surface of the gel, one facing the rear glass surface of the gel. Temperature should be maintained in the range of 17–23°C for gels at room temperature.
   f. Dry the gels on Whatman 3MM paper for 30 min. Autoradiograms are generally exposed for 12–24 hr at −70°C between intensifying screens.

References

Accili, D., Frapier, C., Mosthaf, L., McKeon, C., Elbein, S. C., Permutt, M. A., Ramos, E., Lander, E. S., Ullrich, A., and Taylor, S. I. (1989). *EMBO J.* **8**, 2509-2517.
Bachmann, B., Luke, W., and Hansmann, G. (1990). *Nucleic Acids Res.* **18**, 1309.
Barbetti, F., Gejman, P. V., Taylor, S. I., Raben, N., Cama, A., Bonora, E., Pizzo, P., Moghetti, P., Muggeo, M., and Roth, J. (1992). *Diabetes* **41**, 408.
Casanova, J. L., Pannetier, C., Jaulin, C., and Kourilsky, P. (1990). *Nucleic Acids Res.* **18**, 4028.
Caskey, C. T. (1987). *Science* **236**, 1223-1229.
DiLella, A. G., and Woo, S. L. C. (1987). *In* "Guide to Molecular Cloning Techniques" (S. Berger and A. Kimmel, eds.), Methods in Enzymology, Vol. 152, pp. 447-451. Academic Press, San Diego.
Engelke, D. R., Hoener, P. A., and Collins, F. S. (1988). *Proc. Natl. Acad. Sci. U.S.A.* **85**, 544-548.
Gusella, J. F. (1986). *Annu. Rev. Biochem.* **55**, 831-854.
Hurley, D. M., Accili, D., Vamvakopoulos, N., Stratakis, C., Karl, M., Vamvakopoulos, N., Rorer, E., Taylor, S. I., and Chrousos, G. P. (1991). *J. Clin. Invest.* **87**, 680-686.
Innis, M. A., Myambo, K. B., Gelfand, D. H., and Brow, M. D. (1988). *Proc. Natl. Acad. Sci. U.S.A.* **85**, 9436-9440.
Kadowaki, T., Kadowaki, H., and Taylor, S. I. (1990). *Proc. Natl. Acad. Sci. U.S.A.* **87**, 658-662.
Keohavong, P., and Thilly, W. G. (1989). *Proc. Natl. Acad. Sci. U.S.A.* **86**, 9253-9257.
Kim, H., Kedowaki, H., Sakura, H., Odawara, M., Momomura, K., Takahashi, Y., Miyazaki, Y., Ohtani, T., Akanuma, Y., Yazaki, Y., Taylor, S. I., and Kadowaki, T. (1992). *Diabetologia* **35**, 261-266.
Lerman, L. S., and Silverstein, K. (1987). *In* "Recombinant DNA," Part F (R. Wu, ed.), Methods in Enzymology, Vol. 155, pp. 482-501. Academic Press, San Diego.
McMahon, G., Davis, E., and Wogan, G. N. (1987). *Proc. Natl. Acad. Sci. U.S.A.* **84**, 4974-4978.
Myers, R. M., Lumelsky, N., Lerman, L. S., and Maniatis, T. (1985). *Nature (London)* **313**, 495-498.
Myers, R. M., Sheffield, V. C., and Cox, D. R. (1989). *In* "PCR Technology. Principles and Applications for DNA Amplification" (H. A. Erlich, ed.), pp. 71-88. Stockton Press New York.
Nukiwa, T., Brantly, M., Graver, R., Paul, L., Courtney, M., LeCocq, J. P., and Crystal, R. G. (1986). *J. Clin. Invest.* **77**, 528-537.
Olsen, D. B., and Eckstein, F. (1989). *Nucleic Acids Res.* **17**, 9613-9620.
Orita, M., Iwahana, H., Kanazawa, H., Hayashi, K., and Sekiya, T. (1989). *Proc. Natl. Acad. Sci. U.S.A.* **86**, 2766-2770.
Rossiter, B. J., and Caskey, C. T. (1990). *J. Biol. Chem.* **256**, 12573-12756.
Saiki, R. K. (1989). *In* "PCR Technology. Principles and Applications for DNA Amplification" (H. A. Erlich, ed.), pp. 7-16. Stockton Press, New York.
Saiki, R. K., Bugawan, T. L., Horn, G. T., Mullis, K. B., and Erlich, H. A. (1986). *Nature (London)* **324**, 163-165.
Sheffield, V. C., Cox, D. R., Lerman, L. S., and Myers, R. M. (1989). *Proc. Natl. Acad. Sci. U.S.A.* **86**, 232-236.
Southern, E. M. (1975). *J. Mol. Biol.* **98**, 503-517.
Taylor, S. I., Cama, A., Accili, D., Barbetti, F., Imano, E., Kadowaki, H., and Kadowaki, T. (1991). *J. Clin. Endocrinol. Metab.* **73**, 1158-1163.

25

Techniques to Study Regulation of Gene Expression

Fredric E. Wondisford

I. RNA Analysis
II. Direct Measurement of Transcriptional Rates
III. Indirect Measurement of Transcriptional Rates
   A. Choice of Reporter Gene
   B. Choice of Mammalian Cell
   C. Choice of DNA Transfection Method
   D. Internal Controls for Transfection Efficiency
   E. Analysis of Reporter Gene Assays
   References

Understanding of endocrine regulator systems has been greatly aided by techniques to study gene expression. The following chapter outlines an orderly approach for the analysis of gene expression, with emphasis on transcriptional regulation. Hormonal regulation of the glycoprotein α and β subunit genes of thyrotropin (TSH) by thyroid hormone are described as illustrative of such an analysis. Central to any study of gene expression, though, is a thorough understanding of what factors physiologically regulate the gene of interest. In fact, the techniques described below help explain but should never supplant appropriate physiological data.

I. RNA Analysis

Northern blot and solution hybridization assays are two methods to measure steady-state mRNA levels in cells before and after treatment with any number of hormonal mediators. In most cases, total RNA preparations are adequate for these analyses. The exception is either when the mRNA transcript is of low abundance or when ribosomal RNA cross-hybridizes with the mRNA transcript to be detected. In these cases, analyses of the poly(A) fraction will be necessary. The best, although certainly not the fastest, methods for total RNA preparation use the strong protein denaturant, guanidine isothiocyanate (4 M final concentration), and fractionation of total RNA from genomic DNA on a cesium chloride gradient (Chirgwin *et al.*, 1979). Recently, a modification of this procedure has been described where a single phenol-chloroform extraction replaces the cesium chloride fractionation step (Chomczynski and Sacchi, 1987). This method is superior to the traditional cesium chloride method when small amounts of RNA need to be prepared from a large number of samples. Its disadvantage, though, is that total RNA preparations are more likely to be contained by genomic DNA.

Methods for Northern blot and solution hybridization assays have been described in detail elsewhere (Sambrook *et al.*, 1989). In general, Northern blot analysis is useful for a rapid assessment of mRNA changes in response to various hormonal mediators. Changes in steady-state mRNA levels represent a pretranslational level of regulation (transcription, RNA splicing, transport, stability) and should not be overly interpreted as indicating a transcriptional level of control on gene expression. For example, early studies (Croyle and Maurer, 1984; Gurr and Kourides, 1984; Chin *et al.*, 1985a) concerning the molecular mechanism of thyroid hormone regulation of TSH subunit genes demonstrated that thyroid hormone treatment of hypothyroid mice resulted in a significant reduction in TSH subunit mRNAs (Chin *et al.*, 1985b) (Fig. 1). These studies showed that thyroid hormone inhibited TSH subunit gene expression at a pretranslational level.

Solution hybridization of a uniformly ^{32}P-labeled *in vitro* synthesized RNA probe followed by RNase digestion (RNase protection assay) or hybridization of an end-labeled primer followed by reverse transcription (primer extension assay) is useful when precise quantitation of mRNA levels expressed from a certain gene are required. Moreover, these techniques are necessary, and many times are complementary, when specific mRNA transcripts derived from either alternative start-site utilization or RNA-splicing events are to be quantified. They have been described in detail elsewhere (Sambrook *et al.*, 1984).

II. Direct Measurement of Transcriptional Rates

After establishing that your gene of interest is regulated at a pretranslational level, the next step is to determine if a change in transcriptional rate is responsible for this pretranslational regulation. The most frequently employed method to measure transcriptional rates is the "run-off" assay. Nuclei are isolated from cells and synthesis of previously initiated RNA transcripts is

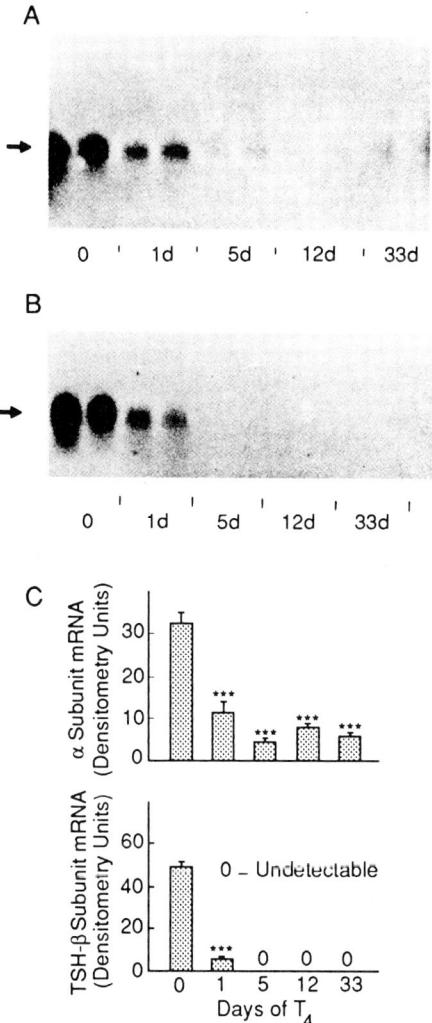

Figure 1 Blot hybridization analyses of total RNAs from thyrotropic tumors in mice treated with T_4 for varying periods of time (α and TSH-β mRNAs). (A and B) Mice bearing thyrotropic tumors were treated with T_4 for 0, 1, 5, 12, and 33 days (d). Animals were killed and total RNA was prepared from the thyrotropic tumors. Fifteen micrograms of total RNA was separated by agarose gel electrophoresis, and blot hybridization analyses were performed using the same blot and labeled α subunit cDNA (A) and TSH-β cDNA (B). Representative autoradiograms of hybridizing bands in A and B corresponding to mRNAs encoding the α and TSH-β of mouse TSH, respectively, are shown. For each time point, the levels of subunit mRNAs were analyzed in two animals and are represented by separate lanes. (C) Hybridizing bands shown in A and B were subjected to semiquantitative analyses by scanning densitometry, and values are expressed in arbitrary densitometric units. Values for four to six animals per group are given as the Mean ± SD.***, $P < 0.001$ vs 0.001 vs controls. (From Chin et al., 1985b, reproduced with permission.)

measured. This is done by radiolabeling RNA transcripts with [^{32}P]UTP. Radiolabeled transcripts are hybridized to membranes containing immobolized plasmid cDNA for the gene of interest. As positive controls in these experiments, plasmids containing cDNAs for other nonregulated "housekeeping" genes are immobilized; as a negative control, plasmid DNA without a cDNA insert is immobilized. After correction for radioactivity recovered from each transcription run-off reaction, hybridization efficiency, and nonspecific hybridization, transcription is expressed as parts per million. An analysis of common α and TSH-β gene transcription by Shupnik *et al.* (1985) revealed that thyroid hormone inhibition of steady-state mRNA levels from these genes was due to changes in gene transcription (Fig. 2). The fold reduction in mRNA levels due to thyroid hormone, in fact, correlated very closely with the fold reduction in mRNA synthesis for both subunits.

III. Indirect Measurement of Transcriptional Rates

The ability to isolate and characterize cis-acting elements responsible for gene regulation is greatly simplified by the use of reporter gene assays. The basic principle of this assay is that changes in rates of transcription can be indirectly measured by changes in the activity of a reporter gene. Various promoter-containing gene fragments are fused to a reporter gene, such as chloramphenicol acetyltransferase (CAT), β-galactosidase (β-GAL), human growth hormone (hGH), or luciferase (LUC) to form chimeric expression vectors. Expression vectors are transfected into mammalian cell cultures; after a defined incubation period, the cell cultures are harvested and reporter gene activity is quantitated. The following are basic considerations before undertaking these types of assays.

A. Choice of Reporter Gene

Histologically, reporter genes were chosen based on the ability of the investigator to measure easily the activity of a protein product of the reporter gene in transfected cell cultures. Each of the reporter genes listed above fulfill this criterion; but of reporter gene activity, LUC activity is clearly the easiest and fastest to obtain (Wet *et al.*, 1987). It is, of course, also important that reporter gene activity is not detected in nontransfected mammalian cells or the contribution of the chimeric expression vector to reporter gene activity will be unknown. CAT, β-GAL, and LUC are genes expressed normally in non-mammalian cells; hGH is only expressed in a subset of human pituitary cells. Finally, the LUC reporter should be employed whenever a weak eucaryotic

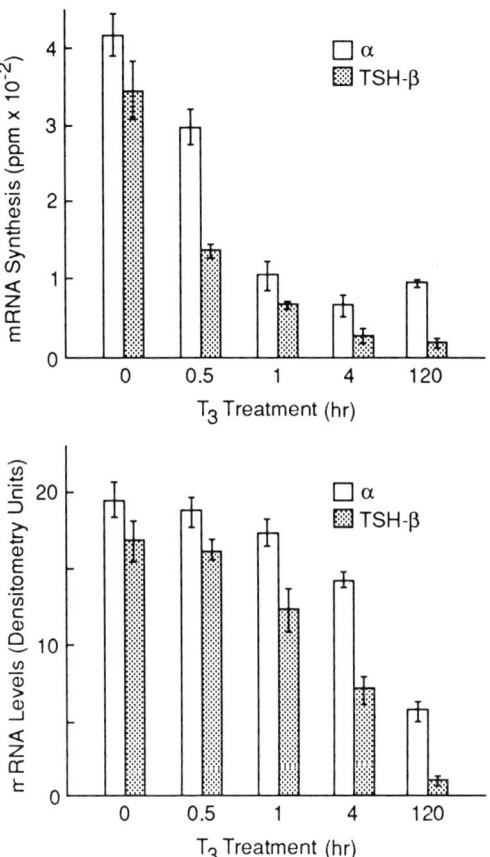

Figure 2 Time course of T_3 inhibition of α subunit and TSH-β mRNA synthesis. (Top) Quantitation of mRNA synthesis. Mice bearing TtT 97 tumors (2/group) were injected with T_3 daily for 5 days. Tumors were excised and nuclei prepared from aliquots of tissue at the indicated times. Isolated nuclei were allowed to synthesize RNA and TSH-β and α subunit mRNA synthesis were determined. Transcription reactions were performed in duplicate. [^{32}P]RNA input was $15-25 \times 10^6$ cpm/hybridization. The bar represents the Mean ± SE for 4 independent determinations/point. (Bottom) Quantitation of cellular α subunit and TSH-β mRNA levels. Total RNA was prepared from tumors and duplicate 10-μg samples were subjected to electrophoresis, diffusion blotting, and hybridization analysis. Densitometry was performed and values are presented as the Mean ± SE for 4 independent determinations/point. □, α; ▨, TSH-β. (From Shupnik et al., 1985, reproduced with permission.)

promoter is to be studied. The LUC assay is the most sensitive reporter gene assay; and unlike other reporter genes, background LUC activity in nontransfected cells is virtually absent.

B. Choice of Mammalian Cell

DNA transfection studies can be performed in most permanent and primary cell cultures. The choice of the cell line depends on the goal of the experiments. Permanent cell cultures are preferable to primary cell cultures for most transfection experiments because they are generally easier to transfect and tend to give more reproducible results. A permanent cell line expressing and displaying normal regulation of the gene of interest is an ideal choice when a model for *in vivo* gene expression is needed. If this type of cell line is not available, then primary cell cultures of the appropriate tissue may be utilized. A major consideration in using primary cell cultures is that this is a heterogenous cell population and this may alter the interpretation of results obtained with these cultures. If the goal of the experiment is to investigate expression of the fusion constructs in the absence of a particular transcription factor, a variety of permanent cell lines are available for this purpose.

C. Choice of DNA Transfection Method

A variety of methods have been described to transfect mammalian cells: calcium-phosphate precipitation, DEAE dextran, electroporation, and lipofection. I generally employ the calcium-phosphate precipitation method first when I am using a new cell line. The calcium-phosphate precipitation and DEAE dextran methods work well in most actively dividing cell cultures. This is because these cell cultures are more efficient at endocytosis of DNA complexes. At best, only 10–20% of cells are transfected within the culture. Lipofection and electroporation are preferable for slowly dividing permanent cell cultures and primary cell cultures. A greater percentage of cells is transfected with these methods but they are disadvantageous because of expense (lipofection) or cell death with the method (electroporation).

D. Internal Controls for Transfection Efficiency

Expression levels for the same vector can vary as much as 5-fold within an experiment and as much as 10-fold between experiments. This variation can be due to many factors such as cell density, cell growth, and transfection efficiency. Obviously, such variation can make it difficult to perform these types of experiments reproducibly. One way to control for these differences is to cotransfect a control expression vector containing a reporter gene (usually β-GAL or hGH) whose expression is controlled by a viral promoter (typically herpes thymidine kinase or Rous sarcoma virus). The rationale for cotransfection is that alterations in expression levels of the test plasmid that are due to the process of transfection itself should also change expression levels of the control plasmid to a similar degree. Thus, by coexpressing a control plasmid, differences in transfection efficiency can be identified, and reporter gene activity can be corrected for differences in transfection efficiency when the data are analyzed.

Protocols for each of the reporter assays have been described previously (Gorman et al., 1982; Wet et al., 1987; Brasier et al., 1989). Below are the current protocols for CAT and LUC assay that we employ in our laboratory. The CAT assay we utilize relies on thin-layer chromatography separation of acetylated and nonacetylated chloramphenicol. Modifications of the standard CAT assay protocol have been described (Sleigh, 1986) to simplify the assay. These modifications involve the use of either radiolabeled acetyl CoA instead of chloramphenicol and a simple extraction step or an enzyme-linked immunosorbant assay (ELISA) of acetylated chloramphenicol. These modified assays are certainly useful and probably preferable, for measurement of CAT activity from relatively strong eucaryotic promotors. When one is measuring CAT activity from weaker eucaryotic parameters, however, the relative sensitivity, cost, and linearity of these modified assays is a concern. Since we routinely measure CAT activity from a weak eucaryotic promoter (TSH-β), we continue to employ the standard CAT assay.

CELL CULTURE PROTOCOL

1. Cell cultures are divided onto either six-well (Costar 3406 or equivalent) or 100-mm plates (Costar 3100 or equivalent), depending on expected activity, on the first day.
2. On the morning of the second day, medium is removed and fresh medium is added (1 ml in each well or 8 ml on a 100-mm plate).
3. Cell cultures are transfected on the afternoon of the second day.
4. On the morning of the third day, medium is removed and the cell cultures are glycerol shocked with 1–2 ml of a solution containing 20% glycerol, 150 mM NaCl, and 20 mM Hepes, pH 7.4, for a defined period of time (1 to 3 min). Fresh medium is added to each culture (1 ml to each well and 8 ml to 100 mm plate) and the diluted glycerol solution is carefully removed (time is important here to

achieve reproducible results). Fresh medium is added with or without hormonal mediators at this time.

5. On the morning of the fourth or fifth day, cell cultures are harvested for either CAT or LUC assays. Medium is saved for hGH assay (Nichols Institute, Capistrano, CA) if this internal standard was used.

CAT ASSAY PROTOCOL (MODIFIED FROM GORMAN ET AL., 1982)

1. Cells are scraped in a buffer containing 150 mM NaCl and 40 mM Tris, pH 7.5 (two scrapings of 0.75 ml each).
2. Cells are collected by centrifugation (2000 rpm in a microcentrifuge for 3–5 min at room temperature).
3. The cell pellet is resuspended in 200–250 μl of lysis buffer (250 mM Tris, pH 7.6) and three cycles of freeze–thaw are performed.
4. The cell suspension is centrifuged (14,000 rpm for 2 min at room temperature), and the supernatant is recovered for CAT assay.
5. Typically 100 μl of extract is assayed but this depends on the total CAT activity. A total of 50 μl of extract is added to 50 μl of a mixture containing 1–4 μl of [^{14}C]chloramphenicol (CPA.754, Amersham Corporation, Arlington Heights, IL), 1–30 μl of 67 mM acetyl CoA (50 mg/ml lysis buffer, volume depends on assay time), and 16-48 μl of lysis buffer.
6. The reaction mixture is incubated from 1 to 16 hr at 37°C. To maintain linerity of the assay during long incubations (>3 hr), excess acetyl CoA must be added. The concentrations of acetyl CoA given above are sufficient for assays from 1 to 16 hr. Assay linearity should be established for any particular assay conditions.
7. The reaction is extracted with 1 ml of ethyl acetate after vigorous vortexing.
8. The upper phase of ethyl acetate is removed (0.95 ml) and dried under vacuum.
9. Residue is resuspended in 10 μl of ethyl acetate and spotted on thin-layer chromatography (TLC) plates.
10. Acetylated and nonacetylated forms are separated by ascending chromatography in a chloroform:methanol mixture (95:5).
11. Autoradiography is performed and the percentage of acetylated chloramphenicol is determined using liquid scintillation counting.

LUC ASSAY PROTOCOL (MODIFIED FROM BRASIER ET AL., 1989)

1. Cells are scraped in 250 µl of a lysis buffer containing 1% Triton X-100, 25 mM glycylglycine, pH 7.8, 15 mM $MgSO_4$, 4 mM EGTA, and 1 mM dithiothreitol (DTT; added just before use).
2. The cell lysate is collected and centrifuged at 14,000 rpm for 2 min at room temperature in a microcentrifuge.
3. Typically, 5 to 200 µl of cell lysate is assayed. All samples are brought to a volume of 200 µl with lysis buffer.
4. A Berthold Model 9501 luminometer is used with two injectors. The first injects 200 µl of assay buffer. Assay buffer is lysis buffer without Triton X-100 but with 2 mM ATP and 15 mM potassium phosphate. After a 10-sec delay, the reaction is started by the addition of 200 µl of 0.5 mM D-luciferin (in lysis buffer). After a 2-sec delay, light emission is measured for 20 sec. Concentration response experiments demonstrate that D-luciferin can be used at a concentration as low as 0.2 mM.

E. Analysis of Reporter Gene Assays

CAT or LUC activity is generally expressed relative to either total cellular protein or expression of an internal standard such as β-GAL or hGH. Some investigators report absolute CAT or LUC values by use of appropriate purified standards within a particular assay. Since most experiments are designed to measure relative expression among many test plasmids, this additional derivation of the data is generally unnecessary.

Indirect measures of transcription using reporter gene assay were performed to determine the cis-acting elements responsible for thyroid hormone inhibition of common α and TSH-β gene transcription (noted above). Various 5' flanking regions of either the common α or TSH-β gene were fused to either the CAT or the LUC reporter gene and transfected into mammalian cells (Carr et al., 1989; Darling et al., 1989; Wondisford et al., 1989; Wood et al., 1989). DNA transfection of cells containing sufficient numbers of nuclear thyroid hormone receptors resulted in thyroid hormone-mediated inhibition of reporter gene activity from either of these promoters. Moreover, by deleting or mutating DNA sequences within these promoters, cis-acting elements responsible for thyroid hormone were located near the "TATA box" in the common α gene and near or downstream of the transcription initiation site in the TSH-β gene. Thus starting from the observation that thyroid hormone reduced steady-state mRNA levels of both the common α and TSH-β gene, the molecular mechanisms responsible for these changes are beginning to be understood.

References

Brasier, A. R., Tate, J. E., and Haeener, J. F. (1989). *BioTechniques* **7,** 1116.
Carr, F. E., Burnside, J., and Chin, W. W. (1989). *Mol. Endocrinol.* **3,** 709.
Chin, W. W., Shupnik, M. A., Ross, D. S., Habener, J. F., and Ridgway, E. C. (1985a). *Endocrinology (Baltimore)* **116,** 873.
Chin, W. W., Muccini, J. A., and Shin, L. (1985b). *Biochem. Biophys. Res. Commun.* **128,** 1152.
Chirgwin, J. M., Przybyla, A. E., MacDonald, R. J., and Rutter, W. J. (1979). *Biochemistry* **18,** 5294.
Chomczynski, P., and Sacchi, N. (1987). *Anal. Biochem.* **162,** 156.
Croyle, M. L., and Maurer, R. A. (1984). *DNA* **3,** 231.
Darling, D. S., Burnside, J., and Chin, W. W. (1989). *Mol. Endocrinol.* **3,** 1359.
Gorman, C. M., Moffat, L. F., and Howard, B. H. (1982). *Mol. Cell. Biol.* **2,** 1044.
Gurr, J. A., and Kourides, I. A. (1984). *Endocrinology (Baltimore)* **115,** 830.
Sambrook, J., Fritsch, E. F., and Maniatis, T. (1989). "Molecular Cloning: A Laboratory Manual." Cold Spring Harbor Lab. Press, Cold Spring Harbor, New York.
Shupnik, M. A., Chin, W. W., Habener, J. F., and Ridgway, E. C. (1985). *J. Biol. Chem.* **260,** 2900.
Sleigh, M. J. (1986). *Anal. Biochem.* **156,** 251.
Wet, J. R., Wood, K. V., DeLuca, M., Helinski, D. R., and Subramani, S. (1987). *Mol. Cell. Biol.* **7,** 725.
Wondisford, F. F., Far, E. A., Radovick, S., Steinfelder, H. J., Moates, J. M., McClaskey, J. H., and Weintraub, B. D. (1989). *J. Biol. Chem.* **254,** 14601.
Wood, W. M., Koa, M. Y., Gordon, D. F., and Ridgway, E. C. (1989). *J. Biol. Chem.* **264,** 14840.

26

Transgenic Animals as a Tool in Endocrinology: Structure/Function Studies of Peptide Hormones Employing Transgenic Mice

John J. Kopchick and Wen Y. Chen

I. Introduction
II. Plasmid Construction and Mutagenesis
III. Transgenic Mouse Production
  A. DNA Preparation
  B. Microinjection and Transgenic Mice Production
  C. Transgenic Mice Detection
  D. Transgenic Mice Used in This Study
IV. IGF-1 Radioimmunoassay
V. Immunoblotting Analysis
VI. Radioreceptor-Binding Assay
VII. Comments
References

I. Introduction

The ability to incorporate exogenous DNA into the germ line of animals and to subsequently develop lines of animals which retain this DNA loosely defines the term "transgenic animal." Recombinant DNA has been inserted into the germ line of many animals including mice, sheep, rabbits, pigs, chickens, and fish (Hammer *et al.*, 1985; Bosselman *et al.*, 1989; Chen and Powers, 1990). Although a variety of technologies have been employed to generate these animals, by far the most commonly used technology has been direct microinjection of the foreign DNA into the pronucleus of fertilized mammalian eggs. This work was developed in the late 1970s and early 1980s (Brinster *et al.*, 1981; Wagner *et al.*, 1981) and is used routinely in many laboratories throughout the world. Other methods employed for the production of transgenic animals include the use of retroviral vectors, embryonic stem cell technology, particle "gun"-mediated gene transfer, and electroporation.

A variety of genes which impact the endocrine status of the organism have been introduced into the germ line of animals and lines of these animals have been generated. Examples of these genes include growth hormone (GH), insulin, insulin-like growth factor-I (IGF-I), a variety of receptor genes, and others (Chao *et al.*, 1985; Magram *et al.*, 1985; Selden *et al.*, 1986; Palmiter *et al.*, 1987; Tremblay *et al.*, 1988; Mathews *et al.*, 1988; Berg *et al.*, 1988). Usually, the gene of choice is injected into the fertilized egg in the form of a fusion gene; that is, a transcriptional regulatory element (TRE) of a different gene is attached or ligated to the protein-coding region of the gene to be studied. Selection of the TRE depends on the type of study to be performed. For example, one may wish to express the gene at relatively high levels in many tissues of the fetus and in the adult animal. A TRE of this type is derived from the mouse metallothionine I gene (McGrane *et al.*, 1991). Alternatively, one may design the experiment such that the gene should be expressed only after birth. In this case the phosphoenol pyruvate carboxy kinase (PEPCK) TRE has been successfully employed (McGrane *et al.*, 1991). Additionally, one may want to express the gene in a tissue-specific manner, for example in the B-acinar cells of the pancreas. In this case the mouse elastase TRE has been used (Berg *et al.*, 1988). Thus, transgene expression as a function of animal development or tissue specificity can vary depending on the TREs utilized for a given gene.

Among the first transgenic animals developed were those which express foreign GH genes (Palmiter *et al.*, 1982; McGrane *et al.*, 1988). In these first studies either the human (h) or rat (r) GH genes were ligated to the metallothionein-I (MT-I) TRE and mice were generated which had incorporated this DNA into their germ cells. These animals could pass the foreign DNA to offspring. When the GH gene was expressed and the protein secreted into the serum of the animals, a large mouse phenotype was produced. Importantly, the enhanced growth phenotype of these mice was not related to gene copy number or the serum levels of GH (Palmiter *et al.*, 1982; McGrane *et al.*, 1988). These exciting results have been reproduced in many laboratories including ours (Kopchick *et al.*, 1990; Chen *et al.*, 1990, 1991a,b,c).

In the remainder of this report we discuss the use of GH transgenic mice as reagents for the study of a variety of growth hormone-related endocrine parameters *in vivo*. The studies detail the methodologies employed to produce mice which express the bovine (b) GH gene or mutated bGH genes and to determine the effect of bGH expression on a variety of endocrine parameters of the animals. The effects of bGH expression in transgenic mice on the animal's growth phenotype, GH receptor levels, IGF-I levels, and pituitary GH are addressed as they relate to the GH cascade. Although we use the growth phenotype and endocrine parameters of transgenic mice as an assay for studies of the biological activities of GH analogs, these general protocols could be applied to most peptide hormones.

II. Plasmid Construction and Mutagenesis

As an example of the use of transgenic animals for endocrine studies, we have evaluated a variety of mutated bGH genes for biological activity when expressed in transgenic mice. The GH gene encodes a protein of 191 amino acids which consists of four α-helices. In this review we focus on the third α-helix (amino acids 109–126) which is encoded between the Tth111I and XmaI restriction endonuclease cleavage sites (Fig. 1). Plasmid, pBGH-10Δ6, is a pBR322-based vector which contains deletions of 1922 bp from the NdeI to BamHI sites in pBR322 and deletions of introns B, C, and D of bGH gene (Fig. 1) was used in these studies. bGH gene mutagenesis is carried out

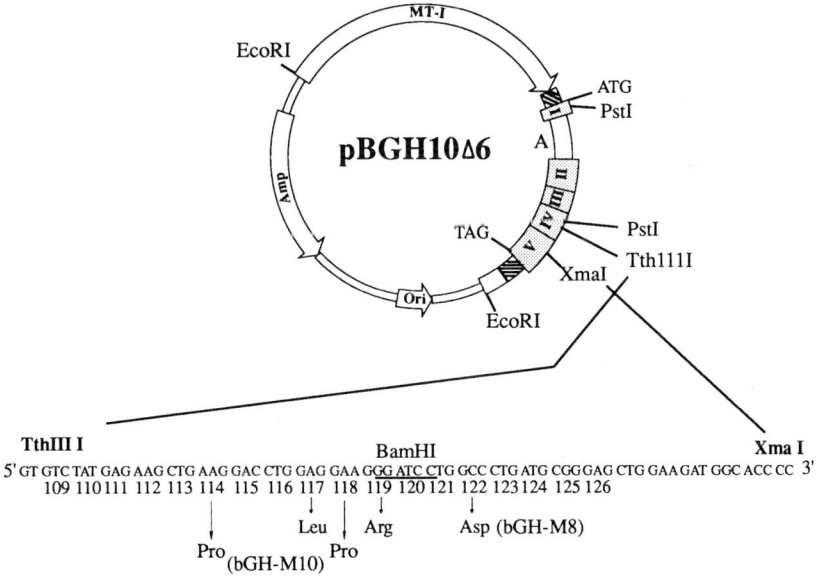

Figure 1 Plasmid, pBGH-10Δ6, is used as the parental vector for site-directed mutagenesis studies. It contains mouse metallothionein-I transcriptional regulatory sequences (MT-I) fused to a modified bGH gene which contains five exons (shaded boxes I–V) and intron A. The pBR322 origin of replication (Ori) as well as the ampicillin-resistant gene (Amp) are indicated. The nucleotide sequence between restriction sites Tth111I and XmaI is shown. A BamHI site is inserted into the plasmid. The changes which generated bGH-M8 and bGH-M10 are also shown.

by oligonucleotide-directed mutagenesis using complementary oligonucleotides to replace the DNA fragment between the *Tth*111I site (found near the 3' end of exon IV) and the *Xma*I site (located near the 5' end of exon V, Fig. 1). In addition to designing mutations in the synthetic oligonucleotides, a silent base-pair change is incorporated to create a unique *Bam*H1 restriction site which simplified plasmid screening procedures (Fig. 1). Other sites could also be introduced, however, one must be sure that the amino acid sequence is not altered.

The nucleotide sequences of the mutated bGH target regions are determined by using the dideoxy chain-termination method with modified T7 DNA polymerase (Sequenase, United States Biochemical; Sanger *et al.*, 1977). An oligonucleotide primer for manual DNA sequencing used in these studies is an 18-mer (5'AAATTTGTCATAGGTCTG3') which is located downstream of the *Xma*I site. This general mutagenesis technique could be used for structure/function studies of any cloned peptide hormone gene.

III. Transgenic Mouse Production

A. DNA Preparation

In order to evaluate mutated gene expression in transgenic mice, one must first design and generate the genes. Following mutagenesis the DNA must be purified prior to microinjection. The DNA used for the production of the different strains of mice included a pBGH-10Δ6-based plasmid which encodes wild-type bGH; pBGH-M8 which encodes the bGH analog with amino acid substitutions at position 117, 119, 122 (E117L, G119R, A122D); and pBGH-M10 which encodes a bGH analog containing proline substitution mutations at positions 114 and 118 (K114P, E118P). The construction of these plasmids has been described (Chen *et al.*, 1990, 1991a). Briefly, plasmids are digested with restriction enzyme *Eco*RI and the linear fragments separated by 1% agarose gel electrophoresis in TAE buffer (40 mM Tris-acetate, 1 mM EDTA, pH 8.0). Following electrophoresis the corresponding DNA fragments to be used for microinjection are electroeluted from the gel slice in a dialysis tubing (Bethesda Research Laboratories, Gaithersburg, MD). After butanol (Kodak) extraction of the ethidium bromide (EtBr) bound to the DNA, the DNA fragments are precipitated in ethanol. The DNA pellets are then dissolved in 1.5 ml of low-salt solution (200 mM NaCl, 20 mM Tris, 1 mM EDTA, pH 7.2). Elu-Tip columns and filters are used to recover the DNA. The final DNA solutions is extracted twice with phenol:chloroform:isoamyl alcohol (25:24:1) and twice with chloroform:isoamyl alcohol (24:1) and then precipitated by ethanol. The DNA pellets are washed twice with 70% ethanol and dried briefly. Microinjection buffer (10 mM Tris, 0.25 mM EDTA, pH 7.5) is used to dissolve the DNA pellets and the concentration of the DNA is determined.

B. Microinjection and Transgenic Mice Production

The procedure for production of transgenic mice by direct microinjection of DNA into the male pronucleus of fertilized mouse eggs has been described (Wagner et al., 1981). Briefly, fertilized eggs at the pronuclear stage are collected from the oviducts of C57BL/6JxSJL hybrid mice. A small drop of culture medium with the fertilized eggs is placed on a microscope slide and covered with paraffin oil. About 10 pl of purified DNA solution (the final concentration of DNA for microinjection is 2–4 μg/ml) is drawn into the injection pipette which is then moved to the drop containing the eggs. A fertilized egg is positioned onto the holding pipette for subsequent injection of DNA solution into male pronucleus. After microinjection, the eggs are removed from the drop of medium, transplanted into the uteri of foster mothers, and carried to term (Wagner et al., 1981).

C. Transgenic Mice Detection

Following birth of the mice, those which carry out the "transgene" are identified by hybridization analysis. Following weaning (approximately 30 days of age), genomic DNA is extracted from a segment of mouse tails. Briefly, 200–400 mg of tail tissue is digested in 0.75 ml of SSTE (1% SDS, 100 mM NaCl, 50 mM Tris, 15 mM EDTA, pH 8.0) containing 0.25 mg/ml proteinase K and 0.125 mg/ml RNase A (Sigma) at 65°C for 12–16 hr. The solutions are extracted once with phenol:chloroform:isoamyl alcohol (25:24:1) and once with chloroform:isoamyl alcohol (24:1). The genomic DNA is then precipitated in 0.1 vol of 3 M NaOAc and an equal volume of isopropanol. After washing with 70% ethanol and brief drying, the DNA pellets are dissolved in 0.5 ml TE (10 mM Tris, 1 mM EDTA, pH 8.0). Hybridization analyses are carried out using Minifold II Slot-Blotter System (Schleicher and Schuell, Keene, NH). Typically, 10 μg of genomic DNA is denatured in 0.2 N NaOH for 1 hr at 65°C and neutralized by addition of equal volume of 2 M NH$_4$OAc, pH 7.4. The samples are then applied to a nitrocellulose membrane (0.2 μm; Schleicher and Schuell, Keene, NH) according to Manifold II instruction. After being baked for 1–2 hr at 80°C in vacuum oven, the nitrocellulose membranes are hybridized to a ^{32}P-labeled bGH-specific probe (PstI–PstI) by standard procedures described elsewhere in this volume. A typical slot blot

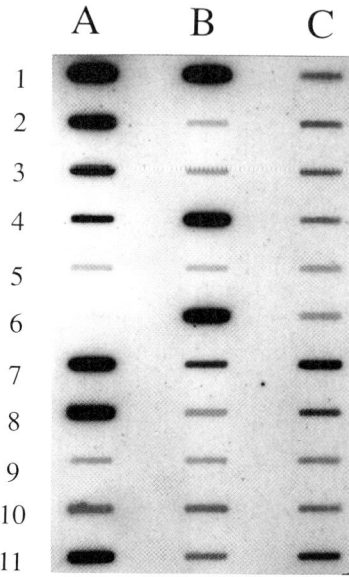

Figure 2 A typical example of slot blot hybridization analysis. (Lane A) No. 5 represents the DNA isolated from nontransgenic mouse, Nos. 1 to 4 represent positive controls prepared with plasmid, pBGH-10Δ6 corresponding to 10, 5, 2, 1 gene copies/cell, respectively. In this case, lane A, Nos. 7, 8, and 11, and lane B, Nos. 1, 4, and 6 represent transgenic mice with approximately 5–10 copies of bGH DNA/cell whereas lane B, No. 7 and lane C, Nos. 7, 8, and 11 represent transgenic mice containing approximately 1 to 2 copies of bGH DNA/cell.

analysis is shown in Fig. 2. Some investigators select a "Southern" hybridization technique for transgenic animal identification; however, we feel that the slot blot procedure takes less time and more animals can be evaluated. Once a transgenic mouse line is established, Southern analysis is used to identify the site of DNA integration.

D. Transgenic Mice Used in This Study

Depending on the endocrine-related gene which is inserted into the mice, one or a few lines of the mice may be established. For the endocrine studies described here, the third generation of dwarf transgenic mice (DTM) from bGH-M8 founder No. 55 and the second generation of wild-type bGH and bGH-M10 transgenic mice are used (Chen *et al.*, 1990, 1991a) (Fig. 3). Each mouse contains one or two copies of transgene with serum bGH levels at approximately 2 μg/ml as determined by immunoblotting analyses (Chen *et*

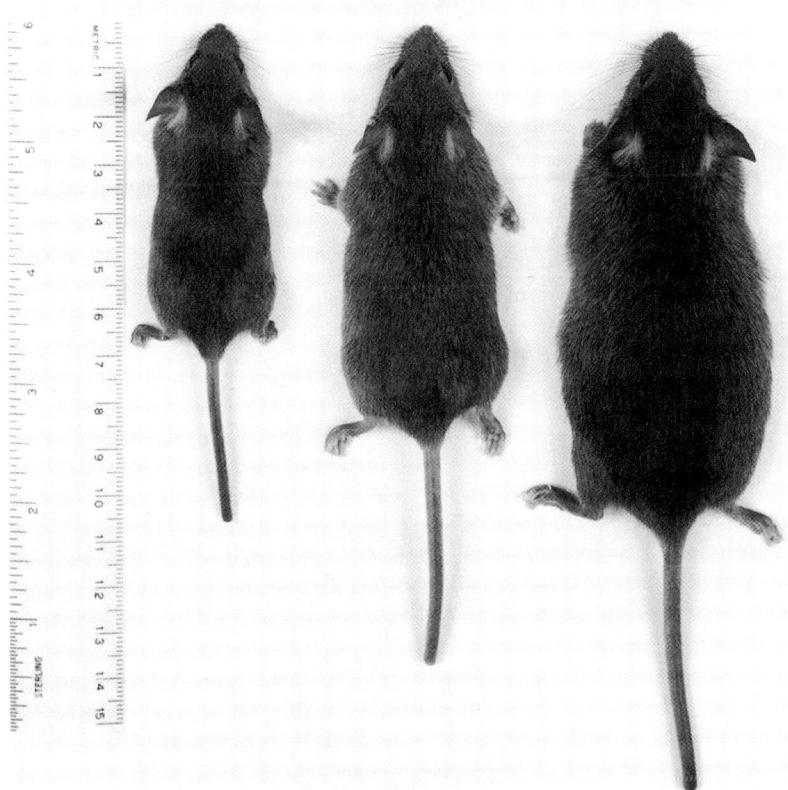

Figure 3 Representative transgenic mice (2-month-old males) which express different bGH analogs. From left: bGH-M8-TG; bGH-M10-TG, which exhibits a phenotype similar to its nontransgenic littermates; and bGH-TG. The mass of the animals is approximately 20, 28, and 42 g, respectively.

al., 1991b). At 2 months of age, the mean growth ratio (mean body weight of transgenic mice per mean body weight of corresponding nontransgenic littermates) for wild-type bGH transgenic mice is approximately 1.7. The value is typical of transgenic mice which express GH genes (Palmiter *et al.*, 1982; McGrane *et al.*, 1988). The mean growth ratio for bGH-M10 mice is 1.0 while the mean growth ratio for DTM is approximately 0.7. There are significant differences ($P < 0.01$) between the body weights of wild-type and bGH-M8 transgenic mice and their nontransgenic littermates as determined by the Student *t* test at all time points. There is no difference between bGH-M10 and their nontransgenic littermates (Chen *et al.*, 1990, 1991a) (Fig. 3).

IV. IGF-I Radioimmunoassay

Since the transgenic mice which expressed wild-type or mutated bGH gene possessed different growth phenotypes, we set out to investigate a variety of GH-related endocrine parameters in these mice. These parameters will certainly change depending upon the endocrine-related gene used in a particular study. The first relationship to be established in this study is that of the growth phenotype of transgenic mice as it related to the total serum IGF-I levels. Serum IGF-I levels in transgenic mice are determined using a heterologous RIA kit (Nichols Institute Diagnostics, San Juan Capistrano, CA) following acid ethanol extraction according to the manufacture's directions. The antibody used in this assay exhibits approximately 2% cross-reactivity with IGF-II (Chen et al., 1991b). A human IGF-I is used as a standard.

Animals are sacrificed and sera from transgenic mice and nontransgenic littermates are collected and analyzed for IGF-I. Serum IGF-I levels from nontransgenic mice are 286 ± 44 ng/ml while serum IGF-I levels from DTM are 119 ± 38 ng/ml which is less than 1/2 the levels of control nontransgenic littermates. Serum IGF-I levels from bGH-M10 mice are 280 ± 53 ng/ml. In contrast, transgenic mice which express the wild-type bGH gene possess IGF-I levels of 501 ± 89 ng/ml, approximately twice that of control animals (Chen et al., 1991b). We examined the serum IGF-I level from both sexes in all three groups of mice. Since no difference is found between sexes within each group, the data are pooled. The results indicate that the IGF-I levels are directly correlated with the growth phenotype (size) of the transgenic mice (Chen et al., 1991b).

V. Immunoblotting Analysis

Expression of elevated levels of bGH or bGH analogs in transgenic mice may affect pituitary levels of GH in these mice. Based on the endocrine feedback loops, one would expected to find lower levels of GH in the pituitaries of animals possessing elevated levels of IGF-I. To examine the pituitary content of GH in the animals, pituitaries are obtained from wild-type bGH, bGH-M8, and bGH-M10 transgenic mice and their nontransgenic littermates of the same age and sex (Chen et al., 1991b). The pituitaries are homogenized in 100 μl of homogenization solution (0.5% Nonidet P-40, 10 mM Tris–HCl, 1 mM EDTA, pH 8.0). The homogenates are briefly centrifuged and the protein concentration of the cytosol is determined by a standard protein assay (Bio-Rad). A total of 100 μg of cytosolic protein (100 μg) is mixed with 3× sample buffer (150 mM Tris, pH 6.8, 10% SDS, 2 M β-mecaptolethanol, 10% glycerol, 0.4% bromophenol blue) and heated at 100°C for

3 min before separation by 12.5% SDS-PAGE. Following electrophoresis, proteins are transferred to nitrocellulose membranes at 100 V constant voltage for 1.5 hr.

Immunoblot detection of bGH is performed by a modified procedure of Viceps-Madore *et al.* (1983). Following protein transfer, the nitrocellulose paper is blocked with 1% gelatin in TBS (50 mM Tris, 0.5 M NaCl, pH 7.5) with gentle agitation for 0.5 hr at room temperature and then washed three times with 0.05% Tween 20 in TBS (5 min/wash). Polyclonal rabbit anti-bGH (1:100 dilution) in 1% gelatin/TBS is added to the nitrocellulose membrane and incubated 12-16 hr at room temperature with gentle agitation. After removing the primary antibody, the nitrocellulose paper is washed three times with 0.05% Tween 20 in TBS and subsequently incubated for 2 hr at room temperature in the presence of a goat anti-rabbit IgG horseradish peroxidase (HRP) conjugate (Boehringer-Mannheim Biochemicals) in 1% gelatin/TBS. Following incubation with secondary antibody, the nitrocellulose membrane is washed three times with 0.05% Tween 20 in TBS.

To visualize the protein bands, the nitrocellulose membrane is incubated for 10 min in a mixture of 50 ml of 0.018% H_2O_2 (v/v) in TBS and 10 ml of methanol containing 30 mg of HRP color development reagent (Bio-Rad). The nitrocellulose membrane is then rinsed with water, air-dried, and photographed. Purified bGH (a gift from Upjohn Co., Kalamazoo, MI) is used to quantify bGH by photographic and densitometric methods (Fernandez and Kopchick, 1990). The above protocol could be used for any peptide hormone for which an antibody is available.

Relative concentrations of GH in the pituitary are determined by spectrophotometric analysis (Fernandez and Kopchick, 1990) using values from control animals as 100%. In wild-type bGH transgenic mice, pituitary GH

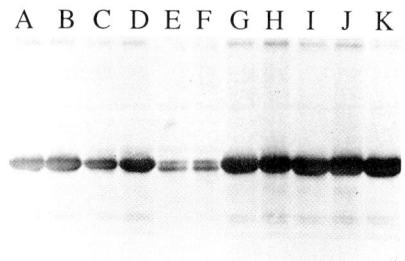

Figure 4 Immunoblot analysis of pituitary gland homogenates from different transgenic mice and their nontransgenic littermates. Each lane represents 100 µg of protein derived from a pituitary of an individual animal. (Lanes A and B) Samples from nontransgenic mice. (Lanes C and D) Samples from bGH-M10 transgenic mice. (Lanes E and F) Samples from bGH transgenic mice. (Lanes G-K) Samples from bGH-M8 transgenic mice, respectively.

levels are found to be decreased by 51 ± 18%. On the other hand, pituitary GH levels in DTM are elevated to 133 ± 16% (Fig. 4). Pituitary GH levels from bGH-M10 transgenic mice are similar to those found in nontransgenic littermates. Thus the levels of GH found in the pituitary varied inversely with the size of the animal and the total serum IGF-I levels. In summary, the large animals (hGH-TG) have elevated levels of IGF-I and decreased levels of pituitary GH. Conversely, the small animals (DTM) possess lower levels of IGF-I and higher levels of pituitary GH.

VI. Radioreceptor-Binding Assay

Increased levels of circulating GH has been shown to upregulate liver GH receptors in mice which expresses GH genes (Herington et al., 1976; Baxter et al., 1982; Posner et al., 1980) as well as in animals which are injected with GH (Baxter and Zaltsman, 1984). As an example of the type of experiments performed on transgenic mice in order to assay receptor regulation, the levels of liver GH receptor in the transgenic animals described above were investigated. The method for preparation of mouse liver microsomal membrane has been described by Smith and Talamants (1987) and is used with minor modifications. Male DTM (bGH-M8-TG, $n = 8$), wild-type bGH (bGH-TG, $n = 4$), and bGH-M10 (bGH-M10-TG, $n = 3$) transgenic mice, and their nontransgenic littermates ($n = 12$), 60–120 days of age, are selected for liver receptor-binding studies. The reason that male mice are used is to decrease the binding of ^{125}I-hGH to lactogenic receptors found in female liver preparations. Liver tissue is homogenized with a Brinkman Polytron in 4 vol (w/v) of 0.3 M sucrose, 10 mM EDTA, 50 mM Hepes, 0.1 mM TPCK, and 1 mM PMSF at pH 8.0. The above step and all the following protocols are carried out at 4°C. The homogenate is centrifuged at 20,000g for 30 min and the supernatant is centrifuged at 100,000g for 1 hr. The pellets are washed once with 10 mM Hepes, pH 8.0, and recentrifuged. The final pellets are suspended in 10 mM Hepes buffer at a concentration of 25 to 50 $\mu g/\mu l$. The suspension is then used for receptor-binding studies.

Competitive binding assays are performed using the following protocol:

1. Microsomal membranes from transgenic or nontransgenic littermates corresponding to 1 mg protein are incubated with ^{125}I-hGH (0.5 ng/ml, sp act = 115 μCi/μg) and with various amount of unlabeled bGH ranging from 1 ng to 1 μg in a total volume of 0.3 ml assay buffer (20 mM Hepes, 10 mM CaCl$_2$, 0.1% BSA, and 0.05% NaN3, pH 8.0).

2. After a 3-hr incubation at room temperature, the reaction is stopped by addition of 1 ml of ice-cold assay buffer followed by centrifugation at 10,000g for 15 min.
3. Membrane pellets then are assayed for radioactivity.

All assays are performed either in duplicate or in triplicate and repeated three to four times.

There is no difference between mean dissociation constants (K_d) of liver membrane preparations from all experimental groups (Chen et al., 1991b). However, the maximal displacement of ^{125}I-hGH from liver membrane preparations for all three types of transgenic mice is greatly increased as compared to those for their nontransgenic littermates. Surprisingly, the levels of GH receptors (total specific binding) in all of the transgenic mice are upregulated independent of the size of the mice.

VII. Comments

One can use a structure/function approach in an attempt to understand the mechanism of GH action or the action of peptide hormones in general. In our laboratory the GH gene has been used which directs expression of biologically active GH. The structure of the protein has been selectively altered by mutating corresponding regions of the gene. The biological function of the gene product has been evaluated by determining the effect of gene expression on various growth parameters in transgenic mice. Overexpression of wild-type bGH gene in transgenic mice results in animals with an enhanced growth phenotype independent of the serum levels of bGH. Animals which express a bGH analog (bGH-M10) do not show an enhanced growth phenotype. This bGH analog has been altered such that the structure of the third α helix of the molecule has been destabilized (Chen et al., 1991a). Also, animals which express the bGH analog (bGH-M8) do not show an enhanced growth phenotype when the serum levels of the analog are relatively low. However, elevated serum levels of the analog (approximately greater than 1 μg/ml) resulted in dwarf transgenic mice (Chen et al., 1990). This observation combined with the fact that the analog binds to GH receptors with an affinity similar to wild-type bGH (Chen et al., 1990) have led us to hypothesize that the GH (ligand) must recognize a second target in order to initiate the signal transduction system for GH activity. Therefore, the combination of GH gene mutagenesis studies along with the ability to establish the biological activity of the GH molecules in transgenic mice has given us an opportunity to elucidate the mechanism of GH action. Furthermore, the ability to generate transgenic mice or other transgenic animals which express peptide hormone genes in a tissue-specific manner or at a specific time during animal development provides a valuable assay for elucidating the biological activity of the respective hormones.

Acknowledgments

We thank Tom Wagner, Jun Yune, and David Wright for their invaluable assistant in the production of transgenic mice. J.J.K. is supported by the State of Ohio's Eminent Scholar Program which includes a grant from Milton and Lawrence Goll.

References

Baxter, R. C., and Zaltsman, Z. (1984). *Endocrinology (Baltimore)* **115**, 2009–2014.
Baxter, R. C., Zaltsman, Z., and Turtle, J. R. (1982). *Endocrinology (Baltimore)* **111**, 1020–1032.
Berg, L. J., Groth, B. F., Ivars, F., Goodnow, C. C., Gilfillan, S., Garchon, H., and Davis, M. (1988). *Mol. Cell. Biol.* **8**, 5459–5469.
Bosselman, R. A., Hsu, R. Y., Boggs, T., Hu, S., Bruszewski, J., Ou, S., Kosar, L., Martin, F., Green, C., Jacobson, F., Nicolson, M., Schultz, J. A., Semon, K. M., Rishell, W., and Stewart, R. G. (1989). *Science* **243**, 533–535.
Brinster, R. L., Chen, H. Y., and Trumbauer, M. (1981). *Cell* **27**, 223–231.
Chao, M. V., Bothwell, M. A., Ross, A. H., Koprowski, H., Lanahan, A. A., Buck, C. R., and Sehgal, A. (1985). *Science* **232**, 518–521.
Chen, T. T., and Powers, D. A. (1990). *Tib Tech.* **8**, 209–215.
Chen, W. Y., Wight, D. C., Wagner, T. E., and Kopchick, J. J. (1990). *Proc. Natl. Acad. Sci. U.S.A.* **87**, 5061–5065.
Chen, W. Y., Wight, D. C., Chen, N. Y., Colman, T. C., Wagner, T. E., and Kopchick, J. J. (1991a). *J. Biol. Chem.* **266**, 2252–2258.
Chen, W. Y., White, M. E., Wagner, T. E., and Kopchick, J. J. (1991b). *Endocrinology (Baltimore)* **129**, 1402–1407.
Chen, W. Y., Wight, D. C., Wagner, T. E., and Kopchick, J. J. (1991c). *Mol. Endocrinol.* **5**, 1845–1851.
Fernandez, E., and Kopchick, J. J. (1990). *Anal. Biochem.* **191**, 268–271.
Hammer, R. E., Pursel, V. G., Rexroad, C. E., Jr., Wall, R. J., Bolt, D. J., Ebert, K. M., Palmiter, R. D., and Brinster, R. L. (1985). *Nature (London)* **315**, 680–683.
Herington, A. C., Phillips, L. S., and Daughaday, W. H. (1976). *Metab. Clin. Exp.* **25**, 341–353.
Kopchick, J. J., McAndrew, S. J., Shafer, A., Blue, W. T., Yun, J. S., Wagner, T. E., and Chen, W. Y. (1990). *J. Reprod. Fertil., Suppl.* No. 41, 25–35.
Magram, J., Chada, K., and Costantini, F. (1985). *Nature (London)* **315**, 338–340.
Mathews, L. S., Hammer, R. E., Brinster, R. L., and Palmiter, R. D. (1988). *Endocrinology (Baltimore)* **123**, 433–437.
McGrane, M. M., deVente, J., Yun, J., Bloom, J., Park, E., Wynshaw-Boris, A., Wagner, T. E., Rottman, F. M., and Hanson, R. W. (1988). *J. Biol. Chem.* **263**, 11443–11451.
McGrane, M. M., Yun, J. S., Moorman, A. F. M., Lamers, W. H., Hendrick, G. K., Arafah, B. M., Park, E. A., Wagner, T. E., and Hanson, R. W. (1991). *J. Biol. Chem.* **265**, 22371–22379.
Palmiter, R. D., Brinster, R. L., Hammer, R. E., Taumbauer, M. E., Rosenfeld, M. G., Birnberg, N. C., and Evans, R. M. (1982). *Nature (London)* **300**, 611–615.

Palmiter, R. D., Behringer, R. R., Quaife, C. J., Maxwell, F., Maxwell, L. H., and Brinster, R. L. (1987). *Cell* **50**, 435–443.

Posner, B. I., Pate, B., Vezinhet, A., and Charrier, J. (1980). *Endocrinology (Baltimore)* **107**, 1954–1960.

Sanger, F., Niklen, S., and Coulson, A. R. (1977). *Proc. Natl. Acad. Sci. U.S.A.* **74**, 5463–5467.

Selden, R. F., Skoskiewicz, M. J., Howie, K. B., Russell, P. S., and Goodman, H. M. (1986). *Nature (London)* **321**, 525–528.

Smith, W. C., and Talamants, F. (1987). *J. Biol. Chem.* **262**, 2213–2219.

Tremblay, Y., Tretjakoff, I., Peterson, A., Antakly, T., Zhang, C. X., and Drouin, J. (1988). *Proc. Natl. Acad. Sci. U.S.A.* **85**, 8890–8894.

Viceps-Madore, D., Cidlowski, J. A., Kittler, J. M., and Thanass, J. W. (1983). *J. Biol. Chem.* **258**, 2689–2696.

Wagner, T. E., Hoppe, P. C., Jollick, J. D., Scholl, D. R., Hodinka, R. L., and Gault, J. B. (1981). *Proc. Natl. Acad. Sci. U.S.A.* **78**, 6376–6380.

27

The Use of Nonmammalian Animal Models in Endocrine Investigation

Ian P. Callard, Lynn Riddiford, and Colin G. Scanes

I. Introduction
II. Vertebrate Models for Hypothalamic Pituitary Interaction and Behavior
   A. Avian Models for Seasonal Breeding/Circadian Functioning
   B. The Teleost Brain: Model for Regulation of Neural Aromatase
   C. Direct Neural Regulation of the Pituitary in Teleost Fish
   D. Hormones and Aggressive Behavior in Quail
III. Vertebrate Reproduction
   A. Reptilian Models for the Regulation of Sex-Determining Genes
   B. Avian and Reptilian Ovaries: Models for Follicular Hierarchies, Follicular Development/Atresia, and Other Aspects of Ovarian Function
   C. The Shark Testis Model for the Study of Stage-Dependent Functions and the Regulation of Spermatogenesis
   D. The Elasmobranch Shell Gland: Model for Steroid Hormone Action
   E. Reptilian Models for the Study of Requirements of Long-Term Sperm Storage
IV. Vertebrate Hormones and Metabolism
   A. Avian and Reptilian Models for the Study of Apolipoproteins and Their Role in Lipid Transport, Metabolism, and Uptake
   B. Chicken Adipose Tissue as a Model for Growth Hormone Action
   C. The Chicken as a Model for Pancreatic Function
   D. Pigeon Crop Sac Gland: A Bioassay for Prolactin and a Model for the Synlactin Hypothesis of Prolactin Action
   E. Atrial Natriuretic Peptides: Fish Models for Biological Actions
V. Invertebrate Models
   A. Models for Neuropeptides Modulating Behavior
   B. Models for Pheromones and Chemoreception
   C. Models for Circadian Rhythms and Photoperiodic Control of Neurohormonal Output
   D. Models for Steroid and Genome Interaction
   E. Models for Hormonal Regulation of Sequential Cell Polymorphism
   F. Models for Programmed Cell Death
   References

I. Introduction

The principle upon which the search for and use of animal models is based was stated by August Krogh, Nobel laureate and Danish physiologist, in a lecture to the 13th International Physiological Congress in Boston (Krogh, 1929): "For a large number of problems there will be some animal of choice, or a few such animals, on which it can be most conveniently studied." This approach has been referred to as the "August Krogh principle" (Krebs, 1975). Most frequently, the identification of model species has come as a natural outcome of the comparative approach, which seeks to establish similarities between organisms and has led to the concept of "connectivity." A high-connectivity model is one in which the body of knowledge is large and has resulted in extensive cross-connection with other species within the "matrix of biological knowledge" (see "Models for Biomedical Research, 1985). Well-known model organisms which are good examples include *Drosophila, Caenorhabditis, Aplysia, Manduca* (insects), *Xenopus* (frog), and *Gallus* (chicken). The concept of connectivity is related to the better known concept of homology which provides the logical rationale by which model organisms may be related through shared evolved characteristics. The connectivity between model organisms extends through phylogenetic time and hence to the human being, thus rationalizing both the models and comparative approach by both potential relevance and basic research. While the concepts of connectivity and the matrix of biological knowledge and their dependence upon homology were recently codified in the NRC publication, these have always been the guiding principles under which comparative biologists have operated in their research and use of "animal models."

Some contributions made to endocrinology by nonmammalian research are listed in Table I, and a large number of models are currently in use (Table II). While we have extensive knowledge of some well-developed models (e.g., avian reproductive tract; amphibian liver) most models have not yet been fully explored or utilized. In some fields, this is due to the adherance to principles which suggest that a species (model) is only useful to the extent that it is closely related (phylogenetically) to the human and thus information derived from the study of such organisms can be readily extrapolated to the human condition. Thus, it is somewhat paradoxical that while some nonmammalian models have contributed in a major way to our understanding of basic concepts (neuronal function, neuroendocrine integration, renal function, development, gene expression) there is still a certain degree of antipathy toward animal models (particularly vertebrates, where the homologies between systems are most obvious) and the comparative approach in the biomedical community. A strong plea for comparative studies in biomedical research, and specifically in comparative endocrinology, was made in "An Evaluation of Research Needs

Table I
Some Contributions to Basic Endocrine Knowledge from Nonmammalian Models

Contribution	Reference	Contribution	Reference
Invertebrates		**Vertebrates**	
Neuroendocrine control of insect molting in gypsy moth larvae	Kopec (1917, 1922)	Nerves secrete hormones (skate, minnow)	Speidel (1919); Scharrer (1928)
Regulation of insect metamorphosis by juvenile hormone in *Rhodnius*	Wigglesworth (1934)	Direct observation of blood flow from hypothalamus to pituitary (frog)	Green and Harris (1947)
Specialized sensory hairs on male silkmoth antennae for female sex pheromone: A model chemosensory system	Schneider (1957)	Electrical properties of neuroendocrine cells in goosefish	Kandel (1965)
Pheromones as means of chemical communication within a species first defined for insects	Karlson and Butenandt (1959); Karlson and Luscher (1959)	Molecular evolution of neurohypophysial hormones	Acher (1974)
Ecdysteroid-induced chromosome "puffing" in *Chironomid* larvae	Clever and Karlson (1960)	Prolactin role in water balance (killifish)	Pickford and Phillips (1959)
Modulation of behavior by neuropeptides	Milburn et al. (1960)	18-hydroxycorticosterone is precursor of aldosterone (bullfrog)	Yulick and Kusch (1960)
Neuroendocrine control of motor programs in the cockroach		The interrenal (adrenal cortical homolog) is pituitary dependent (bullfrog)	Smith (1920)
Egg-laying hormone of *Aplysia*	Kupfermann (1967)	1-α-Hydroxycorticosterone is the natural corticoid of elasmobranch interrenal	Idler and Truscott (1967)
Circadian release of eclosion hormone to trigger moth ecdysis	Truman and Riddiford (1970)	Adenyl cyclase/cAMP implicated as second messenger system for ADH in isolated toad bladder	Orloff and Handler (1967)
Programmed cell death: Insect muscles	Lockshin and Williams (1964)	Calcitonin is a product of ultimobranchial glands of dogfish and chicken	Copp et al. (1967)
Insect growth regulation as a control strategy	Williams (1956, 1967); Salama and Williams (1965)	Thyrotropin-releasing hormone is present in frog blood and skin in high levels	Jackson and Reichlin (1977)
Juvenile hormone prevents metamorphosis		**General Vertebrate**	
Anti-juvenile hormones for precocious metamorphosis	Bowers et al. (1976)	Evolution of pituitary glycoprotein peptide hormones	Licht et al. (1977)
Oocyte maturation in starfish		Evolution of vertebrate neuropeptide families	Sherwood and Parker (1990)
Induction of meiosis by 1-methyl-adenine Maturation-promoting factor	Kanatani and Shirai (1957); Kanatani et al. (1969) Kishimoto and Kanatani (1976)		
Nuclear localization of ecdysteroid receptors in *Drosophila* imaginal discs	Yund et al. (1978)		
Intracellular localization of neuropeptide processing and sorting in *Aplysia*	Fisher et al. (1988)		

Table II
Some Nonmammalian Models Used in Endocrine Research

Model, contribution/utility	Reference	Model, contribution/utility	Reference
Invertebrates		Vertebrates	
		Fish	
Drosophila salivary gland chromosomes for ecdysteroid induction of transcription factors, steroid hormone action study	Ashburner *et al.* (1974); Thummel (1990)	Teleost model for studying the effects of chemicals on female reproductive endocrine function	Thomas (1990)
Drosophila imaginal discs for steroid action, role of hormone rise and fall in coordination of development study	Fristrom and Fristrom (1992)	Sex-changing fish for the study of sex determination and differentiation in vertebrates	Reinboth (1970)
		Amphibians	
Manduca sexta (tobacco hornworm) epidermis for juvenile hormone action in regulation of sequential cellular polymorphism study	Riddiford (1976, 1992)	Mexican leaf frog (*Pachymedusa dacnicolor*) as a model in endocrine research	Bagnara (1990)
M. sexta nervous system for study of programmed cell death regulated by steroids	Truman and Schwartz (1984); Truman (1992a)	Amphibian experimental systems for study of developmental neurobiology and behavioral encodrinology in the clawed frog, *Xenopus laevis*	Kelley and Hayes (1990)
M. sexta prothoracic gland for study of regulation of steroid hormone synthesis and release by a neuropeptide	Sakurai and Gilbert (1990)	Amphibian model system for problems in behavioral neuroendocrinology study	Moore (1990)
Diploptera punctata (cockroach) corpora allata for regulation of juvenile hormone synthesis and release by neurohormones, nervous input, and target tissue factors study	Tobe and Stay (1985)	Frog skin model for ADH action study	Ussing (1952)
		Frog bladder model for study of aldosterone action	Crabbe (1951)
Insect ecdysis for study of neuropeptide-induced behavior, circadian and steroid regulation of neuroendocrine cells	Truman (1992b)	Amphibian oocyte model for study of cellular uptake of macromolecules	Wallace and Dumont (1968)
		Amphibian liver model for study estrogen action	Wallace and Dumont (1968)
Lobster stomatogastric ganglion for study of hormonal control of patterned motor output	Dickinson *et al.* (1990); Turrigiano and Selverston (1989)	Amphibian model for study of steroid-induced oocyte maturation	Zwarenstein (1937)
		Amphibian oocyte model for study of membrane steroid receptors	Beaulieu (1983)

Topic	Reference
Crab pericardial cells for study of neurophysiological aspects of neurohormone release	Stuenkel and Cooke (1988)
Aplysia bag cells for study of neuropeptide processing, sorting, release, and induction of oviposition behavior	Geraerts et al. (1988); Jung and Scheller (1991)
Lymnaea stagnalis (snail) neuroendocrine system for behavior, osmoregulation, growth	Geraerts et al. (1988); Joosse (1988)
Starfish oocyte for study of neuroendocrine control of meiosis	Shirai and Walker (1988)
Moth pheromone-sensitive sensillae for study of chemoreception, especially odorant-binding proteins	Vogt (1987); Vogt et al. (1991)
Drosophila for molecular basis of circadian rhythms study	Hall and Kyriacou (1990)
Vertebrates	
Fish	
Elasmobranch species as models for studies of placental viviparity and its endocrine regulation	Hamlett (1990)
Elasmobranch rectal gland as model for ion transport regulation	Burger and Hess (1960); Shuttleworth (1987)
Endocrine pancreas of teleost fish: A model for interaction of islet hormones?	Plisetskaya (1990)
Teleost Brockman body (principal islet) for insulin studies	Macleod (1922)
Fish and amphibian models for developmental endocrinology	Dickhoff et al. (1990)
Amphibian metamorphosis model for study of thyroid hormone effects and action	Etkin (1963)
Reptiles	
Lizards as models for behavioral endocrinology	Crews (1990)
Iguana iguana: A model species for the studying of ontogeny of behavior/hormone interactions	Phillips (1990)
Salt glands as models for regulation of salt transport	Schmidt-Nielsen and Fange (1958); Templeton et al. (1972)
Birds	
The avian embryo as a model for early developmental endocrinology	Murphy and Clark (1990)
Insulin and insulin-like growth factor-I action in the chick embryo	de Pablo et al. (1990)
The avian ovary: Model for endocrine studies	Bahr (1990)
"Paradoxical" growth hormone secretion in acromegaly: An avian model	Harvey (1990)
Avian bone as model for hormonal regulation of deposition and resorption and osteoporosis	Etches (1986)
Avian reproductive tract for steroid hormone action	O'Malley et al. (1969)
The avian as an animal model for the study of the vitamin D endocrine system	Norman (1990)
Avian songbird model for steroid hormone influence on brain sex differentiation and neuronal plasticity	Nottebohm (1981)
Avial model for study of estrogen regulation of lipid metabolism	Chan et al. (1976)
Avian models for study of hormonal and dietary influences on atherosclerosis	Adelman and St. Clair (1988); Shih (1983)

in Endocrinology and Metabolic Diseases" (U.S. Department of Health and Human Services, PHS, NIH Publ. No. 81-2395, p. 41, 1981): "When viewed superficially, studies of comparative biology may seem esoteric or irrelevant to the human condition. The report of the task force on comparative endocrinology provides numerous examples which attest strongly to the contrary" (attributed to H. A. Bern). Recently, the animal welfare movement has given impetus to the development of alternatives to traditional mammalian systems which has been recognized by the National Research Council through its Institute of Laboratory Animal Resources (ILAR). Nevertheless, more effort should be devoted to the spread of these ideas and their basic relevance within and to the biomedical community. In this chapter we have tried to provide a sampling of some of the models which we believe to be worthy of further consideration by the community of endocrinologists as alternatives to mammals for investigation of specific questions. The compilation is not intended to be comprehensive.

II. Vertebrate Models for Hypothalamic Pituitary Interaction and Behavior

A. Avian Models for Seasonal Breeding/Circadian Functioning

Animals of the temperate zone breed during the spring when feed supplies will be anticipated to be high during the periods of reproduction, care of neonates, and growth of the offspring. Birds are excellent models (as are various hamsters) for examining the physiological basis of this. During the spring, the increasing day length provides an entirely predictable environmental cue of season (more accurately of date). In the spring, the increasing day length stimulates rapid gonadal "growth" (recrudescence) via the hormones of the hypothalamo-pituitary-gonadal axis; this is photoperiodic stimulation of reproduction. Gonadal atrophy occurs in the summer in the presence of day lengths which would be anticipated to be stimulatory; this is referred to as photorefractoriness. Examples of avian species in which seasonal breeding has been intensively studied include the Japanese quail *(Coturnix coturnix Japonica)* (particularly for studies in circadian functioning in photoperiodism) and European starlings *(Sturnus vulgaris)* (particularly for studies on photorefractoriness and circannual rhythms).

1. Photoperiodism and Seasonal Breeding in Japanese Quail

On a short day length, quail have small gonads (approximately 10 mg testes) but on transfer to long day, gonads grow rapidly (in 3 weeks to 2–3 g testes) (Nicholls *et al.*, 1973); plasma concentrations of luteinizing hormone being elevated effectively immediately with exposure to only one long day (Follett *et al.*, 1977). The mechanisms by which photoperiod stimulates reproduction involves interaction of a light-induced signal and circadian rhythm of photosensitivity (see, e.g., Simpson and Follett, 1982); a circadian rhythm is an endogenous rhythm with a period of about 24 hr in the absence of environmental cues but is entrained to exactly 24 hr by environmental cues (e.g., the day/night cycle). The Japanese quail is a very useful model not only for photoperiodism and circadian rhythms but also of the rapid growth of an organ (the gonad).

Photorefractoriness is one mechanism to prevent reproduction when the offspring would have a low chance of survival. Photorefractoriness in European starlings involves reduced synthesis of GnRH (and hence gonadotropins). There is evidence that thyroid hormones and prolactin may be involved in this process (Dawson and Goldsmith, 1983; Dawson, 1989).

Circannual rhythms are analogous to circadian rhythms and as such are endogenous rhythms but with a periodicity of approximately 1 year in the absence of entraining environmental cues. With seasonal cues (e.g., day length changes), the circannual rhythm will be entrained to exactly 12 months. Examples of circannual rhythm of seasonal breeding include European starlings (*S. vulgaris*) held on a 12L:12D light/dark cycle (see, e.g., review in Gwinner, 1990) and also sheep maintained on short day lengths (8L:16D) (Robinson, 1990).

B. The Teleost Brain: Model for Regulation of Neural Aromatase

Aromatase, with other steroidogenic enzymes, is part of the ancient, functionally diverse, and ubiquitous P-450 supergene family (Nebert and Gonzalez, 1987). Expression of the aromatase gene (p450 AROM) in the brain has been demonstrated in all vertebrates down to elasmobranchs (Callard, 1984), suggesting an important function of adaptive value. Since the levels of aromatase in teleost brain are extraordinarily high (about 100 times that of other neural tissues), the teleost brain provides an excellent model system with which to study (a) the basis of the high level of neural expression, (b) its regulation, and (c) the physiologic role of this enzyme as a rate-limiting step in neural responsiveness to androgen. With regard to (a) and (b), the cDNA for goldfish brain P-450 AROM has recently been isolated and characterized and considerable homology to human placental and chicken ovary P-450 AROM cDNA has been noted (Callard *et al.*, 1993). As for the potential physiologic role of the enzyme, Mak *et al.* (1985) and Schlinger and Callard (1989) have demonstrated that the vast majority of aromatase in brain subfractions is located in synaptosomes, from which it is inferred that delivery of estrogen to an adjacent target tissue in a particular fashion resembles that of neurotransmitters/neuromodulators. It remains for future work to determine the relative importance of genomic vs nongenomic mechanisms of action of locally formed neural aromatase (G. V. Callard et al., 1990).

C. Direct Neural Regulation of the Pituitary in Teleost Fish

The adenohypophysis in teleost fishes is unique among vertebrates in that it is directly innervated by neurosecretory fibers so that in a functional sense the median eminence is within the neurohypophysis and thus signals reach target endocrine cells without the intervention of a hypothalmic-pituitary portal system. The usefulness of this unique model has been recently discussed by Peter et al. (1990). In both the pars distalis and pars intermedia, neurosecretory fibers may be separated by as much as two basement membranes from the presumed target cell or may be in direct synaptoid contact (Holmes and Ball, 1974). A variety of studies (see, e.g., Peter et al., 1986; Kah et al., 1986) have emphasized the importance of the innervation of the teleost pituitary by monoaminergic fibers, and these and other surgical studies have demonstrated the uniqueness of the teleost pituitary as a model for studies of the functional anatomy of the vertebrate neuroendocrine system. In addition, direct localization of particular neurohormones within specific pituitary cell types suggests physiological actions. By study of anatomical pathways, it is possible, using this model to associate particular brain sites and transmitters with specific pituitary cell functions.

D. Hormones and Aggressive Behavior in Quail

Behavioral studies in quail have made it possible to test the hypothesis that *in situ* aromatization of androgen regulates the quantity of estrogen available for binding to nuclear receptors and is the rate-limiting step determining the occurrence and magnitude or intensity of a neural response. Such a mechanism would explain gender-based, seasonal, and individual differences in neural responsiveness and sensitivity to a given amount of androgen. Japanese quail are useful models because they form dominance hierarchies ("pecking orders") in nature which is a reflection of individual differences in aggressiveness. In order to establish a clear relationship between circulating androgen and behavioral intensity, a new test was devised for quantifying aggressiveness by the measurement of pecking and locomotor behavior in response to a visual stimulus (Schlinger et al., 1987). Unlike pair-testing and round-robin methods, scores obtained are independent of the sex or behavior of the stimulus bird, remain constant with time despite repeat testing, and can predict the outcome of a single paired fight while avoiding the experiential effects of actual encounters. Results of experiments using this test demonstrated that aggressiveness in males is dependent upon androgen to estrogen conversion and is also a reflection of occupied estrogen receptor levels in limbic brain areas. In addition, individual differences in aggressiveness of mature males were unrelated to plasma androgen levels but were significantly correlated with aromatase activity in the preopticohypothalamic region (Schlinger and Callard, 1989).

III. Vertebrate Reproduction

A. Reptilian Models for the Regulation of Sex-Determining Genes

Many orders of bony fish contain representatives of hermaphroditic and sex-reversed species. These have been reviewed by Reinboth (1975) and are valuable models for endocrine and genetic studies on sex differentiation. Reptiles, which give rise to birds and mammals in phylogeny, are unique in that they exhibit two distinct types of sex-determining mechanisms: genotypic and temperature dependent. Although the primary trigger for gonadal determination now appears to be a "testis-determining" gene, the physiological initiators are not well understood and may be investigated using reptiles in which sex determination may be manipulated by both temperature and steroid hormones (Gutzke, 1990). Thus, in chelonian species (Pieau, 1974; Yntema, 1976) when eggs are incubated at low temperatures ($25 \pm 1°C$ for *Emys orbicularis*) all embryos show a male phenotype; in contrast, when eggs are incubated at $29.5 \pm 0.5°C$, the gonads of all embryos acquire an ovarian structure. It is of interest that this temperature effect is the reverse of that observed in amphibian larvae, in which low temperatures favor female development (Witschi, 1929). Although early models of sex determination favored steroid hormones as the primary physiologic triggers, the validity of the model has been called into question due to the failure of steroids to induce sex reversal in mammals and the observation that testes developed in phenotypically female XY mice with testicular feminization syndrome (tfm) in which abnormal androgen receptors prevent androgen action. However, as pointed out by Gutzke (1990), gonadal steroid administration to developing reptilian embryos can override normal sex-determining mechanisms in species with either genotypic or temperature-dependent sex determination and the embryonic period of steroid sensitivity coincides with the period of normal sex determination. Because of the readiness with which a variety of turtle eggs can be obtained, treated, and manipulated, these species provide excellent models for studies of the physiological mechanisms of sex determination.

B. Avian and Reptilian Ovaries: Models for Follicular Hierarchies, Follicular Development/Atresia, and Other Aspects of Ovarian Function

The structure of the mature ovary of vitellogenic birds and reptiles is broadly similar. Discussion on the avian/reptilian reproductive functioning here is restricted to the chicken. This is because there is vastly more information on this species. Moreover ovulation is daily and at a predicted time. In addition, female chickens (hens) are readily available (they can be purchased as "point of lay") and the requirements for their care in laboratory animal or agricultural settings are well established. For detailed review of the model system, the reader is referred to an excellent article by Etches and Petitte (1990).

1. Ovarian Anatomy

There is a single ovary in the abdominal cavity of the hen. This is readily identified by the large yellow (yolk-filled) follicles. There are three types of preovulatory follicles: the most immature small white follicle (<2 mm diameter), the intermediate small yellow follicles (2–6 mm), and the rapidly growing large yellow follicles (6 to ~ 40 mm). The large follicles form an obvious hierarchy based on size: in order of size there being the F_1 or largest follicle (the next to ovulation) $> F_2$ (next largest) follicle $> F_3 > F_4$, etc. There is no antrum as the oocyte (up to 40 mm in diameter) is filled with yolk (theca and the granulosa). The follicular tissues remaining following ovulation do not form a corpus luteum.

2. Ovarian Functioning

The time of ovulation can be readily predicted (see Etches and Petitte, 1990, and Johnson, 1990, for details). Oviposition (egg laying) occurs approximately 24–26 hr following ovulation during which the egg white and shell are added. For chickens laying eggs every day, the time of oviposition is at the same time each day. For hens laying eggs in sequences interspersed by a missed day, successive oviposition occurs progressively later in the day but in a predictable manner. Ovulation normally occurs approximately 25 min following oviposition. There is a missed day of ovulation which occurs approximately 24 hr prior to the missed day of oviposition.

Thecal and granulosa cells from different follicles can be used to examine steroidogenesis, plasminogen activator, cellular proliferation, and their control (see, e.g., Tilly and Johnson, 1990; Yoshimura and Tamura, 1991). Moreover, small follicles not recruited for further development appear to be a good model for apoptosis and its control (Tilly *et al.*, 1991).

3. Chicken as Germ Line Chimeras

Chick embryos can be manipulated relatively easily and can be utilized to form germ line chimeras (see, e.g., Petitte *et al.*, 1990). Obviously the potential exists for the donor cells to receive transgenes.

C. The Shark Testis Model for the Study of Stage-Dependent Functions and the Regulation of Spermatogenesis

The interaction between Sertoli and germ cells is of paramount importance in the regulation of germ cell progression, but because of the complex organization of the mammalian testis, it is difficult to obtain discrete spermatogenic stages for direct biochemical analysis (Parvinen, 1982). In contrast to the mammalian testis, where the Sertoli cell differentiates from base to apex and simultaneously nurtures three to four successive generations of germ cells, the testis of the dogfish shark *(Squalus acanthias)* displays features which render it an ideal model in which to study these processes (Callard, 1991; DuBois and Callard, 1990). Here there is a cystic mode of spermatogenesis in which a single cohort of Sertoli cells and an isogenic clone of synchronously developing germ cells remain associated throughout development and form distinct anatomical units called spermatocysts. At any given stage of development, the number of germ cells vs Sertoli cells and the germ cell/Sertoli cell ratio for each spermatocyst are predictable. During successive stages of spermatogenesis, spermatocysts (Sertoli/germ cell units) move medially toward the efferent ducts and those spermatocysts in like stages of development are grouped into premeiotic (ZI), meiotic (ZII), and postmeiotic zones (ZIII) that are readily visible in testicular cross sections. A successful method has been developed for isolation and culture of Sertoli cells from different stages as either intact spermatocysts or Sertoli cell monolayers (Dubois and Callard, 1992). The linear arrangement of maturational order, and thus spatial separation of different stages, has facilitated study of structural/functional correlations at each stage of development. Since cysts retain structural integrity in culture and continue to express stage-related growth and differentiation traits, a range of functional attributes associated with particular stages of development can be studied. Despite the wide evolutionary gap between elasmobranchs and mammals, testicular structure shares similarities in the organization of spermatogenic tubules and the close association between germ cells and Sertoli cells which allow transfer of information about the regulation of spermatogenesis from the shark model to the mammal.

D. The Elasmobranch Shell Gland: Model for Steroid Hormone Action

This compound tubuloalveolar gland, found in its most developed form in egg-laying species, such as the little skate, Raja, is an excellent model for steroid hormone-regulated synthesis of collagen-like molecules. The glands are large (8 g) and weight is positively correlated with plasma estradiol titer (Koob et al. 1986). The fully formed egg case is biochemically complex and extraordinarily stable, resistant, and relatively impermeable. Rousaouen (1976; Rousaouen et al; 1976) identified a collagen-type protein, a tyrosine-rich protein, sulfur-rich proteins, phenoloxidase, and peroxidase within the gland. Recent analyses identified six major structural proteins and two enzymatic activities, tyrosine hydroxylase and catechol oxidose, responsible for capsule sclerotization in the little skate (Koob and Cox, 1988, 1993). Initially, these products are stored in cytoplasmic granules and subsequently secreted in proper chronological order for assembly within the lamellae of the gland. Capsule precursors then polymerize *in utero* via a quinone tanning process. (Koob and Cox, 1990).

The current state of knowledge of this structure, its hormone sensitivity, the controlled sequential synthesis, secretion, and polymerization of fibrous protein makes the shell gland an excellent model for hormone regulation of these basic processes. Both *in vivo* and *in vitro* studies are possible because of the availability of tissue culture methodology for elasmobranch tissues. The species itself is readily available off the east coast of the United States and can be maintained in reproductive condition in captivity for long periods of time due to ease of feeding adult animals.

E. Reptilian Models for the Study of Requirements of Long-Term Sperm Storage

Sperm storage is relatively uncommon in laboratory mammals, although retention of sperm with fertilizing capacity has been demonstrated for periods of from 5 months to 1 year in certain bats (Wimsatt, 1942). This phenomenon is common in squamate reptiles (Fox, 1956, 1963; Cuellar, 1966) and chelonians (Hildebrand, 1929; Ewing, 1943). Retention of sperm fertilizing capacity has been demonstrated to last 4 years in turtles (Malaclemys centrata) and 6 years in snakes (Leptodeira annulata; Haines, 1940). Seminal receptacles in the walls of the cranial portion of the oviducts of snakes (Thamnophis) have been described by Fox (1956) and Hoffman and Wimsatt (1972). The adaptive value of sperm storage in situations in which encounters between males and females of a species may be reduced by geographic separation or skewed sex ratios is obvious. While such species would be useful models for the study of factors important to sperm viability in the female tract as well as the phenomenon of capacitation, they have not been exploited in this way, despite the availability of appropriate species. Enough information is now available on the physiology and biochemistry of male reproduction, spermatogenesis, and fertilization in a variety of nonmammalian species, particularly reptiles, to make these species useful and attractive alternates to standard mammalian models (Licht, 1984).

IV. Vertebrate Hormones and Metabolism

A. Avian and Reptilian Models for the Study of Apolipoproteins and Their Role in Lipid Transport, Metabolism, and Uptake

In the human, coronary artery disease is characterized by an imbalance in the relative amounts of the apolipoproteins and their associated lipid classes. The work of Brown and Goldstein (1986) has emphasized the importance of low-density lipoprotein (LDL) cholesterol as the most important component in the pathogenesis of the disease and its sex-linked nature. Although there is a general tendency to assume a protective action of estrogen in coronary artery disease (it is less prevelant in women of reproductive age than in men, and increases to the incidence of males after the menopause), it is probably a more accurate prediction that altered ratios of estrogen, androgen, and progesterone are involved. Thus, estrogen acts to decrease apoB/LDL and to increase apoA1/high-density lipoprotein (HDL) and progesterone and androgens have the reverse effect, possibly contributing to the sex differences. According to the relative amounts of the primary gonadal steroids, conditions may favor (high progesterone, testosterone, low estradiol) or protect against (high estradiol, low progesterone, testosterone) cardiovascular disease.

In oviparous nonmammalian species with large-yolked eggs the massive hepatic synthesis, transport, and uptake of yolk-protein precursors and associated lipids (vitellogenesis) is a normal physiologic event coupled to reproductive cycles in nature. Vitellogenesis is sex dependent, occurring only in females in normal physiologic circumstances, but inducible by estrogens in males. Although most studies focused on the transcriptional regulation of vitellogenin per se, available evidence from birds (Chan *et al.*, 1976) indicates that other lipoproteins are involved. Despite the obvious absence of vitellogenin per se in mammalian species, there are striking similarities between mammalian and nonmammalian systems for the regulation of lipid transport, delivery, and metabolism. Further, known structural homologies between vitellogenin, apolipoprotein B (Perez *et al.*, 1991), and lipoprotein lipase (Baker, 1988; Persson *et al*, 1989) and the involvement of all three proteins in lipid transport and metabolism suggest that they should be considered a functional group, possibly of common evolutionary descent with shared transcriptional/translational control mechanisms (Persson *et al.*, 1989). Recent work (Stifani *et al.*, 1990) showing that vitellogenin and apoprotein B recognize the same oolemma surface receptor in the bird further emphasizes the importance of these models. The value of pigeons as models for the development of classical atherosclerosis when fed a high-cholesterol diet has been recognized for some time (Adelman and St. Clair, 1988).

Attention should also be directed toward reptilian models for investigations of the potential role of sex hormones in lipoprotein metabolism and coronary artery disease (I. P. Callard *et al.*, 1990). Available evidence from the turtle, *Chrysemys picta*, suggests that it may have important advantages over avian species, in particular the presence of a functional luteal phase correlating with elevated progesterone levels. Evidence has also been presented from this species that progesterone and testosterone inhibit estrogen-induced vitellogenesis, probably via hepatic receptors (I. P. Callard *et al.*, 1990).

B. Chicken Adipose Tissue as a Model for Growth Hormone Action

Responses of chicken adipose tissue *in vitro* have potential for to aid the understanding of the mechanism of action of growth hormone. *In vitro* growth hormone (GH) exerts both a lipolytic effect [increasing glycerol release (Campbell and Scanes, 1985)] and insulin-like effects [increasing glucose uptake (Rudas and Scanes, 1983) and exerting an antilipolytic effect by inhibiting glucagon-stimulated glycerol release (Campbell and Scanes, 1987)]. Use of adipose tissue explants from adult male chickens to examine lipolytic and insulin-like (anti-lipolytic) effects of GH has a number of advantages. These include the consistency and rapidity of the response and the absence of need for either hypophysectomized animals for the source of tissue or agents such as theophylline or dexamethasone for full lipolytic responses to be observed. In addition there is evidence that the lipolytic and insulin-like antilipolytic effects of GH involve different mechanisms/structural requirements on GH. For instance the lipolytic but not insulin-like effects of GH are inhibited by cycloheximide, actinomycin D, or verapamil and conversely the insulin-like/antilipolytic but not the lipolytic effects of GH are inhibited by polyamide synthesis inhibitors (Campbell and Scanes, 1988). Moreover there is evidence for different structure requirements for the different GH activities. For instance, while GH preparations from all vertebrates examined (mammals, birds, reptiles, amphibians, and fish) have insulin-like/antilipolytic activity only mammalian and avian GH have lipolytic activity (Campbell and Scanes, 1985; 1987; Campbell et al., 1990, 1991).

C. The Chicken as a Model for Pancreatic Function

Chickens have been used successfully as models for investigations of pancreatic function. Several examples are included below.

1. Chick Embryos. Chicken embryos are useful models to examine the developmental biology of the endocrine pancreas. For instance, somatostatin is observable immunocytochemically at 3 days of embryonic development when the dorsal pancreatic bud is beginning to evaginate from the forgut (Wild, 1979).

2. The Chicken as a Model for Diabetes. The normal chicken has similarities to diabetic mammals including elevated circulatory concentrations of glucose, ketones, and glucagon; depressed circulatory concentrations of bicarbonate and cations; and reduced sensitivity to insulin (reviewed in Scanes and Griminger, 1990).

3. Splenic Lobe of the Pancreas. The splenic lobe of the chicken pancreas has disportionately high concentrations of endocrine cells and thus may have advantages for *in vitro* studies of the endocrine pancreas (see the example of Hazelwood and Cieslak, 1986). It might also be noted that 99% of the pancreas of the chicken can be surgically removed leaving only the splenic lobe. This is followed by rapid growth of splenic lobe with a fivefold increase in weight and pancreatic polypeptide content and increases in splenic lobe insulin and glucagon contents (Cieslak and Hazelwood, 1986). This potentially very useful model of exocrine and endocrine pancreas growth remains almost entirely unexploited.

D. Pigeon Crop Sac Gland: A Bioassay for Prolactin and a Model for the Synlactin Hypothesis of Prolactin Action

Despite the fact that the response of the pigeon crop sac has been the principal biological assay for prolactin since the 1930s, Nicoll (1990) concluded that the crop sac is still "an underappreciated and underutilized model" for examining the mechanism of action of prolactin and other hormones. The crop is a distensible feed storage organ found at the posterior (inferior) end of the esophagus in birds. In pigeons and doves, crop sac epithelium undergoes proliferation in response to prolactin; ultimately the lipid-filled cells are sloughed off as crop milk to feed the young.

1. Prolactin Bioassay

The crop sac prolactin bioassay can be performed in two ways. Prolactin can be injected systemically *or* locally intradermally into either lobe of the crop (administered twice daily for 2 days). The response determined for the systemic crop sac assay is normally wet crop weight. The most used response to prolactin with the local assay is increased mucosal dry weight in the 4-cm-diameter circular section of hemicrop close to the site of injection. The details of this are explained elsewhere (Nicoll, 1967, 1990). Subjective criteria have also been used involving "scored" changes in crop epithelial thickness. An alternative to determining dry crop mucosal weights is to estimate prolactin stimulation of [^{3}H]thymidine incorporation as an index of mucosal epithelial proliferation (Bani *et al.*, 1990). The localized crop sac prolactin assay (using mucosal dry weight as the determined response) is a good bioassay for monitoring the purity of pituitary/placental/biosynthetic lactogens during purification; the assay having acceptable precision and high specificity relative to other pituitary hormones (Nicoll, 1967, 1990). However there is a response to some growth factors [e.g., epidermal growth factor (EGF) and insulin-like growth factor-I (IGF-I) (Nicoll, 1990)].

2. Synlactin and Prolactin Action

There is very strong evidence that the liver (and also perhaps other organs) produce a factor(s), synlactin, which acts as a synergist for the biological activity of prolactin (see, e.g., Anderson *et al.*, 1984; Mick and Nicoll, 1985; reviewed in Nicoll, 1990). Moreover, the release of this synlactin (possibly related to IGF-I) is stimulated by prolactin. At this time, synlactin has not been fully characterized. A corollary to the proposed existence of synlactin is that effects, observed following the systemic administration of prolactin, may be due to prolactin, synlactin, or a combination of both.

The response of the pigeon crop sac gland to prolactin has the potential to be a valuable tool for examining the cellular mechanisms of prolactin action. For instance, the systemic injection of prolactin (200 μg/day for 3 days) has been found not only to increase crop weight and soluble protein content but also to induce specific proteins (Pukac and Horseman, 1984) and specific mRNAs, one by more than 70-fold (Pukac and Horseman, 1987).

Pigeons *(Columbia livia)* have been most extensively employed for prolactin bioassay, investigation of prolactin mechanism of action, and synlactin hypothesis examination. Examples of strain and sources of pigeons which have been used include White Carneaus (e.g. Palmetto Pigeon Plant, Columbia, SC), Silver King and California Kings (e.g., Scott Squab Plantation, Hughson, CA; California Squab Breeders Association, Stockton, CA), and feral pigeons.

E. Atrial Natriuretic Peptides: Fish Models for Biological Actions

As Tables I and II show, fish have provided a number of useful model species, particularly in the field of electrolyte control, and therefore are relevant to the study of cardiovascular function. In the 10 years since the discovery of atrial natriuretic peptide (ANP) by deBold (1982) there has been intense interest in the field spurred by reports of brain atrial natriuretic peptide (BNP), C-type natriuretic peptide (CNP), and ventricular natriuretic peptide (VNP) (Evans and Takei, 1992). These form a peptide family with significant homology between members from different vertebrate species. Extension of functional studies of these peptides to nonmammalian species in classical comparative physiology is likely to prove extremely rewarding for our understanding of the role that these peptides play in the general and tissue-specific regulation of tonicity, peripheral vascular function, and cardiac function.

As pointed out by Evans and Takei (1992), fish, broadly classified, fall into three major groups, the agnathans (lampreys and hagfish), chondrichthyes (sharks, skates, and rays), and osteichthyes (bony fish). Taken together, these species present an array of natural osmoregulatory problems (hagfish, isosmotic to seawater; marine chrondricthyes, hypernatremia and potential hypervolemia; freshwater teleosts, hyponatremia); further, migratory species (anadromous or catadromous) can be readily manipulated in the laboratory (e.g., eels). By appropriate selection of the model one could study the role of ANPs in a variety of physiological circumstances mimicking clinical problems of endocrine/renal origin with cardiovascular sequelae without involvement of pathology or abnormal conditions. Fish are also advantageous from the point of view of the variety of osmotically active surfaces available for study (skin, kidney, bladder, gill, rectal gland). Some fish species may be particularly useful in dissecting out primary and secondary sites of action of the peptides. For example, the aglomerular toadfish allowed direct demonstration for the first time that ANP-induced natriuresis could be produced without changes in GFR. There is also evidence that CNP may be a circulating hormone in the dogfish shark (Evans *et al.*, 1993).

V. Invertebrate Models

A. Models for Neuropeptides Modulating Behavior

1. Insect Eclosion Hormone

Insect eclosion hormone is a 62 amino acid neuropeptide that acts directly on the nervous system to trigger, via a rise in cGMP, a preprogrammed bout of motor output that results in the shedding of the old cuticle (Truman, 1992b). Production of the hormone by two pairs of neurosecretory cells in the brain of moths is continuous, whereas release is governed either by the declining ecdysteroid titer at the end of the molt and/or by a circadian oscillator in the brain. Therefore, this is an excellent model system for studying the effects of a hormone on behavior and of the control of neurohormonal output by a circadian clock which is modulated by a steroid hormone.

2. *Aplysia* Egg-Laying Hormone

Aplysia egg-laying hormone is produced by an isolated group of bag cells and acts on the nervous system to trigger a preprogrammed behavior (Kupfermann, 1967; Geraerts *et al.*, 1988). In this case, the egg-laying hormone is only one of a series of peptides produced by endoproteolytic cleavage of the large precursor. The other peptides serve to prolong the afterdischarge of the bag cells, to modulate other central neurons, and to play other ancillary roles in the egg-laying process. These bag cells have proven ideal for the study of the processing and sorting of these peptides at the electron microscopical level, showing that the different peptides are released from different portions of the cell (Jung and Scheller, 1991).

3. Lobster Stomatogastric Gaglion

The lobster stomatogastric ganglion is probably the best-understood motor pattern generator of any system. Recently, this patterned motor output has been found to be modulated by a variety of neuropeptides such as red pigment-concentrating hormone and cholecystokinin (Turrigiano and Selverston, 1989; Dickinson *et al.*, 1990). This is currently the best system for analysis of the cellular mechanisms by which modulatory substances alter complex neuronal targets.

B. Models for Pheromones and Chemoreception

Insect sex attractants and the complex interactions among the social insects were the first to alert biologists to the importance of chemical communication within a species by pheromones (Karlson and Butenandt, 1959; Karlson and Luscher, 1959). Moreover, the specificity and sensitivity of the sensilla trichodea on the male moth antennae for the female sex pheromone (Schneider, 1957, 1969) allowed the later isolation of an odorant-binding protein and a specific pheromone-degradative enzyme from these hairs (Vogt and Riddiford, 1981). Later work showed that the female antennae contained a different protein for binding general odorants (Vogt et al., 1991). This work was the first in any olfactory system demonstrating the importance of soluble molecules surrounding the nerves in the signal transduction process (Vogt, 1987).

C. Models for Circadian Rhythms and Photoperiodic Control of Neurohormonal Output

Insect studies have provided many basic insights into the nature of circadian rhythms and of the photoperiodic control of release of neurohormones (Saunders, 1982). Importantly, the isolated insect brain can even respond to photoperiod *in vitro* and modulate release of the prothoracicotropic hormone (Bowen et al., 1984) although this system has not been explored in depth as to the nature of the interactions involved. The period (per) mutants of *Drosophila* which have an altered circadian oscillator (Konopka and Benzer, 1971) are however being exploited for a molecular dissection of this circadian oscillator (Hall and Kyriacou, 1990).

D. Models for Steroid and Genome Interaction

1. Midge Salivary Glands

The classic study of Clever and Karlson (1960) showing that ecdysone caused "puffing" (RNA synthesis) of specific loci in the salivary glands of the midge, *Chironomus tentans* within 15 min of application followed by a second series of puffs several hours later demonstrated for the first time that steroid hormones might act directly on the genome. Clever (1964) then showed that protein synthesis was not necessary for the initial puffing response but only for the later one. Later studies with *Drosophila melanogaster* resulted in a hypothesis for ecdysteroid action (Ashburner et al., 1974) in which the initial puffs were thought to code for regulatory proteins that activated the late genes and inactivated the early genes. Isolation of several of the ecdysteroid-activated early genes has recently shown that they encode transcription factors of various types (Thummel, 1990). At least one of these has now been shown to be critical for activating a set of late genes (Guay and Guild, 1991).

2. *Drosophila* Imaginal Discs

Drosophila imaginal discs have proven to be a good model system for the study of steroid hormone action on development. Ecdysteroid *in vitro* causes first pupal, then adult, differentiation of these discs (Fristrom and Fristrom, 1992). These discs proved to have nuclear receptors for ecdysteroid in the absence of hormone (Yund *et al.*, 1978), a "heretical" idea at the time. Moreover, the coordinated series of events induced by ecdysteroid beginning with evagination and morphogenesis of the disc into the adult shape followed by pupal cuticle deposition was shown to depend on initial exposure to ecdysteroid followed by its absence and then a subsequent reexposure (Fristrom *et al.*, 1982; Doctor *et al.*, 1985; Clark *et al.*, 1986). The delay of ecdysteroid-induced events by the presence of the hormone has since been found to be a common theme in the orchestration of development by ecdysteroid in many tissues. One hypothesis currently under study is that ecdysteroid induces transcription factors of varying half-lives, and a positive one of longer duration (Riddiford *et al.*, 1990; Hiruma and Riddiford, 1992). Thus, the genes that these factors regulate are not activated until the hormone itself disappears.

E. Models for Hormonal Regulation of Sequential Cell Polymorphism

Many insect cells display sequential differentiated states as they progress through metamorphosis. The presence of juvenile hormone prevents this progression in response to the ecdysteroids as they cause the molt. The epidermis of the tobacco hornworm *(Manduca sexta)* responds to these hormones *in vitro* (Riddiford, 1976). Recent studies have shown that juvenile hormone has no effect on the initial ability of ecdysteroid to inactivate ongoing gene expression during the molt; rather it only prevents the permanent inactivation of larval-specific genes and the ecdysteroid-induced expression of new genes during the molt (Riddiford, 1992). Also, the presence of juvenile hormone appears to suppress the level of ecdysteroid receptors, particularly in some tissues such as the larval nervous system, and may also influence the isoform of ecdysteroid receptor present (Riddiford and Truman, 1992).

F. Models for Programmed Cell Death

The control of programmed cell death by hormones was first noted for insect muscles during metamorphosis (Lockshin and Williams, 1964). In the past 10 years the insect nervous system has been shown to be a better model system in which to study this effect of ecdysteroids since some cells are affected and others are not (Truman, 1992a). Recent studies suggest that the cells fated to die at adult emergence acquire a different isoform of the ecdysteroid receptor at the outset of adult development and maintain high levels of this receptor until death. Cells that remain in the adult have only low receptor levels.

References

Acher, R. (1974). *In* "Handbook of Physiology, Sect. 7: Endocrinology. The Pituitary Gland and Its Neuroendocrine Control" (R. O. Greep and E. B. Astwood, eds.), Part 1, pp. 119-130. Williams & Wilkins, Baltimore.
Adelman, S. J., and St. Clair, R. W. (1988). *J. Lipid Res.* **29,** 643-656.
Anderson, T. R., Pitts, D. S., and Nicoll, C. S. (1984). *Gen. Comp. Endocrinol.* **54,** 236-246.
Ashburner, M., Chihara, C., Meltzer, P., and Richards, G. (1974). *Cold Spring Harbor Symp. Quant. Biol.* **38,** 655-662.
Bagnara, J. T. (1990). *J. Exp. Zool., Suppl.* No. 4, 145-147.
Bahr, J. (1990). *J. Exp. Zool., Suppl.* No. 4, 192-194.
Baker, M. E. (1988). *Biochem. J.* **255,** 1057-1060.
Bani, G., Sacchi, T. B., and Bigazzi, M. (1990). *Gen. Comp. Endocrinol.* **80,** 16-23.
Beaulieu, E. E. (1983). *Exp. Clin. Endocrinol.* **81,** 3-16.
Bowen, M. F., Saunders, D. S., Bollenbacher, W. E., and Gilbert, L. I. (1984). *Proc. Natl. Acad. Sci. U.S.A.* **81,** 5881-5884.
Bowers, W. S., Ohta, T., Cleere, J. S., and Marsella, P. A. (1976). *Science* **193,** 542-547.
Brown, M. S., and Goldstein, J. L. (1986). *Science* **232,** 34-47.
Burger, J. W., and Hess, W. N. (1960). *Science* **131,** 670-671.
Callard, G. V. (1984). *In* "Metabolism of Hormonal Steroids in Neuroendocrine Structures" (F. Celotti, L. Martini, and F. Naftolin, eds.), pp. 79-102. Raven, New York.
Callard, G. V. (1991). *In* "Oogenesis, Spermatogenesis and Reproduction" (R. K. H. Kihne, ed.), pp. 104-154. Karger, Basel.
Callard, G. V., Schlinger, B., Pasmanik, M., and Corina, K. (1990). *Prog. Comp. Endocrinol.*, 105-111.
Callard, G. V., Drygas, M., and Gelinas, D. (1993). *J. Steroid Biochem. Mol. Biol.* **44,** 541-547.
Callard, I. P., Riley, D., and Perez, L. E. (1990). *J. Exp. Zool., Suppl.* No. 4, 106-111.
Campbell, R. M., and Scanes, C. G. (1985). *Proc. Soc. Exp. Biol. Med.* **180,** 513-517.
Campbell, R. M., and Scanes, C. G. (1987). *Proc. Soc. Exp. Biol. Med.* **184,** 456-460.
Campbell, R. M., and Scanes, C. G. (1988). *Proc. Soc. Exp. Biol. Med.* **188,** 177-181.
Campbell, R. M., Kostyo, J. L., and Scanes, C. G. (1990). *Proc. Soc. Exp. Biol. Med.* **193,** 269-273.
Campbell, R. M., Kawauchi, H., Lewis, V. J., Papkoff, H., and Scanes, C. G. (1991). *Proc. Soc. Exp. Biol. Med.* **197,** 409-415.
Chan, L., Jackson, R. L., O'Malley, B. W., and Means, A. R. (1976). *J. Clin. Invest.* **58,** 368-379.
Cieslak, S. R., and Hazelwood, R. L. (1986). *Gen. Comp. Endocrinol.* **61,** 476-489.
Clark, W. C., Doctor, J., Fristrom, J. W., and Hodgetts, R. B. (1986). *Dev. Biol.* **114,** 141-150.
Clever, U. (1964). *Science* **146,** 794-795.
Clever, U., and Karlson, P. (1960). *Exp. Cell Res.* **20,** 623-626.
Copp, D. H., Cockcroft, D. W., and Kueh, Y. (1967). *Science* **158,** 924-926.
Crabbe, J. (1961). *Nature (London)* **200,** 787-788.

Crews, D. (1990). *J. Exp. Zool., Suppl.* No. 4, 164–166.
Cuellar, O. (1966). *J. Morphol.* **119**, 7–20.
Dawson, A. (1989). *J. Exp. Zool.* **249**, 68–75.
Dawson, A., and Goldsmith, A. R. (1983). *J. Endocrinol.* **97**, 253–260.
deBold, A. J. (1982). *Can. J. Physiol. Pharmacol.* **60**, 324–330.
de Pablo, F., Serrano, J., Girbau, M., Alemany, J., Scavo, L., and Lesnik, M. A. (1990). *J. Exp. Zool., Suppl.* No. 4, 187–191.
Dickhoff, W. W., Brown, C., Sullivan, C. V., and Bern, H. A. (1990). *J. Exp. Zool., Suppl.* No. 4, 90–97.
Dickinson, P. S., Mecsas, C., and Marder, E. M. (1990). *Nature (London)* **344**, 155–158.
Doctor, J., Fristrom, K., and Fristrom, J. W. (1985). *J. Cell Biol.* **101**, 189–200.
Dubois, W., and Callard, G. V. (1990). *J. Exp. Zool., Suppl.* No. 4, 142–144.
Dubois, W., and Callard, G. V. (1992). *J. Exp. Zool., Suppl.* (in press).
Etches, R. J. (1986). *J. Nutr.* **117**, 619–628.
Etches, R. J., and Petitte, J. N. (1990). *J. Exp. Zool., Suppl.* No. 4, 113.
Etkin, W. (1963). *Science* **139**, 810–814.
Evans, D. H., and Takei, Y. (1992). *News Physiol. Sci.* **7**, 15–19.
Evans, D. H., Toop, T., Donald, J., and Forrest, J. N. (1993). *J. Exp. Zool.* **265**, 84–87.
Ewing, H. E. (1943). *Copeia* pp. 112–114.
Fisher, J. M., Sossin, W., Newcomb, R., and Scheller, R. H. (1988). *Cell* **54**.
Follett, B. K., Davies, D. T., and Gledhill, B. (1977). *J. Endocrinol.* **74**, 449–460.
Fox, H. (1963). *Nature (London)* **198**, 500–501.
Fox, W. (1956). *Anat. Rec.* **124**, 519–539.
Fristrom, J. W., and Fristrom, D. (1992). *In* "The Development of *Drosophila*" (M. Bate and A. Martinez-Arias, eds.). Cold Spring Harbor Lab. Press, Cold Spring Harbor, New York. In press.
Fristrom, J. W., Doctor, J., Fristrom, D. K., Logan, W. R., and Silvert, D. J. (1982). *Dev. Biol.* **91**, 337–350.
Geraerts, W. P. M., TerMaat, A., and Vregdenhil, E. (1988). *In* "Endocrinology of Selected Invertebrate Types" (H. Laufer and R. G. H. Downer, eds.), pp. 141–231. Alan R. Liss, New York.
Green, J. D., and Harris, G. W. (1947). *J. Endocrinol.* **5**, 136–146.
Guay, P. S., and Guild, G. M. (1991). *Genetics* **129**, 169–175.
Gutzke, W. N. H. (1990). *J. Exp. Zool., Suppl.* No. 4, 161–163.
Gwinner, E. (1990). *Prog. Comp. Endocrinol.* 632–638.
Haines, T. P. (1940). *Copeia* pp. 116–118.
Hall, J. C., and Kyriacou, C. P. (1990). *Adv. Insect Physiol.* **22**, 221–298.
Hamlett, W. C. (1990). *J. Exp. Zool., Suppl.* No. 4, 129–131.
Harvey, S. (1990). *J. Exp. Zool., Suppl.* No. 4, 195–199.
Hazelwood, R. L., and Cieslak, S. R. (1989). *Gen. Comp. Endocrinol.* **73**, 38–317.
Hildebrand, S. F. (1929). *Bull. U.S. Bur. Fish.* **45**, 25–70.
Hiruma, K., and Riddiford, L. M. (1993). *Intl. J. Insect Morph. Embryol.* (in proof).
Hoffman, L. H., and Wimsatt, W. A. (1972). *Am. J. Anat.* **134**, 71–96.
Holmes, R. L., and Ball, J. N. (1974). "The Pituitary Gland. A Comparative Account." Cambridge Univ. Press, London.
Idler, D. R., and Truscott, B. (1967). *Steroids* **9**, 457.
Jackson, I. M. D., and Reichlin, S. (1977). *Science* **198**, 414–415.
Johnson, A. L. (1990). *Crit. Rev. Poult. Biol.* **2**, 319–346.
Joosse, J. (1988). *In* "Endocrinology of Selected Invertebrate Types" (H. Laufer and R. G. H. Downer, eds.), pp. 89–140. Alan R. Liss, New York.

Jung, L. J., and Scheller, R. H. (1991). *Science* **251**, 1330–1335.
Kah, O., Dubourg, P., Onteniente, B., Geffard, A., and Calas, A. (1986). *Cell Tissue Res.* **238**, 621–626.
Kanatani, H., and Shirai, H. (1967). *Nature (London)* **216**, 284–286.
Kanatani, H., Shirai, H., Nakanishi, K., and Kurokawa, T. (1969). *Nature (London)* **221**, 273–274.
Kandel, E. R. (1965). *Physiol.* **47**, 691–697.
Karlson, P., and Butenandt, A. (1959). *Annu. Rev. Entomol.* **4**, 39–58.
Karlson, P., and Luscher, M. (1959). *Nature (London)* **183**, 55–56.
Kelley, D. B., and Hayes, M. (1990). *J. Exp. Zool., Suppl.* No. 4, 148–149.
Kishimoto, T., and Kanatani, H. (1976). *Nature (London)* **260**, 321.
Konopka, R. J., and Benzer, S. (1971). *Proc. Natl. Acad. Sci. U.S.A.* **68**, 2112–2116.
Koob, T. J., and Cox, D. L. (1988). *Biol. Bull. (Woods Hole, Mass.)* **175**, 202–211.
Koob, T. J., and Cox, D. L. (1990). *J. Mar. Biol. Assoc. U.K.* **70**, 395–411.
Koob, T. J., and Cox, D. L. (1993). *Environ. Biol. Fishes* (in press).
Koob, T. J., Tsang, P., and Callard, I. P. (1986). *Biol. Reprod.* **35**, 267–275.
Kopec, S. (1917). *Bull. Acad. Sci., Cracov., Cl. Sci. Math. Nat., Ser. B* **57**–60.
Kopec, S. (1922). *Biol. Bull. (Woods Hole, Mass.)* **42**, 323–342.
Krebs, H. A. (1975). *J. Exp. Zool.* **194**, 221–226.
Krogh, A. (1929). *Am. J. Physiol.* **90**, 243–251.
Kupfermann, I. (1967). *Nature (London)* **216**, 814–815.
Licht, P. (1984). *In* "Marshall's Physiology of Reproduction" (G. E. Lamming, ed.), Vol. 1, pp. 206–208. Churchill Livingstone, New York.
Licht, P., Papkoff, H., Farmer, S. W., Muller, C. H., Tsui, H. W., and Crews, D. (1977). *Recent Prog. Horm. Res.* **33**, 169–248.
Lockshin, R. A., and Williams, C. M. (1964). *J. Insect Physiol.* **10**, 643–649.
Macleod, J. J. R. (1922). *J. Metab. Res.* **2**, 149–172.
Mak, P., Zenn, R., and Callard, G. V. (1985). *Program 67th Annu. Meet. Endocr. Soc., Baltimore*.
Mick, C. C. W., and Nicoll, C. S. (1985). *Endocrinology (Baltimore)* **116**, 2049–2053.
Milburn, N., Weiant, E. A., and Roeder, K. D. (1960). *Biol. Bull. (Woods Hole, Mass.)* **118**, 111–119.
Moore, F. L. (1990). *J. Exp. Zool., Suppl.* No. 4, 157–158.
Murphy, M. J., and Clark, N. B. (1990). *J. Exp. Zool., Suppl.* No. 4, 177–180.
Nebert, D. W., and Gonzalez, F. J. (1987). *Annu. Rev. Biochem.* **56**, 945–993.
Nicholls, T. J., Scanes, C. G., and Follett, B. K. (1973). *Gen. Comp. Endocrinol.* **21**, 84–98.
Nicoll, C. S. (1967). *Endocrinology (Baltimore)* **80**, 641–655.
Nicoll, C. S. (1990). *J. Exp. Zool., Suppl.* No. 4, 72.
Norman, A. W. (1990). *J. Exp. Zool., Suppl.* No. 4, 37–45.
Nottebohm, F. (1981). *Science* **214**, 1368–1370.
O'Malley, B. W., McGuire, W. L., Kohler, P. O., and Korenman, S. G. (1969). *Recent Prog. Horm. Res.* **25**, 105–160.
Orloff, J., and Handler, J. (1967). *Am. J. Med.* **42**, 757–768.
Parvinen, M. (1982). *Endocr. Rev.* **3**, 404–417.
Perez, L. E., Fenton, M., and Callard, I. P. (1991). *Comp. Biochem. Physiol. B* **100B**, 821–826.
Persson, B., Bengtsson-Olivecrona, G., Enerbach, S., Olivercrona, T., and Jornvall, H. (1989). *Eur. J. Biochem.* **179**, 39–45.
Peter, R. E., Chang, J. P., Nahorniak, C. S., Omeljaniuk, R. J., Sokolowska, M., Shih, S. H., and Billard, R. (1986). *Recent Prog. Horm. Res.* **42**, 513–548.
Peter, R. E., Yu, K.-L., Marchant, T., and Rosenblum, P. (1990). *J. Exp. Zool., Suppl.* No. 4, 84–89.

Petitte, J. N., Clark, M. E., Liu, G., Gibbins, A. M. V., and Etches, R. J. (1990). *Development* **108**, 185–189.
Phillips, J. A. (1990). *J. Exp. Zool., Suppl.* No. 4, 167–169.
Pickford, G. E., and Phillips, J. G. (1959). *Science* **130**, 454–455.
Pieau, C. (1974). *Ann. Embryol. Morphol.* **7**, 365–394.
Plisetskaya, E. M. (1990). *J. Exp. Zool., Suppl.* No. 4, 53–55.
Pukac, L. A., and Horseman, N. D. (1984). *Endocrinology (Baltimore)* **114**, 1718–1724.
Pukac, L. A., and Horseman, N. D. (1987). *Mol. Endocrinol.* **1**, 188–194.
Reinboth, R. (1970). *Mem Soc. Endocrinol.* **18**, 515–543.
Reinboth, R., ed. (1975). "Intersexuality in the Animal Kingdom." Springer-Verlag, New York.
Riddiford, L. M. (1976). *Nature (London)* **259**, 115–117.
Riddiford, L. M. (1993). *Adv. Insect Physiol.* (in press).
Riddiford, L. M., and Truman, J. W. (1992). *Am. Zool.* (in press).
Riddiford, L. M., Palli, S. R., and Hiruma, K. (1990). *Prog. Comp. Endocrinol.* 226–231.
Robinson, J. E. (1990). *Prog. Comp. Endocrinol.* 653–658.
Rudas, P., and Scanes, C. (1983). *Poultry Sci.* **62**, 1838–1845.
Rusaouen, M. (1976). *J. Exp. Mar. Biol. Ecol.* **23**, 267–283.
Rusaouen, M., Pujol, J.-P., Boquet, J., Veillard, A., and Bore, J. P. (1976). *Comp. Biochem. Physiol. B* **53B**, 539–543.
Sakurai, S., and Gilbert, L. I. (1990). *In* "Molting and Metamorphosis" (E. Ohnishi and H. Ishizaki, eds.), pp. 83–106. Jpn. Sci. Soc. Press, Tokyo and Springer-Verlag, Berlin.
Salama, K., and Williams, C. M. (1965). *Proc. Natl. Acad. Sci. U.S.A.* **54**, 411–414.
Saunders, D. S. (1982). "Insect Clocks," 2nd Ed. Pergamon, Oxford.
Scanes, C. G., and Griminger, P. (1990). *J. Exp. Zool., Suppl.* No. 4, 98–105.
Scharrer, E. (1928). *Z. Vgl. Physiol.* **7**, 1–38.
Schlinger, B., and Callard, G. V. (1989). *Endocrinology (Baltimore)* **124**, 437–443.
Schlinger, B., and Callard, G. V. (1989). *Neuroendocrinology* **49**, 433–441.
Schlinger, B., Palter, B., and Callard, G. V. (1987). *Physiol. Behav.* **40**, 343–348.
Schmidt-Nielsen, K., and Fange, R. (1958). *Nature (London)* **182**, 783–785.
Schneider, D. (1957). *Z. Vgl. Physiol.* **40**, 8–41.
Schneider, D. (1969). *Science* **163**, 1031–1037.
Sherwood, N. M., and Parker, D. B. (1990). *J. Exp. Zool., Suppl.* No. 4, 63–71.
Shih, J. C. H., ed. (1983). *Fed. Proc.* **42**, 2475–2552.
Shirai, H., and Walker, C. W. (1988). *In* "Endocrinology of Selected Invertebrate Types" (H. Laufer and R. G. H. Downer, eds.), pp. 453–476. Alan R. Liss, New York.
Shuttleworth, T. (ed.) (1987). *In* "Physiology of Elasmobranch Fishes," pp. 171–199. Springer-Verlag, Berlin.
Simpson, S. M., and Follett, B. K. (1982). *J. Comp. Physiol.* **145**, 391–398.
Smith, P. E. (1920). *Am. Anat. Mem.* **11**.
Speidel, C. C. (1919). *Carnegie Inst. Wash., Publ.* No. 13, 1–31.
Stifani, S., Barber, D. L., Nimpf, J., and Schneider, W. (1990). *Proc. Natl. Acad. Sci. U.S.A.* **87**, 1955–1959.
Stuenkel, E. L., and Cooke, I. M. (1988). *Curr. Top. Neuroendocrinol.* **9**, 124–150.
Templeton, J. R., Murrish, D. E., Randall, E. M., and Mugaas, J. N. (1972). *Z. Vgl. Physiol.* **76**, 255–269.
Thomas, P. (1990). *J. Exp. Zool., Suppl.* No. 4, 126–128.
Thummel, C. S. (1990). *BioEssays* **12**, 561–568.
Tilly, J. L., and Johnson, A. L. (1990). *Endocrinology (Baltimore)* **126**, 2079–2087.

Tilly, J. L., Kowalski, K. I., Johnson, A. L., and Hsuek, A. T. W. (1991). *Endocrinology (Baltimore)* **129**, 2799–28801.
Tobe, S. S., and Stay, B. (1985). *Adv. Insect Physiol.* **18**, 305–432.
Truman, J. W. (1992a). *Exp. Gerontol.* **27**, 17–28.
Truman, J. W. (1992b). *Prog. Brain Res.* **92**, 361–374.
Truman, J. W., and Riddiford, L. M. (1970). *Science* **167**, 1624–1626.
Truman, J. W., and Schwartz, L. M. (1984). *J. Neurosci.* **4**, 274–280.
Turrigiano, G. G., and Selverston, A. I. (1989). *J. Neurosci.* **9**, 2486–2501.
Ussing, H. H. (1952). *Conf. Renal Funct., Trans.* **4**, 1–88.
Vogt, R. G. (1987). *In* "Pheromone Biochemistry" (G. D. Prestwich and G. L. Blomquist, eds.), pp. 385–431. Academic Press, Orlando, Florida.
Vogt, R. G., and Riddiford, L. M. (1981). *Nature (London)* **293**, 161–163.
Vogt, R. G., Prestwich, G. D., and Lerner, M. R. (1991). *J. Neurobiol.* **22**, 74–84.
Wallace, R. A., and Dumont, J. N. (1968). *J. Cell. Physiol.* **72**, 73–90.
Wigglesworth, V. B. (1934). *Q. J. Microsc. Sci.* **77**, 191–222.
Wild, A. (1979). *Gen. Comp. Endocrinol.* **38**, 370–373.
Williams, C. M. (1956). *Nature (London)* **178**, 212–213.
Williams, C. M. (1967). *Sci. Am.* **217**, 13–17.
Wimsatt, W. A. (1942). *Anat. Rec.* **83**, 299–306.
Witschi, E. (1929). *J. Exp. Zool.* **52**, 267–292.
Yntema, C. L. (1976). *J. Morphol.* **2**, 453–462.
Yoshima, Y., and Tamura, T. (1991). *Gen. Comp. Endocrinol.* **84**, 222–227.
Yulick, S., and Kusch, K. (1960). *J. Am. Chem. Soc.* **82**, 6421.
Yund, M. A., King, D. S., and Fristrom, J. W. (1978). *Proc. Natl. Acad. Sci. U.S.A.* **75**, 6039–6043.
Zwarenstein, H. (1937). *Nature (London)* **139**, 112.

28

Proliferation Induced by Growth Factors in Mammalian Fetal Cells

Manuel Benito and Margarita Lorenzo

I. Introduction
II. Isolation and Plating of Cells
III. Establishment of Cell Quiescence
IV. Induction of Cell Proliferation
   A. Determination of DNA Synthesis
   B. Determination of Cell Number and DNA, RNA, and Protein Contents
   C. Analysis of the Cell Cycle
References

I. Introduction

Growth factors can act as positive or negative modulators of cell proliferation and influence differentiation. A readily accessible model to study this is the recruitment of quiescent mammalian cells into the cell cycle under defined conditions in tissue cultures. The general pattern of this process if exemplified by the tyrosine kinase receptor-mediated stimulation of proliferation in cultures of mammalian cells; for instance, the transmembrane tyrosine kinase receptors mediated proliferation in cultured fibroblasts (reviewed in Heldin and Westermark, 1990). The mitogenic response of NIH 3T3 cells, an established cell line from mouse fibroblasts, is considered a two-step process. The first step is the advance of quiescent cells (G_0) to the G_1 phase of the cell cycle by the initiation factors such as PDGF (platelet-derived growth factor), and the second step is the progression of those cells into the S phase of the cell cycle and subsequently commitment to DNA synthesis by the progression factors such as EGF (epidermal growth factor) and IGF-I (insulin-like growth factor-I) (Rozengurt, 1986). Less is known of the mitogenic response of mammalian cells in primary culture. Established cell lines are more widely used. The results obtained with primary cultures, however, reflect more closely the situation *in vivo*.

Hepatocytes isolated from adult liver (Thoresen *et al.*, 1990) or from regenerating liver after partial hepatectomy (reviewed in Michalopoulos, 1990) are able to proliferate *in vitro*. However, during fetal development, growth factors and hormones must play a significant role in regulating the orderly progression of liver growth and differentiation. It follows, therefore, that fetal cells have an intrinsic proliferative capacity which can be stimulated in primary culture under defined conditions. Hepatocytes of fetal rats have been shown to undergo division in primary culture under EGF stimulation (Hoffmann *et al.*, 1989; Molero *et al.*, 1992).

Brown adipocytes (fetal derived) also proliferate and differentiate in primary culture. Despite the limited survival time of the primary cultures, brown adipocyte primary cells have been successfully used to study metabolism as well as gene expression. Brown adipocytes come from a tissue specialized in carrying out thermogenesis and lipid synthesis and are distributed differently than the white adipose tissue. The expression of the uncoupling protein (Porras *et al.*, 1990) and its adrenergic regulation (Porras *et al.*, 1989) have been investigated. Other studies include lipogenic flux (Lorenzo *et al.*, 1988b), the hormonal regulation of lipogenesis (Lorenzo *et al.*, 1988a; and the expression of malic enzyme (Lorenzo *et al.*, 1989; Valverde *et al.*, 1992).

Fetal rat brown adipocyte primary cultures proliferate in response to an individual growth factor such as IGF-I or combinations, such as EGF plus vasopressin plus bombesin, or 10% fetal calf serum (FCS; Valverde *et al.*, 1991). The mitogenic response of these cells to growth factors is quite different from that described for mammalian cell lines, as for example, NIH-3T3 (Randazzo and Jarret, 1990). We have found a high expression of IGF-I receptors in fetal brown adipose tissue (unpublished results), which is consistent with IGF-I being the main individual growth factor involved in fetal brown adipocyte proliferation (Valverde *et al.*, 1991). In addition, IGF-I increases expression of glucose 6-phosphate dehydrogenase (Valverde *et al.*, 1991).

Although this chapter describes specifically a protocol of proliferation with primary cultures of brown adipocytes, the approach followed may be extended to other primary cultures and covers several parameters indicative of cell growth that should be considered in any system to study proliferation.

II. Isolation and Plating of Cells

Fetal brown adipocytes can be isolated from interscapular brown adipose tissue. Fetuses are delivered by rapid hysterectomy on Day 20–22 of gestation. Brown adipocytes are dispersed by a collagenase (Lorenzo et al., 1988a). Obtaining tissue from 22-day-old fetuses (30 mg/animal) is easier than from 20-day-old fetuses (10 mg/animal). Brown adipocytes from 22-day-old fetuses have been employed successfully for metabolic (Lorenzo et al., 1988a,b) as well as enzyme expression studies (Lorenzo et al., 1989; Valverde et al., 1992). Despite the skill needed for dissection of tissue from fetuses younger than 22-days, 20-day-old fetal brown adipocytes are more useful for proliferation studies.

After removal under sterile conditions, the tissue needs to be minced with a pair of fine scissors. The tissue is then placed in a 10-ml sterile bottle containing 100 mM Hepes/4% (w/v) bovine serum albumin (BSA) isolation buffer (Nèchad et al., 1983) containing 2 mg collagenase/ml. The mixture is incubated at 37°C for 30–45 min in a shaking water bath (100 strokes/min). Every 5 min the bottle must be *shaken* for 10 sec, for instance in an automatic mixer, to facilitate tissue digestion. After collagenase treatment the tissue is filtered through a 100-µm-pore-size nylon mesh. The free cells are collected at the bottom of a sterile plastic tube by centrifugation at 80g for 5 min (this g force is sufficient since brown adipocytes do not float) and are washed once with the isolation buffer. The procedure yields, respectively, 3.5×10^6 and 10×10^6 cells/100 mg for 22- or 20-day-old fetal brown adipose tissue. Cell viability, as determined by trypan blue exclusion, is routinely over 90%.

Cells are plated (1×10^6 cells/60-mm-diameter plastic dishes) in 2.5 ml of Eagle's medium modified with Earle's salts (MEM), 1 mM glutamine, and 20 mM Hepes, supplemented with 10% (v/v) FCS, 120 µg/ml penicillin, 100 µg/ml streptomycin, 50 µg/ml gentamycin, and 25 units nystatin/ml. The presence of FCS is essential to allow cells to attach to the plastic surface of the dish. After 4 hr of culture at 37°C under an atmosphere of 6% CO_2 in air with 80% humidity in a cell incubator, cells are rinsed twice with phosphate-buffered saline (PBS). This is to remove any red cells or cell debris remaining in the culture. At this stage, 70% of the initial cells are attached to form a monolayer as observed by inverse light microscopy. This represents about 30% confluence.

III. Establishment of Cell Quiescence

Cells are maintained for 20 hr in a serum-free medium supplemented with 0.2% (w/v) BSA (fatty acid free, e.g., from Sigma) to reduce the intrinsic mitogenic capacity of fetal cells. Propidium iodide staining method followed of flow-cytometric analysis (as described later in this chapter) indicates that

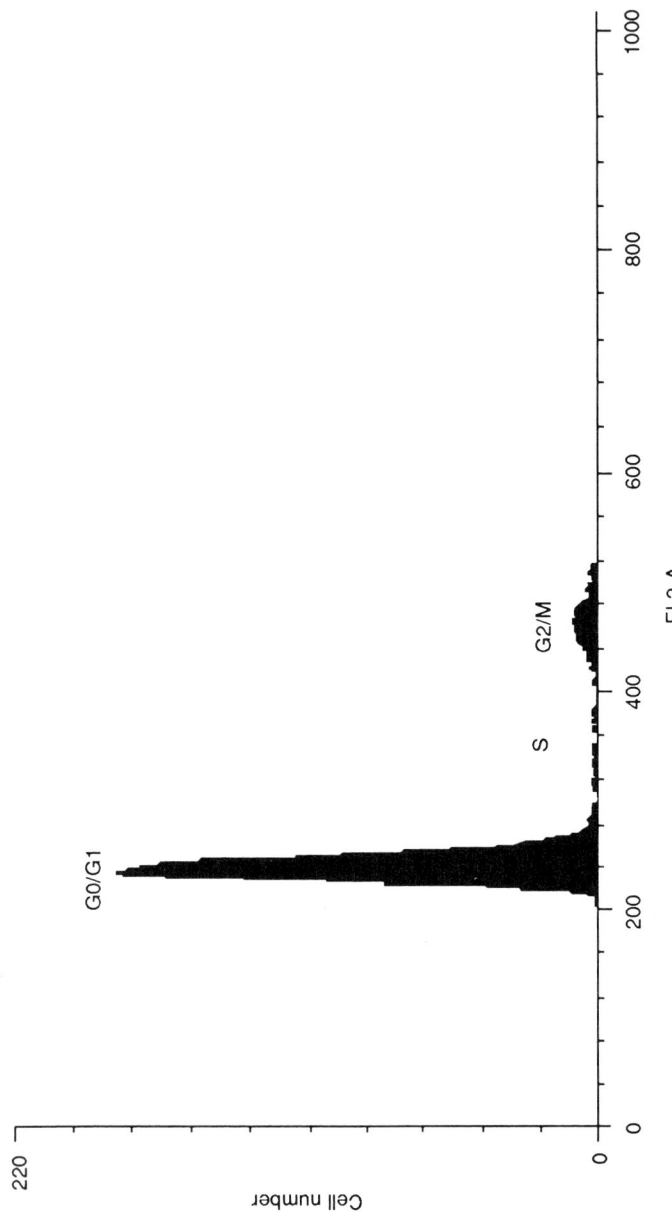

Figure 1 Cell cycle histogram in quiescent fetal brown adipocyte primary cultures. Fetal brown adipocytes from 20-day-old fetuses were plated for 4 hr in 10% FCS–MEM and cultured for 20 hr in 0.2% BSA–MEM. Cell suspensions were stained with propidium iodide and fluorescence intensity per cell (DNA per cell) was analyzed (FL2-^{A}). The histogram represents fluorescence plotted against cell number.

>95% of cells are in the G_0/G_1 phase of cell cycle at this time (Fig. 1). Under these experimental conditions, most of fetal adipocytes maintain a quiescent state. A low cellular density of approximately 2.5×10^4 cells/cm² is found to be suitable for proliferation studies.

The presence of fibroblasts contaminating cultures of primary cells is disturbing scientists working in this field. It is convenient, when dealing with primary cells, to check the phenotype of the cells after the establishment of the culture. Since specific antibodies are available in most laboratories, immunocytochemical detection of specific proteins expressed by primary cells is a good test of the quality of the culture and the degree of fibroblast contaminating. Routine control experiments can be performed to check the purity of the cultures, detecting the presence of the uncoupling protein, a specific mitochondrial protein of brown adipose tissue, by an immunocytochemical technique with rabbit anti-uncoupling protein serum. Dye immunodetection allows uncoupling protein expressing fetal brown adipocytes to be distinguished from fibroblast-like cells, which comprise less than 10% of the total cells.

Brown adipocytes can be further characterized by flow cytometric analysis of their size and endogenous fluorescence (Fig. 2). Cells can be detached from the monolayer with 0.25% trypsin–0.02% EDTA prior to examination on a flow cytometry (model FACScan from Becton–Dickinson). Size signals (FSC-H) are measured in an angle $<10°$ through a filter BP 488/10 nm. Endogenous flavin green fluorescence (FL1-H) was measured through a filter BP 530/30 nm after excitation with a laser at 488 nm. For instance, when we compared brown adipocytes from 20- and 22-day-old fetuses we observed that adipocytes obtained from 20-day-old fetuses were smaller than those from 22-day-old fetuses (Fig. 2A). Fibroblasts show different cytometric characters from brown adipocytes and can be eliminated by sorting the initial cell population. In our cultures their number is routinely lower than 10% of the total cells.

The measurement of flavin fluorescence gives evidence of the metabolic steady state of living cells. Flavins fluoresce when they are in the oxidized state; such fluorescence disappears during reduction (Thorell, 1983). Since brown adipocytes have a high number of mitochondria, the detection of endogenous fluorescence is possible. We observed the presence of two fluorescence peaks in 22-day-old fetal brown cells and a single peak (low fluorescence) in 20-day-old fetal cells (Fig. 2B). The graphical representation of adipocytes size plotted against endogenous fluorescence clearly demonstrates the existence of two population of cells in 22-day-old fetal cells (Fig. 2D), but only a single population in adipocytes from 20-day-old fetuses (Fig. 2C). Thus, 20-day-old fetal brown adipocytes appear to be a more appropriate model for studies of proliferation.

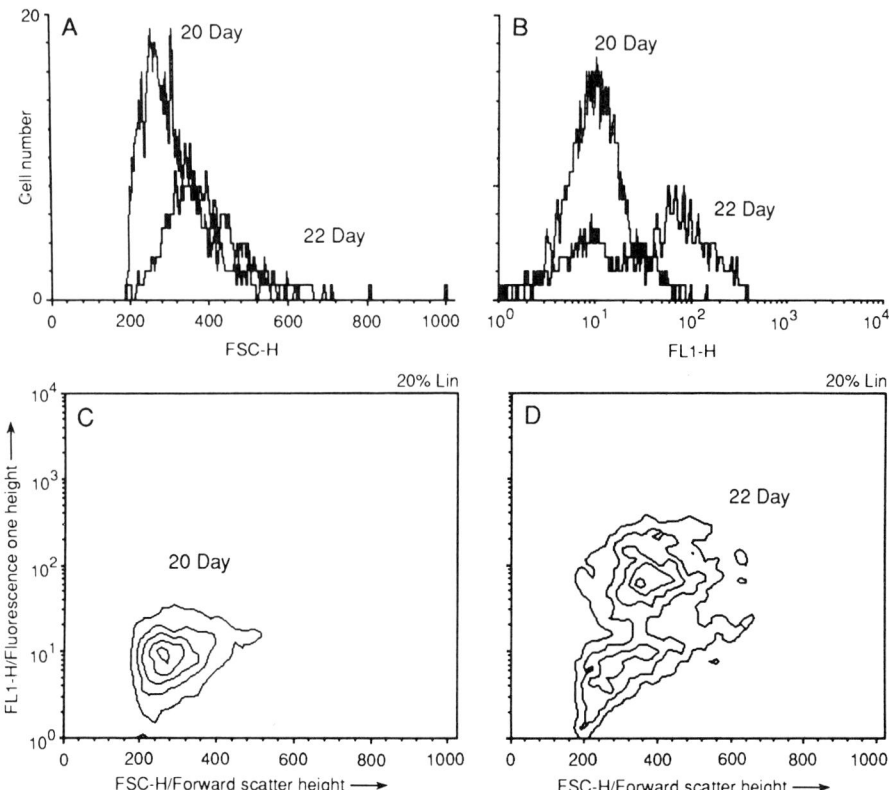

Figure 2 Flow cytometric analysis of size and endogenous fluorescence in fetal brown adipocyte primary cultures. Brown adipocytes from 20- and 22-day-old fetuses were plated for 4 hr in 10% FCS-MEM and cultured for 20 hr in 0.2% BSA-MEM. Cell suspensions were analyzed for size (FSC-H) and endogenous fluorescence (FL1-H). (A) Cell number vs FSC-H. (B) Cell number vs FL1-H. (C) FSC-H vs FL1-H in 20-day-old cells. (D) FSC-H vs FL1-H in 22-day-old cells.

IV. Induction of Cell Proliferation

Quiescent fetal brown adipocytes are used in studies in which proliferation is being determined. These cells (after incubation for 20 hr in a serum-free medium supplemented with 0.2% BSA) are rinsed again with PBS and further cultured for 24, 48, and 72 hr in MEM supplemented with 1% fetal calf serum and various growth factors. In each experiment, wells cultured in the absence of growth factors should be included as a negative control for cellular quiescence. Also cells cultured in the presence of 10% FCS should be included as a positive control for cell proliferation. Culture medium and additives are most effective when changed every 24 hr.

We tested various growth factors, since no evidence of the nature of growth factors involved in fetal brown adipocytes proliferation was available. These included (1) EGF, which influences fetal development in lung (Nielsen, 1989) and liver (Hoffmann et al., 1989); (2) IGF-I and IGF-II; (3) PDGF, which stimulates proliferation in various cell types, including Swiss 3T3 cells (Blakeley et al., 1989); and (4) the amphibian peptide bombesin and vasopressin, which are potent mitogens for Swiss 3T3 cells (Rozengurt and Sinnett-Smith, 1983). The role of cAMP in the regulation of mammalian cell proliferation has been a subject of controversy. β-Adrenergic stimulation of brown adipocyte proliferation in rat and mouse during cold acclimation has been reported (Geloen et al., 1988; Rehnmark and Nedergaard, 1989). However, cAMP inhibits mitogen-induced DNA synthesis in hamster fibroblasts (Magnaldo et al., 1989).

The incorporation of radiolabeled thymidine into DNA is the most widely used method to index cell proliferation. As the rate of DNA synthesis reflects the mitotic activity, the radioactivity of trichloroacetic acid-precipitable material should reflect cell proliferation. This method does not necessarily reflect the cell proliferation rate in all biological models. We suggest that results should be confirmed by counting the number of cells or determining DNA content *and* performing a precise analysis of the phases of cell cycle (see Fig. 3).

A. Determination of DNA Synthesis

DNA synthesis can be determined by [^{3}H]thymidine incorporation during the last 40 hr (continuous labeling) or 4 hr (pulse labeling) of culture (Murat et al., 1990).

Higher differences between the positive and negative control of proliferation are obtained when brown adipocytes are incubated in the presence of [^{3}H]thymidine (0.2 μCi/ml) over the last 4 hr of culture. At the end of the incubation period, cell layers are washed twice with ice-cold PBS, and DNA proteins are precipitated for 20 min on ice with 1 ml of ice-cold 10% (v/v) TCA. After washing twice with absolute ethanol, the precipitate is solubilized in 0.1 M NaOH–2% sodium carbonate–1% SDS. Aliquots are counted (for instance using Cocktail-22-Normascin scintillation liquid from Sharlau). Radioactivity is expressed in dpm after quenching and luminiscence correction (Kontron Betamatic I). Assays may be performed in duplicate dishes from at least five independent primary culture experiments. Results on radioactivity

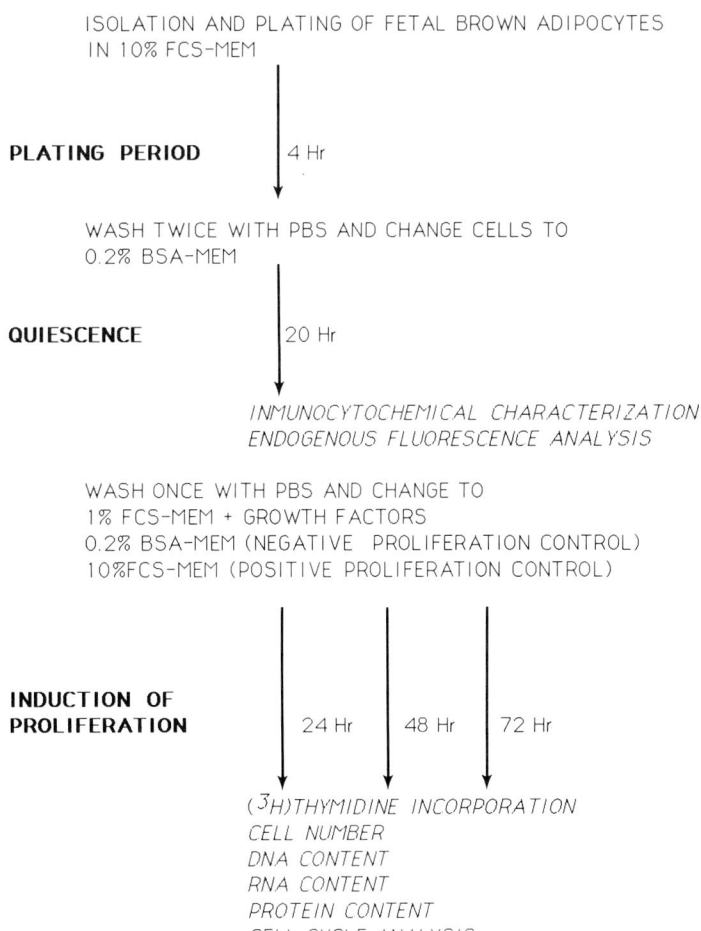

Figure 3 Schematic protocol of proliferation for fetal brown adipocytes in primary culture.

incorporated can be expressed as a percentage relative to the incorporation by the untreated cells.

A time course of [^{3}H]thymidine incorporation into cells in the presence of several growth factors should be performed initially. The maximal effect on DNA synthesis in response to various growth factors in fetal brown adipocyte primary cultures was at 72 hr. In the presence of 10% FCS (positive control) there is a three-fold increase in [^{3}H]thymidine incorporation. To test the mitogenic capacity of growth factors, a dose–response curve is essential. For instance, optimal stimulation of DNA synthesis by IGF-I is reached at 10 ng/

ml concentration (1.4 nM). The magnitude of the significant increase in DNA synthesis induced by IGF-I is higher than that observed with 10% FCS or other growth factors.

B. Determination of Cell Number and DNA, RNA, and Protein Contents

On the basis of the experiments on DNA synthesis, we have chosen adipocytes treated with IGF-I or FCS for further investigation. This included determination of cell number and total cellular DNA, RNA, and protein contents. As positive and negative controls of cell proliferation some dishes cultured in the presence of 10% FCS or in the absence of serum and growth factors, respectively, are included. At the end of the 72 hr of culture under the conditions previously described, cells were rinsed with PBS and specifically collected for each determination. All determinations must be performed in at least five independent experiments with duplicate dishes for each point.

Cell number is determined by flow-cytometric analysis on a FACScan flow cytometer after detaching cells from the monolayer with 0.25% trypsin–0.02% EDTA. DNA content is determined by a fluorimetric method based on the specific binding of the dye 33258-Hoechst (2.5 μg/ml from Sigma) with the adenine–thymidine base pairs of the DNA (Labarca and Paigen, 1980), with calf thymus DNA used as standard. Protein content is determined as described by Bradford (1976), with a reagent available commercially from Bio-Rad, using γ-globulin as standard. RNA content is determined after direct lysis of the monolayer with guanidinium thiocyanate-phenol reagent (RNAzol available commercially from CinnaBiotecx), following the protocol of Chomczynski and Sacchi (1987). The presence of 10% FCS or IGF-I for 72 hr increased cell number, DNA, RNA, and protein contents, as compared with quiescent untreated cells. The analysis of these results indicate a good correlation between DNA content and cell number in all the experimental conditions tested, the DNA content being 28.6 ± 0.8 pg DNA/cell.

C. Analysis of the Cell Cycle

Analysis of the cell cycle is the final (optimal) index of cell proliferation by growth factors. For instance, the percentage of cells in the S and G_2/M phases was determined at 72 hr, after stimulation with IGF-I and 10% FCS, and compared with quiescent cells (Table I). Flow cytometric analysis of the cell cycle is performed in cell suspensions after 0.25% trypsin–0.02% EDTA detachment of cells from the culture dishes. Cell nuclei were stained for

Table I
Analysis of the Cell Cycle in Fetal Brown Adipocyte Primary Cultures[a]

Phase	Addition		
	None	IGF-I	10% FCS
G_0/G_1	96.5	89.2	87.6
S	1.5	7.4	8.0
G_2/M	2.0	3.4	4.5

[a] Cells were cultured for 72 hr in the presence of IGF-I (1.4 nM) and 10% FCS, or in the absence or serum and growth factors. The cell cycle study was performed after propidium iodide staining of nucleus and flow cytometric analysis. Results are expressed as percentage of cells in the phases of the cell cycle.

15 min with ice-cold propidium iodide (125 µg/ml in 0.1% Na_3 citrate) as indicated in the cell cycle test reagent kit (Becton–Dickinson). Propidium iodide is an analogue of ethidium bromide that intercalates in double-stranded DNA helix with a resultant increase in fluorescence. Fluorescence (FL2-A) is measured through a filter BP 572/24 after excitation with a laser at 488 nm, with a double-discriminator module. Integration of the histograms obtained as in Fig. 1 gives us the percentage of cells in the phases of the cell cycle (Table I). Fetal brown adipocytes cultured for 72 hr in a serum-free medium are mainly in the G_0/G_1 quiescent phases of the cell cycle. The presence of IGF-I or 10% FCS for 72 hr increases the percentage of cells both in the S phase and in G_2/M phases compared with quiescent cells (Table I). Thus, both IGF-I and FCS induce the entrance to the cell cycle as well as stimulate the synthesis of DNA, in fetal adipocytes in primary cultures.

Acknowledgments

The authors thank Dr. Angela M. Valverde for carrying out the benchwork on primary cultures and Dr. Cesar Roncero for help in the artwork of the manuscript. We also thank Alberto Alvarez for his technical skill with the flow cytometer.

References

Blakeley, D. M., Corps, A. N., and Brown, K. D. (1989). *Biochem. J.* **258**, 177–185.
Bradford, M. (1976). *Anal. Biochem.* **72**, 248–254.
Chomczynski, P., and Sacchi, N. (1987). *Anal. Biochem.* **162**, 156–159.
Geloen, A., Collet, A. J., Guay, G., and Bukowiecki, L. J. (1988). *Am. J. Physiol.* **254**, 175–182.

Heldin, C. H., and Westermark, B. (1990). *J. Cell Sci.* **96**, 193–196.
Hoffmann, B., Piasecki, A., and Paul, D. (1989). *J. Cell. Physiol.* **139**, 654–662.
Labarca, C., and Paigen, K. (1980). *Anal. Biochem.* **102**, 344–352.
Lorenzo, M., Roncero, C., Fabregat, I., and Benito, M. (1988a). *Biochem. J.* **251**, 617–620.
Lorenzo, M., Fabregat, I., Roncero, C., and Benito, M. (1988b). *Biochem. Soc. Trans.* **16**, 274.
Lorenzo, M., Fabregat, I., and Benito, M. (1989). *Biochem. Biophys. Res. Commun.* **163**, 341–347.
Magnaldo, I., Pouyssegur, J., and Paris, S. (1989). *FEBS Lett.* **245**, 65–69.
Michalopoulos, G. K. (1990). *FASEB J.* **4**, 176–187.
Molero, C., Valverde, A., Benito, M., and Lorenzo, M. (1992). *Exp. Cell Res.* **200**, 295–300.
Murat, J.-C., Gamet, L., Cazenave, Y., and Trocheris, V. (1990). *Biochem. J.* **270**, 563–564.
Nèchad, M., Kuusela, P., Carneheim, C., Bjorntorp, P., Nedergard, J., and Cannon, B. (1983). *Exp. Cell Res.* **149**, 105–118.
Nielsen, H. C. (1989). *Biochim. Biophys. Acta* **1012**, 201–206.
Porras, A., Fernandez, M., and Benito, M. (1989). *Biochem. Biophys. Res. Commun.* **163**, 541–547.
Porras, A., Penas, M., Fernandez, M., and Benito, M. (1990). *In* "Endocrine and Biochemical Development of the Fetus and Neonate" (J. M. Cuezva and A. M. Pascual-Leone, eds.), pp. 147–152. Plenum, New York.
Randazzo, P. A., and Jarret, L. (1990). *Exp. Cell Res.* **190**, 31–39.
Rehnmark, S., and Nedergaard, J. (1989). *Exp. Cell Res.* **180**, 574–579.
Rozengurt, E. (1986). *Science* **234**, 161–166.
Rozengurt, E., and Sinnett-Smith, (1983). *Proc. Natl. Acad. Sci. USA* **80**, 2936–2940.
Thorell, B. (1983). *Cytometry* **4**, 61–65.
Thoresen, G. H., Sand, T., Refsnes, M., Dajani, O. F., Guren, T. K., Gladhaug, I. P., Killi, A., and Christoffersen, T. (1990). *J. Cell. Physiol.* **144**, 523–530.
Valverde, A. M., Benito, M., and Lorenzo, M. (1991). *Exp. Cell Res.* **194**, 232–237.
Valverde, A. M., Benito, M., and Lorenzo, M. (1992). *Eur. J. Biochem.* **203**, 313–319.

29

Expression of Hormones/Growth Factors and Their Receptors in Early Embryogenesis

Flora de Pablo, Gilbert A. Schultz, and Susan Heyner

I. Introduction
II. The Mammalian Preimplantation Embryo
   A. Identification at the RNA Level
   B. Identification at the Protein Level
III. The Chicken Embryo
   A. Identification at the RNA Level
   B. Identification at the Protein Level

I. Introduction

A profound reevaluation of the role of hormones and hormone-like peptides (growth factors and cytokines) in early embryonic development has taken place in recent years, as methods sufficiently sensitive to detect subnanogram concentrations of proteins or a few mRNA molecules per cell have become available. Simultaneous conceptual changes, such as those emerging from the findings that hormones can be produced by nonglandular tissues and that growth/differentiation factors are produced by most cells of vertebrates, have facilitated this advance (De Pablo and Roth, 1990). Therefore, the search for signaling factors and their receptors in the early stages of embryonic development in a variety of experimental models is presently feasible and an area of rapid expansion.

In this chapter, we describe techniques that have been used in our laboratories for the successful analysis of the expression of hormones/growth factors and their receptors in early mammalian and avian embryos. Other standard experimental models in developmental biology such as the fruit fly *Drosophila melanogaster*, the nematode *Caenorhabditis elegans*, and the amphibian *Xenopus laevis* have certain characteristics that may facilitate the study of the function of growth factors in development. For example, in *Drosophila* sophisticated genetics may help elucidate the developmental role of the insulin receptor, widely expressed in embryos (Garofalo and Rosen, 1988). *Xenopus* embryos can be grown synchronously in large quantities, moreover, they can be injected with proteins, mRNAs, and antibodies and, hence, are probably the best alternative for exploring which factors control embryonic induction (Melton, 1991; see also Callard *et al.*, this volume).

Independent of the model used, a strategy to look for any type of hormonal peptide or receptor in embryos should consider a few generally applicable points:

1. For the study of the expression of a hormone/growth factor, the major issue is the sensitivity of the assays employed, since embryonic signaling molecules and their mRNAs tend to be present in extremely low concentrations. For analysis at the peptide level, the most sensitive technique is partial purification of the molecule by chromatographic methods (HPLC, FPLC, affinity chromatography, etc.) followed by a highly specific immunoassay (Scavo *et al.*, 1989). Only if the peptide is expected to be highly localized one can use immunohistochemistry with a reasonable probability of detecting the peptide (De Pablo *et al.*, 1988; see also García-Arrarás, this volume). A new approach using PCR amplification coupled to immunochemistry awaits further confirmation of its applicability (Sano *et al.*, 1992). Immunoblotting and bioassays should be left for the study of more abundant factors or, alternatively, if the starting material is readily available and multiple purification steps allow for high enrichment and high yield of the factor.

For analysis at the mRNA level, the technique of choice is amplification of the reverse-transcribed mRNA using the polymerase chain reaction (PCR) (Serrano *et al.*, 1990; Caldés *et al.*, 1991; Shuldiner *et al.*, 1991; Scavo *et al.*, 1991a). Although strict quantitation of the initial transcript is difficult with this technique, optimization of the PCR conditions should permit remaining in the logarithmic amplification part of the reaction (before a plateau is reached), to obtain semi-quantitative results. Addition of an internal control allows correction for tube-to-tube variations in the reverse transcription and/or the PCR steps (Shuldiner *et al.*, 1991; Deltour *et al.*, 1993). Better quantitation will be perhaps possible when competitive PCR becomes more widely used (Siebert and Larrick, 1992). Less-sensitive, although more adaptable to exact quantitation, is solution hybridization followed by RNase protection analysis (see Adamo *et al.*, this volume). *In situ* hybridization, in our hands less sensitive, can be attempted when the mRNA is thought to be highly localized in certain cells or regions of the embryo (Serrano *et al.*, 1989; see also Bordy *et al.*, this volume).

2. For the study of the expression of receptors, in addition to the requirement of high sensitivity, the technique chosen has to demonstrate high specificity, since many hormonal/growth factor receptors belong to "families" which show certain degrees of interaction with related ligands. At the protein level, autoradiographic techniques permit sensitive detection of highly specific binding (see below). When the starting material is less limited (nonmammalian embryos in general), classical binding-competition studies with membrane preparations can also be performed (Bassas et al., 1988; Scavo et al., 1991b). At the mRNA level, as in the case of the ligands, the approach using PCR is preferred, and the strategy may facilitate the study of several related receptors simultaneously (Scavo et al., 1991a,b).

II. The Mammalian Preimplantation Embryo

The free-living, preimplantation stage of mammalian embryogenesis provides a unique opportunity to study the expression of ligands and receptors that may play roles in the regulation of cellular proliferation and differentiation of the earliest stages of mammalian development. Methods have been developed, during the past 4 decades, for the *in vitro* fertilization of oocytes and the culture of preimplantation embryos (to the stage at which implantation occurs). The ease with which this can be achieved differs considerably among species. The mouse has been the subject of most studies. This is partly due to the ease of recruiting a large number of oocytes (and hence embryos) by superovulation. In addition it is possible to culture embryos from the fertilized zygote to the periimplantation stage (see Hogan et al., 1986, for detailed methods). Taken together with the relatively low cost of maintaining this species and the amount of genetic information that is available, it is not surprising that the mouse serves as a paradigm for early mammalian development.

Although the mouse offers considerable opportunities for research in preimplantation mammalian development, there are also formidable difficulties in the study of the expression of ligands and receptors of polypeptide growth factors during preimplantation embryonic stages. The fertilized ovum contains only 23 pg of mRNA, which declines to approximately 8 pg in the two-cell stage and increases only to 37 pg by the blastocyst stage (reviewed in Schultz, 1986; Heyner et al., 1989a). Thus, if the investigator is interested in the analysis of a message that may be present in low copy number, then the technique of Northern analysis is not sufficiently sensitive. Similarly, the classical biochemical approach to the study of polypeptide growth factors and their receptors (Scatchard plot analysis) is not feasible for the study of early mammalian embryos. The techniques described below provide guidance for the recognition of messenger RNA molecules and the expression of the corresponding proteins in the preimplantation mouse embryo. Evidence is accruing that techniques that are successful in studies of the mouse embryo can be applied with great utility to other species. For example, Watson et al. (1992), using methods developed for study of the mouse embryo, have demonstrated expression of growth factors and their cognate receptors in bovine preimplantation embryos. As these analytical methods become more widely used, we anticipate an exponential increase in information that will lead to both improved methods for mammalian preimplantation embryo culture and a fuller understanding of factors that contribute to successful establishment of pregnancy.

A. Identification at the RNA Level

The use of molecular techniques to identify gene transcripts is one approach to the limitation imposed by paucity of biological material in the mammalian preimplantation embryo. In particular, the PCR-based methodology has proven to be extremely useful (see also Shuldiner et al., this volume). This technique involves the use of two synthetic oligodeoxynucleotides (primers) that flank a specific sequence of DNA. This target sequence is amplified by repeated cycles of heat denaturation of DNA, annealing of primers to complementary sequences, and extension of the annealed primers with a thermostable DNA polymerase. The primers hybridize to opposite strands of the target sequence and are oriented so that DNA synthesis by the enzyme proceeds across the region between the primers, effectively doubling the amount of the bracketed sequence with each successive cycle. This procedure permits selective enrichment and accumulation of a specific DNA fragment by several millionfold from small amounts of an initial template (Saiki et al., 1988).

A powerful modification of the PCR method employs reverse transcriptase (RT) to detect the presence of mRNA transcripts. This has been termed "RNA phenotyping," or "RT-PCR" (Rappolee et al., 1989). This method allows detection of RNA sequences of very small copy number and has been used successfully to detect a number of growth factor transcripts and their corresponding receptors in RNA derived from small numbers of mammalian cells or embryos (Rappolee et al., 1988, 1989, 1990; Heyner et al., 1989a; Hogan et al., 1991; Watson et al., 1992; Schultz et al., 1992). The major features of the method follow.

1. RNA Extraction

Rappolee et al. (1989, 1990) have successfully used the procedure in which cells (one to several hundred eggs or embryos) are lysed in guanidinium isothiocyanate (100 μl) and the mixture is layered over 5.7 M cesium chloride and subjected to ultracentrifugation to pellet the RNA. Full details of this method have been described previously (Berger and Kimmel, 1987; Rappolee et al., 1989). A modification of the Braude and Pelham (1979) microprocedure, in which RNA is extracted from preimplantation mouse embryos by standard phenol/chloroform procedures, has also proven to be satisfactory (see Hahnel et al., 1990; Hogan et al., 1991, for details). In both methods, 5–10 μg of *Escherichia coli* RNA (Boehringer-Mannheim) is included as carrier and final RNA pellets, obtained from ethanol precipitation, are washed twice with 70% ethanol to remove residual salts. The RNA is then dried under vacuum and dissolved in a small volume (approximately 5 μl) of sterile, distilled water for use in the reverse transcription assay.

2. Reverse Transcription

Synthesis of cDNA from polyadenylated mRNA molecules within the RNA preparation can be accomplished by RT in the presence of oligo(dT) primer. Full details of the method are described by Rappolee et al. (1989). A typical reaction can be carried out in a final volume of 10 μl in which the RNA is incubated at 42°C for 1 hr in a mixture containing 20 units of avian myeloblastosis virus or MuLV reverse transcriptase (Molecular Genetics Resources or Boehringer-Mannheim, Indianapolis, IN), 0.2 μg oligo(dT$_{12-18}$) primer (Pharmacia, Inc., Piscataway, NJ), reverse transcription buffer (50 mM Tris–HCl, pH 8.3, 60 mM KCl, 3 mM MgCl$_2$), 1 mM of each of dATP, dCTP, dGTP, and dTTP (Pharmacia), 1 μg of nuclease-free bovine serum albumin, and 5 units of RNasin (Promega, Wisconsin, MD) or RNAguard (Pharmacia). The reaction is then diluted to 30 μl with sterile water, heated to 95°C for 5 min, and chilled on ice to inactivate the reverse transcriptase and denature the RNA/cDNA hybrid. The reverse transcription reaction can also be carried out with a specific 3' (downstream) primer (see Kawasaki, 1990, for details).

3. Polymerase Chain Reaction

The polymerase chain reaction involves two synthetic oligonucleotides (primers), usually 20 to 30 nucleotides in length, that flank a target sequence of DNA. The target sequence is amplified, as described in detail in Shuldiner et al., this volume. As in all PCR assays, the reaction should be optimized for each primer set of interest with respect to magnesium ion, deoxynucleotide triphosphate (dNTP), and oligonucleotide primer concentrations. In general, with the dNTP and primer concentrations used in the RT-PCR procedures of Rappolee *et al.* (1989) magnesium concentrations in the range of 2.5 mM are a good starting point for most assays.

In the RT-PCR procedure, it is common to make RNA preparations from pools of up as many as 100 mouse oocytes or early embryos per sample. After reverse transcription, this is represented by cDNA molecules in a final volume of 30 µl (see above). This cDNA pool can be divided into aliquots such that several different primer pairs complementary to their respective cDNAs can be used simultaneously for PCR analyses within each original preparation. Thus, in practice, in the RT-PCR procedure, 1/5 to 1/10 of the reverse transcription reaction (3 to 6 µl representing cDNA from the equivalent of roughly 10 to 20 mouse oocytes or embryos) is amplified by PCR in a final volume of 50 µl containing 1 unit of *Taq* DNA polymerase (Perkins–Elmer Cetus), PCR buffer (10 mM Tris–HCl, pH 8.3, 50 mM KCl, 2.5 mM $MgCl_2$), 5 µg nuclease-free bovine serum albumin, 1.5 µM of each sequence-specific primer, and 0.6 mM of each dNTP. The mixture is overlayed with mineral oil (Sigma M5904) to prevent evaporation and amplified by 25 to 40 PCR cycles (the number of cycles should be in the lower range for messenger RNAs that are less rare to avoid reaching the plateau in the amplification reaction). Each cycle includes denaturation at 94°C for 1 min, annealing of primers at 55 to 72°C (see below) for 1 to 2 min, and extension at 72°C for 2 min.

When amplification has been completed, the product can be analyzed on a composite agarose gel (3% NuSeive and 1% Seakem agaroses, FMC Corporation, Marine Colloids Div. Rockland, ME, in Tris–borate buffer) or 6 to 10% polyacrylamide gels along with a 1-kb molecular weight DNA ladder (BRL) as marker. A total of 5 to 10 µl of the PCR reaction is used. Gels are stained with ethidium bromide to detect PCR amplification products.

4. Designing Primers

This is discussed in Shuldiner *et al.*, this volume, and reviewed in detail in Rappolee *et al.* (1989). A few essential points include:

   a. oligonucleotide primers between 20 to 30 nucleotides in length, with a T_m near 72°C have proven to yield good specificity in PCR reactions.
   b. primers should have a balanced G/C and A/T concentration and lack secondary structure.
   c. Primers must not be complementary with each other. In particular, 3′

sequence overlaps should be avoided to prevent primer dimer formation.
d. Primers should generally bracket a sequence of 200 to 600 nucleotides (note, however, that longer targets can also be amplified).
e. Primers should be designed to "bracket" an intron, so that amplification from contaminating genomic DNA can be distinguished from the cDNA of interest.
f. Primers should bracket a sequence with a diagnostic restriction site (or a sequence that covers a cDNA insert possessed by the laboratory), so that the identity of the PCR product can be validated by restriction analysis or Southern blot.

Applications RT-PCR methods have been used successfully to classify a number of growth factors and their corresponding receptors (Fig. 1), (Rappolee *et al.*, 1988; Heyner *et al.*, 1989a; Rappolee *et al.*, 1990; Watson *et al.*, 1992; Schultz *et al.*, 1992). In addition, these methods have been used to identify glucose transporter isoforms expressed by early mouse embryos (Hogan *et al.*, 1991).

B. Identification at the Protein Level

In addition to identifying gene transcripts of growth factor ligands and receptors, it is important to confirm expression at the protein level. Several methods to demonstrate expression have been used with preimplantation mammalian embryos; these include indirect immunofluorescence, autoradiography, high-resolution immunoelectron microscopy, and immunoblot analysis.

1. Indirect Immunofluorescence

Indirect immunofluorescence is a rapid means of providing evidence for expression of receptors and ligands at the light microscopic level (see also Smith and Jarett, this volume). It does not give high-quality cellular resolution (due to the geometry of the embryo) but is a very useful technique. It requires antibodies to the receptor and, if available, preimmune serum as a control. If antibodies to the receptor are directed against the extracellular domain, then the procedure can be carried out using living embryos. If the receptor domain is cytoplasmic, or the antibodies are directed against a ligand that is produced by the embryo, then the embryos require fixation and permeabilization. All

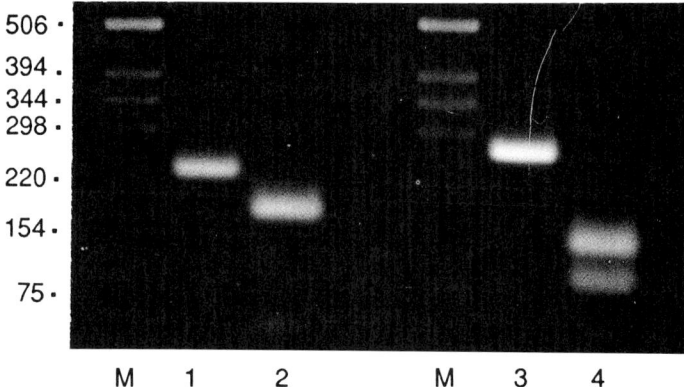

Figure 1 Detection of platelet-derived growth factor (PDGF-αR) and transforming growth factor-β_2 (TGF-β_2) transcripts in bovine blastocysts by RT-PCR. RNA was prepared from a small pool of blastocysts and reverse transcribed with oligo(dT) primer. The reverse transcription product from the equivalent of six blastocysts was used for 40 cycles of PCR in the presence of primers specific for PDGF-αR cDNA (lanes 1 and 2) or TGF-β_2 cDNA (lanes 3 and 4). PCR products were resolved on 2% agarose gels to yield the expected 235-bp product for PDGF-αR in lane 1 and the expected 273-bp product for TGF-β_2 transcripts in lane 3. The PCR primers utilized have been described in Watson et al. (1992). Identity of the PDGF- R PCR product was verified by digestion with *Hin*fI to yield the expected 181- and 154-bp fragments (lane 2) and the TGF-β_2 PCR product was digested with *Sph*I to yield the expected 154- and 119-bp fragments (lane 4). Lanes marked M contain a small-molecular-weight DNA ladder with sizes in base pairs marked on the left side of the figure.

steps that follow are carried out in 50-μl drops under an overlay of silicon (Dow Corning Corp., Midland, MI) or paraffin oil (Squibb, Princeton, NJ) (details of the microdrop preparation are described in Hogan et al., 1986). Reactions are carried out at room temperature.

1. Embryos are treated briefly with acidic phosphate-buffered saline (PBS), pH 2.5, and returned to culture medium for 60 min prior to assay. This step is preferred for living embryos; it is optional otherwise.
2. Embryos are fixed in 3% neutral buffered formaldehyde for 30 min and permeabilized by incubation in 0.1% Tween 20 (Sigma Chemical Co., St. Louis, MO) for 10 min.
3. A blocking step is included for fixed as well as nonfixed embryos. This consists typically of a 30- to 60-min incubation in 10–20% serum from an "innocent bystander" species, such as goat or donkey, in PBS containing 2% bovine serum albumin (BSA, fraction V, Sigma).

4. Embryos are incubated in the primary antibody or control serum (e.g., rabbit anti-receptor antibody; typical dilutions are 1:200 to 1:1000 in PBS-BSA) for 20-30 min. They are washed by passing through three successive microdrops of PBS-BSA, incubated with the second, fluorochrome-conjugated antibody (e.g. rhodamine-conjugated porcine anti-rabbit IgG), and washed as previously, before mounting in a drop of 30% glycerol in PBS. Another important control is the omission of the first antibody, to confirm specificity of the fluorochrome conjugate.
5. Embryos are evaluated under a microscope equipped for fluorescence and are photographed using standard exposure times in order to compare antibody-exposed embryos with controls.

Applications Indirect immunofluorescence has been used to confirm expression of a number of growth factor receptors and ligands in early mouse embryos (Rappolee *et al.*, 1988, 1990; Rosenblum *et al.*, 1986).

2. Autoradiography

Autoradiography permits the investigator to demonstrate that binding of a particular ligand is receptor mediated. It provides evidence for cellular localization and is very useful when there are limited amounts of material. It is time consuming to implement and, if the signal is weak, requires lengthy exposure times.

1. Embryos are flushed at the desired developmental stage from the reproductive tract. They are incubated at 37°C in control medium in microdrops under oil for approximately 1 hr in order to ensure that receptors are unoccupied.
2. Embryos are transferred to microdrops containing labeled ligand (e.g., [^{125}I] insulin) at a physiological concentration (optimal binding is reached at 4 ng/ml of insulin); embryos are incubated for approximately 1 hr, then washed as described above, by passage through microdrops, fixed, dehydrated, embedded in Epon, and sectioned.
3. Immobilization of embryos is achieved using a modification of the method described by Wiley *et al.* (1985). This method allows for the immobilization in a closely grouped, coplanar arrangement that permits ease of sectioning. Embryos are pipetted through drops of cold 1% phytohemagglutinin (Difco Laboratories, Lavonia, MI), placed on a monolayer of confluent fibroblasts on Thermanox plastic coverslips (Nunc, Inc. Naperville, IL), and grouped for convenience in sectioning. Excess fluid is withdrawn and replaced by ice-cold fixative (2.5%

glutaraldehyde in 0.1 M sodium cacodylate buffer, pH 7.4). After fixation (1 hr on ice), the coverslips are washed extensively with buffer, dehydrated, and infiltrated with resin. After polymerization of the resin, 5-μm-thick sections are cut and processed for autoradiography.

4. Sections are placed on glass slides, covered with photographic emulsion [Kodak NTB-2 (Eastman Kodak Co. Rochester, NY) diluted 1:1 with distilled water], and dried overnight. Slides are stored in light-tight boxes containing dessicant and exposed for periods ranging from 1 to 3 weeks at 4°C. Autoradiograms are developed with Kodak D-19 developer, stained with toluidine blue O (0.5% toluidine blue in 0.5% $Na_2B_4O_7$, pH 11), and mounted for microscopic evaluation. Grains are counted under a microscope at 400× magnification. Specific cell-associated counts are obtained by subtracting background grain counts. For controls, the labeled ligand is omitted and coincubated with 1000× excess of the unlabeled ligand. Displacement of the label provides evidence that binding is receptor mediated.

Applications This method has been used successfully, for instance, to demonstrate insulin binding to the cells of the morula-stage mouse preimplantation embryo (Mattson *et al.,* 1988).

3. High-Resolution Electron Microscopy

This approach provides extremely high cellular resolution and permits studies to follow the binding and internalization of growth factors and hormones. The ligand under investigation is labeled with colloidal gold, using the methods described by Smith and Jarett, this volume. Embryos are recovered from the reproductive tract and either subjected to experimental treatment or immobilized immediately, using the methods described above.

1. In order to examine binding and internalization of a particular ligand, embryos of the desired stages are flushed from the maternal tract and cultured briefly so that receptors are unoccupied. Embryos are then cultured in microdrops under oil for an appropriate time (20 – 45 min) in a microdrop overlaid with oil, containing the gold-labeled ligand (Smith *et al.,* 1988) or the control. Initial experiments should be run in order to determine the optimal incubation time so that the events of cell surface binding and internalization may be observed. Specificity of the ligand is demonstrated by incubating embryos with an identical concentration of gold-labeled BSA, or incubating embryos

with an excess of unlabeled ligand, to demonstrate displacement of the label (Heyner et al., 1989b).
2. Following incubation, embryos are washed briefly in ice-cold PBS and then immobilized as described above. As soon as the embryos are firmly adherent, the coverslip is flooded with fixative (2% glutaraldehyde in 0.1 M sodium cacodylate buffer for conventional transmission electron microscopy) and fixed for 1 hr. Following fixation, the cells are washed extensively with cacodylate buffer and osmicated for 1 hr in 2% osmium tetroxide in 0.1 M sodium cacodylate buffer, pH 7.4.
3. The cells and adherent embryos are dehydrated in an ascending series of ethanol, and then passed through three changes of Epon:ethanol (1:2; 2:1; 3:1) of 3 hr each. The final infiltration of resin is achieved by inverting the coverslip, cell-side down, over the top of a BEEM capsule that is filled with Epon and allowing the resin to polymerize for 24 hr at 60°C. Following polymerization, the coverslip is peeled off, leaving the embryos and the cell layer in the resin. For purposes of orientation, 1-μm sections are cut and stained with toluidine blue (0.5% toluidine blue in 0.5% $Na_2B_4O_7$, pH 11). Thin sections are cut, and gold particle counts are made from thin sections across the embryos (Fig. 2).

Applications This method has been successfully applied to demonstrate that preimplantation mouse embryos internalize insulin via coated pits, and at the blastocyst stage, insulin is translocated across the trophectoderm to the inner cell mass (Heyner et al., 1989b).

4. High-Resolution Immunoelectron Microscopy

In this technique, antibody directed against a particular ligand or receptor is labeled and used for analysis. It differs from the previous method in the fixative used and the processing techniques.

1. Embryos are flushed from the reproductive tract or treated experimentally. They are then immobilized on a layer of confluent cells, as described above.
2. Fixation is in BGPA fixative (1% glutaraldehyde, 0.2% picric acid in 154 mM NaCl containing 10 mM $NaPO_4$ and 6 mM eserine, pH 7.4) for 1 hr on ice.
3. Following fixation, embryos are washed extensively in PBS and partially dehydrated with ethanol.

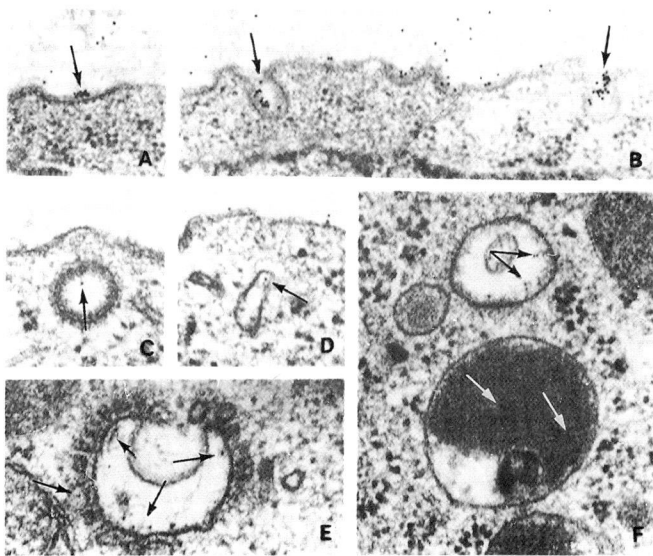

Figure 2 Electron micrographs illustrating binding and internalization of gold-labeled insulin in the blastocyst-stage mouse embryo. Gold-labeled insulin (arrows) clustered over thickened areas of trophoectoderm plasma membrane (A) was internalized in coated pits (B) and transported through the cell in coated vesicular structures (C) or noncoated vesicles (D). Ligand was found in various sizes endosomes (E), multivesicular bodies, and dense bodies (F). (Modified from Heyner *et al.*, 1989a).

4. Embryos are stained with eosin (1 in 70% ethanol) so that they remain visible, before being passed through several changes of LR White resin (London Resin Co., Ltd., London, U.K.).

5. The area of the coverslip containing the embryos is punched out with a paper punch and placed cell-side down in a gelatine capsule filled with LR White resin.

6. The blocks are cured at 55°C for 36–48 hr.

7. Thin sections are cut and mounted on naked nickel grids. The sections are then incubated in a drop of 0.1% BSA in 10 mM phosphate buffer (NaPO$_4$), pH 8.2, for 30–60 min, at room temperature, and then placed in suitably diluted drops of gold-labeled antibody, to be incubated at 4°C overnight.

8. The grids are then washed extensively in phosphate buffer and deionized water, stained with neutralized aqueous uranyl acetate for 5 min, and then washed with deionized water.

9. Sections are then evaluated by electron microscopy for the location of gold particles.

Applications This method has been used to show that blastocysts take up maternal insulin from the reproductive tract in the mouse (Heyner *et al.*, 1989b). It has also been used to show the localization of TGF-α in the mouse embryo (Dardik and G. A. Schultz, personal communication) and to demonstrate the developmental expression and cellular localization of glucose transporter molecules (Aghayan *et al.*, 1992).

5. Immunoblot Analysis

Immunoblot analysis has been used widely in order to identify molecules of interest and to assign molecular weights to mature proteins as well as to post-translational variants. It can be used in the study of preimplantation mammalian development, but, depending on the antibody and the level of protein expression, may require fairly large numbers of embryos.

1. Embryos at the desired developmental stage are flushed from the reproductive tract and cultured for 1 hr.
2. They are washed extensively, resuspended in sample buffer (Laemmli, 1970) containing 5% β-mercaptoethanol (Sigma Chemical Co., St. Louis, MO), and boiled for 3 min.
3. Groups of approximately 200 embryos are subjected to a 12% 1D SDS–PAGE electrophoresis using a minigel system (Bio-Rad, Richmond, CA).
4. The gels are electrotransferred to nitrocellulose membranes overnight at 30 V, 4°C.
5. Following transfer, immunoblot analysis is carried out using either of the two following methods:
   a. Filters are washed for 10 min in distilled water, for 20 min in Tris-buffered saline (TBS: 20 mM Tris, pH 7.4, 150 mM NaCl) and 0.1% Tween 20 (Sigma), and for 30 min in TBS, 5% nonfat dry milk and 0.2% NP-40 (Sigma), at 37°C. Incubation with antibody or the preimmune serum with a 1 : 200 dilution is 2 hr at room temperature in TBS, 5% nonfat dry milk, and 0.2% NP-40. The filters are finally washed for 20 min in TBS and 0.1% Tween 20 and incubated in 0.25 mCi/ml ^{125}I-Protein A (Amersham Corp. Arlington Heights, IL) for 30 min. Filters are washed again and exposed for autoradiography using Kodak X-OMAT AR (Eastman Kodak, Rochester, NY) film at -70°C.

b. Filters are placed in the washing buffer (Tris, 50 mM; NaCl, 150 mM, pH 10.2; Na azide, 5 mM) for 6 hr and for 1 hr in the blocking buffer (washing buffer, 0.05% Tween 20, and 5% nonfat dry milk) at room temperature. They are then incubated with antibody (usually around 1:500), the absorbed antibody control, and the corresponding preimmune serum (1:1000), overnight at room temperature. All antibodies are diluted in washing buffer, containing 0.05% Tween 20. The following morning, the filters are washed for 30 min in washing buffer, with 0.05% Tween 20, prior to incubation with horseradish peroxidase-conjugated goat anti-rabbit IgG (Sigma) at room temperature with a 1:20000 dilution. The filters are finally washed for 50 min in washing buffer, with 0.05% Tween 20, and incubated 1 min in the enhanced chemiluminescence reagents (ECL Western blotting detection system from Amersham). Exposure time to Kodak X-OMAT AR films is 5 to 30 sec depending on the antibody used.

Applications Immunoblot analysis has been used successfully to demonstrate the developmental expression of glucose transporter molecules (Aghayan *et al.*, 1992).

III. The Chicken Embryo

Of all the nonmammalian developmental models, the chicken has the early phases of embryogenesis closest to the human embryo. Chicken development has been well known at the morphological level for many decades (Romanoff, 1960) and the embryo is accessible, independent from maternal influences, from the blastoderm stage until hatching. Other advantages of this system are that large quantities of embryos and specific organs can be obtained for biochemical studies, manipulations are relatively easy, and embryonic microsurgery can be performed in certain areas. The maintenance of chicken embryos in the laboratory is simple and economical. The major disadvantage of this system is the very limited genetic studies feasible, since few inbred lines are available.

A. Identification at the RNA Level

As discussed extensively for mammalian embryos, PCR techniques have allowed studies of the expression of hormones and receptors that would not have been possible by other methods. With chicken embryos the PCR protocols applied are similar to those described for mammalian embryos (for details see Serrano *et al.*, 1990; Caldés *et al.*, 1991; Scavo *et al.*, 1991a). RNase protection assays (Kikuchi *et al.*, 1991) and *in situ* hybridization (Serrano *et al.*, 1989) are clearly less sensitive when using very young chicken embryos and may fail to detect transcripts that are detected by PCR (Serrano *et al.*, 1990; F. de Pablo *et al.*, unpublished observations).

B. Identification at the Protein Level

1. Autoradiography of Whole Chick Embryos to Study Receptors

This type of study can be performed with embryos from the stage of "primitive streak" (about 6 hr of incubation) until neurulation is completed and early organogenesis is underway (about 50 hr of development, stage 14, embryo with 22 somites) (Hamburger and Hamilton, 1951). Older embryos are too thick and the nonspecific binding increases greatly (Girbau *et al.*, 1992). The autoradiography of the whole embryo provides an image that allows gross localization of receptors in different areas of the embryo, although it is not adequate for resolution at the cellular level.

1. Embryos are obtained from fertilized eggs incubated at 38°C in a humidified incubator for the desired period of time. It is important to rotate automatically the embryos every few hours to avoid adherence of the embryonic area to the shell membrane which would make recovery of an intact embryo more difficult.
2. The egg is opened from the pointed end and as much albumen as possible is cut away with scissors.
3. The yolk is then placed in a glass bowl filled with PBS and the embryonic area is identified, usually on the upper side of the yolk (if it is not visible, turn the egg gently with the aid of a perforated spoon).
4. Under a dissection microscope, the embryo is identified in the middle of a more transparent area (area pellucida, Fig. 3). The thicker area around (area opaca) is carefully cut off from the yolk by using Pascheff scissors and fine (N°5) forceps, and put in a culture dish filled with cold PBS, where it is cleared of yolk. Addition of 50 μl of 0.02% phenol red over the embryonic area may help in the identification before dissection.

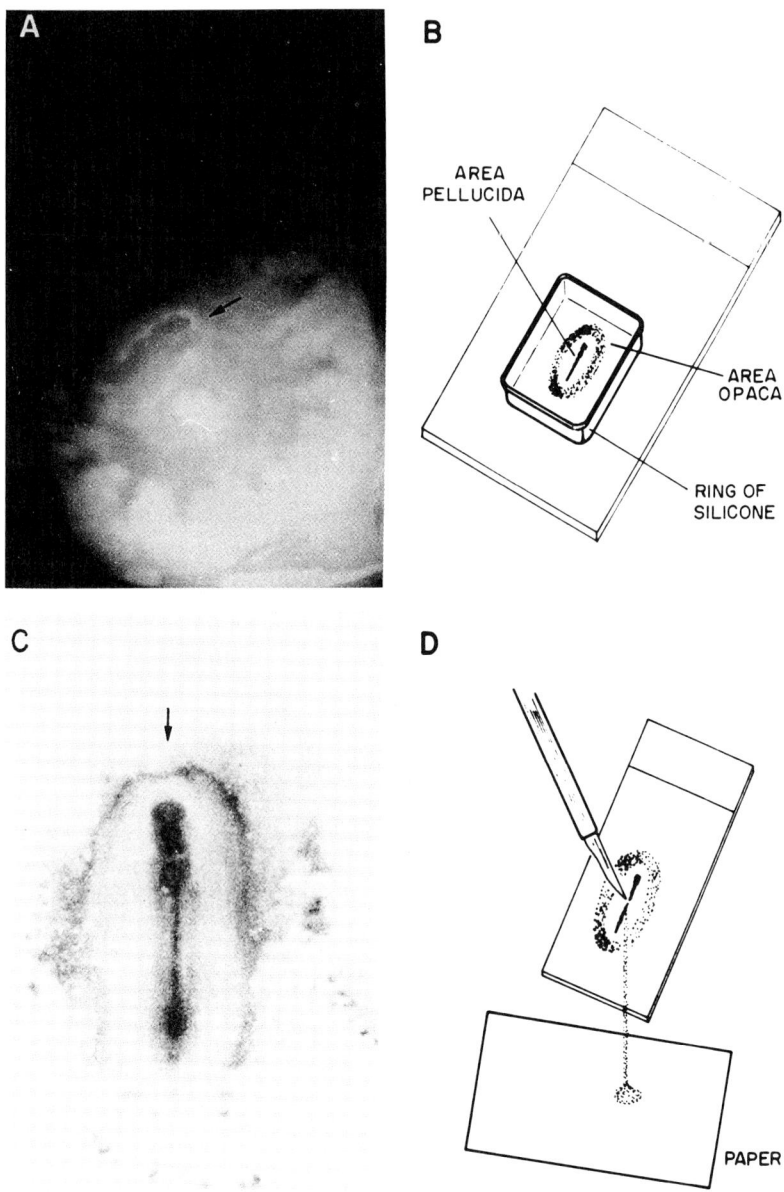

Figure 3 Schematic diagram of a ligand-binding study performed *in situ* in whole-mounted chicken embryos. (A) The chick embryo at approximately 30 hr of incubation (stage 8–9 of Hamburger and Hamilton) is recognized on the animal pole of the yolk. The arrow points to the cefalic end of the embryo (magnification 2.5×). (B) Scheme of a blastoderm and extraembryonic membranes positioned on a glass slide surrounded by a ring of silicone to hold the reagents in

5. The whole blastodisc including the area opaca is trimmed and gently extended on a gelatin-coated glass slide where the stage of the embryo is determined (Hamburger and Hamilton, 1951).
6. The slides are dried in a vacuum desiccator overnight at 4°C and they are then stored at −70°C until binding studies are performed (usually within a few days).
7. For the *in situ* binding study the embryos are brought to room temperature. A silicone ring of 2×2 cm is created around the embryo (Fig. 3) to hold the approximately 300-μl volume of a typical assay and the slides are placed horizontally over a wet filter paper on a tray.
8. The embryos are washed two or three times for 5 min by adding 250 μl of cold PBS (pH 7.8) and then incubated with labeled hormone for 6–8 hr at 4°C in a humid chamber. For insulin and insulin-like growth factor-I (IGF-I) binding the assay buffer contains 50 mM Hepes (pH 7.8), 120 mM Na Cl, 15 mM $NaC_2H_3O_2$, 10 mM glucose, 2.5 mM KCl, 1.2 mM $MgSO_4$, 1 mM EDTA, bovine serum albumin (0.5%, insulin-free from Sigma, A4503), and bacitracin (1 mg/ml).
9. To determine total binding the embryos are incubated in 300 μl assay buffer containing ^{125}I-peptide (approximately 100 cpm/μl or a total of 30,000 cpm per embryo).
10. To obtain nonspecific binding, embryos of similar stage are incubated with the labeled peptide plus an excess of unlabeled peptide (1 to 10 μg/ml). To generate a full competition curve for establishing specificity, intermediate concentrations of unlabeled peptide, or other analogs, can be added to similar embryos (best done with embryos of stages 7 to 11).
11. After incubation is completed, the embryo is washed 10 times, for 1 min, dried with a fan at room temperature for approximately 30 min, and fixed in a vacuum desiccator containing vapors of paraformaldehyde powder, heated at 80°C (in a stove) for two hours.
12. The slides are cooled at room temperature and placed in a X-ray

---

place. (C) Autoradiographic image of ^{125}I-insulin binding to a whole-mounted embryo. Arrow corresponds to cephalic end as indicated in A. (D) Scraping the desired area of the blastoderm from the slide into a paper, the radioactivity can be quantitated (From Girbau *et al.*, 1992; reprinted with permission.).

cassette between filter papers. Exposure to film (Kodak X-OMAT AR) can be for 7 or more days at 4°C to obtain an autoradiogram. Computer color-coded images can be produced from the autoradiogram by an appropriate color-coding densitometric analysis.

13. After obtaining the desired exposure, quantitation of individual embryo binding is possible by scraping the embryo (leaving the extraembryonic membrane on the slide) on a small paper that is counted in a gamma counter tube (Fig. 3).

Applications This technique has permitted macroscopic localization and quantitation of binding and competition studies on a single embryo for insulin receptors and IGF-I receptors (Girbau *et al.*, 1989, 1992).

2. SepPak Fractionation and HPLC for Detection of Polypeptides

To analyze the presence of hormones/growth factors in embryo extracts, embryos are incubated under standard conditions and removed from the shell at the desired age. For embryos up to 2 days of development, the dissection technique is similar to that described for studying receptors, with minor modifications. The egg contents can be placed in an empty large petri dish. The embryonic area is identified and after immobilization of the area with a ring of filter paper, the extraembryonic membranes are cut around this ring. The embryo is isolated from the membranes while maintained in a black-bottom glass dissection cuvette containing a few drops of PBS. Many embryos are collected together after counting (typically, in the order of 100) with estimates of group weight rather than individual weight. For studies with embryos of 3 days of development and older, the dissection can proceed directly over the yolk surface with fine forceps and scissors, cutting through the extraembryonic membranes surrounding the embryo. Then this whole area can be "lifted" with curved forceps and is washed in ice-cold PBS; the embryo is further cleaned off from the membranes while maintained in the dissection cuvette. Groups of 10 or more embryos, depending on the age, are usually pooled together, although individual embryos can be weighed and processed.

1. Embryos are homogenized with a mechanical homogenizer (Polytron or Ultra-Turrax type) for approximately 2 min (three to four pulses of 30 sec) in 5 to 20 vol (w/v) of an acidic extraction medium selected to maximize solubilization of peptides. Different extraction methods may favor recovery vs relative purification, and different peptides may behave differently. Some possible alternative media are

a. 0.5 N HCl alone
b. Acid acetone (80% acetone/0.5 N HCl)
c. 1% NaCl/1 N HCl/1% trifluoroacetic acid (TFA, Pierce, Rockford, IL)/5% formic acid
d. Phosphoric acid/ethanol (90 parts 95% ethanol, 9.5 parts of distilled water, and 0.5 parts of H_3PO_4).

2. The homogenate can be maintained by shaking at 4°C (with the acid acetone method, room temperature is also adequate) for a few hours to overnight and then centrifuged at 3000 rpm for 5 min (or at 1500 rpm for 15 min).
3. The supernatant is collected and either evaporated to 1/10 its original volume or lyophilized.

a. Fractionation by SepPak   C18 SepPak cartridges (Waters Associates, Milford, MA) are very useful to concentrate the peptides in the embryo extract and to remove certain other molecules, such as large binding proteins. Two protocols for preparation of the sample and elution of the bound material that work well are the following:

1. Prepare the cartridge by rinsing with 5 ml of 0.1% TFA in deionized water, followed by 10 ml of acetonitrile (Sigma), and again in 5 ml of 0.1% TFA. Prepare the sample by dilution in 0.1% TFA in water (the volume is not critical but volumes of 2 to 10 ml are adequate) and, if dissociation of binding proteins is desired, maintain the sample at room temperature for 1 hr. Load the extract ($\approx$ 10–15 mg protein/cartridge) at 0.5–1 ml/min, and recycle the flow-through several times. The cartridge is then flushed with 5 ml of 0.1% TFA. The retained proteins are eluted with 3 ml of acetonitrile/20% H_2O/0.1% TFA. The eluate is lyophilized and kept at -20°C until further characterized by HPLC.
2. Prepare the cartridge by rinsing with 5 ml of isopropyl alcohol, followed by 5 ml of absolute methanol and 10 ml of 4% acetic acid. Prepare the sample by reconstitution in water (preferred, if this precartridge sample is also going to be tested for protein content or in any immuno or bioassay). Acidify the sample by dilution in 0.5 N HCl (1:1) and maintain it at room temperature for 1 hr, if dissociation of large proteins is desired. Apply it to the cartridge and recycle as described above. Wash the cartridge with 10 ml of 4% acetic acid. Elute the retained proteins with 3 ml of 100% methanol.

b. HPLC The eluate from embryo extracts passed through SepPak is likely to contain a complex mixture of substances. To further purify the peptide of interest and to obtain some information on its molecular nature, high-performance liquid chromatography is a good choice. To discuss all the possibilities of this powerful technique for purification of proteins is beyond the scope of this chapter (see also Alemany *et al.*, this volume). We and others, have used hydrophobic interaction C18 columns (Vidac TP104 or 10 μm μBondapack), eluted with a linear acetonitrile gradient (Scavo *et al.*, 1989; Robcis *et al.*, 1991; Mesiano *et al.*, 1985).

1. After running a "blank" on the prewashed column (to test in the assay of choice and confirm the absence of contaminating peptides), inject an aliquot of the embryo extract (usually between 0.5 and 2 ml).
2. The column is eluted with a linear gradient from 25 to 65–75% acetonitrile, containing 0.1% TFA throughout, over 40 min at 1 ml/min. If the peptide is concentrated in a few fractions but molecular heterogeneity is suspected, a rerun of selected fractions using a more shallow gradient, for example, from 34 to 40% acetonitrile, will permit increased resolution. Ideally, the peptides used as standard peptides should be run in the HPLC system using an identical, but different, column to avoid residual contamination of the samples. Ultraviolet absorbance measurements permit automatic recording of the profile of total protein eluted.
3. Fractions to be assayed, preferentially in a sensitive radioimmunoassay or radioreceptor assay, are evaporated to dryness and reconstituted in the appropriate buffer (for details see Scavo *et al.*, 1989; Mesiano *et al.*, 1985).

Applications Separation by SepPak and HPLC have been useful for detection and partial purification of insulin, IGF-I and epidermal growth factor in chicken embryo and egg extracts.

References

Aghayan, M., Rao, L. V., Smith, R. M., Jarett, L., Charron, M., Thorens, B., and Heyner, S. (1992). *Development* 115, 305–312.
Bassas, L. I., Lesniak, M. A., Serrano, J., Roth, J., and De Pablo, F. (1988). *Diabetes* 37, 637–644.
Berger, S. L., and Kimmel, A. R. (1987). *In* "Guide to Molecular Cloning Techniques" (S. Berger and A. Kimmel, eds.), *Methods in Enzymology*, Vol. 152, pp. 219–227. Academic Press, Orlando, Florida.
Braude, P. R., and Pelham, H. R. B. (1979). *J. Reprod. Fertil.* 56, 153–158.

Caldés, T., Alemany, J., Robcis, H. L., and De Pablo, F. (1991). *J. Biol. Chem.* **266,** 20786–20790.
Deltour, L., Ledugue, P., Blume, N., Madsen, O., Dubois, P., Jami, T., and Bucchini, D. (1993). *Proc. Nat. Acad. Sci. U.S.A.* **90,** 527–531.
De Pablo, F., and Roth, J. (1990). *Trends Biochem. Sci.* **15,** 339–342.
De Pablo, F., Chambers, S. A., and Ota, A. (1988). *Dev. Biol.* **130,** 304–310.
Garofalo, R. S., and Rosen, O. M. (1988). *Mol. Cell. Biol.* **8,** 1638–1647.
Girbau, M., Bassas, L. I., Alemany, J., and De Pablo, F. (1989). *Proc. Natl. Acad. Sci. U.S.A.* **86,** 5868–5872.
Girbau, M., Gonzalez-Guerrero, P. R., Bassas, L. I., and De Pablo, F. (1992). *Mol. Cell. Endocrinol.* **90,** 69–75.
Hahnel, A. C., Rappolee, D. A., Milan, J.-L., Manes, T., Ziomek, C. A., Theodosiou, N. G., Werb, Z., Pederson, R. A., and Schultz, G. A. (1990). *Development* **110,** 555–564.
Hamburger, V., and Hamilton, H. L. (1951). *J. Morphol.* **88,** 49–92.
Heyner, S., Rao, L. V., Jarett, L., and Smith, R. M. (1989a). *BioEssays* **11,** 171–176
Heyner, S., Smith, R. M., and Schultz, G. A. (1989b). *Dev. Biol.* **134,** 48–58.
Hogan, B. L. M., Lacy, E., and Constantini, F. (1986). "Manipulating the Mouse Embryo." Cold Spring Harbor Lab. Press, Cold Spring Harbor, New York.
Hogan, A., Heyner, S., Charron, M. J., Copeland, N. G., Gilbert, D. J., Jenkings, N. A., Thorens, B., and Schultz, G. A. (1991). *Development* **11,** 363–373.
Kawasaki, E. S. (1990). *In* "PCR Protocols: A Guide to Methods and Applications" (M. A. Innis, D. H. Gelfand, J. J. Sninsky, and T. J. White, eds.), pp. 21–27. Academic Press, San Diego.
Kikuchi, K., Buonomo, F. C., Kajimoto, Y., and Rotwein, P. (1991). *Endocrinology (Baltimore)* **128,** 1323–1328.
Laemmli, U. K. (1970). *Nature (London)* **227,** 680–685.
Mattson, B. A., Rosemblum, I. Y., Smith, R. M., and Heyner, S. (1988). *Diabetes* **37,** 585–589.
Melton, D. A. (1991). *Science* **252,** 234–241.
Mesiano, S., Browne, C. A., and Thorburn, G. D. (1985). *Dev. Biol.* **110,** 23–28.
Rappolee, D. A., Brenner, C. A., Schultz, R., Mark, D., and Werb, Z. (1988). *Science* **241,** 1823–1825.
Rappolee, D. A., Wang, A., Mark, D., and Werb, Z. (1989). *J. Cell. Biochem.* **39,** 1–11.
Rappolee, D. A., Sturm, K. S., Schultz, G. A., Pederson, R. A., and Werb, Z. (1990). *In* "Early Embryo Development and Paracrine Relationships" (S. Heyner and L. M. Wiley, eds.), pp. 11–25. Alan R. Liss, New York.
Robcis, H. L., Caldés, T., and De Pablo, F. (1991). *Endocrinology (Baltimore)* **128,** 1895–1901.
Romanoff, A. L. (1960). "The Avian Embryo." Macmillan, New York.
Rosenblum, I-Y., Mattson, B. A., and Heyner, S. (1986). *Dev. Biol.* **116,** 261–263.
Saiki, R. K., Gelfand, D. H., Stoffel, S., Scharf, S. J., Higuchi, R., Horn, G. T., Mullis, K. B., and Erlich, H. A. (1988). *Science* **239,** 487–494.
Sano, T., Smith, C. L., and Cantor, C. R. (1992). *Science* **258,** 120–122.
Scavo, L., Alemany, J., Roth, J., and De Pablo, F. (1989). *Biochem. Biophys. Res. Commun.* **162,** 1167–1173.
Scavo, L. M., Serrano, J., Roth, J., and De Pablo, F. (1991a). *Biochem. Biophys. Res. Commun.* **176,** 1393–1401.
Scavo, L., Shuldiner, A. R., Serrano, J., Dashner, R., Roth, J., and De Pablo, F. (1991b). *Proc. Natl. Acad. Sci. U.S.A.* **88,** 6214–6218.
Schultz, G. A. (1986). *In* "Experimental Approaches to Mammalian Embryonic Development" (J. Rossant and R. A. Pederson, eds.), pp. 239–265. Cambridge Univ. Press, Cambridge, England.

Schultz, G. A., Hogan, A., Watson, A. J., Smith, R. M., and Heyner, S. (1992). *Reprod. Fertil. Dev.* **4**, 361–371.
Serrano, J., Bevins, C. L., Young, W. S., and De Pablo, F. (1989). *Dev. Biol.* **132**, 410–418.
Serrano, J., Shuldiner, A. R., Roberts, C. T., Jr., LeRoith, D., and De Pablo, F. (1990). *Endocrinology (Baltimore)* **127**, 1547–1549.
Shuldiner, A. R., De Pablo, F., Moore, C. A., and Roth, J. (1991). *Proc. Natl. Acad. Sci. U.S.A.* **88**, 7679–7683.
Siebert, P. D., and Larrick, J. W. (1992). *Nature (London)* **359**, 557–558.
Smith, R. M., Goldberg, R. I., and Jarett, L. (1988). *J. Histochem. Cytochem.* **36**, 359–365.
Strickland, S., Huarte, J., Belin, D., Vassalli, A., Rickles, R. J., and Vassalli, J. D. (1988). *Science* **241**, 680–684.
Telford, N. A., Watson, A. J., and Schultz, G. A. (1990). *Mol. Reprod. Dev.* **26**, 90–100.
Watson, A. J., Hogan, A., Hahnel, A., Wiemer, K. E., and Schultz, G. A. (1992). *Mol. Reprod. Dev.* **31**, 87–95.
Wiley, L. M., Takaki, K. K., and Yamagata, M. (1985). *Gamete Res.* **11**, 51–58.

Index

Adenosine, effects on adenylyl cyclase, 368-369
Adenylyl cyclase
  activation by forskolin and inhibition by adenosine-3'-phosphate, 366
  considerations of enzyme concentrations, reaction times, temperatures and radioactive substrates, 369-374
  determination, 365-388
  protocols for, 374-385
  protocol for data analysis, 386-388
  requirements for enzyme activity, 366-367
Adenylylimidodiphosphate, as inhibitor of nucleoside triphosphate activity, 358
Adipocytes
  considerations on the use of phorbol esters, 352
  proliferation and differentiation, 556
  protocol for isolation and plating, 556-557
Adrenocorticotropic hormone, effect on glycosylphosphatidyl inositol hydrolysis, 395
Adsorption chromatography, use in polypeptide purification, 95
Aequorea Victoria, source of Ca^{++}-binding protein, 408
Aequorin
  fluorescent Ca^{++}-binding protein, 408-409
  protocol for preparation, 410
Affinity chromatography
  in antibody purification, 32-33
  in polypeptide purification 96
  protocol for wheat germ agglutinin in enrichment of receptors, 293-296
Agarose gel electrophoresis, protocol with RNA, 430-432
Aggressive behavior, and hormonal control, Japanese quail model, 536-537
Alkaline phosphatase, in ELISA, 60
Allele-specific oligonucleotide hybridization, protocol, 495-497

Alumina, protocol for separating ATP and cAMP, 381-383
Amino acid copolymers, protocol for tyrosine kinase receptors, 315
Angiotensin II, effect on intracellular free calcium, 417
Antibodies, development and characterization, 67-90
Antibody fractionation
  application to radioimmunoassay, 15-16
  purification, 32-35
Antigenic epitopes, masking/unmasking, 236
Antigens
  antibody access to intracellular proteins, 235-236
  detection of multiple antigens by immunocytochemical electron microscopy, 250
  immunocytochemical electron microscopic localization, 230-232
Apolipoproteins, avian and reptilian models, 541-542
Aromatase, regulation, teleost brain model, 535-536
Ascites, protocol for production of monoclonal antibodies, 86-87
ATP-generating system, for adenylyl cyclase assays, 368-369
Atrial natriuretic peptide, biological actions, fish models, 545
Autophosphorylation
  protocol, 306-310
  receptor tyrosine kinase with *met*, *trk*, and fibroblast growth factor receptors, 303
Autoradiography
  disadvantages for spacial resolution with electron microscopic immunocytochemistry, 228
  ligand blots, 189-191

protocol for demonstration of ligand binding in embryos, 575–576
Avidin, protocol for coating beads, 43

Biotin, protocol for biotinylation of monoclonal antibodies, 43
Blunt end subcloning, protocol, 476–477

Caenorhabditis elegans, model in developmental biology, 567–568
Calcium
  determination of free form in single-cell level, 407–417
  protocol for determination, 410–417
Capillary electrophoresis
  apparatus, 134
  approaches to improve sensitivity, 129
  basic principle, 129–130, 133
  characteristics, 133
  coupled to fluorescence detection in determination of neuropeptides, comparison with radioimmunoassay, 127–128
  detection of neuropeptides, 127–144
  in determination of neuropeptides, comparison to HPLC, 128
CAT assay, *see* Chloramphenicol acetyl transferase assay
Cell cycle, effects of growth factors, 555–564
Cell proliferation
  in fetus, effect of growth factors, 555–564
  quiescent fetal brown adipocytes, induction, protocol for, 560–561
Cell quiescence, protocol for establishment with fetal cells, 558–560
Cell sorting, by flow cytometer, 170–176
Cesium chloride, protocol for RNA isolation, 426–428
Chemiluminescence, in protocol for immunoblotting, 362–363
Chicken embryos, protocol for autoradiographic demonstration of receptors, 581–584
Chickens, as models for pancreatic research, 543
Chloramphenicol acetyl transferase, assay, protocol, 512–513
Cholera toxin, protocol for ADP-ribosylation for G_s-α, 360–361

Circadian rhythms, insect models, 547
Circannual rhythms
  Japanese quail model, 534–535
  starling model, 534–535
Cis-acting elements, isolation by employment of reporter genes, 508
Coelenteramide, chromophore used for calcium determination, 409
Colloidal gold, protocol for preparation of IgG complexes and protein A, G, and AG complexes, 242–245
Complementary DNA, *see* cDNA
Coturnix coturnix Japonica, as model for seasonal breeding/circadian functioning, 534–535
Cryopreservation, hybridomas, 82–84

Denaturing gradient gel electrophoresis, as protocol in determination of DNA/DNA hybrids, 497–500
Developmental Biology, models, 567–586
DNA
  chemical determination, 563
  preparation, method for production of transgenic mice, 518–519
  probe, comparison with riboprobe for *in situ* hybridization, 271–273
  probe labeling, protocol, 435–439
  synthesis, protocol for determination with ^{3}H-thymidine, 561–563
Dowex-50, in protocol for separating ATP and cAMP, 381–383
Down-regulation, protein kinase C by phorbol ester, protocol, 345–346, 352–354
Drosophila melanogaster, as model for development, 548, 567–568

Ecdysone, action as model for steroid action, 547–548
EGF, *see* Epidermal growth factor
Electron microscopy
  immunocytochemical approaches to localization of ligands, receptors, transducers, and transporters, 227–264
  in protocol for mammalian embryos, 576–579
Electrophoresis
  protocol for RNA, 430–432

protocol for separation of IGFBPs by SDS-polyacrylamide gel electrophoresis, 190–192
transfer of protein to nitrocellulose membranes, 189–191
ELISA, *see* Enzyme-linked immunosorbent assay
Embryo
  chicken, protocol for autoradiography of receptors, 581–584
  mammalian, in study of ligand and receptor expression, 569
Embryogenesis, models for study of hormones, growth factors, and receptors, 567–586
End labeling, protocol with oligonucleotides, 438–439
β-Endorphin, sample preparation and derivatization for capillary electrophoresis, 134–135
Enzyme-linked immunosorbent assay
  comparison with radioimmunoassay, 58
  general principles, 56–58
  list of enzymes and substrates, 60
  protocol for evaluation of antisera or monoclonal antibodies, 88–89
  protocol for procedure, 58–63
Epidermal growth factor
  effect on glycosyl phosphatidyl inositol hydrolysis, 395
  as progression factor for cell cycle, 555
Epidermal growth factor receptors
  direct phosphorylation of phospholipase C, 339
  immunocytochemical electron microscopic localization, 253–255
  relationship to phosphorylation of phosphatidylinositol-3α kinase, 302
Epitope mapping, use in specific antibody selection, 35–36
ERK, *see* Extracellular-signal-regulated kinase
Erythrocytes, protocol for protein A-coated cells for reverse hemolytic plaque assay, 113
Escherichia coli, in production of monoclonal antibodies, 68
Extracellular-signal-regulated kinase, *see also* mitogen-activated kinase
  role in signal transduction and employment as substrate, 340

Ferritin-labeled insulin conjugate, in electron microscopic immunocytochemistry, 228
Fixation
  protocols for immunocytochemical electron microscopic studies, 238–240
  protocols for processing of tissues, 208–210
Fluoride, as phosphatase inhibitor, 351
Follicle-stimulating hormone, immunoradiometric assay, 47
Follicular development/atresia, avian and reptilian ovaries as models, 538–539
Forskolin, activation of adenylyl cyclase, 366
Freund's adjuvant, protocol for immunization, 70
Flow cytometry, 157–178
  analysis, 162–165
  capabilities, 158–159
  list of fluorophores, 162
  principle, 160
  separation of somatotroph population, 158
  utility for sorting cells, 169–176
Fluorescamine
  for derivation of neuropeptides, 131
  isothiocianate, use in immunohistochemistry, 213
Fluorescence-activated cell sorting, analysis of cell cycle in adipocytes, 560–564
Fluorescent dyes, in determination of free calcium in cells, 407–417
Fluorophores, in flow cytometry, 162

β-Galactosidase, in ELISA, 60
Galanin, cells, identification by immunohistochemistry, 216–218
Gel filtration
  separation of glucagon and other proglucagon-derived peptides, 148–151
  separation of glucagon-like peptide-1, 153
Gene expression, techniques, 505–514
GH, *see* Growth hormone
Glicentin
  determination by radioimmunoassay, 148–150
  separation by HPLC, 151–153
  structural relationship to proglucagon, 146, 147, 150
GLP-1, *see* Glucagon-like peptide 1
Glucagon, and other proglucagon-derived peptides

cDNA, 146–147
    determination by radioimmunoassay, 147–149
    separation by gel filtration, 153
    separation by HPLC, 151–153
Glucagon-like peptide 1
    separation by gel-filtration, 153
    structural relationship to proglucagon, 146–147, 150
Glucose transporters, immunocytochemical electron microscopic localization, 258–259
Glutaraldehyde-picric acid fixative, protocol for immunocytochemical electron-microscopic study, 239–240
Glycosyl phosphatidyl inositol, labeling with galactose oxidase/NaB^3H_4
    phosphoglycan system, 391–404
    protocol for chemical labeling of glucosamine, 396
    protocol for metabolic labeling, 395–396
    protocol for purification from rat liver, 393–395
GnRH, see Gonadotropin-releasing hormone
Gold-labeled insulin conjugate
    in electron microscopic immunocytochemistry, 229
    protocol for preparation of colloidal particles, IgG complexes, and protein A, G and AG complexes, 242–245
Gonadotropin-releasing hormone, sample preparation and derivation for capillary electrophoresis, 134–135
G-proteins
    activation by aluminum fluoride, 365
    identification and quantification, 357–363
    protocols for ADP-ribosylation by cholera toxin or pertussis toxin, 360–362
    protocol for determination of GTPase activity, 358
    protocol for determination of guanine nucleotide binding with [^{35}S]GTPαS, 358–359
G_i protein, immunocytochemical electron microscopic localization, 256–258
G_q protein, role in activation of phospholipase C, 339
Granulocytes, separation by flow cytometry, 170
GRF, see Growth hormone-releasing factor

Growth hormone
    chicken adipose tissue as model for action, 542–543
    producing cells, see somatotroph
    production by transgenic animals, 515–526
    release, effects of GRF or IGF-I, reverse hemolytic plaque assay, 115–118
Growth hormone-releasing factor
    development of antibodies, 75
    effect on GH release by reverse hemolytic plaque assay, 115–118
(^{35}S)GTPγS, protocol for determination of guanine nucleotide binding to G protein, 358–359
Guanidinium thiocyanate, protocol for RNA isolation, 422–428
Guanine-nucleotide binding assay, protocol, 358–359

Hemolytic plaque assay, see Reverse hemolytic plaque assay
High-performance (pressure) liquid chromatography
    comparison with capillary electrophoresis for determination of neuropeptides, 128
    isolation of embryonic signalling peptides, 568
    in polypeptide purification, 97–98
    protocol for isolation of peptides from developing chicken embryo, 586
    separation of glucagon and other proglucagon-derived peptides, 151–152
Histamine, cross-reactivity of antisera to, by LHRH, 218
Horseradish peroxidase, in chemiluminescence, 362–363
HPLC, see High-performance liquid chromatography
Hybridization
    analysis, in detection of transgenic mice, 519–520
    conditions for, 439–440
    in situ
        complementary uses with immunocytochemistry and ligand-binding autoradiography, 266–267
        controls for specificity, 276–277
        protocol, 280–284

protocol for combination with immunocytochemistry, 284–285
protocol for ^{35}S-labeling riboprobes and alkali hydrolysis, 277–280
Hydrophobic interaction chromatography, in polypeptide purification, 97

IGF, see Insulin-like growth factor
IGFBP, see Insulin-like growth factor binding protein
Immunization
  conjugation of low molecular weight analytes, 69
  development of antibodies, 67–90
  protocol for antisera development, 70–74
  protocol for use of adjuvants, 69–70
Immunoblotting
  protocol, 522–523
  protocol for identification of proteins during ontogeny, 579–580
Immunocytochemical electron microscopy
  controls for specificity, 250–251
  general consideration, 229–237
Immunocytochemistry
  applications to transgene expression, 207
  as method for localizing peptides during embryonic development, 568
  complementary uses with in situ hybridization and ligand-binding autoradiography, 266–267
  double labeling, 207–223
  protocol for combination with in situ hybridization, 284–285
  protocol for fixation, dehydration, embedding and immunolabeling, 238–250
  protocol for use with electron microscopy for studies on mammalian embryos, 577–579
  receptors, transducers and transporters, 227–264
Immunofluorescence
  protocol for use with embryos, 573
  with flow cytometry for pituitary cells identification, 159–160
Immunohistochemistry, controls and specificity, 213–214
Immunoprecipitation
  protocol of isolation of ^{32}P-labeled insulin receptor kinase, 329

protocol for using MARCKS, 349–352
Immunoradiometric assay
  comparison to radioimmunoassay, 4, 29
  data analysis, 47–48
  precision, 45–46
  principle, 10–15, 26–28
  sensitivity, 44–45
  specificity, 46–47
Immunostaining
  protocol for indirect and direct procedures for immunocytochemical electron microscopic study, 245–250
  protocol for pituitary cells, 161–162
Inositol phosphoglycan
  effect on mRNA, 403–404
  insulin-like effects, 400–402
  intracellular mediator, 391
  protocol for purification, 399–400
Insulin
  determination by radioimmunoassay, 145–146
  effect on phosphatidylinositol hydrolysis, 395
  identification of binding and internalization in blastocysts, 578
  localization of hormone-receptor complexes by non-immunologic electron dense complexes, 228–230
  like effects of inositol phosphoglycan, 400–402
  mRNA determination by RT–PCR, 470
  protocols for insulin binding to crude plasma membrane proteins or solubilized receptors, 297–299
  regulation of glycosyl phosphatidylinositol, 392
  relationship of tyrosine phosphatase to signal transduction mechanism, 321–335
Insulin-like growth factor I
  determination by solution hybridization/RNAse protection assay, 443
  effect on GH release by reverse hemolytic plaque assay, 115–118
  progression factor in the cell cycle, 555
  purification by HPLC, 97–98
  mRNA, quantitative by RT–PCR, 47
  radioimmunoassay, interference of IGFBPs, 182
Insulin-like growth factor II, localization of cell of origin, 267

Insulin-like growth factor binding protein
  identification by ligand blotting, 189–193
  localization of cell of origin, 269–270
  mRNA, 202
  protocol binding of ^{125}I-IGF in solution, 184
  protocol for deglycosylation, 196–197
  separation of endogenous IGFs by gel-filtration on charcoal, 186–189
  immunoblotting, 198–202
Insulin-like growth factor receptors
  autophosphorylation, 301–302
  autophosphorylation of wheat germ agglutinin purified receptors, 304–305
  comparison on effects of fixatives on antigen epitopes, 234–235
  immunocytochemical electron microscope localization, 253–255
Insulin receptor
  immunocytochemical electron microscopic localization, 253–255
  isolation, 289–299
  protocol for immunoprecipitation of ^{32}P-labeled autophosphorylation product, 329–330
  protocol for lectin purification, 326–327
  protocol for partial purification by affinity chromatography with wheat germ agglutinin, 293–296
Insulin receptor kinase
  protocol for assay, 327
  protocol for autophosphorylation, 327–329
  as substrate in assay of protein tyrosine kinase, 330–335
Interleukins -1, -2 and -4, effect on glycosyl phosphatidyl inositol hydrolysis, 396
Invertebrate, models for endocrine research, 546–549
Ion-exchange chromatography, in polypeptide purification, 96
3-Isobutyl-3-methylxanthine, inhibitor of phosphodiesterase, 367–368

Japanese quail
  as model for seasonal breeding/circadian functioning, 534–535
  as model for study of hormones and aggressive behavior, 536–537

Lactotrophs
  protocol for immunostaining, 161–162, 174–176
  separation enrichment by flow cytometry, 170–176
LH, see Luteinizing hormone
Ligand-binding, autoradiography, complementary roles with immunocytochemistry and in situ hybridization, 266–267
Ligand blotting, protocol for identification of IGFBPs subunits, 189–196
Lithium chloride, protocol for RNA isolation, 422–426
Luciferase (LUC) assay, protocol, 513
Luminescence, mechanism for determination of free calcium, 409
Luteinizing hormone, release as measured by reverse hemolytic plaque assay, 117
Luteinizing hormone-releasing hormone, sample preparation and derivation for capillary electrophoresis, 134–135
Lymphocytes, separation by flow cytometry, 170

Manduca sexta, use of epidermis in vitro as model for studying effects of developmental related hormones, 548
MARCKS (myristoylated alanine-rich C kinase substrate)
  antibodies to, 349
  phosphorylation by cyclic AMP-dependent kinase, 348
  substrate for protein kinase C, 347
Microdialysis, coupled with push-pull cannula, in isolation of neuropeptides from brain sites, 128
Microinjection
  production of transgenic mice, 519
  protocol for, in determination of free calcium, 412–413
Mitogen-activated protein, see also Extracellular-signal-regulated kinase
  role as substrate in signal transduction, 340
Monoclonal antibodies
  comparison with polyclonal antibodies, 31–32
  comparison with polyclonal antibodies with conformation- dependent epitopes, 234

protocol for ascites production, 86–87
protocol for biotinylation, 43
protocol for development, 74–89
purification, 87–88
Monocytes, separation by flow cytometry, 170
mRNA, see RNA
Mutations, detection in hormone receptor genes, 487–503
Myeloma cells, protocol for use in monoclonal antibody formation, 76–78
Myristoylated alanine-rich C kinase substrate, see MARCKS

Nerve growth factor
  effect on glycosyl phosphatidyl inositol hydrolysis, 395
  localization of cells of origin, 266
Neuropeptides, insect models for behavioral effects, 546–547
Neuropeptide Y
  derivatization, 130–131
  determination by capillary electrophoresis coupled to fluorescence detection, 127–144
  sample preparation and derivatization for capillary electrophoresis, 134–135
Neutrophils, separation by flow cytometry, 170
NGF, see Nerve growth factor
Nicotine, effect on free calcium in chromaffin cells, 414
p-Nitrophenyl phosphate, substrate for ELISA, 60–62
Northern blot analysis, protocol, 432–441
NPY, see Neuropeptide Y

Obelina geniculata, source of calcium-binding photoprotein obelin, 408–409
Oligo(dT)cellulose, protocol for role in isolation of poly(A) RNA, 428–430
Oligonucleotides
  protocol for end labeling, 438–439
  selection as primers for PCR, 465–466
Oocytes, use in study of ligand and receptor expression, 569

Oxytomodulin
  determination by radioimmunoassay, 148–150
  separation by HPLC, 151–152
  structural relationship to proglucagon, 146–147,150

Pancreatic β-cells, separation by flow cytometry, 170
Papaverine, role as inhibitor of phosphodiesterase, 367
Paraformaldehyde
  potassium ferricyanide fixation, protocol for immunocytochemical electron microscopic study, 239–240
  protocol for role in fixing tissue, 208–209
Parathyroid hormone, purification of antibodies to, 32
Parathyroid hormone-related peptide, immunoradiometric assay and protocol, 28, 39–41
PCR, see Polymerase chain reaction
PDGF, see Platelet-derived growth factor
Peptide histidine isoleucine, cross-reactivity of antisera to, role of corticotropin releasing factor, 218
Peroxidase, role in ELISA, 60
Pertussis toxin, protocol for ADP-ribosylation of G_i, 361–362
Phenylephrine, effect on intracellular free calcium, 417
Pheromones, analysis with insect models, 546–547
PHI, see Peptide histidine isoleucine
Phosphatase inhibitors, sodium fluoride as, 351
Phosphatidylinositol-3 kinase, relationship to receptor tyrosine kinase, 301–302
Phosphodiesterase, hydrolysis of cAMP, 367
Phospholipase C
  protocol, 398–399
  role in signal transduction, 339
Phosphotyrosine, determination by Western blotting, 328, 330
Photoperiodism, Japanese quail and European starling as models, 534–535
Pigeon crop sac gland, bioassay for prolactin and synlactin, 543–544
PIT1, localization of cells of origin, 267
Pituitary cells

protocol for dispersion of cells, 111–112, 161
protocol for immunostaining, 161–162, 174–176
Plasmid construction and mutagenesis, consideration for GH producing transgenic animals, 517–518
Platelet-derived growth factor
detection of transcripts in bovine blastocysts by RT/PCR, 574
initiation for the cell cycle, 555
growth factor receptor-direct phosphorylation of phospholipase C, 339
relationship to phosphorylation of phosphatidyl inositol-3 kinase, 302
Point mutations, protocol of detection by allele-specific oligonucleotide hybridization, 495–497
Polyadenylated RNA, protocol for isolation, 428–430
Polyclonal antibodies
comparison with monoclonal antibodies, 31–32
comparison with monoclonal antibodies with conformation dependent epitopes, 234
processing in H- and L- cell types, 146–147, 150
protocol for development of antisera, 70–74
purification, 87–88
structure, 146–147, 150
Polymerase chain reaction, 457–482
applications to endocrine research, 461
direct sequencing of products, 478–479
general protocol, 462–466
optimization, 463–466
protocol for sequencing of amplified cDNA or genomic DNA, 492–495
protocol for subcloning, 474–478
specificity and false positives, 469–462
theory of method, 459–462
Polypeptide purification, strategy, 95
Pressure blotting of Northern blots, protocol for, 434–435
Primer
extension analysis, protocol, 449–454
selection for PCR, 465–466
Probes
advantages of isotopic and nonisotopic labeling, 273–276

comparative advantages for *in situ* hybridization, 271–273
Progesterone, effect on free calcium in hepatocytes, 416–417
Proinsulin, determination by radioimmunoassay, 145–146
Prolactin
bioassay by casein release in hemolytic plaque assay, 123
producing cells, *see* Lactotrophs
release, influenced by TRH by reverse hemolytic plaque assay, 115–118
use of pigeon sac gland as bioassay or to demonstrate synlactin activity, 543–545
Propidium iodine
nuclear stain, 164
role in studying cell cycle, 557
Protein, chemical determination, 563
Protein A
chromatography, fractionation of antibodies, 16
protocol for protein A coated erythrocytes, 113
use in purification of antibodies, 32
use in reverse hemolytic plaque assay, 107
Protein kinase C
isozymes, calcium concentration requirement, 345
protocol for determination of translocation by immunoassay, 345–346
protocol for evaluation of activation, 346–352
protocol for translocation, 342–345
techniques for evaluation, 339–354
use of N-bromosuccinamide fragment of histone as substrate for assay, 342
Protein tyrosine kinase
phosphatase, relationship to signal transduction by insulin, 321–335
protocol for assay, 325–335
Push-pull cannula, coupled with microdialysis as technique in isolation of neuropeptides from brain sites, 128

Radioimmunoassay
application of antibody fractionation, 15–16
comparison with capillary electrophoresis

for the determination of neuropeptides, 127–128
comparison with ELISA, 58
comparison with immunoradiometric assay, 29
determination of insulin and proinsulin, 145–146
glicentin, 148–150
glucagon, 147–149
history and development, 4–5
labeling of ligand, 6
mathematical analysis, 7–9, 18–21
oxytomodulin, 148–150
principle, 5–10
Radioreceptor assay
general considerations, 296–297
principle and advantages, 16–17
protocol for GH assay, 524–525
raf-I protein kinase, role in signal transduction, 341
Random priming labeling, DNA, protocol, 436–438
Receptors, electron microscopic immunocytochemical localizations, 227–263
isolation and binding, 289–299
localization with nonimmunologic electron dense hormone receptor complexes, 228–230
preparation of membranes for vertebrate organs, 291–293
protocols for autophosphorylation and substrate phosphorylation, 306–317
protocol for partial purification by affinity chromatography with wheat germ agglutinin, 293–296
protocol for preparation of membranes from cultured lymphoid cells, 290–291
Receptor tyrosine kinase, analysis of autophosphorylation or substrate phosphorylation, 301–317
Reporter gene, indirect measurement of transcription rates, 508–513
Restriction analysis
protocol for determination of mutation, 488–492
fragment polymorphism use and protocol, 488–492
Reverse hemolytic plaque assay, 107–124
determination of hormone release and mRNA by *in situ* hybridization, 122

principle, 108
protocol, 110–114, 118–119
quantitative aspects, 115–118
simultaneous determination of GH release and intracellular calcium oscillations, 122
validation, 115
Reverse transcription
mRNA levels in embryo, determination by PCR, 568, 570
protocol, 466–469
quantitation by PCR, 469
Riboprobes (cRNA)
comparison with cDNA for in situ hybridization, 271–273
protocol for ^{35}S-labeling, 277–280
Ribosomal protein S6 kinase, as a substrate for kinase, 340
RNA
assessment of quality, 430
chemical determination, 563
general consideration for analysis, 505–506
protocol for argarose/formaldehyde electrophoresis, 430–432
protocol for isolation of poly(A) RNA, 422–430
protocol for Northern blot analysis, 432–441
protocol for solution hybridization, *see* RNAse protection assay
quantitation, 421–454
RNA extraction, general consideration, 422
RNA synthesis, insect model for steroid effects, 547–548
RNase protection assay, protocol, 441–449
RNA template, RNA-specific polymerase chain reaction, 472
RT, *see* Reverse transcription
"Run-off" assay, determination of transcription rate, 506–513

Seasonal breeding, use of Japanese quail or European starlings as models, 534–535
Sectioning, tissue for immunohistochemistry, 211–212
SepPak, protocol for isolation of peptides from developing chicken embryo, 584–585
Separation methods, 93–105

Sequencing, protocol with PCR products, 478–479
Sex determination, reptilian model, 537–538
Shark, testis as model for spermatogenesis, 539–540
Single-stranded conformation polymorphism, protocol to determine mutations, 500–503
Slot blot, protocol, 519–520
Solution hybridization, protocol, 441–449
Somatostatin, identification of cells by immunohistochemistry, 216–218
Somatotroph
  protocol for immunostaining, 161–162, 174–176
  separation (enrichment) by flow cytometry, 170–176
  separation of subpopulations by flow cytometry, 158, 164, 167
Southern blotting
  conditions for determination of mutations, 488–492
  protocol, 490–492
Spermatogenesis, use of shark testis as model, 539–540
Squalus acanthias, use of testis as model for spermatogenesis, 539–540
Staphylococcus aureus, source of protein A, 32
Starling, in study of photorefractoriness and circannual rhythms, 534–535
Steroid action, insect models, 547–548
Sticky-end ligation, protocol, 477–478

Taq DNA polymerase
  employment in PCR, 459
  misincorporation errors, 480–481
TATA box, relative to analysis of reporter genes, 513
TGF α, β_2, see Transforming growth factor α and β_2
^{3}H-Thymidine, employment in determination of DNA synthesis, 561–563
Thyroglobulin, isolation of peptides containing iodotyrosyl residues, 99–105
Thyrotropin
  effect on glycosyl phosphatidyl inositol hydrolysis, 395
  immunoradiometric assay, 42–45
  mRNA, blot hybridization analysis, 507

Thyrotropin-releasing hormone, effect on prolactin release by reverse hemolytic plaque assay, 115–118
Thyroxine (T_4)
  isolation of peptides containing iodinated tyrosyl residues from thyroglobulin, 101–105
  radioimmunoassay, 7
Transcription rate, determination, 506–513
Transcription regulatory elements, use in transgenic animals, 516
Transfection, protocol, 510–512
Transforming growth factor α, immunocytochemical electron-microscopic localization, 251–252
Transforming growth factor β, detection of transcripts in bovine blastocyst by RT/PCR, 574
Transgenic animals
  as tool in endocrine research, 515–526
  mice detection, 519–520
TRH, see Thyrotropin-releasing hormone
Truncated GLP-1, see Glucagon-like peptide 1
Trypanosoma brucei, antibody against surface glycoprotein crossreacts with inositol phosphoglycan, 392
TSH, see Thyrotropin
Two-dimensional electrophoresis, protocol for phosphorylation of MARCKS, 349
Tyrosine kinase receptor
  analysis of autophosphorylation or substrate phosphorylation, 301–317
  phosphorylation of YMXM motif, 304
Tyrosine phosphatase, relationship to signal transduction by insulin, 321–335

Ultrafiltration, use in polypeptide purification, 95
Urease, use in ELISA, 60

Vanadate, inhibition of protein tyrosine phosphatase, 322
Vasopressin, effect on hepatocyte free calcium, 415

Western blotting, protocol for IGFBPs, 198–203

Wheat germ agglutinin, protocol for partial purification of receptors by affinity chromatography, 293–296

Xenopus laevis, as model in developmental biology, 567–568

Yoctomole, definition, 142

Zamboni, protocol for its use in fixing tissues, 209–211
Zeptomole, definition, 142
Zeptomole range, use of capillary electrophoresis to determine neuropeptides, 132